Understanding the Nervous System

IEEE PRESS
445 Hoes Lane, PO Box 1331
Piscataway, NJ 08855-1331

Engineering in Medicine and Biology Society, *Sponsor*

EMB-S Liaison to IEEE PRESS

Murray Eden
National Institutes of Health

Technical Reviewers for EMB-S

Jim Brittain
Georgia Institute of Technology

Kenneth R. Foster
University of Pennsylvania

William Freedman
Drexel University

Thomas J. McGrath
Greater New Haven-State Tech. College

Cedric F. Walker
Tulane University

Charles J. Robinson
University of Pittsburgh

Understanding the Nervous System

An Engineering Perspective

Sid Deutsch

Electrical Engineering Dept.
University of South Florida, Tampa

Alice Deutsch

Bioscreen, Inc., New York

IEEE Engineering in Medicine and Biology Society, *Sponsor*

The Institute of Electrical and Electronics Engineers, Inc., New York

This book may be purchased at a discount from the publisher when ordered in bulk quantities. For more information contact:

IEEE PRESS Marketing
Attn: Special Sales
PO Box 1331
445 Hoes Lane
Piscataway, NJ 08855-1331
Fax: (732) 981-9334

Printed in the United States of America

10 9 8 7 6 5 4

ISBN 0-87942-296-3

IEEE Order Number: PP2915

Library of Congress Cataloging-in-Publication Data

Deutsch, Sid (date)
Understanding the nervous system : an engineering perspective / Sid Deutsch, Alice Deutsch.
p. cm.
Includes bibliographical references and index.
ISBN 0-87942-296-3
1. Neural conduction—Mathematical models. 2. Action potentials (Electrophysiology) 3. Nervous system—Mathematical models. 4. Biomedical engineering. I. Deutsch, Alice. II. Title.
[DNLM: 1. Biomedical Engineering. 2. Models, Neurological. 3. Nervous System—physiology, WL 102 D488u]
QP361.D48 1993
591.1'88—dc20
DNLM/DLC
for Library of Congress

92-49854
CIP

Contents

Preface

In 1987 a book that I wrote with Evangelia M-Tzanakou of Rutgers University, *Neuroelectric Systems*, was published by New York University Press. Sometime before this manuscript was completed it occurred to me that one ought to write a book, about the nervous system, for the lay person.

Of course, many books of this type have been written by life scientists, but a book written by a biomedical engineer would be different. Although some of my relatives and best friends are life scientists, there *are* two cultures. The world in general and the nervous system in particular look quite different to an engineer than they do to a life scientist.

I proceeded to generate a manuscript without equations (or, at least, nothing more complicated than Rate × Time = Distance). It was a challenge, a very enjoyable challenge, to explain the nervous system from an engineering perspective to the intelligent lay person. Alas, the manuscript was unpublishable. Even a literary agent did not help. Lay persons were undoubtedly interested in the nervous system, but not as seen through the eyes of an engineer.

Undaunted, I enlisted the aid of my daughter Alice, who is one of the life scientists that has a Ph.D. in biology. Alice made a semi-infinite number of changes, adding and subtracting so much material that it was only fair to call her a co-author. Unfortunately, this fooled nobody. The revised manuscript remained too heavily flavored by an engineering viewpoint.

Finally, I sent the manuscript to a publisher of *engineering* books—namely, IEEE PRESS. The PRESS's reviewers said, "Put the equations back, and you will have something that is viable!" I went even further—I cheerfully gathered up my favorite homework and exam questions, the residue of 25 years of teaching neuroelectric systems, and added Problems at the end of each chapter. (This includes answers to those exam gems that were formerly carefully guarded.)

One final change was made. One of the reviewers, Murray Eden of NIH, opined that most of the mathematics should be presented in appendix form at the end of each pertinent chapter. I am grateful to Dr. Eden for this excellent suggestion.

As the chapter headings indicate, the book begins with the basic components of the nervous system—with sensory receptors, neurons and their dendrites, and skeletal muscle circuits. This is followed by the more complicated auditory and visual systems. Finally, the climax—the brain—is considered. The 12 chapters correspond to a three-credit biomedical engineering course on the nervous system.

What is the engineering perspective? It is one in which each physiological system is simplified until it can be analyzed mathematically, followed by presentation as a simplified model.

The material in the book has been shaped by a myriad of students and colleagues—at the Polytechnic Institute of Brooklyn (now Polytechnic University) until 1972; at Rutgers University from 1972 to 1979; at Tel Aviv University from 1979 to 1983; and at the University of South Florida (as a "Visitor") since 1983. The manuscript was also modified in accordance with the helpful comments of some six "anonymous" reviewers, in addition to those of Dr. Eden.

I am thankful to the people of USF for their cooperation—especially Tom Smith and Dr. Elias Stefanakos of the Electrical Engineering Department, and Dr. Michael Kovac, Dean of Engineering. Special thanks are also due to the editorial staff of IEEE PRESS, notably Executive Editor Dudley Kay, Production Supervisor Denise Gannon, and Production Editor Karen Miller, without whose expertise this project could not have come to fruition.

Above all, Alice and I are grateful to Ruth, mother and wife, who was one of the "intelligent lay persons" who undertook to read the first few chapters of the first manuscript.

Sid Deutsch
Sarasota, FL

Ever since I can remember, my father always answered my "why" and "how" questions about the world. No matter what he was doing, he always had time to give me a scientific explanation for some phenomenon of our environment. No question was ever answered by that oft-quoted phrase: "Look it up!" I still remember his description of Einstein's theory of relativity that we discussed when I was a preteen. When he came to me with the suggestion that I broaden his engineering interpretation of the nervous system, I was delighted. I saw this as my chance, in turn, to help him understand the whys and hows of the nervous system from the biologist's point of view.

Biology relies so completely on experimentally proven facts. In contrast, I saw the book as a series of gross approximations about various aspects of the nervous system which had as starting points these facts. It was fascinating to me that in this way—using logical mathematical expressions and formulas—one could attempt to explain biological data. I suppose it is the dialectic of starting with some basic observations, developing a theory—in this case often a mathematical model—to explain the facts, and from this, designing better experiments so as to learn yet more facts which in turn will lead to a better mathematical description or model, and so on, until the "absolute" truth about the nervous system is known. It is my hope that this book will serve as a unique point of view about the nervous system that can stimulate both new research and new theories. I also hope that it can lead to a greater understanding of the nervous system by being valuable to both the bioengineer and the neurophysiologist.

Alice Deutsch
New York, NY

1

Excitable Tissue

ABSTRACT. The nervous system includes voluntary and autonomic (sympathetic and parasympathetic) systems. This book is devoted almost entirely to the voluntary system.

The system consists mostly of excitable tissue—sensory receptors, neuron cell bodies, axons, and muscle fibers. If you step on a sharp object, it stimulates sensory receptors that in turn stimulate neurons; the latter send action potentials (APs) via axons to interneurons and motoneurons in the spinal cord. The motoneurons send APs to the appropriate muscles, which contract so as to make you jump off the object.

Some of the activity involves atomic dimensions and, because distances are so small, time intervals are correspondingly small. It takes a factor of about 10^7 to transform atomic distances into dimensions that are familiar to us. For example, most atoms and simple compounds, if magnified by 10^7, turn out to be 3 mm (0.12 in.) in diameter. One centimeter multiplied by 10^7 equals 100 km (62 mi.). In time, one second multiplied by 10^7 is almost equal to 4 months.

Sensory receptors are usually at rest when they are unstimulated. Neuron cell bodies are at rest when they are not generating APs, while muscle fibers and axons are at rest when they are not carrying APs (that is, no APs are propagating along the muscle fiber or axon).

Body tissues are bathed in fluid that has an excess of sodium and chloride ions. Internally, excitable tissue at rest has an excess of potassium and large organic negative ions. This external–internal combination forms a battery that makes the inside of the tissue 60 to 90 mV more negative than the outside. The electric field across the membrane is very high—up to 12,000 V/mm. The physiologically compatible ionic concentrations are maintained by sodium and potassium pumps.

1-1 Introduction

This book is about the nervous system, about how its components function and communicate with each other. The book does not read like a novel. There are drawings, visual images to be stored away in memory along with associated interpretations. There are tables that have to be slowly digested so that your nervous system can store the meanings conveyed by the numbers and by the symbols we call the English and Greek alphabets. To some extent one must learn a new vocabulary, the language of the neurophysiologist. It is a difficult subject [D. Junge, 1981; J. Kline, 1976]. However, familiar words such as "voltage" and "potential" appear throughout because the messages of the nervous system, in addition to relatively slow-speed chemical signals, are transmitted as voltages (and their accompanying currents) [M. V. L. Bennett, 1977].

What are we trying to accomplish in this book? The nervous system from an engineering perspective is quite different from the nervous system as seen by a life scientist. The life scientist is trained to work with living tissue—with the nervous systems of insects, animals, and human patients. He or she painstakingly uncovers the chemical, electrical, and mechanical details behind the behaviors exhibited by these nervous systems. Papers and books written by life scientists are likely to contain page after page of descriptive material. Organic molecules, amino acids, and proteins are frequently mentioned. The engineer, on the other hand, is trained to set up block diagrams and mathematical models. To a biomedical engineer, a model is an approximate representation of a biological system. Most biological systems are nonlinear, so the engineer sometimes comes up with a linear system that is a gross approximation. We try to avoid, of course, models that are excessively inaccurate. We try to generate models that reasonably simplify the biological system while retaining some semblance of accuracy.

With further regard to models and accuracy: Our first encounter with a model, in this chapter, yields the calculated value $C = 0.9444\ \mu\text{F/cm}^2$ for the capacitance of a membrane. The four-place accuracy is somewhat comical because, given uncertainties associated with the dielectric constant and membrane thickness, the real accuracy is probably $\pm 20\%$. Physiologists use the value $C = 1\ \mu\text{F/cm}^2$, and rightly so. For engineering students, however, in this book we give the equation's four-place answer because this allows the reader to check his or her work. The reader has to be reasonably sophisticated in judging the real accuracy of a four-place value. In the same equation, for example, the permittivity of a vacuum is given as 8.854×10^{-12} F/m; this *is* accurate to four places.

Biomedical and clinical engineers greatly contribute to improving the quality of human life by designing (and frequently operating) much of the equipment in a modern laboratory, and by designing prosthetic devices. They have to work with biologists and medical doctors. Unfortunately, engineers and life scientists represent two different cultures. Before they try to communicate with each other, each should learn some of the language of the other's discipline. It is hoped that this book will enable engineers to understand and appreciate some of the problems encountered by neurophysiologists.

Although insect as well as human nervous systems are mentioned earlier, the book actually concentrates on mammalian models. Most of the biomedical engineers who specialize in the nervous system naturally gravitate toward mammalian systems. One should mention, however, that many of the details of an actual neural network have been traced out in the snail *Aplysia*. That creature is almost transparent, thereby exposing its abdominal ganglia to the microelectrode probes of neurophysiologists. Still more remarkable, a voltage-sensitive dye can be applied so that electrical activity becomes *visible*. In a large population of *Aplysia* neurons, where it is only feasible to probe a few with microelectrodes, it is actually possible to see that hundreds are active during, say, the gill-withdrawal reflex [D. Zecevic et al., 1989].

From the chapter headings in this book, it is obvious that the reader is gradually led along various foundation paths until the climax, the brain, is reached in the last chapter (Chap. 12).

What about the relatively new field of artificial neural networks (ANNs)? Well, *real* neural networks are considered throughout the book, but especially in Chapter 11. It is reasonable to also devote a section to ANNs in Chapter 11 (Sect. 11-5; see also Sects. 7-3 and 7-4). First, it is "cost effective" because the same techniques, such as learning, self-organization, associative memory, and parallel processing, are common to real and artificial neural networks. Second, one can argue that many of the networks that are labeled "hypothetical" in this book should more properly be labeled "artificial." Third, people who invent ANNs are forever hopeful that these networks will turn out to be "real." Their publications have been rightly criticized for omitting the "artificial" designation, even when the network has obviously nothing in common with living structures. In short, some "neural networks" are real, some are artificial, and some possess aspects of both of these states.

When we speak of the nervous system we usually think of its glamorous components—those that are involved in the writing of books, music, and equations; those that are involved in the major-league launching of baseballs, tennis balls, and golf balls; and so forth. These activities are controlled by the voluntary or somatic nervous system. (*Somatic* here pertains to the framework of the body.) By contrast, there is also an involuntary, visceral, autonomic nervous system, which actually consists of two systems—the sympathetic and parasympathetic. [The *viscera* are the internal organs of the body, especially those contained within the abdominal (intestinal) and thoracic (chest) cavities. *Autonomic* refers to involuntary, spontaneous activity.] [A. C. Guyton, 1986; T. C. Ruch et al., 1965; T. C. Ruch and H. D. Patton, 1979]

The autonomic nervous system is devoted to "housekeeping" chores that we cannot consciously control. Glands, smooth muscles, and the heart

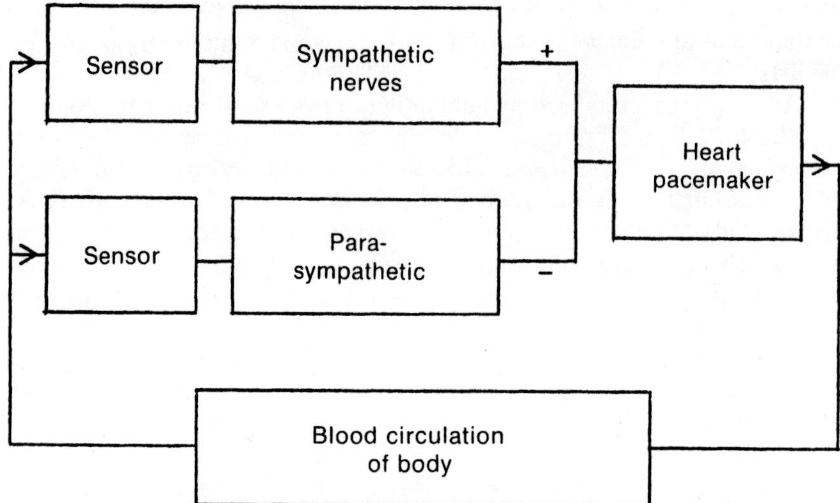

Fig. 1-1 Block diagram showing how the autonomic nervous system—sympathetic and parasympathetic nerves—controls the heart's natural pacemaker via a feedback loop.

are innervated by this system. A typical example, in Fig. 1-1, is provided by the heart's pacemaker, the internal mechanism that determines when the cardiac muscle should contract. The pacemaker is connected to both the sympathetic and parasympathetic systems. When the former is excited, it speeds up the pacemaker; when the parasympathetic fibers are excited, the pacemaker slows down. Each system is part of a feedback loop as shown in Fig. 1-1.

The dual control illustrated in Fig. 1-1 applies to most of the organs innervated by the autonomic system; that is, if the sympathetic system causes a muscle to contract (or relax), the parasympathetic system causes it to relax (or contract). Muscles controlled by the autonomic system are visceral smooth muscles; in contrast, those controlled by the voluntary system are skeletal striated muscles (fibers having transverse stripes). Cardiac and smooth muscles are similar in that they are both stimulated by the autonomic nervous system.

This book is devoted almost entirely to the voluntary, somatic nervous system. The heart is avoided because it is too specialized; entire books are dedicated to this organ alone, including biomedical engineering texts [W. Welkowitz, 1987]. Fortunately, much of the discussion of the voluntary nervous system is almost directly applicable also to the autonomic system.

Another area omitted in this book is that of the electromyogram (EMG). This is a record of the electric response that accompanies a muscle contraction. It forms an important niche in biomedical engineering but, like the heart, it is avoided because it is too specialized. The interested reader can start with the work of Carlo J. DeLuca and his colleagues [R. S. LeFever and C. J. DeLuca, 1982; F. B. Stulen and C. J. DeLuca, 1981]. However, typical feedback loop circuits of skeletal muscles *are* considered in Chapter 8 because the axons to and from muscles are part of the peripheral nervous system. (The brain and spinal cord form the central nervous system.)

At the outset it is appropriate to point out that this book does not regard extrasensory perception as a valid topic in a scientific discussion of the nervous system. Those who are interested in ectoplasm and the like should refer to the *Skeptical Inquirer* magazine (published by the Committee for the Scientific Investigation of Claims of the Paranormal).

[S. D. writes: "How did I, an electrical engineer, only mildly neurotic, get into this business? By way of a biomedical engineering program that some of my colleagues organized at the Polytechnic Institute of Brooklyn in the early 1960s. It seemed to me that much of the nervous system is reminiscent of an electrical communications network. I began to work with neurophysiologists, biologists, and medical doctors, and to read their literature. I gathered material for biomedical engineering courses, especially one on 'Neuroelectric Systems.' And so, like a tadpole turning into a frog, I metamorphosed into a biomedical engineer."]

1-2 Various Types of Excitable Tissue

Much of what we know about the nervous system has been donated by lower animals, such as the squid, to cite one example. This creature has a large nerve fiber or *axon* whose diameter is 0.5 mm (0.02 in.). After the axon is dissected out of the squid (and, incidentally, the animal's minuscule brain is no longer transmitting pain messages along its oversized axon) we can insert a fine wire electrode into the axon and make all kinds of measurements. Besides teaching us a great deal about how the nervous system functions, the experiments reveal that the nervous system of a squid is similar in many respects to that of a human, despite the huge genetic distance between the two species, because of the unified way in which evolution has progressed from animal to animal.

The nervous system consists mostly of *excitable* tissue—tissue that responds to a stimulus. As Luigi Galvani (1737–1798) and Alessandro Volta (1745–1827) discovered, if you suitably apply voltage to the leg of a frog it will jump (if the leg is alive, of course). The muscle is excitable tissue.

It is easy enough to find nonexcitable tissue. Consider a beautiful coleus plant, for example: if the wind blows, it responds by bending over, but that is not what we mean by the "response to a stimulus." Or take a piece of skin that is being grafted from one part of the body to another—it is living tissue, unlike a Band-Aid, but it does not respond to a voltage or any other conventional stimulus.

To be more specific, there are four kinds of excitable tissue: sensory receptors, neuron cell bodies, axons, and muscle fibers. (The branches feeding into a neuron cell body, called *dendrites*, are very much a part of the nervous system, but are omitted from this list because they are passive and unexcitable.) All of these tissues are called into play when you step on a mildly noxious stimulus such as a sharp pebble. Here the sensory receptors are pressure and pain transducers; they respond by generating a voltage called a *receptor potential.* Sensory receptors are usually associated with a neuron, a cell that is specialized in that it, in turn, responds to the stimulus by generating a rapid burst of voltage pulses. (Here a voltage pulse is a relatively high voltage that lasts for a relatively short period of time. A somewhat different type of neuron, which generates a smoothly varying voltage, or *graded potential,* is discussed in subsequent chapters.) [C. F. Stevens, 1979]

The situation is more clearly depicted in Fig. 1-2. The stimulus—in this case, a constant high pressure that results in excessive stress or damage—is accompanied by a pain receptor potential. At first the potential is relatively high, but having "sounded the alarm," it soon decreases to a much lower, steady level. This decrease is called *adaptation.* The neuronal output corresponding to the receptor potential, the 0.1-V high "spikes," are called *action potentials* (APs). If you could hear the neuronal response, it would sound like machine-gun fire; in fact, physiologists frequently amplify the neuronal output and listen to the rat-a-tat-tat while their eyes are occupied in adjusting the electrodes.

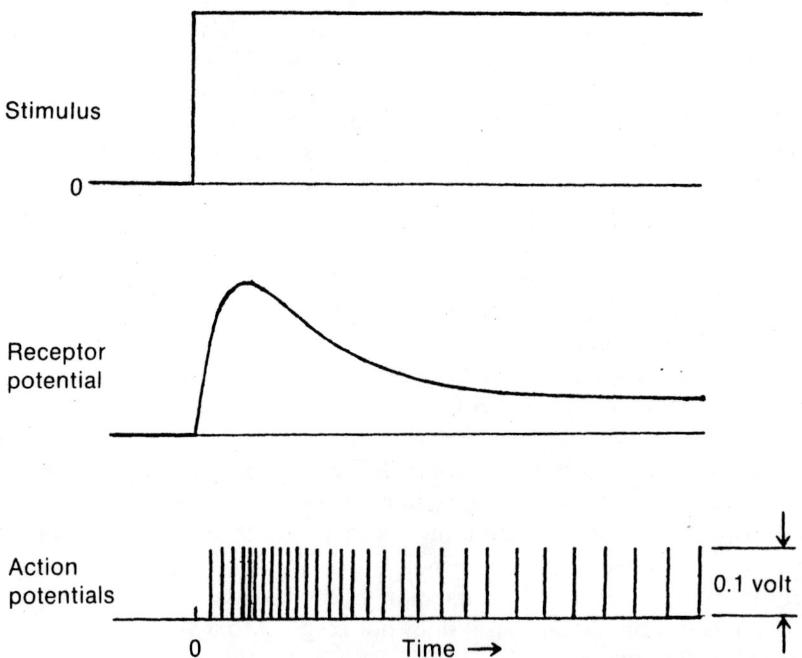

Fig. 1-2 Typical time variations associated with a sudden, steady stimulus. The sensory receptor generates a receptor potential that is relatively high at first, but that rapidly adapts to a much lower, steady level. A neuron that receives information from the sensory receptor generates corresponding 0.1-V APs.

The magnitude of the APs, 0.1 V, has been the same in every animal from before the squid to whatever comes after man, a period of some 500 million years thus far.

Today, voltage labels are familiar to everyone. A cell in the ubiquitous small chemical battery supplies 1.5 V; automobile batteries generate 12 V; home equipment operates on either 110 or 220 V, depending on the country; and so forth. Unfortunately, the early neurophysiologists preferred the word "potential" to "voltage." If you ask for a "1.5-potential battery," you will get an incredulous "What?" even at a Radio Shack store. In this book, however, we will side with the neurophysiologists: a sensory receptor generates receptor potentials, and a neuron generates action or graded potentials.

One part of the body also communicates with another part via chemical messengers, but these are relatively slow. To jump off that noxious pebble you need a high-speed channel. The neuron sends its AP burst along a nerve axon to the base of the spinal cord, as illustrated in Fig. 1-3 (any resemblance here to persons living or dead is purely coincidental). The axon leading to the spinal cord is *afferent*; that is, directed from the periphery of the body inward to the spinal cord or brain. Propagation along an axon is analogous to the transmission of electrical signals along a copper wire. The axon is quite remarkable in that it is an unbroken length of tissue, like a very fine wire, some three feet long for a fiber that runs from the foot of an adult to his or her spinal cord. The particular axon that carries the high-speed "Ouch" message has a diameter of 0.02 mm (0.0008 in.), and the velocity of signal propagation is 120 m/s (400 ft/s or 270 mph).

The afferent axon terminates in many *interneurons*. Figure 1-3 shows three interneurons; the lowest interneuron, with a very short axon, is feeding a *motoneuron*. In each case, when the AP burst reaches the end of an axon, it causes a chemical transmitter to be released across a narrow fluid gap, a *synaptic junction*. At the other side of the gap there is another neuron—an interneuron or motoneuron, as shown. Unfortunately, the drawing is very deceptive. In reality, there are lots of interneurons and lots of synaptic junctions—about 50 times as many interneurons as motoneurons, and the *average* motoneuron receives commands through 5000 synaptic junctions! [M. S. Gordon, 1972]

Chemical transmitters are relatively slow, as noted earlier; typically, they leisurely diffuse across the synaptic gap at a speed of only 2 mm/min (0.0013 in./s). Why, while you are standing on a noxious stimulus, is the body using a slow messenger as part of the signal path? One reason is that the gap is a necessary break in the electrical circuit: without a gap, every footstep would be followed by a twitch as leg muscles are activated. A second reason is that the excitability of the interneuron and motoneuron can be modified by the chemical constituents of the fluid in the gap. For example, in response to danger, the adrenal medullae in the brain release a hormone, epinephrine, into the blood stream. In a few seconds the epinephrine, reaching its target synaptic junctions, increases the excitability of the nervous

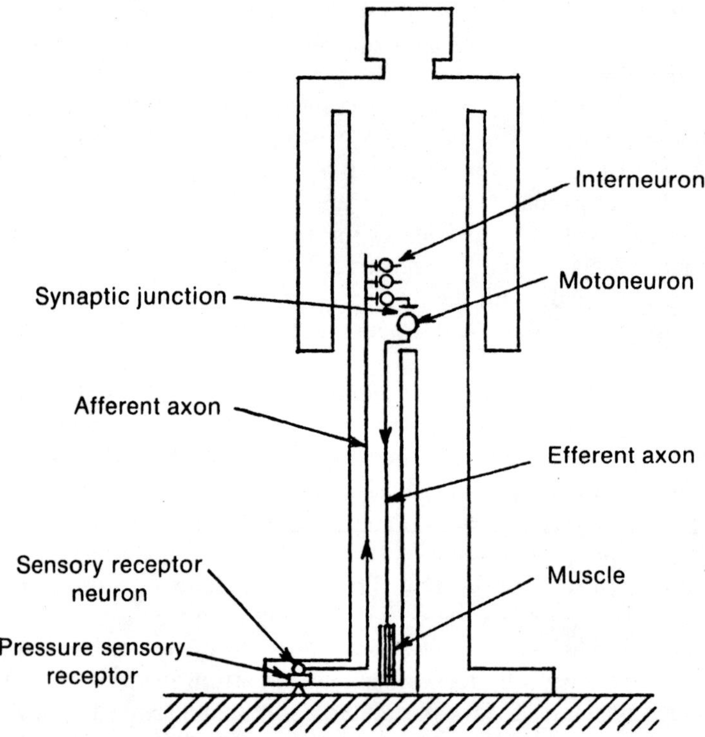

Fig. 1-3 Excitable tissue called into play when a person steps on a sharp pebble. Illustrated is a reflex arc that follows the excitation sequence: sensory receptor to its neuron to afferent axon to interneuron to motoneuron to efferent axon to muscle. The person starts to jump off the noxious stimulus in 0.025 s.

system. (Similarly, you can do amazing things to yourself by adding vodka, hashish, and so forth.)

It turns out, however, that the synaptic gap is so small—only 0.00002 mm (1/1,300,000 in.)—that it only takes 0.6 ms for the local chemical transmitter to bridge the gap! In other words, the synaptic delay turns out to be almost negligibly small.

Interneurons and motoneurons are typical cells in that they contain a nucleus and the biochemical machinery for synthesizing molecules essential for the life of the cell. They respond to the chemical transmitter by generating their own burst of APs. From the motoneuron, these travel three feet or so back down to the leg muscles via another axon. In Fig. 1-3 this axon is *efferent*; that is, directed from the brain or spinal cord to (usually) a muscle.

Finally, the fourth type of excitable tissue has been reached. The muscle responds by contracting. In a well-orchestrated individual, several muscles are involved, and the motoneuron can have over 5000 inputs, as previously mentioned. The net effect is that you promptly jump off the offensive pebble.

Notice that the brain is not part of the "reflex arc" (an involuntary response whose locus, from sensory receptor to motoneuron to muscle, can be loosely depicted as an arc). The afferent axon also synapses with one or more interneurons that lead to the brain, so that you have an awareness of what is going on. But an animal without a brain responds perfectly well to a noxious stimulus such as a pin prick. The brain only slows things down; if you *had* to think about what to do every time you stepped on a sharp object, you would not be able to survive very well in the environment in which your ancestors lived. For the reflex arc of Fig. 1-3, the time elapsed between stepping on a noxious pebble and starting to jump off is 25 ms (8 ms for the AP burst to travel from toe to spinal cord, 8 ms from motoneuron to leg muscles, and 9 ms or so for junction delays). It actually takes somewhat longer because of inertia since you have to reverse the direction of motion.

The main purpose of the preceding discussion is to illustrate how the four types of excitable tissue—sensory receptor, neuron cell body, axon, and muscle—interact when you step on a noxious object. A similar stimulus–response sequence occurs if you touch a hot surface with a finger.

Are you wondering why the signal has to travel so far—six feet—before the reflex arc is completed? Does not "survival of the fittest" dictate a much shorter reflex arc between toe receptors and leg muscles? If the signal could travel directly between them, you could start to jump off the pebble in 6 ms. Apparently the actual delay of 25 ms is short enough, so there was insufficient evolutionary pressure to develop an even faster path.

A few comments about sensory receptors. Children in school are usually taught that there are five types of sensory receptors. Actually, there are many more. It is conjectured that almost all of them transduce or convert a stimulus into a receptor potential [but olfactory (smell) receptors may directly generate action potentials]. An incomplete list of receptors is given in Table 1-1. Eight obvious types are shown plus "visceral." The latter is a catch-all for vitally important "internal" receptors that are not readily apparent, many of which are members of the autonomic nervous system. For example, there are sensors in the brain that maintain body temperature at 37°C (98.6°F); or sensors in the inner ears responsible for maintaining balance; or chemosensors that try to make you breathe again when you upset

TABLE 1-1 A Partial List of Sensory Receptors

Type	Stimulus
Touch	Pressure
Cold	Temperature
Warmth	Temperature
Pain	Excessive stress or damage
Auditory	Pressure
Taste	Chemical
Olfactory	Chemical
Visual	Light
Visceral	Various stimuli

oxygen and carbon dioxide levels by holding your breath. There are sensors for hunger, thirst, urination, defecation, and so forth.

The preceding listing includes many "housekeeping" chores (including those of the autonomic nervous system) that we do not ordinarily think about. Two-thirds of the axons in the body are devoted to these unexotic activities, such as reporting cold, warmth, and pain information. These are slow-conduction fibers—there is no need for a speedy response to any of their primary stimuli. The fibers are small (diameter 0.0003 to 0.0013 mm) and *unmyelinated*; that is, they are relatively simple and lack a thick insulating outer layer made of a substance called *myelin*. On the other hand, the high-speed fibers, such as those in Fig. 1-3, are relatively large (diameter 0.001 to 0.022 mm) and are covered by the relatively thick outer layer of myelin. The victims of multiple sclerosis suffer from deterioration of this myelin cover.

1-3 Of Space and Time

Some pretty small numbers are bandied about in the previous section, such as 1/1,300,000 in. and 0.0006 s. It is impossible for us to conceive of such small distances or time intervals, but we have to at least accept them as realistic values. We have to accept the notion that very important activities—the hustle and bustle of the nervous system, to say nothing of the synthesis of living structures—are executed in the world of microinches and microseconds. The purpose of this section is to "cut us down to size"; an attempt is made to give the reader an appreciation for these small dimensions and time intervals [F. A. Geldard and C. E. Sherrick, 1986].

Consider space (or length) first. How much magnification does it take to make small atoms and molecules "visible"? The answer is 10 million, or 10^7, as illustrated in Fig. 1-4. Magnification by a factor of 10^7 diameters yields the "water molecule" shown in Fig. 1-4(*a*)—about 3 mm in diameter. Its actual diameter is 0.0000003 or 3×10^{-7} mm. This is also the approximate size of most atoms and simple compounds.

Armed with Avogadro's number, it is easy to calculate the size of a molecule. Avogadro's constant, 6.022×10^{23}, is the number of molecules in one mole (M) of a substance. A mole is the amount of a substance that has a mass in grams numerically equal to the molecular weight. For water, as an example, the molecular weight is 18.016, so that 18.016 g of water contains 6.022×10^{23} molecules. The specific gravity of water (its weight in grams per cubic centimeter) is 1 (ignoring minor temperature and pressure effects). Then 18.016 cm^3 of water contains 6.022×10^{23} molecules, or each molecule is equivalent to a *cube* whose edge dimension is 3.104×10^{-8} cm. It is customary to express these small dimensions in angstrom (Å) units: 1 Å = 10^{-8} cm, so that the diameter of a water molecule is around 3 Å.

A factor of 10^7 is represented on the map of Fig. 1-4(*b*). A good atlas gives the scales to which each map is drawn. Accordingly, three scales are listed in Fig. 1-4(*b*): 1 to 10,000,000; 1 inch = 158 statute miles; and 1 cm

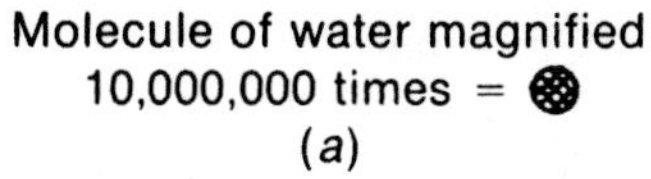

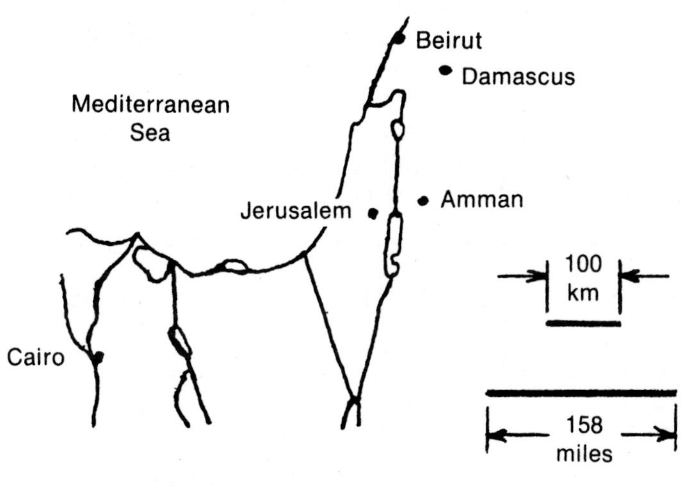

Fig. 1-4 Two illustrations of scale changes by a factor of 10^7 diameters. (*a*) A water molecule "appears" to be a 3-mm sphere. (*b*) Lines 100 km and 158 mi long on a map correspond to 1 cm and 1 in., respectively. Key: One centimeter equals 100 km; one inch equals 158 mi; scale factor = 10,000,000.

= 100 km. The equivalences are based upon 10^7 in. = 830,000 ft = 158 mi, and 10^7 cm = 100,000 m = 100 km. The map shows the countries along the southeast coast of the Mediterranean Sea. To cover the 1-cm equivalent you would have to drive 100 km/hr (62 mph, a bit fast for that part of the world) for one hour; to cover the one-inch equivalent you would have to drive 53 mph for 3 hr (possible on I-80 in the United States, but not in the Middle East).

The mere thought of such a trip by automobile, of traveling 10^7 in., is exhausting. It is perhaps the best way to appreciate what it means to magnify by a factor of 10^7, as in Fig. 1-4(*a*), in order to "make visible to the naked eye" a molecule of water. (For many technical reasons, this is only a thought experiment; even if it were possible to magnify by 10^7 diameters, a small atom or molecule would not look like a well-defined sphere.)

Equally incomprehensible is the factor of 10^7 in the time domain. What can conceivably happen in 10^{-7} s? When cosmologists talk about what was happening 10^{-7} s after the "Big Bang" that started the universe, aren't they being rather silly? Not any more than you are when you watch a television program. It turns out, strictly by coincidence, that happenings on the picture tube take place in 10^{-7} s!

To illustrate, imagine our unnamed pebble-stepper of Fig. 1-3 as he or she would appear on a TV screen, as depicted in Fig. 1-5(*a*). The TV image is of course formed by a single small spot of light. To use an analogy borrowed from the nervous system, the spot of light is the response of a fluorescent coating on the inside face of the picture tube to the stimulus of being hit by a high-speed beam of electrons. To avoid unnecessary complications, the comments that follow are restricted to black-and-white TV. (In the early days of TV, partly because of "national pride" and partly

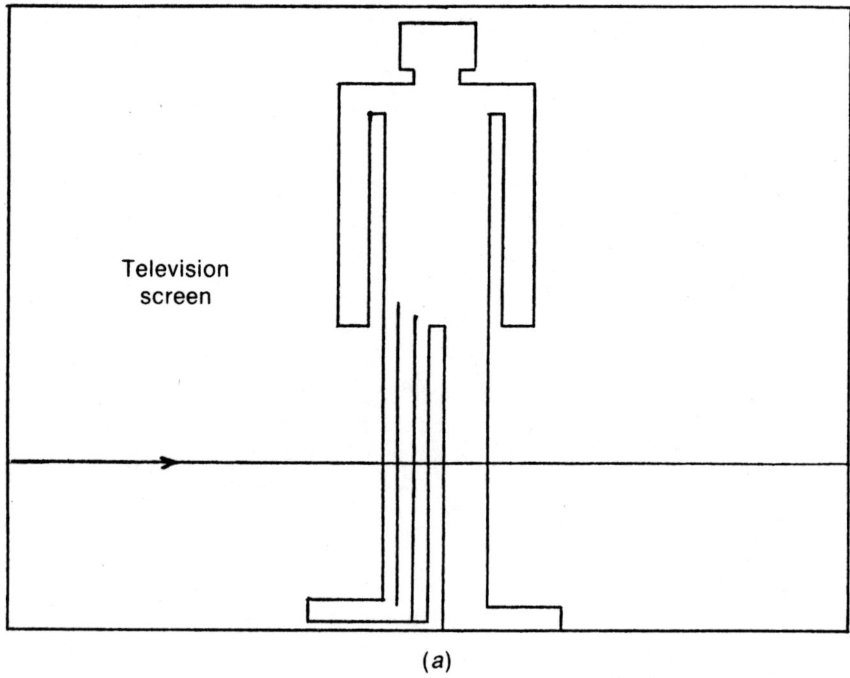

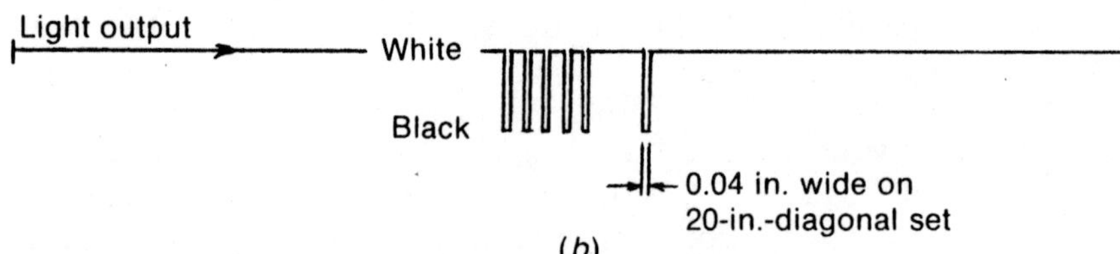

Fig. 1-5 (*a*) Somewhat simplified view of the "pebble-stepper" of Fig. 1-3 as he or she would appear on a television screen. (*b*) Light output corresponding to a single scanning line crossing the legs in (*a*). On a standard TV picture in the United States each of the six lines is approximately 10^{-7} s wide.

because of existing manufacturing facilities, each major participant insisted on preserving its own system, with the result that there are minor differences from one country to the next. Our comments are therefore restricted to TV in the United States.)

The spot of light moves rapidly from left to right and downward almost exactly as our eyes do when reading printed material (the differences can be ignored for our purposes here). If you step up close to any TV picture that is in focus, you can see the individual lines made by the spot of light as it scans from left to right (the internal structure of some tubes also yields vertical lines). In order to give the subjective impression of continuous motion, the TV picture has to be scanned very rapidly; the entire picture is covered in 1/30 s. There are 525 lines (but only about 490 lines are visible because the picture needs borders, for technical reasons, similar to those of a printed page). A *single* line, such as the line crossing the legs of our pebble-stepper in Fig. 1-5(*a*), is generated in around 500×10^{-7} s. As the beam moves from left to right it is responsible for the light emission illustrated in Fig. 1-5(*b*). The picture is white except where the legs and axons are crossed; here the beam generates "black" (that is, it does not generate anything at all). The television bandwidth is 4 MHz (after Heinrich R. Hertz, 1857–1894), which, of course, represents a cycle width of 0.25 μs. The narrowest vertical black line has a width of one-half cycle, 0.125 μs, or approximately 1×10^{-7} s. On a 20-in. tube, the narrowest black line is 0.04 in. wide, as shown.

In other words, each of us is witness to something that is happening in 10^{-7} s. We can even become hostile witnesses if the picture is slightly defective. Each of those lines has to be precisely generated and controlled in 10^{-7} s. The engineers and physicists who invented all of this are able to tame the electron because of exactly the same device that enables us to watch a TV picture: the unbelievably fast lateral motion is captured on a cathode ray tube, where a visible trace is left that can be examined at leisure. On a 20-in. diagonal picture the beam scans each line—16 in. wide—in 500×10^{-7} s as mentioned before. That corresponds to a "reading" speed of 17,000 mph. If you think that is fast, reflect upon the fact that it only amounts to 4.7 mps, which is "standing still" compared to the velocity of the electrons that are hitting the screen; *they* are traveling at 60,000 mps. Little wonder that the fluorescent material lights up when it is hit by such a projectile. But this in turn is much slower than the speed of light and radio waves in a vacuum—186,000 mps.

Moving in the other direction, how long a time is 10^7 s? It is 170,000 min, or 2800 hr, or 116 days, or a bit under 4 months.

The main point we wish to make is that very important events in the nervous system can occur at short distances and short time intervals. We live out our conscious lives in time intervals of seconds, minutes, and hours; we walk at a speed of around 3 mph. A "thought" that lasts for several seconds is the outcome of millions of *cortical* potentials chasing around inside the brain. A pair of *electroencephalographic* (EEG) electrodes attached to the head shows an unending stream of irregular discharges. It does not do much good to "think simple" during an EEG examination; the electrical signal remains incomprehensible. (Its magnitude is around 10 μV.)

[S. D. writes: "I once was approached by a patient who insisted that certain people were 'listening in' to his inner thoughts by analyzing his EEG signals, and that they were able to do this without even using electrodes. Alas, after many years of sophisticated monitoring of brain waves, one can only conclude that 'millions of cortical potentials are chasing around inside the brain'; the nature of the signal radically changes during trauma or epileptic seizures and during sleep, but a dream remains the private property of its originator."]

One final analogy may be helpful. Imagine that you are in a satellite looking down upon a large city. You cannot see individual people, of course. But if thousands of bodies march on the Capitol building, you may be able to see a blob, the integrated outcome of many individual images. Similarly, a pair of scalp electrodes cannot "see" individual brain potentials on the other side of relatively thick skull bone. What it does see is the momentary movement of thousands of signals that happen to be "marching" in the same direction, but this direction changes many times in one second.

1-4 Excitable Tissues at Rest

Excitable tissues are not in a constant state of excitation. There is a time for excitation and a time for relaxation. In this section we will look at excitable tissues at rest. Recall that there are four kinds of excitable tissue: muscle fibers, sensory receptors, neuron cell bodies, and axons. How does each behave at rest?

A muscle that is not contracting is, obviously, at rest. What is not so obvious is that many muscles in the body contract without our being aware of it. Physiologists list three varieties of muscle; as mentioned in the Introduction, they are skeletal, cardiac, and smooth. Skeletal muscles are attached to bones by means of tendons, and they move these bones when they contract; contraction occurs in response to action potentials (APs) that arrive either via a reflex path, as in Fig. 1-3, or via a "voluntary" path from the brain. Cardiac and smooth muscles, as agents of the autonomic nervous system, are outside our conscious levels of perception. We are not aware of heart contractions, or of the smooth muscle peristaltic movements associated with many of the tubular structures in the body such as blood vessels and the digestive tract.

Walking is another example of muscle contraction that we usually take for granted. It involves a complex scenario in which many muscles contract simultaneously as well as in sequence, which is why walking is an excellent exercise. Not only leg muscles are used; back muscles have to contract so as to maintain proper posture, or you would topple over regardless of what the leg muscles are doing. Although all of these muscles are skeletal, we do not ordinarily think about, or consciously direct, any of them. The entire walking cycle is a mixture of an inherited sequence and a learned, environmentally modified series of AP discharges [R. Johansson, M. Magnusson, & M. Akesson, 1988; R. S. Lakes et al., 1981; B. E. Maki, P. J. Holliday, & G. R. Fernie, 1987].

Sensory receptors are usually at rest when they are unstimulated (possible exceptions are visual receptors that respond to a decrease of light). As is the case with muscles, there is usually a lack of awareness of sensory stimulation. Referring to the "partial list of sensory receptors" in Table 1-1:

1. Some touch receptors are always being stimulated (except for a nude astronaut floating inside his or her space vehicle), since gravity forces us to sit, or stand, and so forth. One can argue that most of the touch receptors rapidly adapt, so receptor response ends after a few seconds if the person remains motionless.

2. Cold and warmth receptors are presumably inactive if a clothed person is at "room temperature," 20°C or 68°F. The presumption is probably incorrect. It is very difficult to monitor individual receptors of any kind because they are relatively small, but in the few instances in which receptors have been penetrated by a microelectrode without damage, one finds a relatively large "receptor potential." Practically all of this voltage is due to chemical differences between the inside of the receptor and the fluid bathing the cells; it is similar to the voltage produced by any chemical battery. It is the *change* in receptor potential that is the message. At room temperature a heat sensor in an index finger generates a potential, say, of 10 mV. When the finger touches a hot surface the potential increases to 10.2 mV. The 0.2-mV change is the stimulus for the associated neuron that accordingly emits a train of APs for transmission to synaptic junctions in the spinal cord.

3. It is easy to conceive of a situation in which there is minimal stimulus for cold, warmth, pain, auditory, taste, olfactory, and visual receptors. The person has to be lying completely relaxed on a water bed in a sound-absorbing room, totally insulated from extraneous noises, in pitch darkness, at 20°C. Experimental psychologists are fond of such experiments in sensory deprivation. Beyond a feeling of sympathy for graduate students or patients who are subjected to this form of torture, we are happy to avoid the subject by declaring it to be beyond the province of this book.

4. There is at least one group of "visceral" receptors that is never "at rest": the *carotids* (pronounced KA-ROT'IDS) are the two major arteries in the neck that carry blood to the head. In their walls are located pressure sensors (also called *baroreceptors*) that monitor arterial pressure; as the pressure increases, the receptor potential likewise increases. This activates a reflex system so as to maintain approximately constant arterial pressure in the brain. If you suddenly stand up from a lying position, for example, there is a rapid drop in pressure simply because the head is suddenly much higher than the heart, whereas the two were originally at the same level.

The decreased flow of blood to the brain can result in dizziness or a fainting spell. In that event the carotid baroreceptors cause vasoconstriction (you turn pale because blood flow to the skin is constricted), increased cardiac rate, and increased cardiac contraction so as to increase the arterial pressure and restore normal blood flow to the brain.

It would seem to be simple enough to discover when a neuron is at rest—this category of excitable tissues does not generate action potentials if it is at rest. But here, also, there is a caveat. Many neurons generate a slow discharge called *unstimulated, spontaneous activity*. It is easy to see the advantage of spontaneous activity: any stimulus, no matter how small, will then result in an increased rate of AP discharge. A resting neuron, on the other hand, may be a sleeping giant that requires a large stimulus before it wakes up. The evidence is that different neurons have different threshold levels. Some are characterized by a spontaneous discharge and high sensitivity to weak stimuli, while others are recruited only by relatively strong stimuli.

The last on our list of excitable tissues, axons, are passive transmission pathways in the sense that, like muscles, they do not originate any signals. An axon is at rest when it is not carrying an action potential. (There is, however, an ongoing relatively slow longitudinal movement of nutrients along the axon.)

Regardless of which kinds of excitable tissue we are considering, all of them are substantially "at rest" in between APs. An AP disturbance is over in a few milliseconds; after it has recovered from an AP the tissue is approximately at rest.

The situation at an atomic level is illustrated in Fig. 1-6. This is a highly magnified, highly simplified cross section of the membrane surrounding a neuron cell body, muscle fiber, or (unmyelinated) axon. How highly magnified? Recall from Fig. 1-4(*a*) that a water molecule has a diameter of 3×10^{-7} mm. The membrane in Fig. 1-6 is 75×10^{-7} mm thick—equivalent to only 25 water molecules side-by-side. How can such a thin membrane hold the contents of a cell in place? Well, the cell is small, so on a comparable basis the membrane thickness is not unreasonable. A soap bubble is made out of liquid—mostly water—whose molecules attract each other with sufficient force so as to form a thin membrane. The membrane of a living cell is similar except that it is made out of hydrophobic layers of molecules.

The fluid outside the cell—the external fluid—is mostly salt water, a sodium chloride solution. This is "blood" after one removes red blood cells, white cells, and other floating components. These normally constitute about 40% of the blood volume. The remaining 60% is a complex mixture known as plasma. The sodium and chloride ions of Fig. 1-6 are the major ionic components of blood plasma. As implied by the $-$ and $+$ signs, a chloride ion is a chlorine atom that is negatively charged because it has "borrowed"

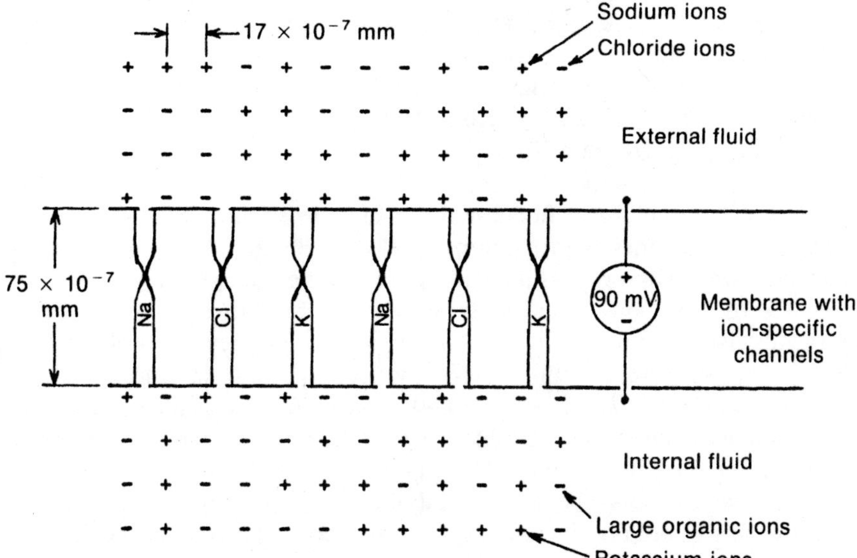

Fig. 1-6 Highly magnified and highly simplified cross section of membrane surrounding a neuron, muscle fiber, or unmyelinated axon in the resting state. At this moment the transmembrane channels are closed, but the sodium channel "pump" has created an excess of + ions in the external fluid and − ions in the internal fluid. This results in a transmembrane potential of, say, 90 mV. (In reality, the pumps are constantly working to maintain the concentration gradient.)

an electron from a sodium atom; the latter, which now lacks an electron, becomes a positive ion.

The fact that the external fluid features salt water is sometimes cited as evidence that life originated in the sea. This argument does not hold water because, after billions of years of evolution, there is little chance that the original constituents would be preserved. Although the life-originated-in-the-sea theory is not necessarily true, it appears that evolution has given us plasma that contains salt water because sodium chloride, in combination with everything else, is a viable solution. A practical consequence is that in the modern treatment for cholera a quart or two of mostly salt water is administered (oral rehydration therapy); this effects a remarkable lowering of the fatality rate.

The inside of a fiber or cell is an extremely complex structure. If we only concentrate on the ionic component, it mostly consists of relatively large molecules that have "borrowed" an electron from a potassium atom. The internal ionic component mostly consists of large negative organic ions and positive potassium ions as shown in Fig. 1-6. (An appreciable potassium deficit can result in dire consequences.)

[S. D. writes: "I once came across a patient who was hallucinating; the medical doctor laughingly explained that it was 'nothing serious, merely potassium unbalance.' The next day, after intravenous correction of the unbalance, the patient had no recollection of the previous day's events; it was like a forgotten dream."]

How do the cells and fibers manage to keep sodium chloride out and potassium in? By means of "pumps." When cells and fibers die, sodium chloride moves in and potassium leaks out until the voltage across the membrane is zero. For living tissue, the energy to operate the pumps comes from the metabolism (chemical breakdown, burning, or oxygenation) of the food we eat. The membrane is punctuated by sodium, chloride, and potassium channel proteins [R. D. Keynes, 1979]. Some of these protein molecules carry sodium ions *into* the excitable tissue (called *passive transport* because energy is not required), while others pump sodium ions *out* of the tissue (called *active transport* because metabolic energy is required). Similarly, there are passive potassium carriers and active potassium pumps. (It appears that the same protein is both a sodium and potassium pump, removing sodium ions as it returns potassium ions. To simplify matters, however, we will regard the protein as two separate units.) In the resting state these protein channels can be considered to be closed, as shown in the figure. One of the great simplifying assumptions of neurobiology was that each of the small ions—sodium, chloride, and potassium—has its own channel proteins, and that each operates largely independently of the others. The behavior of each channel is determined by local environmental conditions. By assuming independent channels, Alan L. Hodgkin and Andrew F. Huxley were able to decipher many of the details of tissue excitation; for this they were awarded a Nobel Prize in 1963 (shared with John C. Eccles).

Returning to Fig. 1-6: The major pump is that of the sodium channel. It forcefully ejects sodium ions, carrying them from the inside to the outside. The net result is an excess of sodium ions (+) in the external fluid and a corresponding excess of large organic ions (−) in the internal fluid. This is illustrated as follows: Along the membrane in the external fluid side there are seven sodium (+) ions, but only five chloride (−) ions. On the opposite side, along the membrane, there are seven large organic (−) ions, but only five potassium (+) ions. Since opposite charges attract, the excess sodium and large organic ions try to recombine. In addition, because the ions are in constant motion (the kinetic theory of heat), there is a thermodynamic tendency for each ion species to diffuse across the membrane until its concentration on each side is the same. One can measure the "force" of attraction because it corresponds to a voltage that is analogous to the voltage of a chemical battery. For the excitable tissue of Fig. 1-6 the voltage is 90 mV, as symbolized by the voltmeter that bridges across the membrane.

The 90 mV turns out to be a very impressive value. The voltage is low compared to the 1500 mV of the usual chemical battery, but the distance involved is minuscule. What is important is the electric field, which is the voltage divided by distance. This comes to 12,000 V/mm! Students are always shocked to learn these "facts of life." When the cell dies, the sodium and other ions move across the membrane until the voltage disappears.

How can the membrane withstand 12,000 V/mm? Because it happens to be a pretty good insulator, comparable to organic liquids such as aliphatic, aromatic hydrocarbons that are used for electrical insulation. (One of the best inorganic insulators, mica, can withstand 150,000 V/mm). All cells in general display a transmembrane voltage, because dissimilar internal and external ionic environments convert them into chemical batteries, but the voltage is usually less than 90 mV. For red blood cells it is about 10 mV, internal negative. (In general, mainly because of greater Na^+ concentration outside, the internal potential of a cell is negative.)

Table 1-2 lists ion concentrations (in micromoles/cubic centimeter, including water molecules) for mammalian muscle cells at rest. This table is discussed more fully later on in the chapter. For now the following point is made: In any small volume of fluid, the number of + ions must be equal to the number of − ions. Notice in Table 1-2, accordingly, that + and − ion totals are the same in the external fluid as in the internal fluid. The spacing between ions in the fluid is around 17×10^{-7} mm as shown in Fig. 1-6. This is an *average* value; the ions are not neatly arranged as shown, but instead are in constant random thermal motion in accordance with the kinetic theory of heat. Similarly, the channels are not evenly spaced as shown, but also experience a certain degree of lateral motion in the membrane, which is not, after all, a rigid structure.

TABLE 1-2 Ion Concentrations for Mammalian Muscle Cells at Rest[a]

	External Fluid[b]	Internal Fluid	Ratio Ext/Int
Water Molecules	55,200	55,200	1
Sodium ion (+)	145	12	12
Potassium ion (+)	4	155	1/39
Other ions (+)	5		
Hydrogen ions (+)	3.8×10^{-5} (pH = 7.4)	13×10^{-5} (pH = 6.9)	1/3
Positive ion total	154	167	
Chloride ion (−)	120	4.1	29
Bicarbonate ion (−)	27	8	3.4
Organic ions (−)		155	0
Other ions (−)	7		
Negative ion total	154	167	

[a] The concentrations are in micromoles/cubic centimeter, including water molecules.

[b] The "external fluid" column represents mammalian blood plasma in general.

We will frequently have occasion, in subsequent chapters, to refer to the *capacitance* of the membrane. Capacitance is given by

$$C = \frac{\epsilon_0 K A}{l}, \tag{1-1}$$

where $\epsilon_0 = 8.854 \times 10^{-12}$ in MKS units,
K = dielectric constant,
A = area,
l = thickness of insulation.

In order to calculate C, one must know the dielectric constant, K. Experimental evidence indicates that C is approximately 1 μF/cm^2; in fact, Hodgkin and Huxley employed this value, and the "1" appreciably simplified their calculations. If we substitute $C = 1\ \mu\text{F/cm}^2$ into Eq. (1-1), we get $K = 8.47$. This seems to be a bit high for a lipid bilayer; in this book, therefore, the value $K = 8$ is assumed for the dielectric constant of a membrane. For the unmyelinated membrane it yields $C = 0.9444\ \mu\text{F/cm}^2$. Another reason for choosing a simple numerical value for dielectric constant rather than capacitance is that K is a natural constant whereas the farad is a man-made unit.

Having calculated C, we can now calculate the resistivity of an unmyelinated membrane. This calculation is based on the experimental observation that a typical axon at rest responds to a transient disturbance as if it has a time constant of $RC = 4$ ms. Then the membrane leakage resistance is given by $R = 0.4235\ \Omega/\text{m}^2$. To translate this into resistivity we use the general formula

$$R = \frac{\rho l}{A}, \tag{1-2}$$

where ρ = resistivity. Solving for ρ,

$$\rho = \frac{RA}{l} = 5.647 \times 10^9\ \Omega\cdot\text{cm}. \tag{1-3}$$

This value fully justifies the previous statement that the membrane at rest "happens to be a pretty good insulator." Actually, evolution has produced some fibers, notably horizontal-cell dendrites in the retina, that seem to have a much higher membrane resistivity than the value given in Eq. (1-3). The 4-ms time constant for a typical fiber at rest is reasonable, since it allows a return to baseline in around 12 ms following a disturbance.

To summarize the most important points: In excitable tissue at rest, channel proteins force an unequal distribution of ions such that in the external fluid there are mostly sodium and chloride ions, while in the internal fluid there are mostly potassium and large organic ions. The sodium pump creates an excess of positive ions in the external fluid and negative ions in the internal fluid so that the voltage across the membrane is between 60 and

90 mV. When the tissue dies, ions migrate across the membrane until the voltage is zero.

1-5 Ion Concentrations of Excitable Tissues at Rest

The body is a mechanical as well as electrochemical engine. By the latter we mean that electric force fields interact with chemicals, the materials out of which the body is constructed. To illustrate one interaction, suppose that we start with a sodium (+) ion combined with a chloride (−) ion that forms a molecule of sodium chloride. Now imagine that the ions are physically moved apart: an electric field forms between them because of the attraction between − and + charges. Try visualizing the field as "lines of force," like stretched rubber bands, trying to pull the ions toward each other; this is, in fact, the way an electric field is shown. In effect, it is the work done in pulling apart oppositely charged ions that results in the 90-mV potential of Fig. 1-6.

Interactions of the body's electrochemical engine can become very complicated, especially if large and complex organic molecules are involved; fortunately, the happenings in the nervous system constitute a drama in which the three simple ions of Fig. 1-6 are the chief members of the cast. The pursuit of these ions can become boring to an engineer; nevertheless, one final step is justified before we abandon chemistry. Figure 1-6 is oversimplified; it is important to look more closely, quantitatively, at the ion concentrations of excitable tissues at rest.

Table 1-2 contains a list of the principal ions in mammalian muscle cells. Besides the four ions of Fig. 1-6, an additional negative ion, bicarbonate (HCO_3^-), is present in a significant amount. (The calcium ion, Ca^{++}, is sometimes very important, but not in this particular case [R. R. Llinas, 1982].) The "external fluid" column represents mammalian blood plasma in general. The internal fluid is the liquid component inside a cell or muscle that comes in contact with the membrane of the cell or fiber. With quantitative differences, the "internal fluid" column of Table 1-2 is also applicable to all mammalian neurons and axons at rest.

The "concentration" of water, in micromoles/cubic centimeter, is calculated as follows: 18.016 cm^3 of water constitutes one mole of water. Its concentration is therefore 1/18.016 M/cm^3, or 55,500 μM/cm^3. The external fluid of Table 1-2 contains $2 \times 154 = 308$ μM/cm^3 of ions, leaving $\cong$55,200 μM/cm^3 of water molecules, as listed. One can regard it as water that is 99.4% pure. The internal fluid leads to a similar characterization.

All of the ions of Table 1-2 have the same valence; specifically, each has gained or lost one electron. In that event the total number of positive ions, 154 outside and 167 inside, has to equal the total number of negative ions as shown. The internal fluid contains about 8% more ions than the external fluid.

We see that the membrane channel proteins of Fig. 1-6 are actually programmed so that the internal fluid contains some sodium and chloride ions and the external fluid includes some potassium ions. In their research, Hodgkin and Huxley elucidated the programming of each channel protein by artificially changing the transmembrane voltage and observing the corresponding behavior of each channel species. Although they worked with the squid, the results turned out to be qualitatively applicable to mammalian neurons and axons as well [H. Mino and K. Yana, 1989].

Table 1-2 also includes the pH value for the muscle fluid environments. The pH is based on the hydrogen ion concentration: pH $= -\log_{10}$ of H ion concentration in millimoles/cubic centimeter. The concentration is extremely small, only 3.8×10^{-5} μM/cm^3 in the "external fluid" column of Table 1-2, and 13×10^{-5} μM/cm^3 in the "internal fluid" column. Nevertheless, despite its small value, the hydrogen ion concentration is of utmost importance because the reactions of organic molecules are usually a function of pH value. If the pH is much below 7, the fluid is acidic; if much above 7, it is basic. From Table 1-2 we see that the internal fluid is more acidic than the external fluid.

The chapter ends with a look at what determines the resting potential. Although ions are transported across the membrane by protein molecules that bridge the membrane, the membrane potential is determined by energy considerations, ion concentrations, and the rates at which ions cross the membrane. The potential is given by the Goldman or "constant-field" equation. The three ion species that contribute are sodium, potassium, and chloride; since each of these is univalent, the equation is somewhat simplified (if Ca^{++} ions are important, for example, one must include a valence coefficient) [I. Levine, 1983]:

$$V = \frac{k_B T}{q_e} \log \left\{ \frac{P_{Na}[Na]_e + P_K[K]_e + P_{Cl}[Cl]_i}{P_{Na}[Na]_i + P_K[K]_i + P_{Cl}[Cl]_e} \right\}. \qquad (1\text{-}4)$$

(In this book, unless otherwise indicated, "log" stands for the *natural* logarithm.) Here the kinetic energy of motion of an ion is matched with the electric energy change when it crosses the membrane:

k_B = Boltzmann's constant, 1.3806×10^{-23}, J/K,

q_e = electron charge, 1.6022×10^{-19}, C,

$[Na]_e$ = external concentration of sodium ions,

$[Na]_i$ = internal concentration of sodium ions, and so forth,

P_{Na} = membrane permeability to sodium ions, and so forth,

T = temperature (310 K for mammalian cells, so that $k_B T/q_e = 0.02671$).

Notice that positive external and negative internal ion concentrations are in the numerator of the log term; positive internal and negative external ion concentrations are in the denominator. We illustrate with numerical values based on Bernhard Katz (1966). Data are given for a squid axon at 6.3°C (so that $k_B T/q_e = 0.02407$). Permeability values are relative, but the concentrations are in micromoles/cubic centimeter:

$$\begin{aligned} P_{Na} &= 0.07, \quad [Na]_e = 455, \quad [Na]_i = 72 \\ P_K &= 1.8, \quad [K]_e = 10, \quad [K]_i = 345 \\ P_{Cl} &= 0.8, \quad [Cl]_e = 540, \quad [Cl]_i = 61. \end{aligned} \tag{1-5}$$

Substituting into Eq. (1-4), $V = -57$ mV. The main contributions to the -57-mV resting potential come from external sodium and chloride ions, and from internal potassium and chloride ions.

REFERENCES

M. V. L. Bennett, Electrical transmission: A functional analysis and comparison to chemical transmission, in *Handbook of Physiology*, ed. E. R. Kandel. Bethesda, Md.: American Physiological Society, 1977.

F. A. Geldard and C. E. Sherrick, Space, time and touch, *Sci. Am.*, vol. 255, pp. 90–95, July 1986.

M. S. Gordon, *Animal Physiology: Principles and Adaptations,* 2nd ed. New York: Macmillan, 1972.

A. C. Guyton, *Textbook of Medical Physiology*. Philadelphia: Saunders, 1986.

R. Johansson, M. Magnusson, and M. Akesson, Identification of human postural dynamics, *IEEE Trans. Biomed. Eng.*, vol. BME-35, pp. 858–869, Oct. 1988.

D. Junge, *Nerve and Muscle Excitation,* 2nd ed. Sunderland, Mass.: Sinauer, 1981.

B. Katz, *Nerve, Muscle, and Synapse*. New York: McGraw-Hill, 1966.

R. D. Keynes, Ion channels in the nerve-cell membrane, *Sci. Am.*, vol. 240, pp. 126–135, March 1979.

J. Kline, ed., *Biological Foundations of Biomedical Engineering*. Boston: Little, Brown, 1976.

R. S. Lakes, K. Korttila, D. Eltoft, A. DeRose, and M. Ghoneim, Instrumented force platform for postural sway studies, *IEEE Trans. Biomed. Eng.*, vol. BME-28, pp. 725–729, Oct. 1981.

R. S. LeFever and C. J. DeLuca, A procedure for decomposing the myoelectric signal into its constituent action potentials, Part 1; with A. P. Xenakis, Part 2, *IEEE Trans. Biomed. Eng.*, vol. BME-29, pp. 149–164, March 1982.

I. Levine, *Physical Chemistry*. New York: Academic Press, 1983.

R. R. Llinas, Calcium in synaptic transmission, *Sci. Am.*, vol. 247, pp. 56–65, Oct. 1982.

B. E. Maki, P. J. Holliday, and G. R. Fernie, A posture control model and balance test for the prediction of relative postural stability, *IEEE Trans. Biomed. Eng.*, vol. BME-34, pp. 797–810, Oct. 1987.

H. Mino and K. Yana, A parametric modeling of membrane current fluctuations with its application to the estimation of the kinetic properties of single ionic channels, *IEEE Trans. Biomed. Eng.*, vol. 36, pp. 1028–1037, Oct. 1989.

T. C. Ruch et al., *Neurophysiology,* 2nd ed. Philadelphia: Saunders, 1965.

T. C. Ruch and H. D. Patton, eds., *Physiology and Biophysics—The Brain and Neural Function*. Philadelphia: Saunders, 1979.

C. F. Stevens, The neuron, *Sci. Am.*, vol. 241, pp. 54–65, Sept. 1979.

F. B. Stulen and C. J. DeLuca, Frequency parameters of the myoelectric signal as a measure of muscle conduction velocity, *IEEE Trans. Biomed. Eng.*, vol. BME-28, pp. 515–523, July 1981.

W. Welkowitz, *Engineering Hemodynamics: Application to Cardiac Assist Devices,* 2nd ed. New York: New York Univ. Press, 1987.

D. Zecevic, J.-Y. Wu, L. B. Cohen, J. A. London, H.-P. Hopp, and C. X. Falk, Hundreds of neurons in the *Aplysia* abdominal ganglion are active during the gill-withdrawal reflex, *J. Neurosci.*, vol. 9, pp. 3681–3689, 1989.

Problems

1. Look up the atomic weight and specific gravity of platinum. Assuming that the atom has the shape of a cube, find its edge dimension, d, in angstrom units. [Ans.: $d = 2.474$ Å.]
2. Given an axon that has a diameter of 1 μm and length of 1 cm, find (a) the capacitance from Eq. (1-1) and (b) the membrane leakage resistance from Eq. (1-2). (c) Verify that $RC = 4$ ms. [Ans.: (a) $C = 296.7$ pF; (b) $R = 13.48$ MΩ.]
3. Given Fig. 1-6, assume that the membrane is the insulation separating two capacitor plates. Using $q = CV$, (a) find the bound charge, in electrons/square centimeter, of the "internal fluid" plate. (b) Find the average distance between these electrons. [Ans.: (a) 5.306×10^{11} electrons/cm^2; (b) 137.3 Å.]
4. Repeat Table 1-2, but add columns giving the number of water molecules or ions per cubic centimeter. [Ans.: $33{,}200 \times 10^{18}$ water molecules/cm^3.]
5. (a) Verify that the average spacing between ions in Fig. 1-6 is 17 Å. (b) Find the average spacing between ions of *like* charge. [Ans.: (b) 22 Å.]
6. Verify the pH values given in Table 1-2.
7. Find the internal potential of human blood cells using Eq. (1-4). Assume that the external fluid concentrations of Table 1-2 are correct, and that $[\mathrm{Na}]_i = 20$, $[\mathrm{K}]_i = 140$, and $[\mathrm{Cl}]_i = 80$. (a) Assume that the membrane permeability values of Eq. (1-5) are correct. (b) Assume that the permeability of the chloride ion is infinite. [Ans.: (a) $V = -38.9$ mV; (b) $V = -10.8$ mV.]
8. Given the squid axon resting potential calculation following Eq. (1-5), the measurement shows that $V = -60$ mV rather than -57 mV. Assuming that the given $P_{\mathrm{Cl}} = 0.8$ is inaccurate, find the correct P_{Cl}. [Ans.: $P_{\mathrm{Cl}} = 0.117$.]

2

Sensory Receptors

ABSTRACT. Each of 15 sensory receptors is considered. The typical receptor generates a receptor potential in response to a stimulus and it is associated with a neuron.

In response to a step stimulus, the receptor potential rises rapidly to a peak and then slowly adapts to a steady level.

In response to a ramp stimulus, the receptor potential slowly rises at first. Thereafter, depending on how fast the stimulus rises, the receptor potential can continue to rise; or it can fall back because of adaptation.

In response to an oscillatory stimulus, the receptor potential is also oscillatory. There is a resonant frequency for which the peak-to-valley magnitude of the receptor potential variation is maximum. Below this frequency the output variation decreases because of adaptation; above this frequency it decreases because the receptor cannot follow the rapid stimulus variations.

2-1 Various Sensory Receptors

In Table 1-1 there is a list of eight types of peripheral receptors plus a single row titled "visceral." In this chapter we can more leisurely look at the visceral receptors; accordingly, the eight peripheral categories are combined with seven visceral types to create the (still incomplete) list of 15 receptors in Table 2-1.

Some brief and oversimplified comments on each of these sensory receptors follow.

Touch: There are several different varieties of touch receptors, depending on location. Those located in internal regions of the body are more properly called *pressure* rather than touch receptors. A receptor very prev-

TABLE 2-1 A List of 15 Sensory Receptors

Type	Stimulus	Location
Touch	Pressure	Skin
Cold	Temperature	Skin
Warmth	Temperature	Skin
Pain	Damage to tissue	Mostly in the skin
Auditory	Pressure	Inner ears
Taste	Chemical	Tongue
Olfactory	Chemical	Nose
Visual	Light	Retinas
Temperature	Change from 37°C	Brain
Vestibular	Acceleration	Inner ears
Chemoreceptor	Chemical	Arteries
Baroreceptor	Pressure	Arteries
Kinesthetic	Angle	Joints
Spindle	Stretch	Skeletal muscles
Tendon organ	Force	Tendons

alent in the finger tips, and exquisitely sensitive to vibration, is the *Pacinian corpuscle*. It is a favorite experimental candidate because it is relatively large [J. A. B. Gray and M. Sato, 1953; F. Grandori and A. Pedotti, 1980]. Most pressure transducers rapidly adapt, and in effect only respond to *changes* in pressure. For most of the peripheral touch receptors, the potential curve of Fig. 1-2 drops back to the initial baseline. For example, as you read this, are you aware of your wrist watch, or of the wallet in your back pocket? If you rotate your wrist you may be able to momentarily feel the watch; if you squirm in your seat, you may be able to sense the wallet, but one can never be sure without feeling it by hand. Adaptation evolved to free the brain from a perpetual cacophony of trivia, but that was in the days before pickpockets.

Cold: Pick up a massive metallic object at room temperature, such as an electric clothing iron, and hold it against your arm. The flatiron is at 20°C (68°F), while your body is at 37°C (98.6°F). Heat passes from your body to the flatiron, so that it feels "cold." If you use a massive piece of wood it will not feel particularly cold because wood is a poor conductor of heat. If you use a small metallic object, like a coin, it will only feel cold for a few seconds because it rapidly warms up to body temperature.

Warmth: If you heat the electric iron up to 42°C (107.6°F) and hold it against your arm, it will feel hot since heat flows from the iron to your body. If you heat it to 45°C (113°F) you will feel pain. Tissue damage starts to occur, and you will be literally cooking your arm. You can tolerate much higher temperatures from a large piece of wood because it is a poor conductor of heat. It starts out at 50°C, say, but the surface in contact with your skin is rapidly cooled down by the body until it is only slightly above body temperature.

Pain: Why do we feel pain when tissue is damaged? One theory has it that many chemicals are released when cells break down, and two of the substances—histamine and bradykinin—stimulate pain receptors. Another noxious stimulus is lactic acid, which results from insufficient oxygenation; in this case, the tissue "damage" is reversible. If pain receptors respond to a chemical stimulus, it is reasonable to assume that there are chemicals capable of inhibiting pain [R. Melzack, 1990].

Auditory: Sound vibrations in air result in vibrations that are transmitted via the ear drum, middle ear bones, and cochlear fluid to the *basilar membrane* in each cochlea; the latter is mechanically coupled to *hair cell* receptors. Although the hair cells respond to movement, the primary cause of the movement is cochlear fluid pressure change. Hair cells are unbelievably sensitive; they can respond to vibratory movement amplitudes comparable to atomic dimensions. By the same token, they can be damaged by the loud, unnatural noises encountered in a modern environment.

Taste: There seem to be four different taste receptors: taste buds for sour, salty, sweet, and bitter substances. Examples of primary substances are, respectively, hydrochloric acid, sodium chloride, sucrose, and quinine. Each of the receptors responds broadly to many different chemicals, so the designations sour, salty, sweet, and bitter are only nominal. Our conjecture is that the brain "looks at" the relative signal from each of the four different receptors; this corresponds to a point in a three-dimensional "taste space" somewhere in the brain. The various locations in this map are associated with our genetic and learned responses to food.

Olfactory: The sense of smell is similar to that of taste, and the two are closely linked in the brain (and also in physiology textbooks). Because of the difficulty of making measurements in the olfactory system, it is not definitely known what the primary substances are. Our assumption is that there are four or more different receptor types, and the relative signal from each corresponds to a point in a three-dimensional olfactory space. The brain associates each of these locations with our genetic and learned responses to volatile substances in the environment.

Visual: Here we are on firmer ground because the retina is relatively accessible, and a few animals such as the mudpuppy salamander have retinal cells that are large enough to accept microelectrodes. There are three different types of color receptors in the human eye. Each type broadly responds to visible light (as opposed to infrared and ultraviolet light). The color receptors are called *cones* because of their shape, and their nominally maximum responses are to red, green, and blue light, respectively. But color is a luxury that requires relatively high light intensity; in addition to the cones the retina contains *rods,* rod-shaped cells with which we see in dim light. The subjective rod impression is that of light and dark areas without color.

Temperature: The receptors for body temperature are a group of *thermostat* cells in the hypothalamus region of the brain. We do not know how they sense temperature. If their temperature falls below 37°C, they directly or indirectly release chemical messengers, hormones such as epinephrine, into the bloodstream. (The 37°C reference level is increased if defense against an infection calls for fever conditions.) This action has four consequences: (1) blood flow to the skin is blocked (vasoconstriction) so that evaporative cooling via sweating stops and the skin plus underlying fat acts as an insulating blanket; (2) the hair stands on end (piloerection or "goose pimples"), which tends to trap air and increase the insulating properties of the skin; (3) blood flow to the muscles is increased so that they tend to contract and do additional work, thus increasing internal production of heat; and (4) many muscles contract in a vibratory fashion (shivering), which further increases internal heat production. Conversely, if the temperature of the thermostat cells rises above 37°C, the skin sweats, hair lies down, muscle tone decreases, and we feel lethargic [I. B. Mekjavic and J. B. Morrison, 1985].

Vestibular: These are balance and acceleration receptors in the inner ear. A person who lacks functioning equilibrium mechanisms in both ears will fall over if he or she is blindfolded (so that visual cues are cut off). Whenever a normal person is starting or stopping (linear acceleration) or rotating the head (angular acceleration), the vestibular apparatus generates signals that help maintain equilibrium. If you rotate continuously for a few seconds and then suddenly stop, the angular acceleration receptors cannot respond as they should to this unnatural and childish maneuver, with the result that you feel "dizzy" [L. R. Young, 1966].

Chemoreceptors: In several strategic locations, notably walls of the aorta and carotid arteries, these receptors monitor blood chemistry. The aorta is the main artery leaving the heart, and the carotids are the two major arteries in the neck that carry blood to the head. The chemoreceptors respond to oxygen, carbon dioxide, and pH. When corrective action is required, the chemoreceptor response appropriately causes respiratory activity (breathing rate and depth) to be modified.

Baroreceptors: These are pressure sensors, also located in the walls of the aorta and carotid arteries. They are mentioned in Section 1-4 as an example of a sensory receptor that is never at rest. Arterial pressure rises and falls with each heart beat, such as from a maximum of 120 to a minimum of 80 millimeters of mercury (mmHg), but it is an ever-present stimulus. The baroreceptor outputs rise and fall in synchronism with the pressure variations.

Kinesthetic: The joints between bones are the sites of kinesthetic receptors. Each response is a function of the angle formed by the two bones.

The brain builds up its sensations of bodily position with the help of these responses. Without them we would have to look at our arms and legs each time we move them to be certain they end up in the desired locations.

Spindle: Embedded in skeletal muscles are receptors called *spindles* because of their shape. The *desired* degree of muscle contraction stimulates the receptor, while the *actual* amount by which the muscle has contracted inhibits the receptor. The net response is a function of the *difference* between desired and actual contractions. Why such a complicated system? Because the contraction of a muscle depends greatly on fatigue, on its exhaustion of molecules that supply energy, and on the accumulation of waste products. The actual contraction can fall far short of the desired motion. The signal from the spindle receptor is fed to the muscle's motoneuron, and continues to prod it, like a peasant walking behind a reluctant donkey, until the muscle's contraction meets the desired specification.

Tendon Organ: These are force transducers embedded in skeletal muscle tendons. Without them a muscle is capable of contracting with sufficient force to tear its tendons, or even rip one of them away from its bone. These are also called *Golgi receptors,* after Camillo Golgi (1844–1926). When the Golgi receptor senses excessive tendon tension, it inhibits the muscle's motoneuron, which in turn decreases muscle contractile force to a safe level.

This completes our brief comments on the list of 15 receptors of Table 2-1. There certainly is a wide variety of receptor types, and the list is not complete. But there are some unifying elements: in each case, with the exception of olfactory and possibly some visceral types, the receptor is associated with a neuron body and its axon. The olfactory receptor seems to be a neuron whose membrane responds directly to airborne chemical stimuli by generating action potentials (APs). With the exception of some of the visceral receptor–neuron pairs that release hormones directly into the bloodstream, all of the other receptors generate a *receptor potential.* This is also discussed in Section 1-2. In the next section we stimulate a sensory transducer and consider its receptor potential.

2-2 Receptor Response versus Stimulation

The 15 receptors of Table 2-1 are all members of the same family—they all respond in similar fashion to similar time variations of their respective stimuli. We can therefore represent all receptors by a single model—the block diagram of Fig. 2-1.

In the two sections of Fig. 2-1, the first block converts the stimulus σ into an internal, inaccessible, and perhaps hypothetical voltage, V_i. In general, the conversion is nonlinear, as indicated by the waveform sketches in Fig. 2-1, because V_i eventually begins to saturate as σ increases. The second block operates linearly upon V_i so that the graded output, V_r, shows limited high-frequency response and also suffers from adaptation, as indicated by the waveform sketch. These effects can be modeled by an equivalent *C, G,*

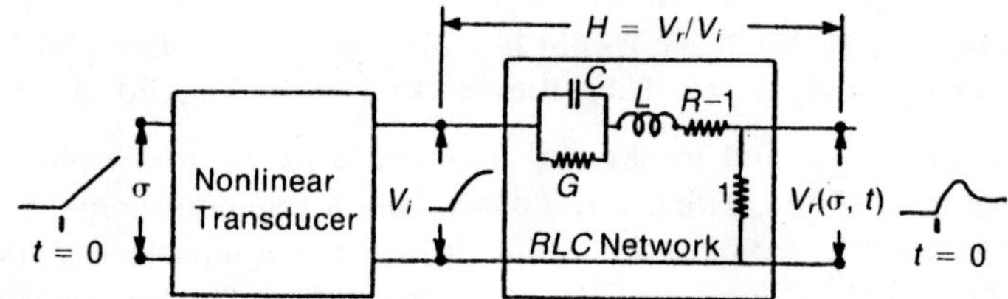

Fig. 2-1 Model of a sensory receptor. σ is the applied stimulus. V_i is an internal, inaccessible voltage, and V_r is the graded receptor potential output. The relation between σ and V_i is, in general, nonlinear. The relation between V_i and V_r is, however, linear. The *RLC* network represents finite rise time as well as adaptation. (From Deutsch and Tzanakou, *Neuroelectric Systems*, New York Univ. Press, New York, 1987.)

L, and *R* network, as shown in the block. The transfer function of this linear network is defined as output/input, or

$$H = \frac{V_r(s)}{V_i(s)}, \tag{2-1}$$

where $s = j\omega$ [N. Balabanian, 1958; E. R. Glaser and D. S. Ruchkin, 1976; V. Z. Marmarelis, 1989].

We can focus on a single representative receptor and examine its response to certain standard stimuli. Our favorite receptors are those that measure cold or warmth. The temperature senses are unique in that everyone is familiar with the unit of measurement—either °C or °F—and, furthermore, is accustomed to taking reasonably accurate readings. In fact, a difference of one degree in the way two people read a thermometer can result in a heated discussion.

It is somewhat more convenient to work with "warmth" rather than "cold" because a "cold" stimulus entails a drop in temperature. If this is represented on a sheet of graph paper, it appears as a negative-going effect. It is more natural to use a positive effect—a *rise* in temperature—as the embodiment of stimulus. We will therefore use *warmth* as the typical representative sensory receptor.

There are a semi-infinite number of ways to stimulate a receptor. Only three of them, however, are sufficient to illustrate the main characteristics of transducers in general: the step, ramp, and oscillatory stimuli.

The step response of an idealized warmth receptor is depicted in Fig. 2-2(*a*). With regard to the receptor potential output waveform, there are many ways to fit this curve to an algebraic expression, but the most convenient is one that embodies exponentials because the Laplace transform of $\epsilon^{-\alpha t}$ is a particularly simple form, $1/(s + \alpha)$. Accordingly,

$$V_r = V_p\left[k - \exp\left(-\frac{t}{\tau_1}\right) + (1 - k)\exp\left(-\frac{t}{\tau_2}\right)\right], \tag{2-2}$$

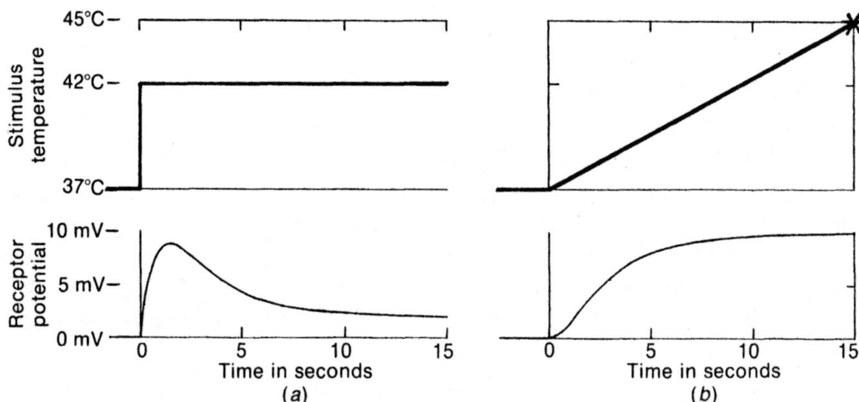

Fig. 2-2 For an idealized warmth receptor, stimulus temperature, and corresponding receptor potential output. (*a*) Step stimulus. (*b*) Ramp stimulus. The receptor "burns out" at 45°C, denoted by the * symbol.

where V_p = amplitude of the step supplied by the internal V_i source,
kV_p = final value after adaptation, as $t \to \infty$,
τ_1 = time constant of the leading edge,
τ_2 = adaptation time constant.

Applying the Laplace transformation to Eq. (2-2), we get

$$V_r(s) = V_p\left(\frac{k}{s} - \frac{1}{s + \dfrac{1}{\tau_1}} + \frac{1 - k}{s + \dfrac{1}{\tau_2}}\right). \tag{2-3}$$

We can continue in the preceding manner, but the algebraic expressions become increasingly complicated. You can follow more easily if a specific numerical situation is illustrated. In Fig. 2-2(*a*), accordingly, we have a step stimulus of temperature which, at "zero time," abruptly rises from 37°C to 42°C, and is maintained constant at 42°C thereafter. The corresponding receptor potential is given by

$$V_r = 0.02(0.1 - \epsilon^{-1.25t} + 0.9\epsilon^{-0.4t}). \tag{2-4}$$

Comparing Eq. (2-4) with Eq. (2-2), we evidently have $V_p = 0.02$, V = 20 mV, $k = 0.1$, the final value after adaptation = 2 mV, the time constant of the leading edge is $\tau_1 = 0.8$ s, and the adaptation time constant is $\tau_2 = 2.5$ s. The waveform rises rapidly to a peak of 8.814 mV at 1.464 s, and then slowly adapts to its final value.

The response curve of Fig. 2-2(*a*) is, alas, an invention. It has not been possible to measure the output potential of a warmth receptor. This unfortunately holds true for most of the receptors in Table 2-1 because they are too small and/or too inaccessible. Some of them are so closely associated with a neuron body and its axon that it is not possible to clearly separate receptor from neuron; in that event the putative receptor potential is synthesized from the train of neuron-generated action potentials. This is the case with cold and warmth receptors. The millivolt and second values in Fig. 2-2(*a*) are "reasonable." The curve is based on a simple equation that allows us to also predict the response to ramp and oscillatory stimuli. Furthermore, all of us are familiar with the step curve through prior experience: when we plunge a hand into the baby's bath water, at first it seems to be quite hot, but after a few seconds of adaptation it feels just right.

Applying the Laplace transformation to Eq. (2-4), and combining terms over a common denominator, there results

$$V_r(s) = \frac{0.02(0.89s + 0.05)}{s(s + 1.25)(s + 0.4)}. \tag{2-5}$$

It is all-important to find the transfer function, H, because with it we can calculate the response V_r to any input V_i if the system is linear. For the preceding numerical example, $V_i(s) = 0.02/s$, so that Eq. (2-1) yields

$$H = \frac{0.89(s + 0.05618)}{(s + 1.25)(s + 0.4)}. \tag{2-6}$$

The second standard stimulus, a ramp, is depicted in Fig. 2-2(*b*). The temperature rises linearly from 37°C at "zero time" to 45°C in 15 s. At the latter point the receptor "burns out"—it is literally destroyed by excessive temperature. (Although 45°C burnout is assumed in this chapter, this actually depends upon how long the 45°C is maintained.)

The receptor potential waveform depends on how nonlinear the transducer happens to be. If it is linear, so that V_i is a ramp, one can simply integrate the step response, Eq. (2-4), to get the ramp response. Figure 2-2(*b*) illustrates a special but more realistic case: the transducer saturates so that, in response to a ramp stimulus

$$\sigma = mt, \tag{2-7}$$

the inaccessible internal voltage is given by

$$V_i = a(1 - \epsilon^{-b\sigma}). \tag{2-8}$$

Notice that as $\sigma \to \infty$, $V_i \to a$. Substituting Eq. (2-7) for σ,

$$V_i = a(1 - \epsilon^{-bmt}). \tag{2-9}$$

This is the rising exponential waveform sketched in Fig. 2-1. Its Laplace transform appears as

$$V_i(s) = \frac{abm}{s(s + bm)}. \tag{2-10}$$

The receptor potential curve of Fig. 2-2(*b*) depicts a special case: the potential levels off, approaching 10 mV before burnout occurs. What makes this a special case? Two opposing effects exactly cancel each other, as follows.

As the stimulus ramp increases from 37°C to 45°C, the receptor potential "tries" to follow the stimulus. After a stimulus rise of a few degrees, however, saturation starts to set in. We begin to feel subjectively a burning sensation of maximum intensity because of the activity of noxious thermal receptors. The warmth receptor potential is no longer proportional to the stimulus rise. At the same time, the receptor potential tends to droop downward as the transducer adapts, as in Fig. 2-2(*a*). In Fig. 2-2(*b*) the rising and drooping tendencies exactly cancel, and the receptor output smoothly approaches 10 mV.

All sensory receptors, in general, display a saturation effect as the stimulus increases. A point is reached at which further stimulus increase does not result in further receptor-potential increase; the sensory receptor is "putting out" the maximum of which it is capable.

Another difference between Fig. 2-2(*a*) and (*b*) is demonstrated in the vicinity of zero time. For a step stimulus the receptor potential rises steeply. For a ramp stimulus, on the other hand, the potential gently rises from its initial zero level because $dV_r/dt\,|_{t=0} = 0$. If the temperature rise of Fig. 2-2(*a*) represents a massive metallic object resting against your arm, you would almost immediately feel the sudden increase in temperature. For Fig. 2-2(*b*) it would take a second or two before you felt the object warming up, and after 15 s it would start to sting painfully.

The special case of Fig. 2-2(*b*) corresponds to the cancellation of the zero in the numerator of Eq. (2-6) by the pole in the denominator of Eq. (2-10). In the latter equation let $bm = 0.05618$. Since $V_r(s) = HV_i(s)$, multiplication yields the ramp response transform

$$V_r(s) = \frac{0.89abm}{s(s + 1.25)(s + 0.4)}$$

$$= 1.78abm \cdot \frac{0.5}{s(s^2 + 1.65s + 0.5)}. \qquad (2\text{-}11)$$

The inverse transform is given by the Laplace pair

$$x(s) = \frac{bs + \alpha^2 - \beta^2}{s(s^2 + 2s\alpha + \alpha^2 - \beta^2)}$$

$$x(t) = 1 - \left(\frac{\beta + \alpha - b}{2\beta}\right)\epsilon^{-(\alpha-\beta)t} - \left(\frac{\beta - \alpha + b}{2\beta}\right)\epsilon^{-(\alpha+\beta)t} \qquad (2\text{-}12)$$

With $\alpha = 0.825$, $\beta = 0.425$, and $b = 0$, this yields for the ramp response

$$V_r = 1.78abm(1 + 0.4706\epsilon^{-1.25t} - 1.4706\epsilon^{-0.4t}). \qquad (2\text{-}13)$$

The receptor potential of Fig. 2-2(*b*) is plotted with the reasonable value $1.78abm = 10$ mV.

The third example of a stimulus is the oscillatory time variation of Fig.

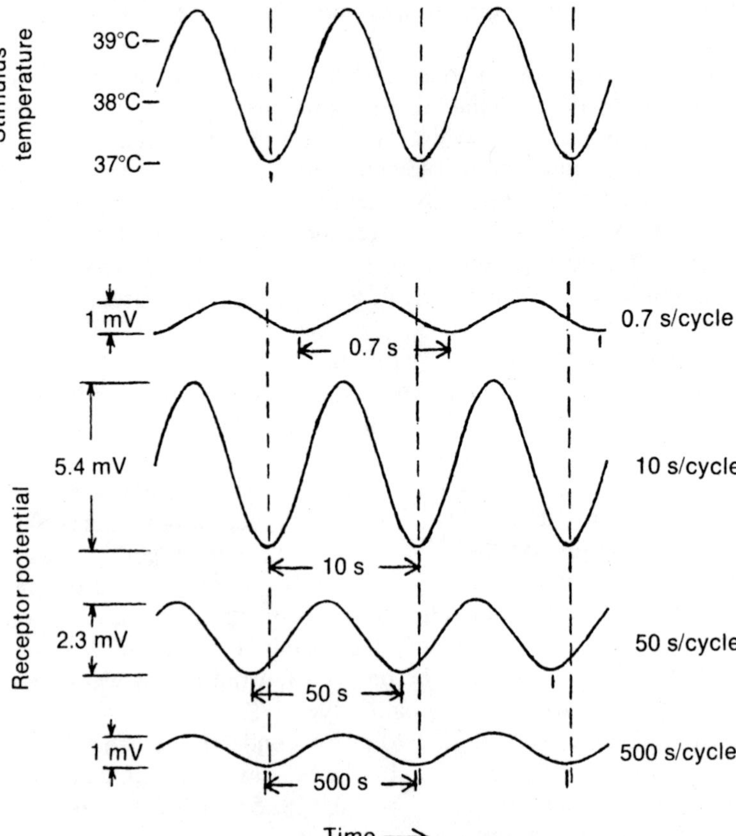

Fig. 2-3 For the idealized warmth receptor of Fig. 2-2(*a*)—Oscillatory stimulus temperature and corresponding receptor-potential outputs for various stimulus frequencies assuming that the system is linear—the dashed lines indicate when the stimulus reaches minimum temperature (37°C). Notice that each receptor-potential waveform has a different time scale corresponding to the change in stimulus frequency.

2-3. Here the temperature goes from a minimum of 37°C to a maximum of 39.5°C and back over and over again, following a sine wave.

Is the oscillatory stimulus a legitimate example? It is sometimes applied by a neurologist in testing a part of the nervous system [J.-F. Baland, E. R. Godaux & G. A. Cheron, 1987; R. J. Peterka, 1981; R. E. Poppele and C. A. Terzuolo, 1973; D. L. Sherrill and C. D. Swanson, 1981]. When you walk (or jog or run), no less than four sensory receptor systems receive an oscillatory stimulus: (1) the pressure sensors in your legs report back to the rest of the nervous system each time your foot touches the ground. Can you imagine walking if the pressure transducers in your feet were disabled? It would be quite impossible because you would not know when to shift the weight of the body from one foot to the other. (2) The vestibular apparatus in your inner ears receives an oscillatory stimulus as you shift from one foot to the other. You cannot maintain balance while walking if this sensory system is defective. (3) How do you know how large a step to take when walking? The kinesthetic receptors send signals back that indicate the status of the leg joints. (4) The stretch receptors receive an oscillatory stimulus as each muscle contracts and relaxes in accordance with the way it is orchestrated for normal walking.

Other examples of oscillatory stimulus occur in the (1) auditory system, because sound consists of the oscillatory longitudinal motion of air molecules; (2) olfactory system, because odors are sensed when we inhale, but not when we breathe out; (3) visual system, which is frequently subjected to noticeable flicker by modern lighting arrangements; and (4) baroreceptors as arterial pressure rises and falls in synchronism with the heart beat.

Returning to Fig. 2-3, the four curves shown below the "stimulus temperature" curve represent the responses of the warmth receptor to various stimulus conditions. Visualize the stimulating object, if you will, as a metallic plate resting against your arm. It has hot and cold water supplies circulating through tubes, with valves electronically operated such that the rate at which the heating cycle occurs can be adjusted.

In Fig. 2-3 it is assumed that the stimulus is relatively weak, so that the transducer is *linear.* To get the response to this steady-state sine wave, substitute $j\omega$ for s in H of Eq. (2-6):

$$H(j\omega) = \frac{0.89(0.05618 + j\omega)}{(0.4 + j\omega)(1.25 + j\omega)}. \tag{2-14}$$

The amplitude response is then given by

$$|H(j\omega)| = 0.89\sqrt{\frac{0.05618^2 + \omega^2}{(0.4^2 + \omega^2)(1.25^2 + \omega^2)}}, \tag{2-15}$$

while the phase response is

$$\theta = \arctan\frac{\omega}{0.05618} - \arctan\frac{\omega}{0.4} - \arctan\frac{\omega}{1.25}. \tag{2-16}$$

Also, in order to draw the waveforms of Fig. 2-3, it is convenient to know the time delay:

$$TD = -\frac{\theta}{\omega}. \tag{2-17}$$

One must be careful to convert $\theta°$ to θ radians. A frequently useful form is given by $TD = -\theta°/(360f)$.

Figure 2-3 illustrates four stimulus frequency conditions, but the f values are so low that the periods are given instead: 0.7, 10, 50, and 500 s/cycle. Various calculations are summarized in Table 2-2. The $|H(j\omega)|$ values of Table 2-2, which are dimensionless, have been multiplied by 10 to

TABLE 2-2 Summary of Various Hypothetical Calculated Values for the Sinusoidal Temperature Stimulus of Fig. 2-3. The amplitude in millivolts peak-to-valley is assumed to be 10 times $|H(j\omega)|$. The output leads the input if the time delay is negative

Period s/cycle	ω rps	$\|H(j\omega)\|$	Amp mV p-v	θ degrees	*TD* seconds
0.7	8.976	0.09811	0.9811	−79.88	0.1553
10	0.6283	0.5388	5.388	≅0	≅0
50	0.1257	0.2326	2.326	42.73	−5.935
500	0.01257	0.1024	1.024	10.23	−14.21

get the reasonable amplitudes of Fig. 2-3 (in millivolts). Some comments follow.

Consider first the receptor potential if the stimulus oscillates at a 0.7-s/cycle rate; that is, it takes 0.35 s for the temperature to rise from 37°C to 39.5°C and 0.35 s for it to drop back down to 37°C. The temperature oscillates so rapidly that the receptor can hardly follow. We are all familiar with the television beam that scans so rapidly that it seems to be a constant source of luminance. For the 0.7-s/cycle stimulus of Fig. 2-3, the receptor potential is not quite constant; there is a relatively small output of around 1-mV peak-to-valley, as shown.

The small ticks underneath the curves of Fig. 2-3 locate the lowest extremes of the potential excursions. For the 0.7-s/cycle curve, the potential peaks lag 0.1553 s behind the stimulus peaks. In other words, the receptor cannot respond instantaneously; this is also illustrated by the rounded leading edge of the potential curve of Fig. 2-2(*a*).

Next, look at the 10-s/cycle curve of Fig. 2-3. Now it takes 5 s for the temperature to rise and 5 s for it to drop back. It happens that 10 s/cycle is optimum for our warmth receptor. It *is* able to follow these temperature changes. The receptor potential has a peak-to-valley height of around 5.4 mV, and one can say that the receptor "resonates" at 10 s/cycle. Every receptor has its own resonance frequency. When we rapidly sniff in response to a suspicious odor, we may be trying to stimulate the olfactory system at its optimum rate. The visual system seems to resonate, very approximately, at 10 Hz; we are maximally sensitive to flicker at around this frequency. A bright, high-contrast flicker near this rate can induce a seizure in an individual who is predisposed to epilepsy. The brain's "alpha rhythm" (see Chap. 12) has a frequency between 8 and 13 Hz in human adults. It is reasonable to assume that there is a connection between the maximum sensitivity to flicker and the alpha rhythm frequencies.

Next we move to the 50-s/cycle curve. It is so slow that appreciable adaptation occurs, so the receptor potential is only around 2.3-mV peak-to-valley. As the ticks indicate, the receptor potential peaks occur 5.935 s *before* the stimulus peaks. How can this be possible? How can the effect precede the cause by 5.9 s? It is an illusion entirely due to adaptation. The stimulus takes such a long time to decrease from 39.5°C to 37°C, 25 s, that the potential bottoms 5.9 s early. The potential then starts to increase because it is "trying" to get to the *average* level between the positive and negative extremes. But just when the potential is about to reach the average level, the stimulus starts to increase toward 39.5°C, and the potential is destined to follow, reaching its positive peak 5.9 s before that of the stimulus. And so forth.

The last curve, with 500 s/cycle, is so slow—8.3 *minutes* per cycle—that the receptor completely adapts. In Fig. 2-2(a) a temperature rise of 5°C eventually results in a receptor potential of 2 mV; proportionally, for the 500-s/cycle curve, a rise of 2.5°C peak-to-valley results in a receptor potential of around 1-mV peak-to-valley. The ticks show that the potential excursions reach their extreme values 14.21 s *before* those of the stimulus (remember that, for this curve, the time scale is 10 times slower than that of the 50-s/cycle curve).

To summarize: In general, when sensory receptors are subjected to an oscillatory stimulus, there is a so-called resonant frequency for which the peak-to-valley magnitude of the receptor potential is maximum. If the receptor is stimulated at a much higher frequency, it cannot follow the variations, and the peak-to-valley magnitude of the potential decreases. If the receptor is stimulated at a much lower frequency, adaptation takes place and, again, the peak-to-valley magnitude of the potential decreases.

The chapter ends with a determination of the C, G, L, and R values of Fig. 2-1 for those who wish to construct the linear portion of the warmth sensory receptor. First, invert the H of Eq. (2-6):

$$\frac{1}{H} = \frac{s^2 + 1.65s + 0.5}{0.89s + 0.05}. \tag{2-18}$$

Next, by long division,

$$\frac{1}{H} = 1.124s + 1.791 + \frac{1}{2.168s + 0.1218}. \tag{2-19}$$

By inspection, the values are $L = 1.124$ H, $R - 1 = 0.791\ \Omega$, $C = 2.168$ F, and $G = 0.1218$ S $= 1/(8.209\ \Omega)$. For those who are skeptical, calculate H using these values and verify that Eq. (2-6) is obtained. And for those who are really ambitious, construct the circuit and subject it to various V_i input waveforms. (More convenient circuit values can be obtained by applying appropriate frequency and impedance transformations.)

REFERENCES

N. Balabanian, *Network Synthesis*. Englewood Cliffs, N.J.: Prentice-Hall, 1958.

J.-F. Baland, E. R. Godaux, and G. A. Cheron, Algorithms for the analysis of the nystagmic eye movements induced by sinusoidal head rotations, *IEEE Trans. Biomed. Eng.*, vol. BME-34, pp. 811–816, Oct. 1987.

E. R. Glaser and D. S. Ruchkin, *Principles of Neurobiological Signal Analysis*. New York: Academic Press, 1976.

F. Grandori and A. Pedotti, Theoretical analysis of mechano-to-neural transduction in Pacinian corpuscle, *IEEE Trans. Biomed. Eng.*, vol. BME-27, pp. 559–565, Oct. 1980.

J. A. B. Gray and M. Sato, Properties of the receptor potentials in Pacinian corpuscles, *J. Physiol.*, vol. 122, pp. 610–636, 1953.

V. Z. Marmarelis, Signal transformation and coding in neural systems, *IEEE Trans. Biomed. Eng.*, vol. 36, pp. 15–24, Jan. 1989.

I. B. Mekjavic and J. B. Morrison, A model of shivering thermogenesis based on the neurophysiology of thermoreception, *IEEE Trans. Biomed. Eng.*, vol. BME-32, pp. 407–417, June 1985.

R. Melzack, The tragedy of needless pain, *Sci. Am.*, vol. 262, pp. 27–33, Feb. 1990.

R. J. Peterka, Determination of otolith afferent response parameters using small amplitude sinusoidal roll and pitch tilts, *IEEE Trans. Biomed. Eng.*, vol. BME-28, pp. 624–630, Sept. 1981.

R. E. Poppele and C. A. Terzuolo, Myotatic reflex: Its input-output relation, *Science*, vol. 159, pp. 743–745, 1973.

D. L. Sherrill and G. D. Swanson, Minimum phase considerations in the analysis of sinusoidal work, *IEEE Trans. Biomed. Eng.*, vol. BME-28, pp. 832–834, Dec. 1981.

L. R. Young, Vestibular control system, in *Biological Control Systems* (NASA Cr-577), ed. L. Stark et al. Washington, D.C.: National Aeronautics and Space Administration, Sept. 1966.

Problems

1. Given a touch receptor for which, in the step response of Eq. (2-2), $V_p = 15$ mV, $kV_p = 3$ mV, $\tau_1 = 1.25$ ms, and $\tau_2 = 5$ ms, plot the step response. Find the peak value and when it occurs. [Ans.: $V_r = 8.263$ mV at $t = 2.682$ ms.]
2. Carry out the step-by-step derivation of Eq. (2-11) using algebraic expressions.
3. Given $x(t)$ of Eq. (2-12), apply $\mathscr{L}(\epsilon^{-\alpha t}) = 1/(s + \alpha)$ and show that it leads to the given $x(s)$.
4. Carry out the step-by-step derivation of Eq. (2-13).
5. Given a touch receptor for which, in the step response of Eq. (2-2), with millivolt and millisecond units, we have $V_p = 15$, $kV_p = 3$, $\tau_1 = 1.25$, and $\tau_2 = 5$. Now the input stimulus is a ramp. (a) Find the transfer function $H = V_r(s)/V_i(s)$. (b) The transducer is linear so that V_i is a ramp, $V_i = 0.25t$. Find V_r and plot the ramp response for $0 < t < 20$ ms. (c) The transducer is nonlinear, and fits the special case in which the pole of Eq. (2-10) cancels the zero of H, with $abm = 0.25$. Find V_r and plot the ramp response for $0 < t < 20$ ms. (d) How do the curves of (b) and (c) compare as $t \to 0$? [Ans.: (a) $H = 0.64(s + 0.05)/(s^2 + s + 0.16)$. (b) $V_r = 0.6875 + 0.05t + 0.3125\epsilon^{-0.8t} - \epsilon^{-0.2t}$. (c) $V_r = 1 + 0.3333\epsilon^{-0.8t} - 1.3333\epsilon^{-0.2t}$. (d) As $t \to 0$, both curves $\to 0.08t^2$.]
6. Derive all of the $|H(j\omega)|$, $\theta°$, and TD values of Table 2-2.
7. (a) Derive the algebraic expression for the transfer function, H, of Eq. (2-3). (b) Rearrange the terms so as to display the zeros and poles; that is, each term should be of the form $(s + \alpha)$. (c) Write the algebraic expression for the amplitude

response, $|H(j\omega)|$. (d) Write the algebraic expression for the phase shift, θ. (e) Write the algebraic expression for the time delay, TD. [Ans.: (a) $H = \{k + s[\tau_2 - \tau_1(1 - k)]\}/[(1 + s\tau_1)(1 + s\tau_2)]$.]

8. Given a touch receptor for which, in the step response of Eq. (2-2), with millivolt and millisecond units, we have $V_p = 15$, $kV_p = 3$, $\tau_1 = 1.25$, and $\tau_2 = 5$. Now the input stimulus is a relatively weak steady-state sine wave, so that the transducer is linear. (a) Find $|H(j\omega)|$ and plot the amplitude response for $1 < f < 1000$ Hz. Use $\log_{10}$ scales for both $|H(j\omega)|$ and f. (b) Find $\theta°$ and plot the phase response for $1 < f < 1000$ Hz. Use $\log_{10}$ scale for f. (c) Find $|H(j\omega)|$, θ, and TD at $f = 1$ Hz; (d) at θ_{max}; (e) at $|H(j\omega)|_{max}$; (f) at $f = 1000$ Hz. [Ans.: (a) $|H(j\omega)| = 0.64\sqrt{(0.05^2 + \omega^2)/[(0.2^2 + \omega^2)(0.8^2 + \omega^2)]}$ (krps units). (b) $\theta = \arctan(\omega/0.05) - \arctan(\omega/0.2) - \arctan(\omega/0.8)$. (c) $TD = -13.65$ ms; (d) $f = 12.86$ Hz, $\theta_{max} = 30.48°$; (e) $f = 62.07$ Hz, $|H(j\omega)|_{max} = 0.6451$; (f) $TD = 0.2260$ ms.]

9. Given the network values of Eq. (2-19)—$L = 1.124$ H, $R - 1 = 0.791\ \Omega$, $C = 2.168$ F, and $G = 0.1218$ S—show that they yield the given transfer function, H, of Fig. 2-1.

10. It is claimed, in Eq. (2-19), that $1/H = sL + R + 1/(sC + G)$, where L, R, C, and G are the network components of Fig. 2-1. Find the algebraic expressions for L, R, C, and G and show that they yield the transfer function, H.

3

Generation of the Action Potential

ABSTRACT. An axon is a relatively poor conductor, so that a graded potential is severely attenuated after traveling less than a millimeter. A different scheme is used for all long-distance transmission. The graded potential is translated into a series of voltage spikes, called *action potentials* (APs), each 100 mV high, that are continually regenerated. The number of spikes per second is roughly proportional to *change* in the original graded potential, so what we have is "frequency modulation": the AP frequency constitutes the message. The AP discharge is characterized by a considerable amount of random "noise," because it is initiated by ions that are constantly buffeted about in accordance with the kinetic theory of heat. The noise is not all bad; it is partly responsible for nonstereotyped behavior, an important ingredient of "creativity."

During an AP, sodium ions enter the axon while potassium ions leave. This migration is subsequently undone as the ions are pumped back via the expenditure of metabolic energy.

3-1 The Need for Action Potentials

Now that we have sensory information, what do we do with it? The information is in the form of graded receptor potentials, such as those of Figs. 2-2 and 2-3. The message has to be transmitted from the sensory receptor to its intended target. There is no exotic way, beyond a "wire" covered by insulation, in which this electrical signal can be transmitted. In the case of a telephone line, the wire is of course made out of an excellent conductor such as copper. The analogous axon is filled with axoplasm, a fluid that is a conductor by virtue of its positively charged potassium ions and negatively charged protein molecules. The latter are relatively large, complex molecules.

Alas, the voltage that represents the nervous system's message is attenuated very rapidly by an axonal conductor. The situation is quantitatively depicted in Fig. 3-1(*a*). The axon, with a diameter of 0.3 μm (0.0003 mm), is the smallest unmyelinated fiber to be found in the human body. (As described in Chap. 1, the fiber has a 75-Å-thick insulating membrane rather than a relatively thick myelin outer membrane.) The axon causes the potential (V) to fall to 37% of its original value (V_{in}) in only 0.15 mm. The distance to the 37% point is called the *length constant*; it gives one an idea about how far the signal can travel before it becomes unusable. The exponential decay of V is described by

$$V = V_{in} \exp\left(-\frac{x}{\lambda_0}\right), \tag{3-1}$$

where λ_0 is the length constant and x is the distance along the axon.

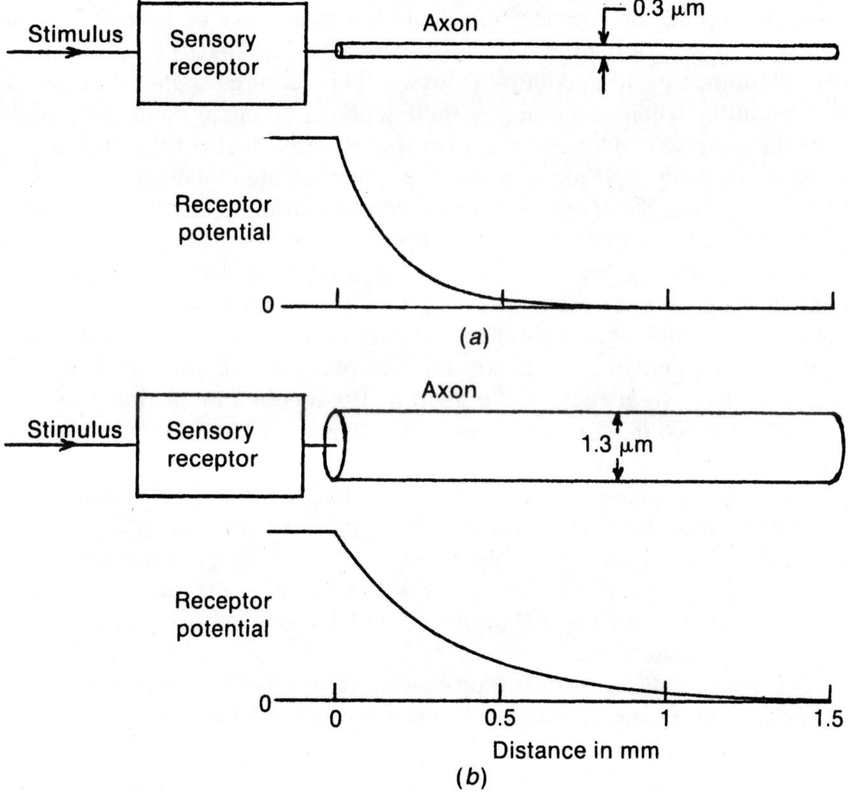

Fig. 3-1 How an idealized unmyelinated axon attenuates the receptor potential output of a sensory receptor. (*a*) If the diameter is 0.0003 mm (the smallest mammalian axon). (*b*) If the diameter is 0.0013 mm (the largest mammalian axon).

The length constant is proportional to the square root of axon diameter. The largest unmyelinated axon is illustrated in Fig. 3-1(*b*); with a diameter of 1.3 μm, it is approximately four times as large as the 0.3-μm fiber, and its length constant, 0.33 mm, is larger by a factor of 2.

The nonregenerating axon was an evolutionary dead end. A length constant of 0.33 mm is viable for a small organism, a few millimeters long at the most. The length constant could be increased to 1 mm, say, by increasing the diameter of its axon by a factor of 10, but it is not feasible to fill up a small creature with large-diameter fibers. Improving the conductivity of axoplasm would help; a really dramatic increase in length constant, by a factor of 8500, would take place if axoplasm could be replaced by copper. For a 1.3-μm axon this would yield a length constant of 2.8 m (9.3 ft). Iron is not as good as copper, but it would suffice to make a midget who literally has nerves of steel and a magnetic personality. It happens that the human nervous system filled with axoplasm is now able to make printed circuit lines that are 0.3 μm wide!

Evolution, not able to replace its ionic conductor with a solid metallic conductor, opted for regeneration. This is not as easy as it sounds. One method for regenerating the receptor potential of Fig. 3-1(*b*) is to amplify it every 0.3 mm, say, to make up for losses. This requires sophisticated transistor circuitry when it is done by the telephone company, and it certainly is not the method employed by the nervous system. Instead, the latter uses an "all-or-nothing" technique: when the receptor potential reaches a threshold level, such as 8 mV, the sensory receptor neuron "fires," generating a 100-mV action potential (AP). The message is not that the neuron fired; it is *how many times a second* it fired. With a receptor potential increase of 10 mV it fires 10 times per second, say; with a potential increase of 29 mV it may fire 50 times per second, as illustrated by an example in the next section. This is known as *frequency modulation*: the receptor potential modulates the firing frequency of the neuron. Frequency modulation was thus invented over 500 million years ago, when relatively large animals began to evolve. It sets no distance limit on the transmission of nervous information. When the voltage on each small section of an axon exceeds 8 mV, that section of the axon fires, sending its 100-mV spike on to the next section where it, in turn, is regenerated. The frequency is preserved: if a proximal section of the axon fires 50 times per second, the distal end of the *same* axon will, after allowing time for the AP to propagate from one end to the other, also fire 50 times per second.

It is interesting to speculate on how all of this evolved, but speculation here is dangerous and is best left to life scientists who are experts on such matters.

3-2 From Sensory Receptor to Neuron

We can examine the generation of an AP from many different observation points. We can descend to the molecular level and pursue ions as they diffuse across the excitable membrane. Some of this ion chasing is done at the end of the chapter, but in the present section generation is considered from a much simpler point of view: The neuron is analogous to a loaded gun; when a critical threshold level is reached, the "gun" explosively generates an AP.

Generation from this point of view is illustrated in Fig. 3-2. The APs are traced out by the heavy lines, which extend from −90 mV to +10 mV, a total height of 100 mV. The drawing is magnified in the vertical direction in order to better show what is going on at the base of the AP.

The drawing pertains to the neuron of a sensory receptor–neuron pair. In Fig. 3-2(*a*) the sensory receptor potential is −80 mV; in Fig. 3-2(*b*) it is −61 mV. We see that some new and perhaps provocative language has been introduced. We have a firing voltage and an extinguishing voltage. It happens that electrical engineers have been using for many years devices that "fire" and "extinguish" in exactly the same sense as that of a neuron. Their common usage calls for "voltage" rather than "potential," so Fig. 3-2 is a hybrid, a mixture of voltages and potentials.

Let us follow the heavy line in Fig. 3-2(*a*). At zero time the internal fluid of the neuron is at −90 mV (see Fig. 1-6). The potential gently rises to meet the receptor potential at −80 mV, but a critical threshold is reached at −82 mV, when the time is at 0.1 s. The neuron "fires"; the internal

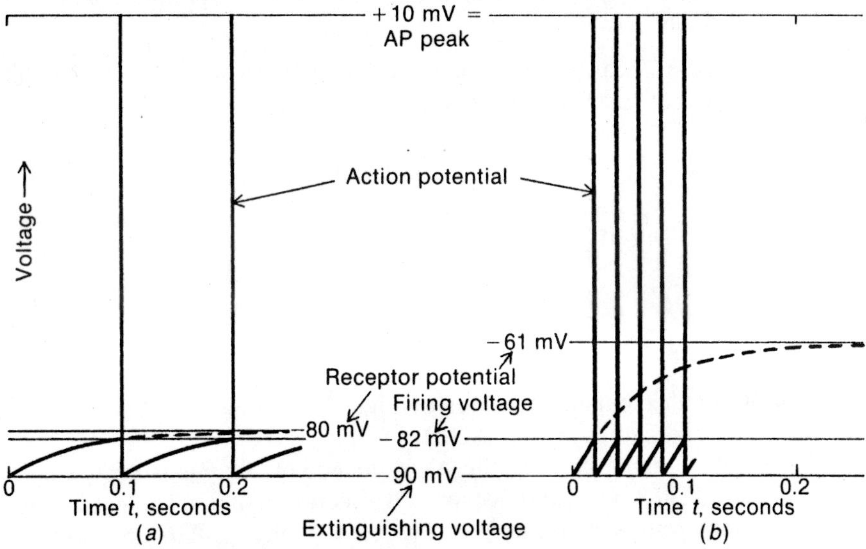

Fig. 3-2 Idealized "all-or-nothing" model for the generation of an action potential by the neuron of a sensory receptor–neuron pair. Starting at the −90-mV extinguishing level, the voltage gently rises toward the receptor potential with a time constant of 0.062 s, but the neuron fires when the firing level at −82 mV is reached, generating a 100-mV AP spike. (*a*) Receptor potential = −80 mV, firing frequency = 10 Hz. (*b*) Receptor potential = −61 mV, firing frequency = 50 Hz.

potential rapidly rises to +10 mV, then rapidly falls to −90 mV. The latter is called the *extinguishing* level because the neuron's "catastrophic" behavior terminates.

On the time scale of Fig. 3-2 the rise and fall are represented by a single vertical line. It looks like a "spike" and the AP is, in fact, commonly known as a *spike*.

The cycle repeats, so that in Fig. 3-2(a) the AP has a period of 0.1 s, or a frequency of 10 Hz.

In Fig. 3-2(b), where the internal potential of the neuron is approaching a receptor potential of −61 mV, it goes from the extinguishing to firing levels in 0.02 s, so that the AP frequency is 50 Hz. The figure therefore illustrates how a stimulus that causes the receptor potential to increase from −80 mV to −61 mV is translated into an AP frequency change of from 10 to 50 Hz.

The voltage exponential of Fig. 3-2 is given by

$$V = V_r - (V_r - V_E)\exp\left(-\frac{t}{\tau_0}\right), \tag{3-2}$$

where V_r = receptor potential,
V_E = extinguishing voltage,
τ_0 = exponential's time constant. (In Fig. 3-2, the time constant is 0.062 s.)

At $t = 0$, the V_r cancels and we are left with $V = V_E$. At $t = 1/f$, which is the reciprocal of the firing frequency, substitution in Eq. (3-2) yields

$$V_F = V_r - (V_r - V_E)\exp\left(-\frac{1}{f\tau_0}\right). \tag{3-3}$$

Solving for the firing frequency,

$$f = \frac{1}{\tau_0 \log\left(\dfrac{V_r - V_E}{V_r - V_F}\right)}. \tag{3-4}$$

Substitution of the numerical values of Fig. 3-2 will show that f = 10 Hz in Fig. 3-2(a) and 50 Hz in Fig. 3-2(b).

It is well to remember that the AP sequence of Fig. 3-2 is not a single, isolated event. The body contains some 10^{12} neurons (it is fashionable nowadays to compare this with the U.S. National Debt in dollars and, like the National Debt, the estimate of the total number of neurons has been increasing by an order of magnitude every few years). At any particular instant of time, many of the neurons can be generating APs; in the awake state, therefore, millions of APs can occur each second. It has been traditional to probe life via the heartbeat because one can feel and hear the systolic pulse, but it is not at all a fundamental unit. Contraction of the heart muscle is itself the product of action potentials. A more reliable measure of life is the graded potential or action potential, which can be seen with modern electronic instruments. An electroencephalographic (EEG) scalp signal is mostly due to graded potentials in the brain; the lack of an EEG signal marks "brain death," the end of life.

3-3 Trains of Action Potentials

One of the greatest milestones in humanity's understanding of natural phenomena was reached via the kinetic theory of heat. At an atomic and molecular level the universe is in constant motion. In a gas or liquid, an atom or molecule is like an elastic rubber ball that flies through space in a straight line until it strikes another atom or molecule; the two then rebound and start on new paths. As temperature is increased, the average velocity of the "rubber balls" increases. For these tiny particles, gravity has only a small effect at room temperature. The flying particles do not fall down; they continue forever to rebound from each other. In a solid, they vibrate about an average position, pulling and pushing their neighbors without breaking loose and wandering away. In a liquid or gas they wander away from each other in the phenomenon known as *diffusion*.

In an aqueous suspension of pollen, Robert Brown in 1827 observed, by looking through a microscope, the continuous, haphazard, zigzag movements of the pollen grains as they were struck by surrounding molecules of water. It took 50 years before the correct explanation was recognized, and another 28 years (1905) before a quantitative theory was developed, notably by Albert Einstein.

What does this have to do with the nervous system? In an axon, the water molecules, ions, and membrane molecules are in constant random thermal motion. The precise horizontal lines that represent extinguishing voltage, firing voltage, and receptor potential in Fig. 3-2 are only approximations. The axon membrane consists, in part, of protein molecules that carry Na^+ ions into the axon (passive transport because energy is not required), and other protein molecules that pump Na^+ ions out of the axon (active transport because metabolic energy is required). When the firing level is reached, protein molecules carry Na^+ ions into the axon; this "depolarizes" the membrane, initiating an AP [M. S. Gordon, 1972]. At the extinguishing level the protein channels close, ending the AP. The opening and closing of the channels is affected by the jostling of the atoms that make up the protein molecules, so there can be considerable randomness in the resulting train of APs. Instead of the clocklike precision of Fig. 3-2 we can have a noisy sequence such as that of Fig. 3-3(*a*). (Although "noise" was originally an auditory adjective, the word is applied now to any undesired departure from a perfectly ordered system.) For protein molecules, many of the biochemical reactions at body temperature can go to completion only because of the previously cited jostling of atoms.

Depending upon the receptor potential and many other factors, some sensory receptor neurons are much noisier than others. The train of APs shown in Fig. 3-3(*a*) is similar to one observed by Nelson Kiang and his colleagues (1965) from the unstimulated, spontaneous spikes generated by auditory neurons of the cat. The sequence shown here corresponds to the following idealized experimental conditions: the microelectrode was in place for 1300 s (22 min) and detected 65,000 APs. The *average* discharge frequency, therefore, was 50 spikes per second, the same as in Fig. 3-2(*b*). At this frequency the *average* interval between discharges is 20 ms, but the very small sample of 10 intervals shown in Fig. 3-3(*a*) displays the following sequence in milliseconds: 4, 15, 9, 26, 6, 20, 12, 37, 5, and 17. Certainly anything but clocklike. Why are there not more automobile accidents if the nervous system is made up of noisy elements like that of Fig. 3-3(*a*)? Mainly because thousands or millions of neurons are involved in any activity such as driving a car, and the randomness of individual neuron discharges "averages out" because of the large numbers of neuron cell bodies, axons, muscle fibers, and so forth. Nevertheless, even an experienced typist makes mistakes, and the noise inherent in individual neuron discharges is an ever-present and sometimes dangerous phenomenon.

One can also say that evolution yielded neurons that duplicate each other so as to reduce noise effects because it was not able to evolve copper wiring.

The AP train of Fig. 3-3(*a*) demonstrates an additional characteristic besides randomness: the minimum interval is 4 ms. This is called the *refractory* period. Following the refractory period, each section of axon membrane that produced an AP pumps out sodium ions equal to those that entered during that AP (and it pumps back in potassium ions equal to those that left the neuron cell body or axon or muscle fiber). The refractory period sets an upper limit on how rapidly the excitable tissue can discharge. If the refractory period is 4 ms, the maximum discharge frequency is the reciprocal, 250 APs per second. In response to a relatively high receptor potential, the neuron of Fig. 3-3(*a*) could theoretically fire at a constant rate of 250 times per second, *without noise*.

Figure 3-3(*b*) illustrates, for the same neuron, an in-between case. The

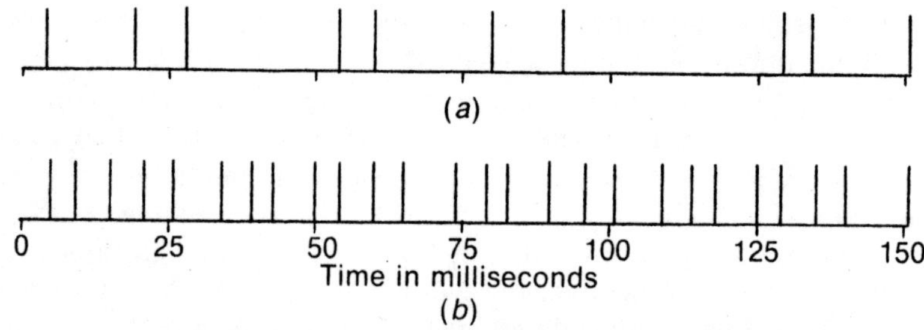

Fig. 3-3 Small samples of an idealized "noisy" train of APs similar to the spikes generated by auditory neurons of the cat. The total sample time was 1300 s, with a refractory period of 4 ms. (*a*) Unstimulated, spontaneous emission. The average discharge frequency is 50 spikes/s. (*b*) Stimulated, with the receptor potential such that the average discharge frequency is 167 spikes/s.

receptor potential is such that the average discharge rate is 167 spikes/s. The total sample time, 1300 s, is the same as before, but now a total of 218,000 APs was detected. Comparing Fig. 3-3(*a*) and (*b*), two differences stand out. First, the frequency of Fig. 3-3(*b*) is of course much higher than that of Fig. 3-3(*a*). Second, visual inspection shows that the discharge train of Fig. 3-3(*b*) is much less noisy than that of Fig. 3-3(*a*). What is the reason for this? The first 10 intervals are, in milliseconds, 5, 4, 6, 6, 5, 8, 5, 4, 7, and 4. The departures from 4 ms are much less than in Fig. 3-3(*a*). Higher receptor potentials therefore result in less noisy APs. At a sufficiently high receptor potential, every period is 4 ms wide and there is no noise at all. In this sense, the 4-ms refractory period is a stabilizing influence.

The noisiness of the nervous system has important implications beyond the occasional mistake made by a typist or even fatal error made by a motorist. It is a good illustration of how mutations operate in evolution so as to originate new species. A perfect system without noise can survive very well under constant environmental conditions that introduce a minimum of competition; the real world, however, demands an imperfect but robust noisy system that can adjust itself to ever-present change.

For example, in the spontaneous discharge 10-interval sequence of Fig. 3-3(*a*) the average period is 20 ms and the widest period is 37 ms. But in the total 22-min experiment there were 65,000 discharge intervals and, although the average period was 20 ms, there were some very wide intervals. The five widest intervals, in milliseconds, were 181, 170, 164, 159, and 156. Figure 3-4 shows the sequence containing the widest interval, 181 ms. The latter stands out like a "sore thumb," of course, and one would normally reject the data containing this sequence in the belief that, mysteriously, something went wrong. On the contrary, the 181-ms interval is a perfectly natural rare event to be expected from an unstimulated auditory neuron of a cat if it has an average frequency of 50 Hz, refractory period 4 ms, and it discharges for 22 min.

The spike sequence of Figs. 3-3 and 3-4 is known as a Poisson point process because Simeon Denis Poisson (1781–1840) made sense out of these random events [A. V. Holden, 1976; W. D. O'Neill, J. C. Lin & Y.-C. Ma, 1986; A. Papoulis, 1965; E. Parzen, 1962; D. L. Snyder, 1975; M. C. Teich, 1989; X. Yu and E. R. Lewis, 1989]. Consider the Poisson point process of Fig. 3-5(*a*), which depicts 15 discharge intervals during 318 ms. (The refractory period is zero in this example.) The *time intervals* between spikes,

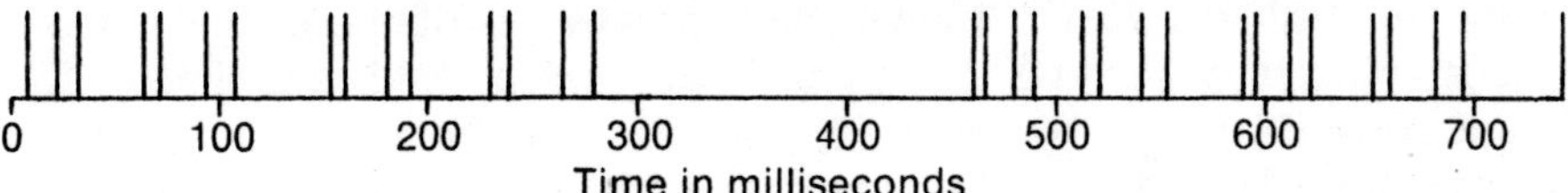

Fig. 3-4 Small sample of the same sequence of APs that gave rise to the spontaneous discharge of Fig. 3-3(*a*), except that the sample now includes the rare event in which, out of the total of 65,000 APs, the interval between one spike and the next was 181 ms.

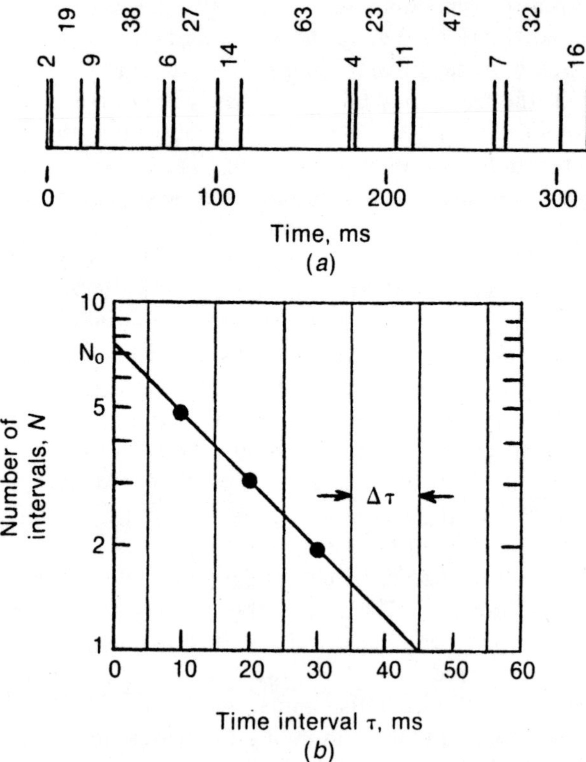

Fig. 3-5 Idealized example of a spontaneous neuron discharge that obeys a Poisson point process in accordance with $N = 7.5\epsilon^{-45\tau}$ (units in seconds), where N is the expected number of intervals and τ is the time width of each interval. (*a*) Train of action potentials containing 15 intervals, with corresponding τ values as indicated. (*b*) The interval histogram using bins $\Delta\tau = 10$ ms wide. $N_0 = 7.5$ is the $\tau = 0$ intercept. The data are summarized in Table 3-1. (From Deutsch and Tzanakou, *Neuroelectric Systems*, New York Univ. Press, New York, 1987.)

τ, are 2, 19, 9, . . . ms wide, as indicated. If we tabulate the 15 intervals, listing them in order of decreasing size, we get Table 3-1. Here the M number identifies intervals according to their (decreasing) size place. Poisson's contribution was to derive the "random point process distribution," and to place the intervals into time bins that are $\Delta\tau$ wide, as depicted in Fig. 3-5(*b*). The value $\Delta\tau = 10$ ms is illustrated. The first bin runs from 5 to 15 ms, and is centered on $\tau = 10$ ms. The second extends from 15 to 25 ms and is of course centered on $\tau = 20$ ms, and so forth. On the right in Table 3-1 is listed the bin into which each of the 15 intervals falls. According to the table, $M = 1$, with $\tau = 63$ ms, falls into the 60-ms bin; $M = 2$, with $\tau = 47$ ms, falls into the 50-ms bin. At the other end, the five intervals from $M = 9$ to 13, with values ranging from 14 to 6 ms, all fall into the 10-ms bin. The M

TABLE 3-1 The 15 Intervals of Fig. 3-5(a) with Corresponding M Numbers and Bins

M Number	Time Interval (τ, ms)	Bin Centered on (ms)
1	63	60
2	47	50
3	38	40
4	32	30
5	27	30
6	23	20
7	19	20
8	16	20
9	14	10
10	11	10
11	9	10
12	7	10
13	6	10
14	4	—
15	2	—
	t_T = 318	

Note: Intervals versus M numbers are calculated from Eq. (3-5). The M number identifies intervals according to their decreasing size place. The "Bin" column lists the nearest bin into which each interval falls (centered on 10, 20, . . . ms). And t_T is the total sample time.

= 14 and 15 intervals fall into the bin between 0 and 5 ms, but this is a half-bin and can be ignored (it is sometimes facetiously referred to as a "has-been"). The intervals versus M number in Table 3-1 are given by

$$M = \frac{N_0}{\lambda \ \Delta\tau} \epsilon^{-\lambda\tau} \tag{3-5}$$

(which is derived in Appendix A3-1), where N_0 = the $\tau = 0$ intercept on the N axis of Fig. 3-5(b), and λ is a constant. (For the *special case* of zero refractory period, λ is the average AP frequency.)

The reason for considering the Poisson point process in some detail is that *all* AP sequences are, to some extent, noisy. Nevertheless, most of the published data show APs with clocklike precision; Kiang et al. (1965) is a notable exception. The noise displayed by auditory neuron APs (Chap. 9) is a central feature that one cannot ignore; in fact, the central auditory processor in the brain normally gets a variable-interval neuronal signal. The important point with regard to Fig. 3-5(b) is that the number of intervals, N, that are *expected* to fall into each bin, is described by

$$N = N_0 \epsilon^{-\lambda\tau}. \tag{3-6}$$

Figure 3-5 is drawn for $N = 7.5\epsilon^{-45\tau}$ (units in seconds). At τ = 10 ms we approximately get N = 5 (the M = 9 to 13 intervals of Table 3-1); at τ =

20 ms we approximately get $N = 3$ (the $M = 6$ to 8 intervals); and at $\tau = 30$ ms we approximately get $N = 2$ (for $M = 4$ and 5). The sample is far too small to yield accurate values, especially for $M = 1$, 2, and 3.

Figures 3-3 and 3-4 show only small samples of their complete spike trains. For the complete sequence of Figs. 3-3(*a*) and 3-4, we have $f_{avg} = 50$ spikes/s and $t_T = 1300$ s for a total of $\eta = 65{,}000$ action potentials, and $\tau_{min} = 4$ ms. The data have not been placed into bins, but it is easy to evaluate $N_0/\lambda\ \Delta\tau$. First, from Eq. (3-18), we get $\lambda = 62.5$ Hz; then from Eq. (3-15), $N_0/\lambda\ \Delta\tau = 65{,}000\epsilon^{1/4}$. Substituting into Eq. (3-5), and solving for τ, there results

$$\tau = 0.016 \log\left(\frac{65{,}000}{M}\right) + 0.004. \tag{3-7}$$

For the smallest interval, which corresponds to $M = 65{,}000$, Eq. (3-7) obviously gives the τ_{min} value of $\tau = 4$ ms. For the widest interval we can expect, which corresponds to $M = 1$, Eq. (3-7) gives 181 ms, as displayed in Fig. 3-4. For the next widest interval we can expect, with $M = 2$, Eq. (3-7) gives 170 ms, and so forth. In other words, out of the 65,000 AP discharges, of which Fig. 3-4 is a small sample, we expect an interval 181 ms wide, another 170 ms wide, and so forth. The point is that, if we wait long enough, the rare event *will* occur.

As another illustration of a rare event, suppose that every morning you toss an "honest" coin into the air 10 times and record which way it landed: H for heads and T for tails, such as HHHTTHT · · · , and so forth. You do this for two years and 10 months. At the end of this time you will have 1024 head–tail sequences. The chances are that one of them will be the rare event in which ten heads occur in a row: HHHHHHHHHH. The chances are that another sequence will be the rare event in which ten tails occur in a row: TTTTTTTTTT. Each of these has one chance of appearing out of the $2^{10} = 1024$ days in which you tossed the coin into the air 10 times in a row. There is no guarantee that ten Hs and ten Ts in a row will ever occur, and they may even occur more than once, but we can *expect* each of these rare events to occur once.

In cell reproduction, which is also a somewhat "noisy" system because the atoms are constantly jostling each other, some of the mistakes are called *mutations*. If we wait long enough, several reproducible mutations will occur simultaneously under favorable circumstances, giving birth to a change in gene expression or phenotype. To hasten the process, biologists routinely work with millions of bacterial cells that can replicate and divide every 20 min, and the desired mutation is recognized by the cell's resistance to a drug or some such strategy. With a smaller population of interacting individuals in a mammalian species that reproduces over years rather than minutes, one may have to wait 10,000 years for a significant evolutionary change to occur.

Similarly, in the nervous system, not every rare event is associated with a deleterious combination such as a typist's error or a fatal accident. If one's energies are deliberately focused, some rare events can give rise to new ideas, a process sometimes called *creativity* or *serendipity*. And remember that each of us has won a lottery, the rare uniting of a particular sperm with an egg cell.

3-4 The Hodgkin–Huxley Action Potential

Up to this point in the chapter the AP is a narrow spike. In the present section we look at it through a "microscope" that reveals that it is about 2 ms wide, as shown at the top of Fig. 3-6. (Voltage is shown relative to the resting potential.)

The AP depicted in this figure is from a model by Alan L. Hodgkin and Andrew F. Huxley, as elucidated in a series of experiments reported in 1952; they were subsequently awarded a Nobel Prize in 1963, as mentioned in Section 1-4 [see also A. L. Hodgkin and B. Katz, 1949; S. A. Talbot and U. Gessner, 1973].

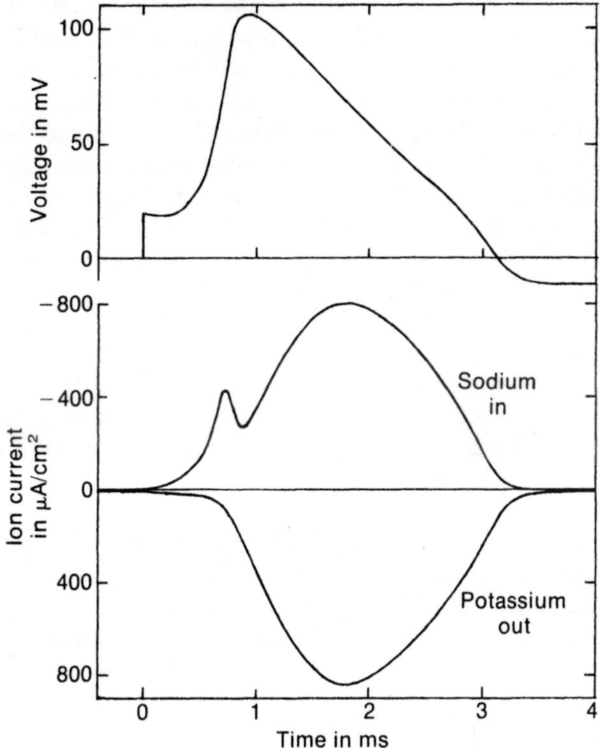

Fig. 3-6 The AP spike and associated transmembrane ion currents as seen through a "microscope" that reveals that they are about 2 ms wide. The curves are from a model described by Alan L. Hodgkin and Andrew F. Huxley in which a +20-mV trigger is applied at zero time to the giant axon of the squid. Voltage is shown relative to the resting potential. Current values are for 1 cm^2 of membrane surface area.

Hodgkin and Huxley worked with the unmyelinated axon of the squid because it is a 0.5-mm giant axon through which one can pass a fine wire. Another advantage is that after it is dissected out of the animal it can survive for hours in sea water at 6.3°C (43.3°F). Hodgkin and Huxley made a large number of observations in which they applied various voltages to the fine wire inside the axon, thus forcing it to assume various points on the AP curve. They changed the chemical composition of the sea water in order to reveal the separate roles played by sodium, potassium, and chloride ions. The research was done in the days before transistors and electronic computers were widely available. The measurements utilized sophisticated vacuum-tube circuits that were designed by Kenneth S. Cole in 1949 (see his book published in 1972), and calculations were laboriously done on mechanical calculators.

Hodgkin and Huxley were very lucky, because it turned out that the movement of ions during an AP is not a disorganized hodgepodge in which charged particles rush through holes in the axonal membrane. Instead, the 75-Å-thick membrane is bridged by protein molecules that are unique for each ion species. For an axon (or excitable tissue in general), because of the concentration gradients and transmembrane potential (see Sect. 1-5), Na^+ ions "try" to diffuse into the axon, while K^+ ions "try" to diffuse out. Chloride ions are practically at equilibrium and only play a minor role during an action potential, while negatively charged protein molecules in the axoplasm are too large to "escape." As pointed out in Section 3-3, some protein molecules carry Na^+ ions into the axon (passive transport because energy is not required), while others pump Na^+ ions out of the axon (active transport because metabolic energy is required). Similarly, some protein molecules carry K^+ ions out of the axon (passive transport), while others pump K^+ ions into the axon (active transport). For every Na^+ ion that enters the axon during membrane depolarization, a membrane protein molecule subsequently captures a Na^+ ion and uses energy to carry it outside. This is known as the *sodium pump*. Similarly, a *potassium pump* uses metabolic energy to return "escaped" K^+ ions. To a physicist, the process is reminiscent of "Maxwell's Demon," which presides over the tiny opening between two gas chambers and decides to let certain atoms or molecules through while blocking others.

For the AP of Fig. 3-6, effect follows cause as follows (the t values refer to the time scale of Fig. 3-6):

1. At $t = 0$: A +20-mV voltage pulse is momentarily applied to the wire inside the axon. (The wire is substituting for a neuronal initiator of APs.) The pulse duration is sufficient to completely charge the membrane capacitance. Immediately afterwards, at $t = 0^+$, the voltage source V becomes an open circuit and the applied current I goes to zero.
2. At $t = 0.4$ ms: The membrane becomes depolarized by the 20-mV

pulse, thereby causing an abrupt increase in permeability to Na^+ ions.

3. At $t = 0.6$ ms: The internal voltage of the axon rises sharply.
4. At $t = 0.8$ ms: The potassium ion carriers in the membrane carry K^+ ions out of the axon.
5. From $t = 1$ to 3.2 ms: The internal voltage gradually decreases.
6. Some 3.2 ms after the initial trigger, the AP has substantially ended. (The word "trigger" here denotes a voltage or current level that initiates a burst of activity.)

During the AP, the sodium carrier proteins convey Na^+ ions into the axon in accordance with the "sodium in" curve of Fig. 3-6 (passive transport). We define current that flows *out* of the axoplasm as "positive"; therefore, the "sodium in" current of Fig. 3-6 is negative, as shown. (Later on, not shown, is the positive current corresponding to the pumping-out [active transport] of these Na^+ ions.) The peculiar dip in the curve is easily explained: this coincides with the peak of the AP curve, where the actual voltage across the membrane is so low that very little pressure exists to drive the sodium ions into the "waiting arms" of its transport proteins.

The potassium carrier proteins convey K^+ ions out of the axon in accordance with the "potassium out" curve of Fig. 3-6 (passive transport). (This is a positive current; later on, not shown, is a negative curve corresponding to the pumping-in [active transport] of the K^+ ions.) To the left of the sodium dip the Na^+ current in is much greater than the K^+ current out (resulting in a rapid rise in voltage), but to the right of the dip the two currents are approximately equal and opposite; here the net current crossing the membrane is a relatively small excess of potassium ions (resulting in a slow drop in voltage). We can only assume that the preceding scenario is the best that evolution has been able to do; perhaps, on another earth, a more efficient AP has evolved (or will evolve on this earth given another 500 million years or so).

Hodgkin and Huxley used the equivalent circuit of Fig. 3-7 for the short length of squid axon. The axon consists of a conductor (axoplasm) surrounded by an insulating membrane. The "cable" is immersed in sea water, which is a good conductor. All of the action takes place in the membrane which, it turns out, is an imperfect and variable (nonlinear) insulator. The g_{Na} is the equivalent conductance of the membrane to sodium ions; g_K is the conductance to potassium ions; and g_{Cl} is the nominal conductance to chloride ions (actually, it is a catchall, the equivalent conductance to all ions except sodium and potassium). The arrows next to the *I*s indicate positive directions of current flow. The 115-, 12-, and 10.6-mV voltage sources are equivalent values at rest (the axon is unstimulated, with $V = 0$). At rest, a small sodium ion (Na^+) current (0.15% of the 803-μA maximum of Fig. 3-6) flows into the axoplasm as if propelled by 115 mV in series with g_{Na}; a small potassium ion (K^+) current (0.5% of the 837-μA maximum) flows out of the axoplasm as if propelled by 12 mV in series with g_K; and chloride *ions* (Cl^-) flow out of the axoplasm (equivalent to chloride *current* flowing into the axoplasm) as if propelled by 10.60 mV in series with g_{Cl}. These "at rest" currents add up to a small longitudinal ion flow, along the axon, either toward or away from the neuron body [J. H. Schwartz, 1980]. Whatever unbalance this causes may be restored during the next action potential, which is, after all, a relatively violent event compared with the quiet "at rest" happenings.

Hodgkin and Huxley assumed a membrane capacitance of 1 $\mu F/cm^2$, as shown in Fig. 3-7. This is only an approximate value, as discussed in connection with Eq. (1-1), but the "1" appreciably simplified their calculations. Interestingly enough, the value 1 $\mu F/cm^2$ is reasonably accurate for *all* unmyelinated fibers, from that of the squid to the human nervous system.

Hodgkin and Huxley were able to fit their experimental data to a few relatively simple equations. The main body of their work is summarized in the following six equations, Eqs. (3-8) to (3-13):

As defined in Fig. 3-7,

$$I = I_C + I_{Na} + I_K + I_{Cl}, \tag{3-8}$$

or, matching term-for-term,

$$I = C\frac{dV}{dt} + g_{Na}(V - 115) + g_K(V + 12) + g_{Cl}(V - 10.60). \tag{3-9}$$

They defined three *activity coefficients*, *m*, *h*, and *n*, as follows (the units are microamperes [μA] and millisiemens [mS] per square centimeter of membrane surface area; millivolts [mV]; and milliseconds [ms]):

$$g_{Na} = 120\, m^3 h, \qquad g_K = 36 n^4, \qquad g_{Cl} = 0.3. \tag{3-10}$$

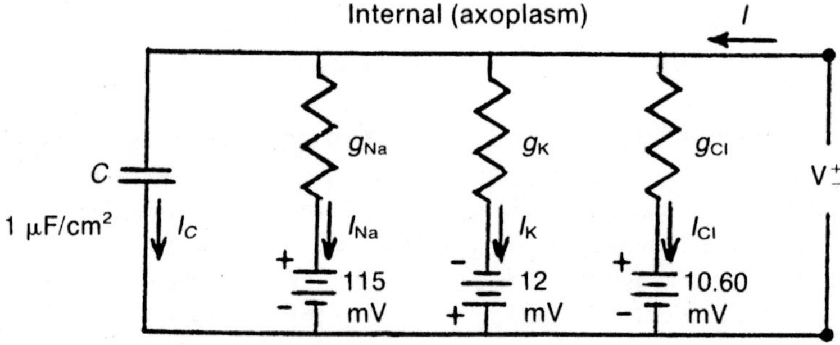

Fig. 3-7 Equivalent circuit of a short length of the squid axon according to the Hodgkin–Huxley model: I_{Na}, I_K, and I_{Cl} are the ion currents that flow in response to the equivalent battery in series with conductance *g*; *C* is the membrane capacitance and I_C is its charging (or discharging) current; *V* is an external voltage applied to the wire inside the axon; and *I* is the resulting electron current. If the external source is an open circuit, *V* is the membrane potential and $I = 0$.

The activity coefficients are related to voltage and time as follows:

$$\frac{dm}{dt} = \frac{0.1(25 - V)}{\epsilon^{0.1(25-V)} - 1}(1 - m) - 4m\epsilon^{-V/18}, \tag{3-11}$$

$$\frac{dh}{dt} = 0.07\epsilon^{-V/20}(1 - h) - \frac{h}{\epsilon^{0.1(30-V)} + 1}, \tag{3-12}$$

and

$$\frac{dn}{dt} = \frac{0.01(10 - V)}{\epsilon^{0.1(10-V)} - 1}(1 - n) - 0.125n\epsilon^{-V/80}. \tag{3-13}$$

The preceding equations represent the distillation of a tremendous amount of painstaking experimental work. It was necessary to "curve fit" the activity coefficients and their derivatives to the experimental data, which included the isolation of sodium, potassium, and chloride currents in addition to the action potential itself. It may appear to an engineer that the preceding curve-fitting is not worthy of a Nobel Prize, but the truth is that it enabled life scientists and biomedical engineers to calculate the activity coefficients; from this, the conductance values; and finally, the sodium, potassium, and chloride currents as a function of time during an action potential [J. E. Mann, N. Sperelakis & J. A. Ruffner, 1981; M. S. Spach and J. M. Kootsey, 1985].

Table 3-2 contains the first few calculated values for the waveforms of Fig. 3-6 as given by the Hodgkin–Huxley equations. (The derivatives are replaced by deltas since dt is actually replaced by $\Delta t = 0.01$ ms. Larger

TABLE 3-2 The First Few Calculated Values for the Waveforms of Fig. 3-6 as Given by the Hodgkin–Huxley Equations

	$V = 0, t < 0$	$V = 20$ mV, $t = 0^+$	$V = 19.86$ mV, $t = 0.01$ ms
m	0.05293	0.05293	0.05954
h	0.5961	0.5961	0.5946
n	0.3177	0.3177	0.3184
g_{Na} mS	0.01061	0.01061	0.01506
g_K mS	0.3666	0.3666	0.3702
g_{Cl} mS	0.3	0.3	0.3
I_{Na} μA	−1.220	−1.008	−1.432
I_K μA	4.400	11.73	11.80
I_{Cl} μA	−3.180	2.820	2.780
dm/dt 1/ms	0	0.6602	0.6406
dh/dt 1/ms	0	−0.1499	−0.1478
dn/dt 1/ms	0	0.07702	0.07616
dV/dt mV/ms	0	−13.55	−13.14

Note: Values are for 1 cm^2 of membrane surface area. The $V = 0, t < 0$ column gives values during the resting state. The $V = 20$ mV, $t = 0^+$ column gives values immediately after a +20-mV pulse is applied to the wire inside the axon. The last entry, $dV/dt = -13.55$ mV/ms, indicates that 0.01 ms later the potential will be $20 - 0.1355 = 19.86$ mV; this is the basis for the last column.

values for Δt may lead to excessive inaccuracies.) Some significant calculated values are: I_{Na} and I_K peak at -803 and 837 $\mu A/cm^2$, respectively, at $t = 1.8$ ms; I_{Cl} peaks at 29 $\mu A/cm^2$ at $t = 0.9$ ms; the net current, $-C(dV/dt)$, peaks at $t = 0.71$ ms with $I_{Na} = -417$, $I_K = 83$, $I_{Cl} = 19$, and $I_{net} = -315$ $\mu A/cm^2$.

Regardless of whether or not the preceding process is efficient, the reader is reminded that millions of these APs are generated each second as sodium and potassium transmembrane protein carriers act out their biochemical roles. It is appropriate to ask, then: Are graded potentials and action potentials the physical embodiment of the so-called "human spirit"?

An important question that is easier to answer is the following: Given the smallest fiber, whose diameter is 0.0003 mm, does a single AP appreciably change the internal sodium concentration of this axon? The normal internal Na^+ concentration, 12 $\mu M/cm^3$, is given in Table 1-2. Multiplying this by Avogadro's constant, there are 7.226×10^{18} Na^+ ions/cm^3. Multiplying this by the volume of the fiber, we get 51.08×10^7 Na^+ ions in a 1-mm length of fiber. But the area under the sodium-in curve of Fig. 3-6 shows that, in a single AP, 1.405 $\mu C/cm^2$ enter the axon. Dividing by the electron charge, this is equivalent to 8.769×10^{12} Na^+ ions/cm^2. Multiplying this by the area of the axon membrane, we get 8.265×10^7 Na^+ ions entering a 1-mm length of fiber in a single AP. This represents a 16% change in internal sodium concentration, and should be regarded as significant because it can disrupt normal axon function. In fact, it is probably the reason why axon diameters are not less than 0.0003 mm. The maximum discharge frequency of such a small fiber is 50 APs per second, which is equivalent to 20 ms between discharges. One can picture an AP lasting for 2 ms, followed by the fiber feverishly pumping out the Na^+ ions (and pumping in lost K^+ ions) during the next 18 ms in order to restore normal ion concentrations. As usual, we are totally unaware of all of this "behind the scenes" activity.

If the preceding calculation is repeated for a large fiber, it turns out that the change induced by a single AP is inversely proportional to diameter. For a giant squid axon, 0.5 mm in diameter, the change in Na^+ ion concentration due to a single AP is 0.01%, completely negligible.

A3-1 A POISSON POINT PROCESS

The total number of spike discharges, η, is given by the summation of N values. In general, the summation runs from $\tau = 0$ to $\tau = \infty$. If there is a refractory period τ_{min}, however, the summation (integration) runs from $\tau = \tau_{min}$ to $\tau = \infty$ [S. Deutsch and E. M-Tzanakou, 1987]:

$$\eta = \frac{1}{\Delta\tau} \int_{\tau_{min}}^{\infty} N \, d\tau. \tag{3-14}$$

Substituting Eq. (3-6) for N, we get

$$\eta = \frac{N_0}{\lambda \, \Delta\tau} \exp(-\lambda\tau_{min}). \tag{3-15}$$

The total sample time, t_T, is given by the summation of $N\tau$ values. The summation (integration) is given by

$$t_T = \frac{1}{\Delta\tau}\int_{\tau_{\min}}^{\infty} N\tau \, d\tau, \tag{3-16}$$

and the substitution of Eq. (3-6) for N yields

$$t_T = \frac{N_0}{\lambda^2 \, \Delta\tau}(1 + \lambda\tau_{\min}) \exp(-\lambda\tau_{\min}). \tag{3-17}$$

The average frequency is, evidently, the total number of AP spikes divided by the total sample time, or

$$f_{\text{avg}} = \frac{\eta}{t_T} = \frac{\lambda}{1 + \lambda\tau_{\min}}. \tag{3-18}$$

Finally, there is a special reason for listing the intervals in Table 3-1 in order of decreasing size. The M number, in that event, is the partial summation of N values using τ rather than $\tau_{\min}$ for the lower limit in Eq. (3-14). This integration easily yields Eq. (3-5): $M = N_0\epsilon^{-\lambda\tau}/(\lambda \, \Delta\tau)$.

In Fig. 3-5 we have $N_0 = 7.5$, $\lambda = 45$ Hz, and $\Delta\tau = 0.01$ s. Substituting into Eq. (3-5) and solving for τ, we get

$$\tau = \frac{-\log(0.06M)}{45}.$$

This was used to calculate the τ intervals corresponding to the M numbers of Table 3-1.

REFERENCES

K. S. Cole, *Membranes, Ions and Impulses*. Berkeley, Calif.: Univ. of California Press, 1972.

S. Deutsch and E. M-Tzanakou, *Neuroelectric Systems*. New York: New York Univ. Press, 1987.

M. S. Gordon, *Animal Physiology: Principles and Adaptations,* 2nd ed. New York: Macmillan, 1972.

A. L. Hodgkin and A. F. Huxley, A quantitative description of membrane current and its application to conduction and excitation in nerve, *J. Physiol.,* vol. 117, pp. 500–544, 1952.

A. L. Hodgkin and B. Katz, The effect of sodium ions on the electrical activity of the giant axon of the squid, *J. Physiol.,* vol. 108, pp. 37–77, 1949.

A. V. Holden, *Models of the Stochastic Activity of Neurones*. New York: Springer-Verlag, 1976.

N. Y. S. Kiang, T. Watanabe, E. C. Thomas, and L. F. Clark, *Discharge Patterns of Single Fibers in the Cat's Auditory Nerve*. Cambridge, Mass.: M.I.T. Press, 1965.

J. E. Mann, N. Sperelakis, and J. A. Ruffner, Alteration in sodium channel gate kinetics of Hodgkin-Huxley equations applied to an electric field model for interaction between excitable cells, *IEEE Trans. Biomed. Eng.*, vol. BME-28, pp. 655–661, Sept. 1981.

W. D. O'Neill, J. C. Lin, and Y.-C. Ma, Estimation and verification of a stochastic neuron model, *IEEE Trans. Biomed. Eng.*, vol. BME-33, pp. 654–666, July 1986.

A. Papoulis, *Probability, Random Variables, and Stochastic Processes.* New York: McGraw-Hill, 1965.

E. Parzen, *Stochastic Processes.* San Francisco: Holden-Day, 1962.

J. H. Schwartz, The transport of substances in nerve cells, *Sci. Am.*, vol. 242, pp. 152–171, Apr. 1980.

D. L. Snyder, *Random Point Processes.* New York: Wiley, 1975.

M. S. Spach and J. M. Kootsey, Relating the sodium current and conductance to the shape of transmembrane and extracellular potentials by simulation: Effects of propagation boundaries, *IEEE Trans. Biomed. Eng.*, vol. BME-32, pp. 743–755, Oct. 1985.

S. A. Talbot and U. Gessner, *Systems Physiology.* New York: Wiley, 1973.

M. C. Teich, Fractal character of the auditory neural spike train, *IEEE Trans. Biomed. Eng.*, vol. 36, pp. 150–160, Jan. 1989.

X. Yu and E. R. Lewis, Studies with spike initiators: Linearization by noise allows continuous signal modulation in neural networks, *IEEE Trans. Biomed. Eng.*, vol. 36, pp. 36–43, Jan. 1989.

Problems

1. The length constant is given by $\lambda_0 = 2.910\sqrt{d}$, where d is the axon diameter (in centimeters). Find the λ_0 of a squid axon if $d = 0.5$ mm. [Ans.: $\lambda_0 = 6.5$ mm.]
2. The length constant of an unmyelinated axon is given by $\lambda_0 = 32.54\sqrt{d/\rho}$, where d is the axon diameter and ρ is the resistivity of the axoplasm (in centimeters). Find λ_0 for a 1.3-μm axon if (a) $\rho = 125\ \Omega\cdot\text{cm}$; (b) the axoplasm is replaced by copper wire; (c) the axoplasm is replaced by iron wire. [Ans.: (a) 0.33 mm; (b) 2.83 m; (c) 1.19 m.]
3. For Fig. 3-2, the time constant is 0.062 s, the firing level is −82 mV, and the extinguishing level is −90 mV. Find the receptor potential corresponding to an AP frequency of (a) 20 Hz; (b) 100 Hz. [Ans.: (a) −76 mV; (b) −36 mV.]
4. For Fig. 3-2, the time constant is 0.062 s, the firing level is −82 mV, and the extinguishing level is −90 mV. Plot the AP frequency as a function of receptor potential as the latter varies from −80 mV to zero.
5. Carry out the step-by-step derivation of Eq. (3-15) from Eq. (3-14), and Eq. (3-17) from Eq. (3-16).
6. For a Poisson point process, express interval τ as a function of f_{avg}, η, τ_{min}, and M. {Ans.:

$$\tau = [(1/f_{avg}) - \tau_{min}] \log(\eta/M) + \tau_{min}.\}$$

7. For the Poisson point process of Fig. 3-3(*b*), given $f_{avg} = 167$ Hz, $\eta = 218{,}000$, and $\tau_{min} = 4$ ms: (a) express interval τ as a function of order number M; (b) find the widest interval to be expected. [Ans.: (b) $\tau_{max} = 28.6$ ms.]
8. Every morning you toss a coin into the air three times and record which way it lands. You do this for eight days. (a) On how many of these days can you

expect to get three Hs? (b) Two Hs? (c) One H? (d) No Hs? [Ans.: (a) 1; (b) 3; (c) 3; (d) 1.]

9. Find the values for Table 3-2 at $t = 0.02$ ms. [Ans.: $V = 19.73$ mV, $I_{Na} = -1.945$ μA, $I_K = 11.86$ μA, $I_{Cl} = 2.739$ μA, $dV/dt = -12.65$ mV/ms.]
10. Find the values for the third column of Table 3-2 if $\Delta t = 0.05$ ms rather than 0.01 ms. [Ans.: $V = 19.32$ mV, $I_{Na} = -4.289$ μA, $I_K = 12.05$ μA, $I_{Cl} = 2.617$ μA, $dV/dt = -10.38$ mV/ms.]
11. Find the percent change in internal sodium ion concentration due to a single action potential for the largest unmyelinated mammalian axon, $d = 1.3$ μm. [Ans.: 3.7%.]

4

Propagation of the Action Potential

ABSTRACT. There are two types of regenerating nerve fibers: unmyelinated and myelinated.

Unmyelinated fibers are generally small (diameter up to 0.0013 mm or 0.00005 in.) and have a slow speed [up to 1.5 meters/second (m/s) or 3.2 mph]. Two-thirds of the fibers in the body are of this type; they are used for chores for which low speed is adequate, such as the reporting of pain and temperature.

A regenerating axon fires (starts to generate an action potential [AP]) when the voltage across the membrane rises above a threshold level such as 8 mV. The AP propagates like an ocean wave; the conduction speed of an unmyelinated fiber is determined by the forward speed of the wavefront at the 8-mV level.

Myelinated fibers are relatively large (diameter up to 0.022 mm or 0.0009 in.) and have a high speed (up to 130 m/s or 290 miles per hour [mph]). They are used for the retrieval and processing of information where high speed is vital.

A myelinated fiber is high speed because its membrane is relatively thick, but it is interrupted by nodes at which the AP is regenerated. The conduction speed is determined by the time taken for the AP to reach the 8-mV level from one node to the next. This time is approximately constant, 17 μs, regardless of nerve fiber diameter.

Nonregenerative or weakly regenerative sections of nerve fiber are reflected in lower-than-normal conduction speeds; this is important clinical evidence for nerve dysfunction, such as in multiple sclerosis.

4-1 In an Unmyelinated Fiber

In Chapter 3 the action potential (AP) is stationary. In Fig. 3-6, for example, the AP is the voltage inside a short section of squid axon with respect to the outside fluid. Following a +20-mV trigger, the voltage rises rapidly 100 mV above the resting potential and then slowly falls back, completing the AP maneuver in around 2 ms.

In the present chapter the AP propagates; that is, it fulfills its intended *raison d'être* by conveying a nervous-system message for an unlimited distance along a regenerating axon—an axon that continually compensates for losses so that the AP suffers no decrease in amplitude [J. W. Cooley and F. A. Dodge, 1966; A. Finkelstein and A. Mauro, 1963; J. J. B. Jack, D. Noble & R. W. Tsien, 1975; B. Katz, 1966; K. N. Leibovic, 1972; A. Saito and A. Noguchi, 1981; A. C. Sanderson and R. J. Peterka, 1983; A. C. Scott, 1977].

Figure 4-1 illustrates the process in an unmyelinated fiber; that is, in an axon filled with axoplasm that has a very thin membrane, only 75 Å thick (75×10^{-7} mm), so that it is relatively easy for ions to be moved across the membrane. A specific numerical example is depicted, which has the following specifications:

Diameter d = 0.944 μm = 0.000944 mm,

Velocity of propagation v = 1.23 meters/second (m/s)

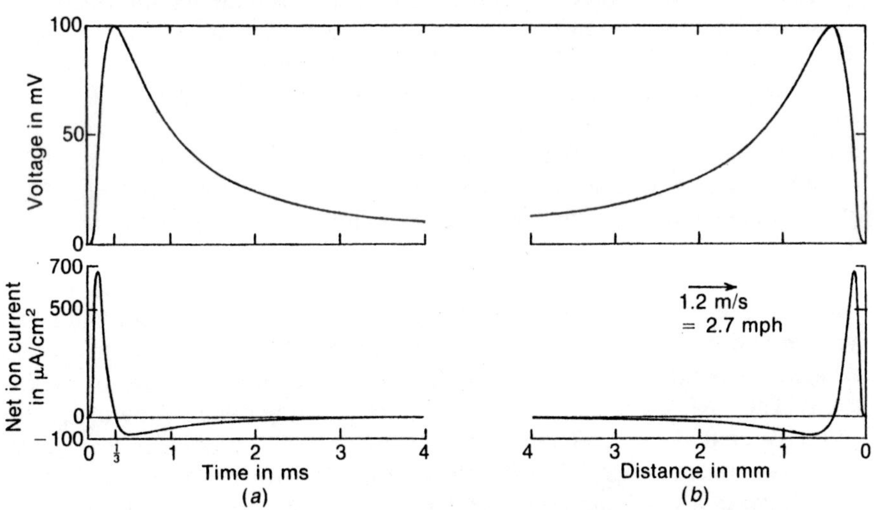

Fig. 4-1 Idealized action potential and accompanying net ion current on an unmyelinated fiber 0.94 μm in diameter. (*a*) Changes with time as seen at a particular location. (*b*) Changes with distance as "photographed" at a particular time. The curves are mirror images of each other (except for different horizontal scales).

Because unmyelinated fiber diameters range from 0.3 to 1.3 μm, the fiber of Fig. 4-1 is a valid representative.

Figure 4-1(*a*) shows the AP waveform, similar to that of Fig. 3-6. The latter is the Hodgkin–Huxley AP, whereas the AP of Fig. 4-1(*a*) is based on a less accurate but much simpler model known as the *impulse response of an* RC *cable*. (The characteristics of the *RC* cable are considered below.) Also shown in Fig. 4-1(*a*) is the net ion current accompanying (or, to be more truthful, causing) the AP. The net ion current is the "sodium in" current of Fig. 3-6 minus the "potassium out" current minus the "chloride out" current. In Section 3-4 it was convenient to use the Hodgkin–Huxley sign for current: $I_{\text{net}} = -C\,dV/dt$. In this chapter, however, it is convenient to use $I_{\text{net}} = +C\,dV/dt$.

In connection with Fig. 3-6 it is pointed out that "to the left of the sodium dip the Na^+ current in is much greater than the K^+ current out (resulting in a rapid rise in voltage)." To the right of the dip "the net current crossing the membrane is a relatively small excess of potassium ions (resulting in a slow drop in voltage)." Besides the previously mentioned sign reversal, the net ion current of Fig. 4-1 is not exactly the same as that of Section 3-4, because the curves in the present chapter are calculated using simple but approximate mathematical expressions rather than the relatively complicated Hodgkin–Huxley formulas, Eqs. (3-8) to (3-13).

To record the AP curve of Fig. 4-1(*a*), we insert a microelectrode into the axon at a convenient location and wait. The voltage will be traced out on an oscilloscope. Soon the voltage starts to rise as an AP approaches in the contiguous section of axon. The starting point is labeled "zero time." When the voltage reaches 8 mV, at time $t = 0.081$ ms, the sodium, potassium, and chloride transmembrane protein carriers enable ions to enter and leave in accordance with the "net ion current" curve. The voltage reaches a peak of 100 mV at $t = 0.33$ ms and then starts to drop. Because the action takes place at a much faster pace than the eye can follow, it is mandatory to record the curves in some permanent form so that they can be studied at leisure.

The Hodgkin–Huxley description of an action potential is much too complex if one wishes to derive general expressions for the velocity of propagation and other functions of diameter. Simplification is based on the observation that all APs are reminiscent of the impulse response of an *RC* cable. The "*RC* cable" is, in fact, a natural starting point as the model for an unmyelinated fiber. The cable conductor is axoplasm, which has appreciable resistance. Measurements show that a resistivity of $\rho = 125\ \Omega\cdot\text{cm}$ ($1.25\ \Omega\cdot\text{m}$) is a reasonable value for mammalian axoplasm, and this is the value employed in this book. Actual values can depart appreciably from 125 Ω·cm, however, so one must expect that calculated results can be gross approximations.

The central conductor of an electrical cable has inductance as well as resistance. The inductance of a nerve fiber can be neglected, however, because it is "swamped out" by axoplasmic resistance, and the signal frequencies are relatively low.

The cable insulation, the 75-Å-thick membrane, is far from perfect, as is abundantly revealed by Fig. 3-7. Here the cable is shown with a capacitance of 1 $\mu F/cm^2$, but the gs stand for appreciable shunt leakage conductance. Nevertheless, to a first approximation, the fiber behaves as if the effect of shunt leakage is small compared to that of shunt capacitance.

Capacitance is given by Eq. (1-1), which is repeated here for convenience:

$$C = \frac{\epsilon_0 KA}{l}. \tag{4-1}$$

The value 8 is used in this book for the dielectric constant of membrane material, as explained in connection with Eq. (1-1). For surface area A = 1 cm^2 of membrane, with l = 75 Å, Eq. (4-1) yields C = 0.9444 $\mu F/cm^2$ (0.009444 F/m^2).

To a first approximation, an unmyelinated fiber can be represented as the infinitely long RC cable of Fig. 4-2, where the ΔRs stand for series axoplasm resistance and the ΔCs stand for shunt membrane capacitance. Although this is a passive, nonregenerating model, it is valid for deriving, in the following, many of the characteristics of a regenerating fiber.

It is easier to use infinitesimally small branches in Fig. 4-2, so that the deltas can be replaced by derivatives, and we have, from Ohm's law,

$$\frac{dV}{dx} = -IR. \tag{4-2}$$

(The negative sign is appropriate because voltage decreases as distance x increases.) Similarly, from Ohm's law,

$$\frac{dI(s)}{dx} = -sCV(s), \tag{4-3}$$

where $s = j\omega$. Eliminating I from these simultaneous equations,

$$\frac{d^2V(s)}{dx^2} = sRCV(s). \tag{4-4}$$

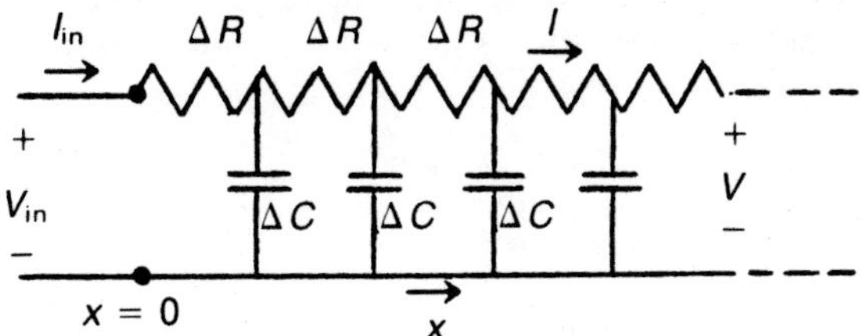

Fig. 4-2 The passive, nonregenerating unmyelinated fiber as an infinitely long RC cable consisting of infinitesimally small series resistances, ΔR, and shunt capacitances, ΔC.

The general solution to this equation is

$$V(s) = c_1\epsilon^{-x\sqrt{sRC}} + c_2\epsilon^{x\sqrt{sRC}} \tag{4-5}$$

as can be easily verified by substituting for V in Eq. (4-4). Because V must vanish as x approaches infinity, however, we get $c_2 = 0$. Because $V = V_{in}$ at $x = 0$, we get $c_1 = V_{in}$. This leaves

$$V(s) = V_{in}(s)\epsilon^{-x\sqrt{sRC}}. \tag{4-6}$$

For a V_{in} impulse of area A_0 V·s, $V_{in}(s) = A_0$ so that

$$V(s) = A_0\epsilon^{-x\sqrt{sRC}}. \tag{4-7}$$

The waveform corresponding to Eq. (4-7) is a standard form that is given in inverse Laplace transform tables as

$$V = \frac{A_0x}{2t^{3/2}}\sqrt{\frac{RC}{\pi}}\exp\left(-\frac{x^2RC}{4t}\right). \tag{4-8}$$

If a voltage impulse is applied to the input end of the RC cable of Fig. 4-2, it produces an "action potential" that is rapidly attenuated and broadened as it propagates to the right. At what value of x will the impulse response look like a bona fide AP? What values for A_0 and RC are reasonable in Eq. (4-8)? To answer these questions, a "genuine" AP is defined as follows [and it is illustrated by the voltage waveform of Fig. 4-1(a)].

(a) The peak amplitude is 100 mV. Differentiating Eq. (4-8), we find that this peak is located at time

$$t_p = \frac{x^2RC}{6}. \tag{4-9}$$

(b) The width of the curve is 2 ms. It is difficult to define the "width" of a curve that has a long trailing edge; nevertheless, visual inspection suggests that the reasonable width of an AP is given by a point at one-fourth of the peak height. Some numerical simplification occurs later if the point is at $V = 23.75$ mV rather than at 25 mV.

Requirement (a) is satisfied by substituting $V = 100$ mV and Eq. (4-9) into Eq. (4-8). Requirement (b) is satisfied by substituting $V = 23.75$ mV and $t = 2$ ms into Eq. (4-8). From this set of simultaneous equations we get

$$A_0 = 216.2 \text{ mV·ms} = 216.2 \times 10^{-6} \text{ V·s}, \tag{4-10}$$

independent of fiber diameter, and

$$x_0 = \sqrt{\frac{0.002}{RC}} \quad \text{(MKS units)}, \tag{4-11}$$

where the 0 subscript denotes the special location, along x, at which a 100-

mV, 2-ms AP is observed, and the 0.002 in the numerator is the width of the AP. This initial section of artificial fiber can be called the *equivalent action potential generator cable length* (EAPGECL).

Incidentally, because the *RC* cable has zero shunt leakage conductance, charge is preserved because it cannot leak away. As the impulse propagates, it decreases in height and broadens, but the area under the curve remains constant. In other words, the area under the voltage versus time waveform in Fig. 4-1(*a*) is 216.2 mV·ms.

Finally, substituting Eqs. (4-10) and (4-11) into Eq. (4-8), we get the expression for this idealized AP:

$$V = \frac{2.728 \times 10^{-6}}{t^{3/2}} \exp\left(-\frac{0.0005}{t}\right) \qquad \text{(MKS units).} \qquad (4\text{-}12)$$

The net ion current curve is given by $C\, dV/dt$. Using $C = 0.9444\ \mu\text{F/cm}^2$,

$$I = \frac{1.288 \times 10^{-15}}{t^{7/2}} (1 - 3000t) \exp\left(-\frac{0.0005}{t}\right) \qquad \text{(MKS units).} \qquad (4\text{-}13)$$

This curve, in Fig. 4-1(*a*), peaks at $t = 0.1225$ ms; the peak value is 676 μA/cm^2. This is about twice as large as that of the Hodgkin–Huxley curves of Fig. 3-6, for which the peak value $I_{net} = -315\ \mu\text{A/cm}^2$ is given in Section 3-4. The reason for the discrepancy is that the leading edge of the AP, in Fig. 4-1, is about twice as steep as that of the Hodgkin–Huxley AP; this is one of the inaccuracies introduced by using tractable formulas rather than the Hodgkin–Huxley equations.

Since current lines must be continuous, the *I* path is completed longitudinally through axoplasm, then it returns across the membrane in the form of pumped ions and in the fluid surrounding the nerve fiber.

The curves of Fig. 4-1(*b*) show voltage and current versus *distance*. In general, the distance curves can be quite different from the time curves. The unmyelinated axon is a special case, however, in which the AP propagates without any change in shape or amplitude. For this special case the distance and time curves are mirror images of each other (except for different horizontal scales), as can be seen in Fig. 4-1.

To obtain the AP curve of Fig. 4-1(*b*), we choose a convenient location and wait with an imaginary magic camera. The camera is able to "see" the voltage inside the axon along a 4-mm-wide section of axon. Soon an AP comes into view; it is rushing to the right at a speed of 1.2 m/s, which corresponds to a reasonable walking speed, 2.7 mph. (To get miles per hour from meters/second, multiply by 2.237.) When the leading edge of the AP reaches the right side of the 4-mm section, at a location labeled "zero distance," we take a photograph. (The magic camera is also able to photograph the net ion current curve.) The AP photograph shows the leading edge rising to 8 mV at distance = 0.1 mm. The voltage reaches a peak of 100 mV at $x = 0.4$ mm, and so forth.

Now let your imagination run wild based on the AP curve of Fig. 4-1(b). Imagine that the AP in each axon lights up, and that you can see these APs. The human body would then be seen with thousands of APs, each 2½ to 5 mm long, rushing along unmyelinated fibers at a speed of around 1½ to 3 mph, with afferent fibers carrying APs toward the spinal cord and brain, and with efferent fibers carrying them in the opposite direction.

[S.D. writes: "A model of the nervous system in which nerve trunks were portrayed in this way was displayed at the Man and Life Pavilion in Montreal at Exposition 1967. Unfortunately, some of the circuits did not operate properly, and the sponsors were fearful of a complete nervous breakdown. The company for which I worked sent me to Montreal to repair the system. I sat behind the control panel, studied some of the circuits, and was able to make a few minor changes. But time ran out and I had to return home, leaving behind a monster in schizophrenic condition."]

4-2 Conduction Speed

One of the important neurological tests is a measurement of conduction speed. If the velocity of propagation is lower than normal, it is an indication of nervous system dysfunction. In this section, accordingly, some of the pertinent aspects of conduction speed are examined with the aid of Fig. 4-3.

In Fig. 4-3(a) we have the normal, regenerating axon of Fig. 4-1. An AP is shown at two locations: at distance = 0.2 mm and at 0.7 mm. Since this AP propagates at a speed of 1.2 mm/ms, the time required to travel the intervening 0.5 mm is given by $t = 0.5/1.2 = 0.4$ ms. The two AP curves are accordingly shown 0.4 ms apart.

In Fig. 4-3(b), on the other hand, the axon of Fig. 4-1 is *nonregenerating*. As indicated by arrows, the waveforms are shown at the distance = 0.2-mm, 0.3-mm, and 0.7-mm locations. The example is chosen on the basis that x_0 has a convenient value, 0.2 mm. This is the x location at which, starting with an impulse applied to the RC cable at $x = 0$, we get a 100-mV-high, 2-ms-wide AP. According to Eq. (4-11), therefore, RC has the value 5 s/cm^2 for the axon of Figs. 4-1 and 4-3.

It is easy to find the axon's diameter. Using a resistivity of $\rho = 125$ Ω·cm for axoplasm, R is given by

$$R = \frac{\rho l}{A} = \frac{1.592}{d^2} \qquad \text{(MKS units)} \qquad (4\text{-}14)$$

while C, from Eq. (4-1), is given by

$$C = 0.02967d \qquad \text{(MKS units)} \qquad (4\text{-}15)$$

so that

$$RC = \frac{0.04722}{d} \qquad \text{(MKS units).} \qquad (4\text{-}16)$$

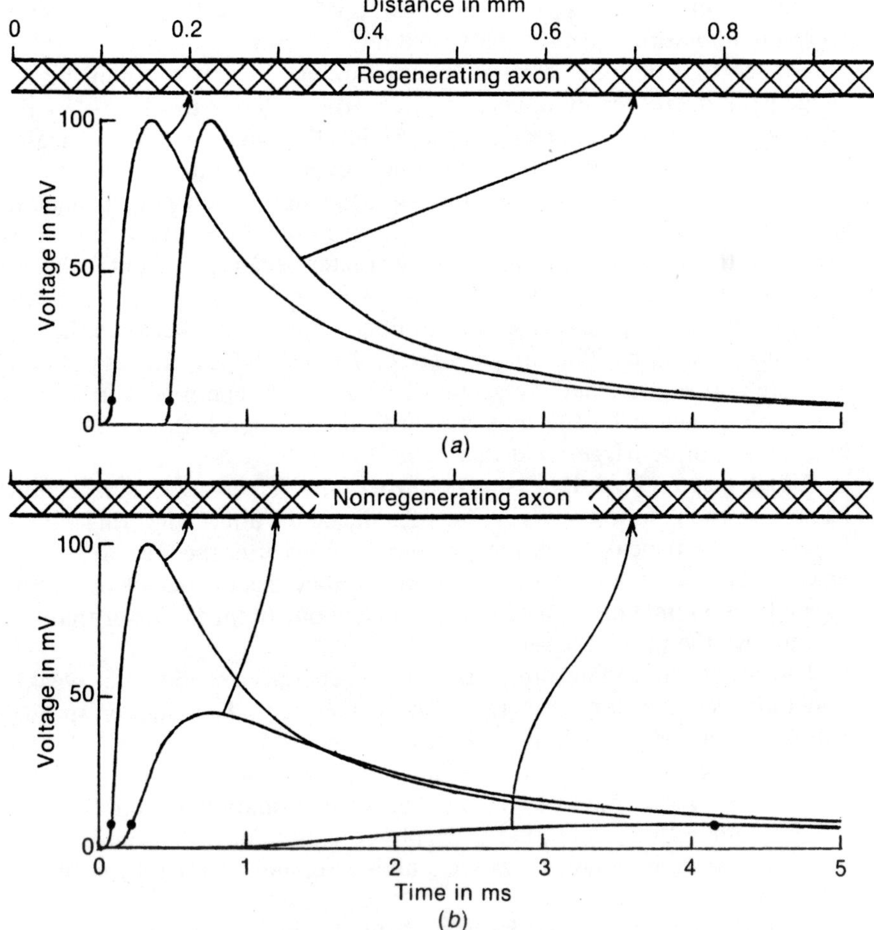

Fig. 4-3 Idealized propagation of the AP. (*a*) Along a normal axon 0.94 μm in diameter. As indicated by arrows, the AP is shown at two locations, 0.5 mm apart in distance, which is equivalent to 0.4 ms apart in time. (*b*) Along a nonregenerating axon. As indicated by arrows, the waveforms are shown at the distance = 0.2-, 0.3-, and 0.7-mm locations. Starting with a normal 100-mV AP at the 0.2-mm point, the AP decreases to 44 mV at the 0.3-mm point, and to 8 mV at the 0.7-mm point.

For Figs. 4-1 and 4-3, where $RC = 50{,}000$ s/m^2, Eq. (4-16) yields $d = 0.9444$ μm.

The numerical values just given are substituted into Eq. (4-8) to plot the curves of Fig. 4-3(*b*). We get

$$V = \frac{0.01364x}{t^{3/2}} \exp\left(-\frac{12{,}500x^2}{t}\right) \qquad \text{(MKS units).} \qquad (4\text{-}17)$$

In addition to the AP at $x = 0.2$ mm, Eq. (4-17) is used to plot the nonregenerating impulse responses at $x = 0.3$ and 0.7 mm.

We see that nonregeneration is catastrophic. In only 0.1 mm the AP voltage peak decreases by a factor of 2.25, from 100 mV to 44 mV. By the time it reaches the 0.7-mm location the AP has flattened out like a pancake, decreasing to 8 mV. (The actual situation is even more unfavorable. In the present chapter, for the sake of simplicity, the shunt leakage resistance of the membrane is neglected, but in Chap. 5, on dendritic summation, it is shown that the attenuation is appreciably greater because of shunt leakage.) Since 8 mV is the threshold level at which triggering can occur, Fig. 4-3(*b*) illustrates the limiting case: a section of fiber 0.5 mm long, between the 0.2- and 0.7-mm locations, fails to regenerate. The AP is nevertheless able to propagate across the defective section because, although it is reduced to a mere 8-mV pancake by the time it gets to the 0.7-mm location, a normal 100-mV AP can be triggered at this point.

A high price is paid, however, because of the defective 0.5-mm section. The dots on each of the curves in Fig. 4-3 indicate the 8-mV triggering points. We see that the pancake curve triggers an AP 4 ms after the normal AP was triggered. The corresponding conduction speed across the defective section is given by 0.5 mm/4 ms = 0.12 mm/ms, just about 10 times slower than the speed across the normal 0.5-mm section!

The conclusion, therefore, is that nonregenerative sections, or weakly regenerative sections, are reflected in lower-than-normal conduction speeds. Even worse, a single excessively long nonregenerative section will disable the entire axon. It could be a 1-m-long axon running from the toe to the base of the spinal cord, but a single defective section only 0.5 mm long can have the same effect as cutting a telephone wire. For a 0.3-μm fiber the maximum allowable nonregenerative length is 0.3 mm; for a 1.3-μm fiber it is 0.6 mm.

What determines the conduction speed, 1.2 m/s, for the fiber of Figs. 4-1 and 4-3? It is the speed with which the AP wavefront advances. We are all familiar with an advancing wavefront as we stand at the ocean's edge. The corresponding electrical wavefront is depicted in Fig. 4-4. To give the illusion of motion, the wave is shown at three closely spaced instances of time—at 71, 81, and 91 *microseconds* after zero time—and also much later at time = 333 μs. These are similar to photographs taken at the water's edge as a wave advances. The waves of Fig. 4-4 are photographs taken with our imaginary magic camera, which can "see" the voltage inside the axon.

The various wavefronts are given by Eq. (4-17) by letting x vary as t takes on the values 71, 81, 91, and 333 μs.

How do the wavefronts of Fig. 4-4 differ from that of Fig. 4-1(*b*)? In the latter, a "photograph" is taken of a normal, regenerating fiber. To find the velocity of propagation, however, we have to capture the advancing wavefront *before* it triggers regeneration; therefore, Fig. 4-4 describes a nonregenerating fiber. The speed of the AP is determined by what is happening at the 8-mV triggering level. To make this more visible, the region

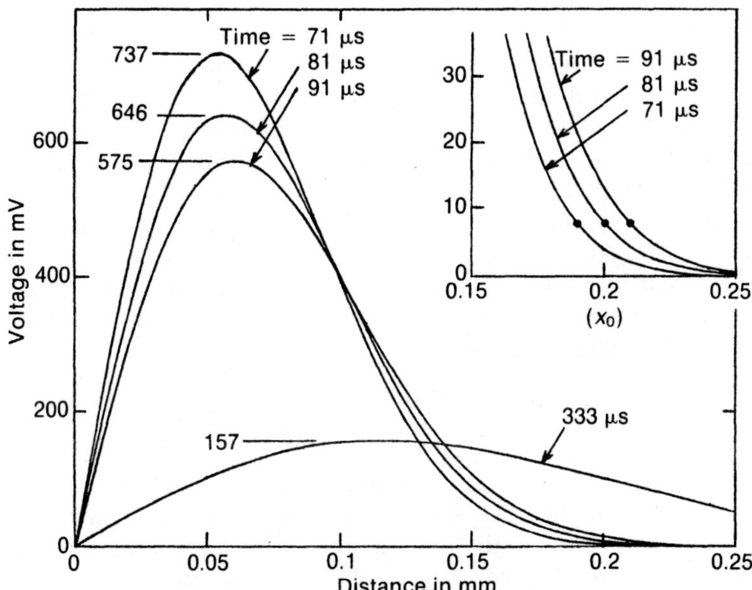

Fig. 4-4 Idealized AP as a wavefront propagating on a nonregenerating axon 0.94 μm in diameter. At time = 0 the left end of the fiber, at distance = 0, is "zapped" with a very narrow voltage spike of suitably enormous amplitude. The voltage rapidly falls, as shown by "photographs" taken at time = 71, 81, 91, and 333 μs after the initial zap. An electrode located at the distance = 0.2-mm point (x_0) sees a normal 100-mV AP. To show the advancing wavefront at the 8-mV triggering level, the region in the vicinity of x_0 is shown magnified, by a factor of 10 in the vertical direction, in the inset.

around distance = 0.2 mm is shown magnified, by a factor of 10 in the vertical direction, in the inset. The three dots indicate the 8-mV level. You can see that the dots advance 0.02 mm in 0.02 ms; we get speed = 0.02/0.02 = 1 mm/ms = 1 m/s. We have been claiming a conduction speed of 1.2 m/s for this fiber, but there are minor discrepancies because all of these values are based on various approximations. The important point is that the advancing wavefront in Fig. 4-4 serves to illustrate the pretriggering phenomena—the events just before AP activity is initiated.

In Fig. 4-4 we "zap" the left end of the artificial cable, at $x = 0$, with a very narrow voltage spike. The area under the spike is 216.2 mV·ms. Since the artificial cable does not regenerate, the voltage falls rapidly. At time = 71 μs after the initial "zap" the peak voltage has decreased to 737 mV; in 20 μs more it has decreased to 575 mV. At time = 333 μs the wavefront has pancaked down to 100 mV at the 0.2-mm point, as shown in Fig. 4-4 and also by the AP of Fig. 4-1(b). At time = 2 ms (not shown) the wavefront is almost completely dissipated, down to 24 mV at the 0.2-mm point, and so forth. The reason for the preceding strategy is that it provides an AP that can be described by a simple mathematical expression, Eq. (4-17). Start with an impulse and let it decay and broaden until it is 100 mV high and 2 ms wide; this is the impulse response of the *RC* cable at $x = 0.2$ mm. In other words, an electrode at the 0.2-mm point would see the voltage rise and fall exactly as it does in a "genuine" AP, so here is the place to install our magic camera in order to take photographs of the wavefront as it propagates to the right.

It is useful to derive an algebraic expression for the velocity of propagation. The velocity at which the wavefront advances is dx/dt, with V held constant. Applying this to Eq. (4-8), there results

$$v = \frac{x}{2t}\left(\frac{x^2RC - 6t}{x^2RC - 2t}\right). \tag{4-18}$$

But we are only interested in the velocity in the vicinity of x_0, since it is here that a 100-mV, 2-ms AP will appear. Also, we are only interested in a specific value of time, t_t, the time at which triggering occurs, because V then reaches the 8-mV level. If typical numerical values for x_0 and t_t are introduced into Eq. (4-18), it turns out that $x_0^2RC - 6t_t \cong x_0^2RC - 2t_t$, so one can cancel these terms to leave

$$v = \frac{x_0}{2t_t}. \tag{4-19}$$

In Fig. 4-4, as the inset shows, $t_t = 81$ μs. Since $x_0 = 0.2$ mm, Eq. (4-19) yields $v = 1.23$ m/s.

What is the velocity as a function of diameter? The trigger time, $t_t = 0.081$ ms, is independent of diameter. From Eqs. (4-11) and (4-16), however, $x_0 = 0.2058\sqrt{d}$ (MKS units) so that

$$v = 1270\sqrt{d} \qquad \text{(MKS units)}. \tag{4-20}$$

In words, the velocity is proportional to the square root of diameter.

4-3 In a Myelinated Fiber

We have described the AP of Fig. 4-1 as "rushing to the right at a speed of 1.2 m/s." It may be "rushing" compared to the speed with which chemical messengers are transported by the bloodstream, but it is painfully slow if one is standing on a pointed object. It is obvious that there must be tremendous evolutionary pressure for high-speed nerve conduction. In fact, we have already seen, in connection with Fig. 1-3, a velocity of 120 m/s (270 mph); the axons in that figure are *myelinated.*

Myelin is a relatively thick fatty layer that surrounds the axon. Instead of a membrane 75 Å thick we have, in effect, a membrane that is 2000 Å or more thick. Regeneration of the AP becomes impossible with such a thick membrane. Regeneration is based on movement of sodium and potassium ions across the axon membrane; the 2000-Å membrane is a barrier through which nothing penetrates. Because of decreased capacitance, however, the thick membrane yields a substantial increase in conduction speed [P. Morell and W. T. Norton, 1980]. A comparable effect is observed with a telephone wire: given a wire with very thin insulation, increasing the insulation thickness results in faster transmission of information. Of course, the thicker insulation may cost more, but it will also withstand higher voltages without breaking down.

The evolutionary compromise for the thick, nonregenerating myelin sheath is to provide periodic nodes; here the 75-Å basic membrane is exposed to the external interstitial fluid, as illustrated in Fig. 4-5. The axon looks like a string of sausages because the diameters are exaggerated; in reality, however, the nodes (called *nodes of Ranvier*) are some 100 outer diameters apart. For the largest mammalian fiber, which has a diameter of 0.022 mm,

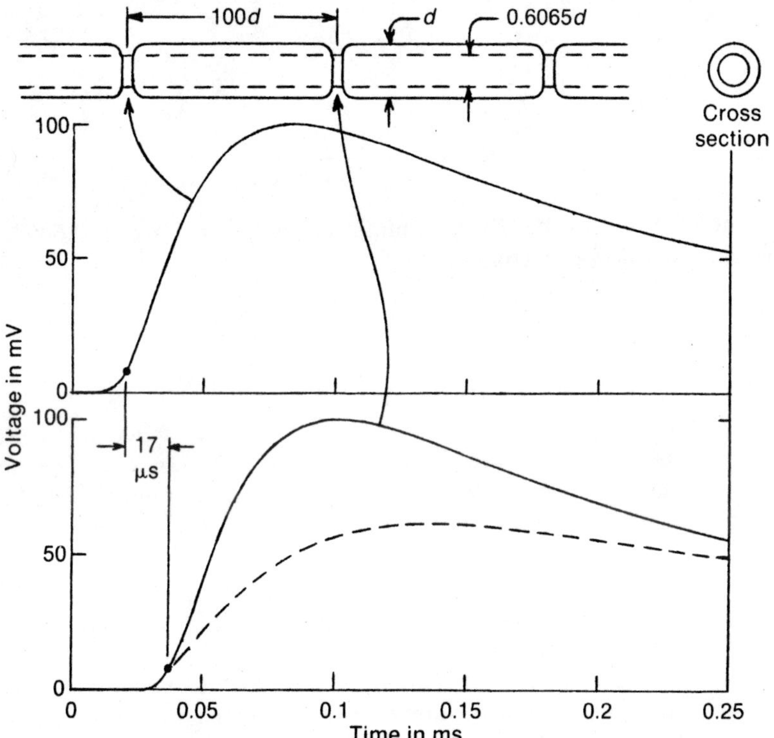

Fig. 4-5 Idealized AP of a myelinated fiber. For convenience, the diameters are exaggerated; actually, the nodes are 100 outside diameters apart. As indicated by arrows, the waveforms are shown at two successive nodes. The upper waveform depicts the AP at the first node. The AP would theoretically appear as the dashed curve at the next node, but when it reaches 8 mV, as indicated by the • dots, this node fires and a new AP is generated as shown. The time from one • dot to the next is 17 μs, regardless of diameter.

the nodes of Ranvier occur approximately 2.2 mm apart; for the smallest fiber, with a diameter of 0.001 mm, they are around 0.1 mm apart. On Fig. 4-5 they are drawn four diameters apart for convenience in showing the thick membrane, and this leads to the somewhat misleading "string of sausages" analogy [R. Fitzhugh, 1962].

The myelin thickness has evolved so as to approximately minimize RC. For a coaxial structure of outside diameter d, capacitance is given by

$$C = \frac{2\pi\epsilon_0 K}{\log(1/\alpha)}, \tag{4-21}$$

where α is the ratio of cable conductor diameter to outside diameter d. The resistance is given by

$$R = \frac{4\rho}{\pi\alpha^2 d^2}, \tag{4-22}$$

so that

$$RC = -\frac{8\epsilon_0 K\rho}{d^2\alpha^2 \log \alpha}. \tag{4-23}$$

Differentiation shows that RC is minimized when $\alpha = 1/\sqrt{\epsilon} = 0.6065$. This is the value used in this chapter. It yields

$$RC = \frac{3.851 \times 10^{-9}}{d^2} \quad \text{(MKS units)}. \tag{4-24}$$

The AP is regenerated at the nodes of Ranvier; the process is almost exactly the same as it is on an unmyelinated fiber. As indicated by arrows in Fig. 4-5, the waveforms are shown at two successive nodes. The narrow nodes, compared to the uninterrupted membrane of an unmyelinated fiber, apparently result in a considerably narrower AP. Whereas the unmyelinated-fiber AP is 2 ms wide, the myelinated-fiber AP is only 0.5 ms wide. The spatial decrease at the nodes is accompanied by a time decrease that is all to the good insofar as conduction speed is concerned.

For myelinated fibers the velocity of propagation is given by a simple relationship: in second units, it is six million times the diameter [H. S. Gasser, 1941]. For example, for the smallest fiber (0.001 mm) the speed is 6000 mm/s, or 6 m/s; for the largest fiber (0.022 mm) it is 132 m/s. The body has naturally evolved so that the fibers for which speed is most important, those involved with monosynaptic reflex arcs (see Chap. 8), have the largest diameter. Since space is at a premium in any organism, smaller fibers are employed wherever possible. This has all been fine-tuned, of course, by millions of years of evolution. The diameter of a fiber directly tells us how vital the conduction speed of that particular fiber is for survival of the organism. In a mammal, only one-third of the fibers are myelinated; two-thirds are slow-speed (less than 1.5 m/s), small-diameter (up to 0.0013 mm), unmyelinated fibers used for chores for which low speed is adequate, such as the reporting of pain and temperature. (Temperature information is also transmitted in the autonomic nervous system.) There surely is an important message here: only a small hierarchy of large-diameter carriers is engaged in the high-speed retrieval and processing of "exotic" information.

Although regeneration at the nodes is similar to that of an unmyelinated fiber, the conduction speed is determined by a much simpler mechanism than the advancing wavefront of Fig. 4-4. The upper waveform in Fig. 4-5 depicts the AP at a particular node of a myelinated fiber. To get $V(t)$ we can substitute into Eq. (4-8). The AP is one-fourth as wide as that of an unmyelinated fiber, so the area under the curve, A_0, is 54.05 mV·ms. In Eq. (4-11) we replace 2 ms by 0.5 ms:

$$x_0 = \sqrt{\frac{0.0005}{RC}} \qquad \text{(MKS units)}, \tag{4-25}$$

and RC is given by Eq. (4-24). Substituting into Eq. (4-25),

$$x_0 = 360.3d \qquad \text{(MKS units)}, \tag{4-26}$$

and Eq. (4-8) then yields

$$V = \frac{3.409 \times 10^{-7}}{t^{3/2}} \exp\left(-\frac{0.000125}{t}\right) \qquad \text{(MKS units)}. \tag{4-27}$$

Since the AP propagates without the benefit of regeneration, it would appear as the *dashed* curve at the next node. The attenuation of the AP in Fig. 4-5 is the same as that of the lower curves in Fig. 4-3 except that, in the latter illustration, the deterioration is allowed to reach a catastrophic level, decreasing from 100 mV to 8 mV. From one node to the next of a myelinated fiber, however, the decrease is only from 100 mV to 61 mV, as shown by the dashed curve in Fig. 4-5. This curve is shown dashed because it never actually appears in a normal fiber; as soon as the propagating AP reaches 8 mV, as indicated by the • dots, the node fires and a new AP is generated.

The dashed curve is obtained by substituting into Eq. (4-8); A_0 and RC are unchanged, but the location, from Eq. (4-26), is at $x = 460.3d$. We get

$$V = \frac{4.356 \times 10^{-7}}{t^{3/2}} \exp\left(-\frac{0.000204}{t}\right) \qquad \text{(MKS units)}. \tag{4-28}$$

In Eqs. (4-27) and (4-28), solving for the time at which V reaches 8 mV and triggering occurs, we respectively get $t_t = 20.36$ μs for the upper curve of Fig. 4-5, and $t_t = 37.23$ μs for the dashed curve. Therefore, the delay in triggering from one node to the next is 17 μs. The conduction speed is therefore 100 diameters/17 μs = 6,000,000 diameters/s.

The 17-μs node-to-node transit time is the same for *all* myelinated fibers, regardless of diameter, if the node-to-node distance is 100 outside diameters. It is most interesting that here is a vital activity taking place in a time interval measured in millionths of a second. For a 0.022-mm fiber, where the nodes are 2.2 mm apart, try to visualize the message as it hops from one node to the next in 17 μs to give a conduction speed of 2.2 mm/0.017 ms = 132 m/s (almost 300 mph). Although it may appear that this is practically instantaneous, experience shows that it is better to wear shoes when stepping on thumb tacks.

In connection with the loss of AP voltage along a nonregenerating unmyelinated fiber as illustrated in Fig. 4-3, it is stated in Section 4-2 that the maximum tolerable length for a 0.0013-mm fiber is 0.6 mm. What do the calculations reveal for a myelinated fiber? That the maximum nonregenerating length is around 900 diameters, or nine node-to-node lengths. For a 0.001-mm fiber, therefore, the maximum nonregenerating length is 0.9 mm; in other words, over this length a 100-mV AP will pancake down to 8 mV, but it can be restored at the next node. If the defective length is greater than 0.9 mm, the fiber suffers fatal interruption of its signal. For a 0.022-mm fiber the maximum nonregenerating length is 20 mm (over ¾ in.). Because of its thicker "membrane," a myelinated fiber is better than an unmyelinated fiber of the same diameter with regard to the ability to transmit across a nonregenerating section.

If the AP *is* able to "jump" across a defective section 900 diameters long, the abnormal delay in triggering caused by the loss of AP voltage results in a time delay of 1 ms in traversing the section. This is equivalent to a conduction speed of 900 dia/1 ms = 900,000 dia/s. Since this is 6.7 times slower than the normal conduction speed of 6,000,000 dia/s, the conclusion is the same as it is for unmyelinated fibers: "Nonregenerative sections, or weakly regenerative sections, are reflected in lower-than-normal conduction speeds."

In multiple sclerosis the reduction in speed can be disastrous, but it is not due to nonregenerative nodes. In multiple sclerosis the myelin sheath disintegrates (probably because of an autoimmune attack), exposing the "bare" axon with its 75-Å-thick membrane. Consider what this forced conversion to an unmyelinated fiber can do to what was originally a high-speed (132-m/s) fiber: The original diameter is 22 μm, but demyelination leaves a 13.3-μm bare fiber. An unmyelinated fiber of this diameter has a conduction speed of 4.6 m/s, only 3.5% of the original value. Since this is the effect upon the largest myelinated fiber, where high speed is vital, the result can be disastrous, as previously noted. For smaller, less important fibers the reduction in speed is less severe: Demyelination reduces a 6-m/s, 1-μm fiber to a diameter of 0.6 μm; the latter, as an unmyelinated fiber, has a conduction speed of 1 m/s. Here multiple sclerosis yields a speed that is 16% of its original value. As one should expect, the amount of dysfunction depends on many factors besides the diameter of the fiber and the length that is demyelinated [W. I. McDonald, 1977; M. Rasminsky, 1978; D. Regan, R. Silver & T. J. Murray, 1977; K. B. Roberts, P. D. Lawrence & A. Eisen, 1983].

4-4 "Photograph" of a Myelinated Fiber AP

An AP is shown in Fig. 4-1(*a*), at a particular location, as it changes with time. In Fig. 4-1(*b*) it is shown, at a particular time, as it changes with distance. The latter is a "photograph" taken with a magic camera that can "see" the voltage inside the axon. The two APs are mirror images of each other (except for unequal horizontal scales), because the axon is unmyelinated so that the AP propagates without attenuation or change of shape.

The corresponding photograph of a myelinated fiber is not at all a mirror image of the AP as it changes with time. Instead, as shown in Fig. 4-6, we have a bizarre creature that is a good candidate in the "truth is stranger than fiction" category. The photograph is that of a 0.014-mm fiber (for which the EAPGECL is 5 mm), with the AP flying to the right at 184 mph. A "flying bird" analogy is apt, with the bird beating its wings at each node location in order to restore height. The leading edge of the AP in Fig. 4-6 has just reached the node labeled "0," and this also defines the distance = 0 point. At node "1" the voltage has not reached the 8-mV threshold, so nothing much has happened. But at node "2" the threshold has been exceeded; the node has fired and the voltage is rising toward 100 mV. All of the nodes to the left of "2" have already fired, with node "5" just reaching 100 mV. To the left of this point the AP is ending, with voltages slowly decreasing toward zero.

The reason for the zigzag, of course, is that the voltage falls while it is in "free flight" or "gliding" from one node to the next. The upper edge of the zigzag traces out a mirror image of the AP of Fig. 4-5, while the lower edge traces out a mirror image of the dashed curve of Fig. 4-5.

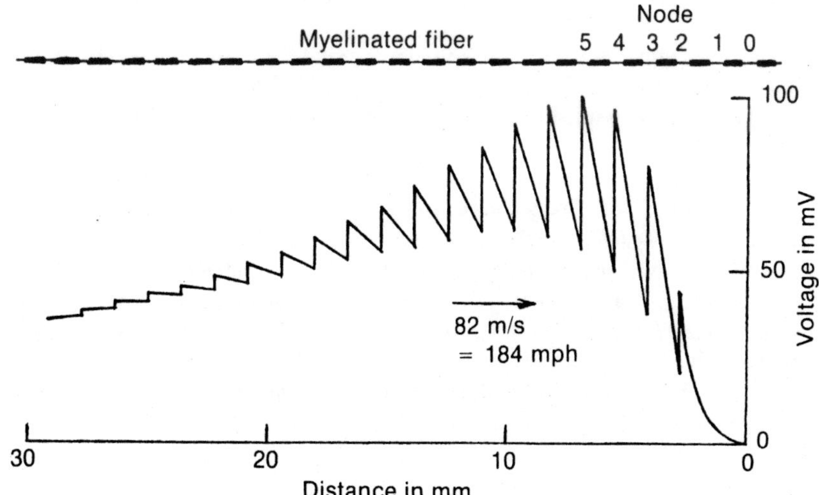

Fig. 4-6 Idealized AP of a myelinated fiber 14 μm in diameter. The curve shows changes with distance as "photographed" at a particular time.

The zigzag is actually not as severe as shown in Fig. 4-6. For example, according to the drawing, node "5" shows a vertical jump from 57 to 100 mV, but this is technically impossible because it implies that the left and right sides of the node are disconnected from each other. Instead, current feeds to the left in a *backward* direction, fueled by the 43-mV jump, so that the bottom of each sawtooth is actually rounded like a U rather than sharp-pointed like a V. (Propagation in a backward direction is called *antidromic*, while the forward direction is *orthodromic*. An axon can conduct equally well in either direction, and neurophysiologists frequently use antidromic propagation in making nerve conduction measurements. Some of the nerve conduction during epileptic seizures may be antidromic.) We show straight sawtooth lines in Fig. 4-6 to save time in making the drawing, but only the curve between nodes "0" and "2" has been accurately cast.

According to the distance scale in Fig. 4-6, a single AP on this myelinated fiber, which is a representative example, is 80 mm (3 in.) long. Again, let your imagination run wild and imagine that the AP in each axon lights up, and that you can see these APs. The human body would then be seen with thousands of dotted APs (the zigzags), each from 5 to 100 mm long, flying along myelinated fibers at a speed of 15 to 300 mph, with afferent fibers carrying APs toward the spinal cord and brain, and with efferent fibers carrying them in the opposite direction.

REFERENCES

J. W. Cooley and F. A. Dodge, Digital computer solutions for excitation and propagation of the nerve impulse, *Biophys. J.*, vol. 6, pp. 583–599, 1966.

A. Finkelstein and A. Mauro, Equivalent circuits as related to ionic systems, *Biophys. J.*, vol. 3, pp. 215–237, 1963.

R. Fitzhugh, Computation of impulse initiation and saltatory conduction in a myelinated nerve fiber, *Biophys. J.*, vol. 2, pp. 11–21, 1962.

H. S. Gasser, The classification of nerve fibers, *Ohio J. Sci.*, vol. 41, pp. 145–159, 1941.

J. J. B. Jack, D. Noble, and R. W. Tsien, *Electric Current Flow in Excitable Cells*. Oxford: Oxford Univ. Press (Clarendon), 1975.

B. Katz, *Nerve, Muscle, and Synapse*. New York: McGraw-Hill, 1966.

K. N. Leibovic, *Nervous System Theory*. New York: Academic Press, 1972.

W. I. McDonald, Pathophysiology of conduction time in central nerve fibers, in *Visual Evoked Potentials in Man: New Developments*, ed. J. E. Desmedt. Oxford: Oxford Univ. Press (Clarendon), pp. 427–437, 1977.

P. Morell and W. T. Norton, Myelin, *Sci. Am.*, vol. 241, pp. 88–118, May 1980.

M. Rasminsky, Physiology of conduction in demyelinated axons, in *Physiology and Pathobiology of Axons*, ed. S. G. Waxman. New York: Raven Press, pp. 361–376, 1978.

D. Regan, R. Silver, and T. J. Murray, Visual acuity and contrast sensitivity in multiple sclerosis-hidden visual loss, *Brain*, vol. 100, pp. 563–579, 1977.

K. B. Roberts, P. D. Lawrence, and A. Eisen, Dispersion of the somatosensory evoked potential (SEP) in multiple sclerosis, *IEEE Trans. Biomed. Eng.*, vol. BME-30, pp. 360–364, June 1983.

A. Saito and A. Noguchi, Dynamic analysis for an active nerve pulse of the single unmyelinated fiber, *IEEE Trans. Biomed. Eng.*, vol. BME-28, pp. 812–816, Dec. 1981.

A. C. Sanderson and R. J. Peterka, *Neural Modeling and Model Identification. CRC Critical Rev. Bioeng.* 1983.

A. C. Scott, *Neurophysics*. New York: Wiley, 1977.

Problems

1. A dc voltage V_{in} is applied to a nonregenerating unmyelinated fiber. The fiber can be represented as an infinitely long *RG* cable consisting of infinitesimally small series resistances, ΔR, and shunt conductances, ΔG. Find $V(x)$. [Ans.: $V = V_{in} \exp(-x\sqrt{RG})$.]
2. For the *RC* cable model of Fig. 4-2, find $I(s)$. [Ans.: $I = A_0\sqrt{sC/R} \exp(-x\sqrt{sRC})$.]
3. Equation (4-8) can be normalized by substituting $t = x^2RCt_n/4$ and $V = 4A_0V_n/(x^2RC)$. Make these substitutions, derive V_n as a function of t_n, and plot the normalized "action potential." [Ans.: $V_n = \pi^{-1/2}t_n^{-3/2} \exp(-1/t_n)$.]
4. Carry out the step-by-step derivation of Eq. (4-9) from Eq. (4-8). Suggestion: Let $t = 1/u$, and find the peak of $V(u)$.
5. Carry out the step-by-step derivation of Eqs. (4-10) and (4-11) from Eq. (4-8).
6. Carry out the step-by-step derivation of Eq. (4-12) from Eq. (4-8). Plot the curve.
7. Carry out the step-by-step derivation of Eq. (4-13) from Eq. (4-12). Plot the curve.
8. For the nonregenerating unmyelinated fiber of Fig. 4-3(*b*): (a) find $V(t)$ at $x = 0.5$ mm; (b) find V_{max} and the time at which it occurs; (c) plot the curve. [Ans.: (a) $V = 6.82 \times 10^{-6}/t^{3/2} \times \exp(-0.003125/t)$ (MKS units); (b) 16 mV at 2.083 ms.]
9. Because of disease, a section l_{nr} of unmyelinated fiber becomes nonregenerating. What is the maximum length of the defective section, as a function of diameter, that will allow propagation to resume past the section? [Ans.: $l_{nr} = 0.52\sqrt{d}$ (MKS units).]
10. Carry out the step-by-step derivation of Eq. (4-18) from Eq. (4-8). Suggestion: Start with the log of Eq. (4-8).
11. For the nonregenerating unmyelinated fiber of Fig. 4-4: (a) find $V(x)$ at $t = 0.2$ ms; (b) find V_{max} and the distance at which it occurs; (c) plot the curve. Find the wavefront velocity at (d) $x = 0.05$ mm; (e) $x = 0.12$ mm; (f) $x = 0.2$ mm. [Ans.: (a) $V = 4822x \exp(-6.25 \times 10^7 x^2)$ (MKS units); (b) 261.6 mV at 0.089 mm; (d) 0.4886 m/s; (e) −0.45 m/s; (f) 0.25 m/s.]
12. An unmyelinated fiber has a diameter of 0.7 μm. Find (a) R; (b) C; (c) RC; (d) x_0; (e) v. [Ans.: (a) 32490 MΩ/cm; (b) 207.7 pF/cm; (c) 6.746 s/cm²; (d) 0.1722 mm; (e) 1.058 m/s.]
13. Show that RC of a myelinated fiber, given in Eq. (4-23), is minimized if $\alpha = 0.6065$.
14. Carry out the step-by-step derivation of Eqs. (4-27) and (4-28) from (4-8), and show that $\Delta t_t = 16.87$ μs.
15. A myelinated fiber has a diameter of 7 μm. Find (a) R; (b) C; (c) RC; (d) x_0; (e) v. [Ans.: (a) 882.9 MΩ/cm; (b) 8.901 pF/cm; (c) 7.859 ms/cm²; (d) 0.2522 cm; (e) 42 m/s.]

16. At some abnormally small diameter, the conduction velocity of unmyelinated and myelinated fibers is the same. Find the diameter and velocity. [Ans.: d = 0.045 μm, v = 0.27 m/s.]
17. Because of disease, a section l_{nr} of myelinated fiber becomes nonregenerating. What is the maximum length of the defective section, as a function of diameter, that will allow propagation to resume past the section? [Ans.: $l_{nr} = 910d$ (MKS units).]
18. A myelinated fiber has a diameter d. Derive an expression for the new velocity of propagation divided by the original velocity if the fiber is demyelinated because of multiple sclerosis. [Ans.: Ratio = $0.0001648/\sqrt{d}$ (MKS units).]
19. Given the myelinated fiber of Fig. 4-6, with v = 82 m/s: (a) find $V(x)$ for the upper boundary of the curve; (b) find $V(x)$ for the lower boundary; (c) plot the curves with x increasing to the left, as in Fig. 4-6. [Ans.: (a) $V = 0.0002531x^{-3/2} \times \exp(-0.01025/x)$; (b) $V = 0.0003235x^{-3/2} \exp(-0.01673/x)$ (MKS units).]

5

Dendritic Summation

ABSTRACT. The input architecture of many neurons consists of a tree-like structure called a *dendritic tree*. The branches gather in as many as 100,000 messages, coming from the axons of preceding neurons, via synaptic and electrical gap junctions. The branches are nonregenerating unmyelinated fibers, so signals are rapidly attenuated. As the dendritic potential (DP) propagates toward the neuron body (the trunk of the tree), it gains time width as it loses voltage height. The neuron body is excited if many action potentials (APs) arrive more or less simultaneously at the periphery, but junctions located closer to the neuron body suffer less attenuation, so fewer of them have to be excited simultaneously to stimulate the neuron. The latter, in effect, gathers in all of the dendritic inputs and fires, as usual, when the net summation (after attenuation) exceeds the threshold. If the neuron is a type that generates a graded potential, the output is simply proportional to the net summation of attenuated DPs.

There is a very important class of *inhibitory* inputs; because their inputs are relatively weak, inhibitory junctions are located on or close to the neuron body. In effect, inhibitory inputs raise the firing threshold of the neuron.

The outputs resulting from several simple combinations of input signals are examined in detail.

5-1 Junctions

In Chapter 2 the output of a sensory receptor is a graded potential—a small voltage that increases and decreases in accordance with the stimulus to the transducer. In Chapters 3 and 4 neurons generate and propagate action po-

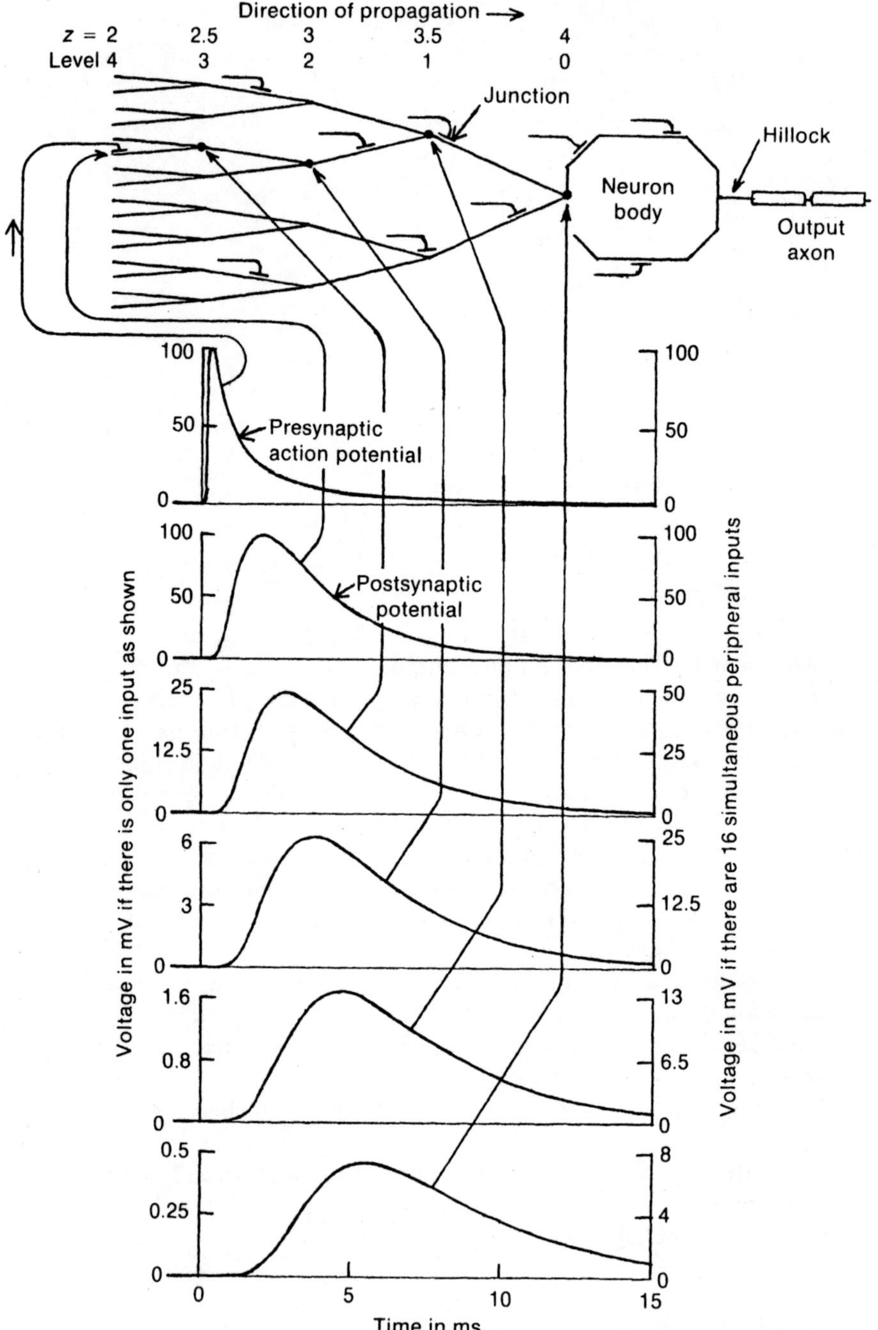

Fig. 5-1 An idealized, particular model of a dendritic tree. The waveforms depict the action that results from APs arriving at the periphery (level 4). The left-hand scale corresponds to a *single* AP arriving at the sixth branch down from the upper

tentials (APs). Also, there are neurons whose outputs consist of graded potentials. What happens to all of these messages of the nervous system?

In general, there are four targets, as listed below.

1. Action potentials can be directed to a skeletal muscle. The APs always result in contraction.

2. Action potentials, and in some cases graded potentials (GPs), can be directed to cardiac and smooth muscles and to the visceral organs. The AP or GP results in release of a chemical transmitter, which in turn stimulates the target tissue. For example, if the heart muscle is stimulated by a sympathetic nerve, adrenaline is released, which causes the heart muscle to increase its rate of contraction; if the heart is stimulated by a parasympathetic neuron, the AP results in the release of acetylcholine, and the cardiac muscle decreases its rate of contraction. Another example, if the pancreas is stimulated by a parasympathetic neuron, acetylcholine is released and the pancreas will increase its production of digestive enzymes; if the pancreas is stimulated by a sympathetic neuron, which may happen during times of stress to an individual, adrenaline is released and this will inhibit the production of pancreatic enzymes.

3. Action potentials can be directed to the neurosecretory cells. The AP results in the release of neurohormones. For example, neurons in the hypothalamus send axons to the nearby posterior pituitary. An AP will cause the secretion of substances that stimulate or inhibit the secretion of hormones by the pituitary.

4. Graded potentials and APs can terminate on the *dendrite* of another neuron. The second neuron then processes the message in accordance with the neuron's function in the scheme of things. Some of the details of this processing are discussed in this chapter.

A dendrite, as depicted in Fig. 5-1, is a treelike structure. (One usually calls this a *dendritic tree*, but use of the word "tree" here is redundant.) It is customary to show signals traveling from left to right; in order to agree with this convention, the tree is shown lying on its side. The dendrite's purpose is to gather messages—anywhere from a few to as many as 100,000—that arrive via either *synaptic* or *electrical gap* junctions. To avoid cluttering the diagram, only a single "presynaptic AP" is shown feeding a

---

edge, at time = zero. The postsynaptic potential (PSP) is 100 mV high and 4 ms wide. As the dendritic potential (DP) propagates to the right, it loses height and gains width until, at the neuron body (level 0), it is 0.46 mV high and 7.5 ms wide. The right-hand scale corresponds to a situation in which *each* of the 16 peripheral branches simultaneously receives an AP; this yields 16 times as much DP height (7.4 mV) at the neuron body. z is distance measured in length–constant units.

junction; it is at the periphery of the dendrite on the left. A few junctions are pictured without input connections, but it should be understood that a large dendrite can be covered with as many as 100,000 of these junctions. Usually, the neuron body (soma) also receives junctions [R. E. Kalil, 1989; H. D. Patton, 1965].

A typical synaptic junction has a diameter of 1 μm (10,000 Å). As pointed out in Section 1-2, the synaptic gap is around 200 Å wide. When an AP reaches the synaptic junction, it may cause the release of an excitatory chemical transmitter such as acetylcholine (ACh). Calcium plays a vital role in the sequence of events [M. Ito, 1984]. The ACh molecules "leisurely" diffuse across the junction, stimulating transmembrane protein carriers at the receiving end of the gap to transport sodium ions into the dendrite. Assuming that the events that follow are similar to those of an axonal AP, there is also some potassium ion current out of the dendrite. As the "net ion current" curve in Fig. 4-1 shows, at first the sodium in current far exceeds the potassium out current; this imbalance results in a positive or excitatory "postsynaptic potential" (PSP), as shown in Fig. 5-1. The ion current curves in Fig. 3-6 show the individual sodium and potassium contributions.

In Fig. 5-1 the chemical transmitter results in a PSP within the dendrite that has a peak height of 100 mV (although this is subject to modification by components in the fluid bathing the nerves). In Section 1-2 a calculation is carried out that shows that it takes 0.6 ms for the chemical transmitter to diffuse across the gap; the buildup of voltage then begins but, as shown in Fig. 5-1, it takes about 2 ms for the PSP peak to occur after the presynaptic AP gives birth to it. The chemical transmitter molecules do not march across the 200-Å gap in unison, but instead straggle across, spread out in space and time. As a result the PSP is around 4 ms wide, whereas its parent AP may be 2 ms or less in width.

We surmise that one purpose of the 200-Å gap is to provide access for different chemical messengers, such as hormones in the bloodstream, to stimulate or inhibit neurons. They can modify the excitability of the dendritic terminal of the gap (an effect also mediated by alcohol and certain drugs).

A major milestone in neurophysiology was the discovery that there are certain chemical transmitters (and chemical messengers) that *decrease* the excitability of the dendritic terminal. This vitally important phenomenon is called *inhibition*. An example of an inhibitory transmitter is gamma aminobutyric acid (GABA). These molecules stimulate the protein carriers to transport potassium ions out of the dendrite while blocking sodium ions. Since the potassium out current allows negative charge to accumulate inside the axon, the net ion current curve of Fig. 4-1 becomes a *negative* spike, and the corresponding inhibitory PSP is *negative*. The effect is relatively weak; whereas an excitatory PSP can have an amplitude of 100 mV, the amplitude of an inhibitory PSP is only around 10 mV.

The preceding discussion refers to APs that cause the release of ACh or GABA when they arrive at a junction. What if the neuron generates a graded potential rather than APs, as in the retina, where distances between neuron cell bodies are small? The electron microscope reveals synaptic junction structures in the retina that are filled with chemical transmitters [R. H. Masland, 1986]. We conclude from this that chemical transmitters carry graded potential messages, as well as action potential messages, across synaptic junctions to the dendrites on the other side of the gap.

In living systems, alas, nothing is simple. There are junctions in which the 2-ms time delay between a presynaptic AP and the PSP peak may be intolerable. There are excitatory junctions where the body never has a need to alter excitability via infusions of epinephrine or the like from the bloodstream. In these cases, there is no need for a 200-Å gap; the latter can be replaced by a "tight" or electrical gap junction in which an axon terminal of the first neuron and dendrite of the second neuron are so tightly intertwined that the electrical signal gets across to the dendrite although there is a finite but negligibly small gap. An electrical gap junction can transmit an AP or graded potential, but it cannot elicit an inhibitory effect; for that reversal we need molecules such as GABA diffusing across a conventional synaptic junction.

In addition to axon-to-dendrite and axon-to-soma junctions, there appears to be a relatively small population of axon-to-axon and dendrite-to-dendrite junctions.

5-2 The Dendritic Tree

For convenience, the tree in Fig. 5-1 is labeled as having levels 4, 3, 2, 1, and 0. Level 4 is the *periphery* of the tree, while level 0 is the *neuron body*, or *soma*. One can think of signals that are propagating to the right as also moving to a lower level. The tree funnels the junction messages, modified, into the neuron body at the right. The neuron has a single output axon, which can subsequently give birth to collateral branches—anywhere from a few to as many as thousands. The neuron can generate graded potentials or action potentials. In Fig. 5-1 the neuron generates APs, and its output axon is myelinated.

The dendritic branches, neuron body, and output axon are all part of a single cell. As shown, the initial section of output axon, adjacent to the neuron body, is called the *hillock*. It is here that the AP is generated when the voltage inside the hillock exceeds a threshold level, such as 8 mV. (All voltages are given with respect to the resting potential.)

The organism would be ill-served by a dendrite that simply gathers in thousands of these axon messages and indiscriminately passes them on to the neuron body [A. F. Kohn, 1989]. Since a PSP is 4 ms wide, only 250 of them arriving in each second would be sufficient to flood the neuron with an approximately steady and meaningless stream. With 10,000 junctions, a total of 250 PSPs/s amounts to, for each synaptic junction, an average of 0.025 spikes/s, or a single AP every 40 s. Instead, as Fig. 3-3 illustrates, an axon can be carrying an unstimulated, random, *messageless* AP discharge of 50 spikes/s. In that event, only five of these inputs could saturate the neuron with nonsense. (A sarcastic reader may suggest that this is exactly what happens sometimes.)

To function as a useful device, the dendrite has to *attenuate* each PSP. The tree accordingly consists of *nonregenerating*, *unmyelinated* fibers. The characteristics of a passive *RC* cable are derived in Chapter 4 in connection with Fig. 4-2, but this model is inadequate for a dendrite branch. In Chapter 4 the shunt leakage conductance G of the unmyelinated membrane is neglected in order to simplify the analysis. In the present chapter, where the nonregenerating fiber is relatively long, excessive error is introduced if G is discarded. Accordingly, in the equivalent circuit of Fig. 5-2, shunt ΔGs are included.

An analysis of the dendrite as an *RCG* cable network is given in Appendix A5-1. We will try to apply various input signals or "messages" to the dendritic network so as to examine its characteristics as a summing device.

The tree in Fig. 5-1 is that of a motoneuron in the spinal cord. It is of medium size, with perhaps 10,000 input junctions. It is due to Willis Rall, who made certain reasonable simplifying assumptions in his model of a dendritic tree. Rall's most useful assumption is that the characteristic resistance (R_0) of the equivalent *RCG* cable ensemble remains constant, independent of the number of branches [S. Deutsch, 1983; W. Rall, 1964, 1969, 1977]. It is as if we take a coaxial cable, cut it in half longitudinally to form the two branches between levels 0 and 1, cut each of these branches in half to make the four branches between levels 1 and 2, and so forth. Between levels 3 and 4 we have 16 relatively thin branches, each having a characteristic resistance of $16R_0$, but giving an ensemble value of R_0 if all of the branches are connected in parallel.

In Table 5-1 are summarized the characteristics for the PSP and four dendritic potentials (DPs) of Fig. 5-1 (right-hand scale). The $V(z, t)$ expression is given by Eq. (5-11) using the numerical value for z, where z is distance expressed in terms of length constants; the time of the peak, t_p, is given by Eq. (5-12); substituting this into the $V(z, t)$ equation yields the peak height, V_p; the area is given by Eq. (5-15); and the curve width t_w by Eq. (5-16). According to the model, if 16 simultaneous PSPs start out at the periphery, each with a height of 100 mV and width of 4 ms, they reach the neuron body

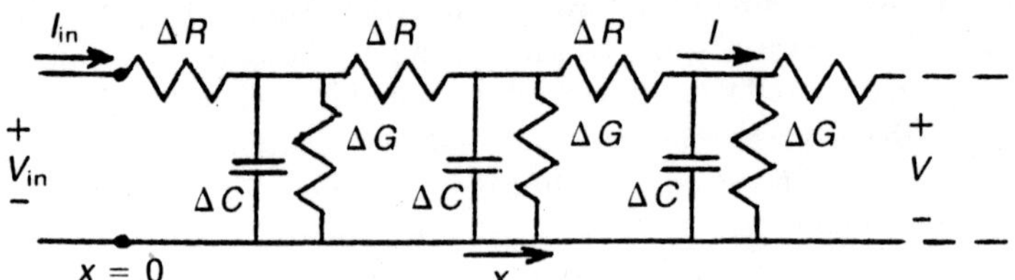

Fig. 5-2 The nonregenerating unmyelinated dendritic fiber as an infinitely long *RCG* cable consisting of infinitesimally small series resistances, ΔR; shunt capacitances, ΔC; and shunt leakage conductances, ΔG.

TABLE 5-1 Idealized Characteristics of the Postsynaptic and Dendritic Potentials

Level	z^a	t_p (ms)	V_p (mV)	Area (mV·ms)	t_w (ms)	$V(z, t)$ (mV, ms units)
4	2.0	2.00	100	413	4.13	$V(2, t) = \frac{3446}{t^{3/2}} \exp\left[-\left(\frac{4}{t} + \frac{t}{4}\right)\right]$
3	2.5	2.83	49.0	251	5.12	$V(2.5, t) = \frac{4307.5}{t^{3/2}} \exp\left[-\left(\frac{6.25}{t} + \frac{t}{4}\right)\right]$
2	3.0	3.71	25.3	152	6.01	$V(3, t) = \frac{5169}{t^{3/2}} \exp\left[-\left(\frac{9}{t} + \frac{t}{4}\right)\right]$
1	3.5	4.62	13.5	92	6.83	$V(3.5, t) = \frac{6030.5}{t^{3/2}} \exp\left[-\left(\frac{12.25}{t} + \frac{t}{4}\right)\right]$
0	4.0	5.54	7.4	56	7.59	$V(4, t) = \frac{6892}{t^{3/2}} \exp\left[-\left(\frac{16}{t} + \frac{t}{4}\right)\right]$

[a] z is distance expressed in terms of length constants.

Note: The PSP is launched at level 4, $z = 2$ of Fig. 5-1. The DPs result as the PSP propagates to the right. This is in response to 16 APs that simultaneously arrive, at $t = 0$, at the periphery terminal junctions. It is assumed that each PSP has a height (V_p) of 100 mV that occurs at peak time t_p, and the curve width (t_w) is around 4 ms. The areas under the waveforms are given in mV·ms units. The waveform equations are given by $V(z, t)$.

with a height of 7.4 mV and width of 7.6 ms. The $V_p = 7.4$-mV value is approximately sufficient to trigger an action potential at the neuron hillock.

If the neuron is a type that generates a graded potential, the output is simply proportional to the net summation of attenuated DPs.

Table 5-2 gives a listing of branch diameter, length, characteristic resistance, and surface area (πdl); also, R, C, and G are given *for the entire branch* rather than per centimeter. The neuron body is included as if it is part of the RCG cable, with a diameter of 15.87 μm and R_0 of 7.3 MΩ. The diameter may be reasonable, but the R_0 value is of little meaning. On the other hand, although a conventional transmission line yields reflections at

TABLE 5-2 Idealized Characteristics of the Branches of the Dendritic Tree of Fig. 5-1

z	Number of Branches	Dia (μm)	Length (mm)	R_0 (MΩ)	Area (μm²)	R (MΩ)	C (pF)	G (nS)
2 to 2.5	16	2.50	0.230	117.2	1807	58.58	17.06	4.266
2.5 to 3	8	3.97	0.290	58.59	3614	29.29	34.13	8.533
3 to 3.5	4	6.30	0.365	29.30	7227	14.65	68.26	17.07
3.5 to 4	2	10.00	0.460	14.65	14455	7.323	136.5	34.13
soma	1	15.87	—	7.324	—	—	—	—
Row ratio =	2	$2^{2/3}$	$2^{1/3}$	2	2	2	2	2

Note: The entire tree is determined by the following three assumptions: (a) the branch from $z = 3.5$ to 4 has a diameter of 10 μm; (b) each branch length is $\lambda_0/2$ (half its length constant); (c) the ensemble R_0 (characteristic resistance) is constant. The surface area of each branch is given in μm² units. Axoplasm resistance R, membrane capacitance C, and shunt leakage conductance G values are given for each branch.

an R_0 discontinuity, the waveforms of Fig. 5-1 show that the *RCG* cable losses are so high that reflections can be ignored. (In fact, when the neuron fires, its AP feeds back antidromically as a weak, rapidly decaying spike.)

Because of the constant-R_0 assumption, as we go from one row to the next higher row in Table 5-2, diameter decreases by a factor of $2^{2/3}$; length decreases by a factor of $2^{1/3}$; characteristic resistance doubles; and surface area is halved. The *total* surface area of each row remains constant at 28,900 μm^2. For the four-stage tree the total surface area is 116,000 μm^2 (0.116 mm^2). Since a junction has a diameter of 1 μm, or area of around 1 μm^2, in theory 100,000 of them could fit on the tree. Therefore, although it may be difficult to visualize, signals from 10,000 different prior neurons *can* be received by the tree of Fig. 5-1. In addition, as previously noted, many junctions may be located directly on the neuron body.

From Table 5-1 one can conclude that "position is everything in life" for a dendritic junction. The tree is democratic with respect to messages coming in at a particular level, but axons that carry more important APs or graded potentials have to terminate closer to the neuron body, at the lower levels of the tree. It is impossible for this location dominance to be predetermined by an individual's genetic makeup. The amount of information required to specify locations for each of the 10,000 junctions of a single neuron is enormous, and a blueprint for the entire nervous system is certainly beyond the capabilities of the DNA molecule. Instead, location dominance is gradually modified as the organism ages. In fact, the most popular theory concerning learning and memory proposes that "learning" is embodied in the growth and/or movement of dendritic junctions, and that "memory" resides in the locations and strengths of dendritic junctions. (Note that some of the collateral branches of an incoming axon can terminate on more than one junction of the *same* dendrite.)

In humans about one hour is required before short-term memory is consolidated into long-term form. We know from accident victims that a massive trauma such as a severe blow to the head, or electroconvulsive shock, may partially or fully erase memories of events that occurred during the previous hour, but it has no effect on events experienced before that and putatively stored as synaptic or electrical gap junctions. The preceding is cited as evidence that it takes one hour for a junction to grow; as we have calculated, there is plenty of room for new junctions, 100,000 if need be on the dendrite of Fig. 5-1. Nevertheless, anatomical evidence for a one-hour growth is lacking and, obviously, difficult to obtain in vivo.

There is *no* evidence that junctions, or the information stored in junctions, can migrate along dendritic branches. Nevertheless, this is attractive as a theory because it explains "forgetting." As new junctions form at lower levels of the tree, old junctions have to be displaced toward higher levels until they literally fall off the peripheral end of the tree. If this were not so, the tree would soon become saturated with junctions. In other words, the dendrite has to "forget" as much as it "learns"; "forgetting" is a normal activity that accompanies "learning"; a person who cannot learn (remem-

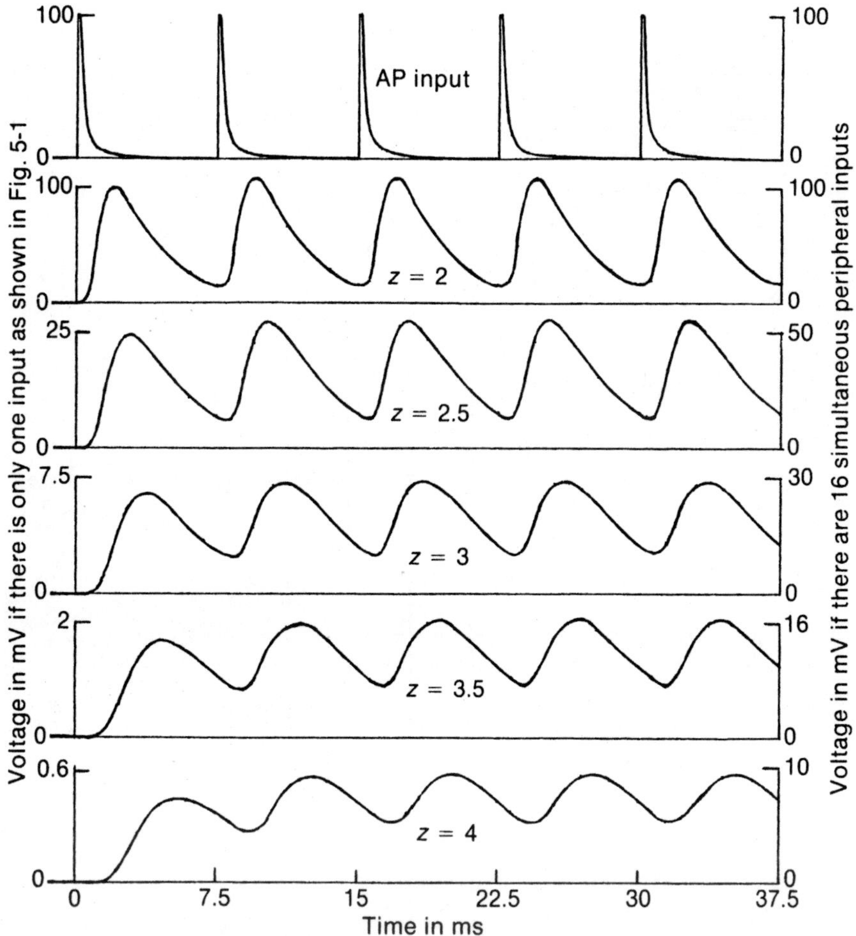

Fig. 5-3 Idealized waveforms for the model of Fig. 5-1 if the input is a repetitive burst of APs, in this case 7.5 ms apart (133 per second) so that appreciable integration occurs. The left-hand scale corresponds to a single AP burst arriving at the periphery, starting at time = zero. The right-hand scale corresponds to a situation in which *each* of the 16 peripheral branches simultaneously receives the AP burst.

ber) recent events is sometimes able to remarkably recall distant events; you cannot teach an old dog new tricks, but at least it does not forget the old ones; and so forth. Perhaps this is a good way to design robots. (A section devoted to "Memory" appears in Chap. 12.)

Figure 5-1 illustrates two very special and simple sets of input conditions: that of a single peripheral input, and that of 16 simultaneous peripheral inputs. In general, of course, the input message sequence can be a very complicated function of dendritic location and of time, leading to a correspondingly complicated neuronal response. One can program a computer to

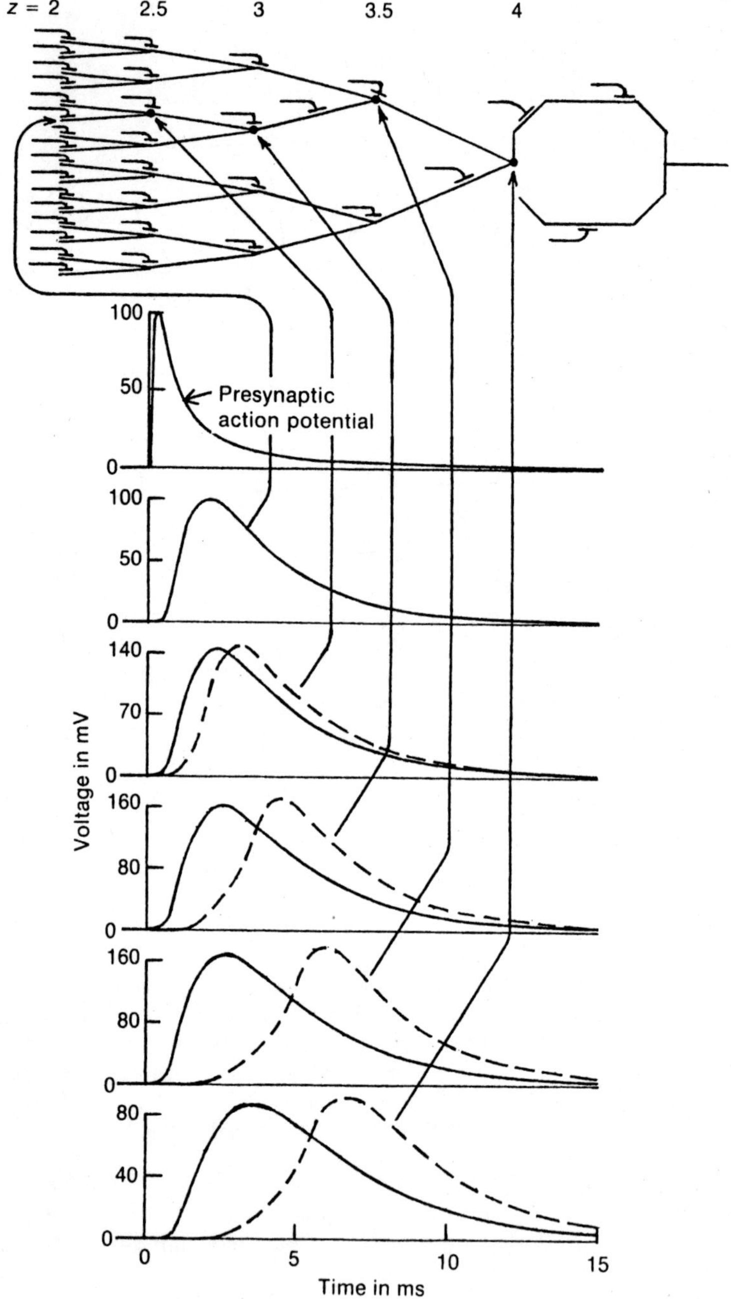

Fig. 5-4 Idealized waveforms for the model of Fig. 5-1 if there is massive excitation. The solid set of curves corresponds to a situation in which every peripheral branch and every node simultaneously receives an AP at time = zero. The dashed

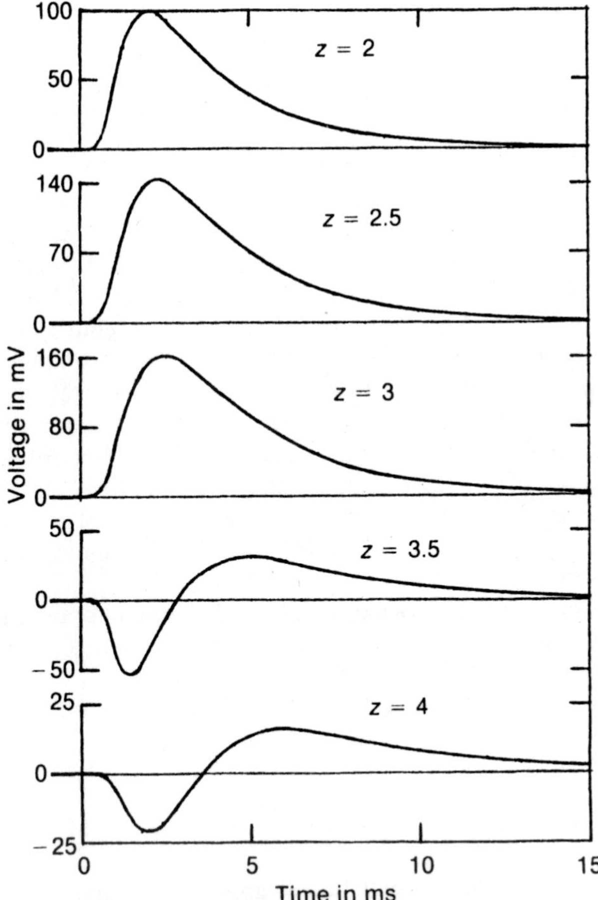

Fig. 5-5 Idealized waveforms for the model of Fig. 5-1 if the PSPs at the two nodes at $z = 3.5$ are *inhibitory*, of height 92 mV, while all of the other nodes and every peripheral branch simultaneously launch excitatory PSPs, of height 100 mV.

calculate the approximate response to any arbitrary set of input messages. There is much to be learned from these procedures beyond the conclusion that the system is complex [A. Pellionisz and R. Llinas, 1977]. Instead of results taken from a computer, however, in Appendix A5-2 a few simple input configurations, from which one can get a "feel" for dendritic summation, are examined. The input messages are illustrated in Figs. 5-3, 5-4, and 5-5.

←

set of curves corresponds to a situation in which the APs at $z = 2$, 2.5, 3, and 3.5, respectively, arrive at time = 0, 1, 2.5, and 4 ms.

5-3 Effect of Inhibitory Postsynaptic Potentials

One of the examples of dendritic summation (and subtraction) is the set of curves of Fig. 5-5. Here we illustrate a process that is discussed in Section 5-1: Inhibition.

Although it may be weak, an inhibitory PSP can have a profound effect in depressing neuronal activity. The reason for this is easily demonstrated by modifying Fig. 3-2. In the latter drawing we have a neuron with an extinguishing (resting) potential of −90 mV and a firing (threshold) level of −82 mV. In Fig. 3-2(*a*) a graded receptor potential of −80 mV is associated with a firing rate of 10 spikes/s; in Fig. 3-2(*b*) a receptor potential of −61 mV results in a firing rate of 50 spikes/s. An inhibitory stimulus is equivalent to *raising the threshold level* (or raising the firing voltage). If the threshold level is raised 10 mV, to −72 mV, Fig. 3-2 becomes modified as shown in Fig. 5-6. The neuron stops firing altogether in Fig. 5-6(*a*) because the receptor potential of −80 mV is below the threshold level. In Fig. 5-6(*b*), as the voltage gently rises from −90 mV toward the receptor potential at −61 mV, it takes much longer to reach the −72-mV threshold, resulting in a 3-to-1 reduction in firing rate.

As a check on the drawing, in Chap. 3 the action-potential frequency is derived in Eq. (3-4):

$$f = \frac{1}{\tau_0 \log\left(\dfrac{V_r - V_E}{V_r - V_F}\right)}.$$

Substituting the values of Fig. 5-6(*b*), $f = 16.6$ Hz, which has a period of 0.06 s and agrees with Fig. 5-6.

Of what use is inhibition? It will appear over and over again in the chapters that follow. For now, however, we would like to cite one example that is pertinent to something we have already considered. In Fig. 1-3 a person steps on a sharp pebble, initiating a sequence in which a motoneuron is excited, which in turn stimulates a muscle that contracts so as to get the foot to jump off the noxious object. But muscles usually come in pairs: a flexor is paired with an extensor. Obviously, it does not make sense to stimulate both of them at the same time. The nervous system is therefore "wired" so that, if a flexor is excited, its paired extensor is inhibited and vice versa. In order to retain simplicity in Fig. 1-3 the afferent axon is shown leading only to a single motoneuron. Actually, the afferent axon reaches many motoneurons; motoneurons that feed synergistic muscles are excited, while motoneurons that feed antagonistic muscles are inhibited.

Because it is ten times weaker than an excitatory PSP an inhibitory junction, in order to be effective, has to be located on, or very close to, the neuron body. If it is *on* the body, an inhibitory PSP has the effect of raising the threshold level, as previously noted. It may not raise the level by 10 mV because the neuron body has a large volume compared to the region under

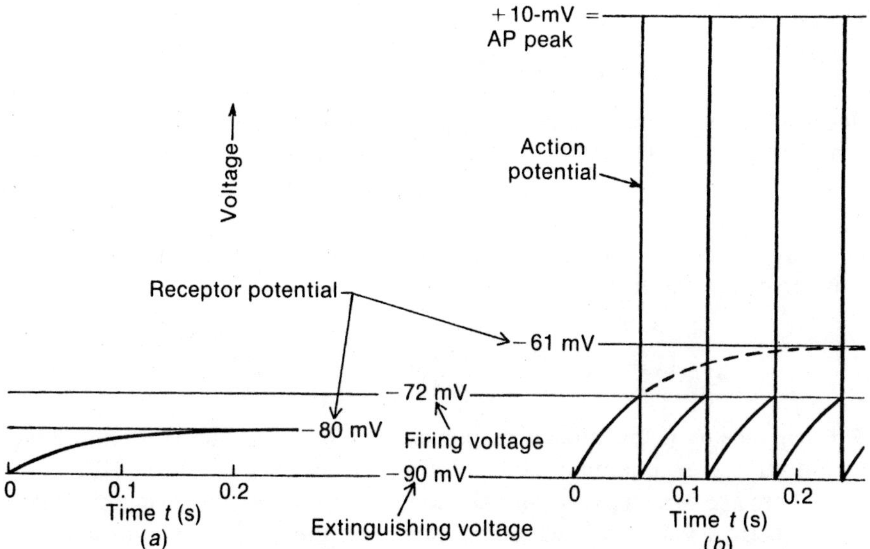

Fig. 5-6 How Fig. 3-2 is modified if the neuron receives an inhibitory input that raises its threshold level from −82 mV to −72 mV. (*a*) Receptor potential = −80 mV, the neuron fails to fire. (*b*) Receptor potential = −61 mV, firing frequency = 16.6 Hz compared to previous value of 50 Hz.

a single junction, but many inhibitory PSPs simultaneously arriving will block the generation of APs or greatly reduce their frequency.

The lesson to be learned here is that an inhibitory PSP arrives first, because its junction is located close to or on the neuron body, but sustained inhibition requires a continuous barrage of inhibitory inputs. Obviously, a sufficient number of inhibitory junctions must be active to subtract from the positive DPs arriving at the hillock.

A5-1 A DENDRITE AS AN RCG CABLE NETWORK

It is easier to use infinitesimally small sections in Fig. 5-2, so that the deltas can be replaced by derivatives and we have, from Ohm's law,

$$\frac{dV}{dx} = -IR. \tag{5-1}$$

Similarly, from Ohm's law,

$$\frac{dI(s)}{dx} = -(G + sC)V(s), \tag{5-2}$$

where $s = j\omega$. Eliminating I from these simultaneous equations,

$$\frac{d^2V(s)}{dx^2} = R(G + sC)V(s). \tag{5-3}$$

The general solution to this equation is

$$V(s) = c_1\epsilon^{-x\sqrt{R(G+sC)}} + c_2\epsilon^{x\sqrt{R(G+sC)}}. \tag{5-4}$$

Because V must vanish as x approaches infinity, we get $c_2 = 0$. Because V is an impulse of area A_0 V·s at $x = 0$, we get $c_1 = A_0$, so that

$$V(s) = A_0\epsilon^{-x\sqrt{R(G+sC)}}. \tag{5-5}$$

The waveform corresponding to Eq. (5-5) is a standard form that is given in inverse Laplace transform tables as

$$V = \frac{A_0 x}{2t^{3/2}}\sqrt{\frac{RC}{\pi}}\exp\left[-\left(\frac{x^2RC}{4t} + \frac{Gt}{C}\right)\right]. \tag{5-6}$$

This happens to be the impulse response of an RC cable, Eq. (4-8), multiplied by a decaying exponential, $\exp(-Gt/C)$.

There are two steps we can take to make Eq. (5-6) more tractable. First, use numerical values whenever possible, including mV, ms, and cm units rather than MKS units. Second, express distance in terms of length constants.

The length constant is associated with the dc response of the RCG cable. In Eq. (5-4), let $C = 0$ since it is irrelevant for a dc input, and let $V = V_{in}$ at $x = 0$. We then get

$$V = V_{in}\epsilon^{-x\sqrt{RG}}. \tag{5-7}$$

This is an exponential decay. From an analogy with the time constant in $\exp(-t/RC)$, we have a length constant λ_0:

$$\lambda_0 = \frac{1}{\sqrt{RG}}, \tag{5-8}$$

where λ_0 is the distance at which V decays to 36.8% of its initial value. (For the RC cable of Fig. 4-2, where $G = 0$, the length constant is infinite because there is no dc decay.)

To express distance in terms of length constants, let

$$x = \frac{z}{\sqrt{RG}}. \tag{5-9}$$

If $z = 1$, x is one length constant away from $x = 0$; if $z = 2$, x is two length constants away; and so forth.

In Section 1-4, the value $C/G = 4$ ms is reported as the time constant of a typical unmyelinated fiber. For the tree of Fig. 5-1, $C/G = 4$ ms yields reasonable answers. However, because dendrites come in a wide variety of sizes and topological configurations, depending on the number of input junctions and location in the organism, there may be wide departures from the $C/G = 4$-ms value [C. Koch, T. Poggio & V. Torre, 1982; G. M. Shepherd, 1974, 1978].

The substitution of $z/\sqrt{RG}$ for x and 4 ms for C/G into Eq. (5-6) leaves

$$V = \frac{A_0 z}{\sqrt{\pi}\, t^{3/2}} \exp\left[-\left(\frac{z^2}{t} + \frac{t}{4}\right)\right] \quad \text{(ms units).} \qquad (5\text{-}10)$$

The next step is a plot of $V(t)$ for $z = 1, 2, 3, \ldots$. This reveals that the impulse response of the *RCG* cable, at $z = 2$, has a width of approximately 4 ms, which is correct for a PSP. (It is the postsynaptic potential curve in Fig. 5-1.) To get a peak amplitude of 100 mV, we need $A_0 = 3054$ mV·ms. (For the unmyelinated fiber of Chapter 4, where there is no shunt leakage, an impulse area $A_0 = 216.2$ mV·ms is sufficient.) With $A_0 = 3054$ mV·ms in Eq. (5-10), there results

$$V = \frac{1723z}{t^{3/2}} \exp\left[-\left(\frac{z^2}{t} + \frac{t}{4}\right)\right] \quad \text{(mV, ms units).} \qquad (5\text{-}11)$$

Differentiating Eq. (5-11), we find that the peak is located at time

$$t_p = \sqrt{9 + 4z^2} - 3 \quad \text{(ms units).} \qquad (5\text{-}12)$$

Notice that, at $z = 2$, the peak occurs at exactly 2 ms.

The characteristic resistance of a cable is its input resistance if the length is infinite. In Eqs. (5-1) to (5-4), solving for input I with $C = 0$, we get

$$R_0 = \sqrt{\frac{R}{G}}. \qquad (5\text{-}13)$$

The characteristic resistance is a dc specification. Since the impulse response of the *RCG* cable entails ac behavior, what does the exercise of repeatedly cutting cables in half longitudinally do to the characteristic *impedance*? It is shown in Table 5-2 that, with each cut, R per branch doubles, C is halved, and G is halved. The C/G ratio remains constant at 4 ms, RC likewise remains constant, but the characteristic impedance doubles.

To get the waveforms of Fig. 5-1, we "zap" an *RCG* cable with an impulse of area 3054 mV·ms at the input end, $z = 0$. The impulse rapidly falls in height and broadens as it propagates to the right. At $z = 2$ length constants from the input, the impulse has a height of 100 mV and width of 4 ms. At this point it has reached level 4 in Fig. 5-1. Now, another simplifying and reasonable assumption is that *each branch has a length of* $z/2$ [E. W. Pottala, T. R. Colburn & D. R. Humphrey, 1973]. In other words,

Level =	4	3	2	1	0
corresponds to z =	2	2.5	3	3.5	4,

and

$$z = \frac{8 - \text{Level}}{2}, \qquad \text{Level} = 8 - 2z. \qquad (5\text{-}14)$$

From here on, we use the mathematically more convenient z designation for dendritic location rather than the Level designation.

Equation (5-11) calls for a scenario in which all 16 peripheral terminals at $z = 2$ receive the PSP waveform. In that event the waveforms at $z = 2.5, 3, \ldots$ are dimensioned by the *right-hand* scale in Fig. 5-1.

In Chapter 4 the area under a nonregenerative curve remains constant as it decays because there is no shunt leakage. In the present chapter, however, the area exponentially decays in accordance with

$$\text{Area} = A_0\epsilon^{-z} \tag{5-15}$$

[S. Deutsch and E. M-Tzanakou, 1987]. The area is useful because, dividing area by peak amplitude, we get a reasonable equivalent width, t_w, for these decidedly nonrectangular curves:

$$t_w = \frac{\text{Area}}{V_p}. \tag{5-16}$$

The history of a *single* PSP deposited at the periphery is given by the left-hand scale of Fig. 5-1. At the $z = 2.5$ bifurcation the propagating DP meets zero contribution from the other incoming branch, with the net result that there is a 2-to-1 reduction in height (to 24.5 mV) compared to that of the right-hand scale (49 mV). This 2-to-1 reduction is repeated at every bifurcation, with the net result that the DP reaching the neuron body has a height of only 7.4/16 = 0.46 mV. The contribution of a single peripheral PSP is therefore minuscule.

Consider next the physical dimensions of the tree. Rall assumed that each branch between $z = 3.5$ and 4 has a diameter of 10 μm. This, plus the $\lambda_0/2$ and constant-R_0 assumptions, determines the entire tree. Using a resistivity of 125 Ω·cm for axoplasm, we have previously calculated in Eq. (4-14) that

$$R = \frac{159.2}{d^2}\ \Omega/\text{cm of cable} \qquad \text{(cm units)}. \tag{5-17}$$

For G, it is convenient to use Eq. (1-3), which gives 5.647×10^9 Ω·cm as the resistivity of membrane insulation. Then, using cm units,

$$G = \frac{A}{\rho l} = 7.418 \times 10^{-4}d \quad \text{S/cm of cable} \tag{5-18}$$

where d is, of course, the diameter of the branch. Substituting Eqs. (5-17) and (5-18) into Eq. (5-13), there results

$$R_0 = \frac{463.2}{d^{3/2}} \qquad (\Omega,\ \text{cm units}) \tag{5-19}$$

and, from Eq. (5-8),

$$\lambda_0 = 2.910\sqrt{d} \qquad \text{(cm units)}. \tag{5-20}$$

If $d = 10\ \mu m$, R_0 of the $z = 3.5$ to 4 branch is 14.65 MΩ and the branch length, $\lambda_0/2$, is 0.46 mm.

A5-2 RESPONSE TO VARIOUS INPUT MESSAGES

A few simple input signal combinations are considered below. We have found that it is a great help, to *our* dendrites, at least, if the input message is listed as a sort of computer instruction, including an address, as follows:

z value	Node or branch number	Time in milliseconds	PSP peak in millivolts

For example, the vertical scale on the left side of Fig. 5-1 is associated with:

z value	Node or branch #	Time in ms	PSP peak in mV
2	06	0.0	100

That is, the message is addressed to the periphery, at the sixth branch down from the upper edge in the diagram, delivered by an AP that arrives at time = zero, and the PSP has a peak of 100 mV.

The vertical scale on the *right* side of Fig. 5-1 is associated with the input messages:

z value	Node or branch #	Time in ms	PSP peak in mV
2	01	0.0	100
2	02	0.0	100
⋮	⋮	⋮	⋮
2	16	0.0	100

This wastes paper by printing many lines where one will do; instead, the abbreviated form:

z value	Node or branch #	Time in ms	PSP peak in mV
2	01 to 16	0.0	100

stands for a message that is addressed to $z = 2$, at each of its 16 peripheral branches, delivered by APs that arrive at time = zero, and the PSPs have a peak of 100 mV.

[S.D. writes: "During an examination I once asked the students some questions about the propagation of a message that is delivered at:

z value	Node or branch #	Time in ms	PSP peak in mV
2.35	06	0.0	100

This resulted in utter confusion. I had to remind the class that junctions can occur *anywhere* on the dendrite, and z value 2.35 is a junction located 70% of the distance downstream between levels $z = 2$ and $z = 2.5$."]

Consider next Fig. 5-3. We are again trying to "kill two birds with one stone" by using two different vertical scales. The left-hand scale is associated with:

z value	Node or branch #	Time in ms	PSP peak in mV
2	06	0.0	100
2	06	7.5	100
2	06	15.0	100
2	06	22.5	100
2	06	30.0	100

The drawing shows the integration that takes place if APs follow each other at a rapid pace, in this case 7.5 ms apart, or 133/s. The first waveform shows the train of APs at the periphery, at the sixth branch down from the upper edge, as in Fig. 5-1. The resulting PSPs show very little integration; that is, each PSP almost ends before the next one begins. According to Table 5-1, however, the DPs get continuously wider as they propagate toward the neuron body. There is an increasing tendency for one DP to run into the next. By the time the DPs reach the $z = 4$ value, after a short transient period we get a fairly steady level plus a 133-Hz ripple.

In Fig. 5-3 the first parts of each waveform, between the 0- and 7.5-ms points, are identical to those of Fig. 5-1 between the 0- and 7.5-ms points. At the neuron body, for example, the first peak is at 0.46 mV, the same as in Fig. 5-1, but integration of several DPs eventually yields a peak value of 0.59 mV.

The scale at the right side of Fig. 5-3 is associated with:

z value	Node or branch #	Time in ms	PSP peak in mV
2	01 to 16	0.0	100
2	01 to 16	7.5	100
2	01 to 16	15.0	100
2	01 to 16	22.5	100
2	01 to 16	30.0	100

Since we now have 16 peripheral inputs, the output at the neuron body is 16 times as great as for the left-hand scale. The peak value after steady-state conditions are reached is $16 \times 0.59 = 9.4$ mV. This should be sufficient to cause the neuron to fire at a high rate almost continuously ("almost" because of ever-present noise). If the neuron is a type that generates a graded potential, the waveform shown in Fig. 5-3 could itself be the graded potential.

Mathematically, the waveform at $z = 2$ in Fig. 5-3 (right-hand scale) is given by (ms units)

$$V(t) = V(2, t) + V(2, t - 7.5)\,|_{t>7.5} + V(2, t - 15)\,|_{t>15} + V(2, t - 22.5)\,|_{t>22.5} + V(2, t - 30)\,|_{t>30}, \quad (5\text{-}21)$$

where $V(2, t)$, derived from Eq. (5-11), is also given in Table 5-1. Similarly, the waveform at $z = 4$ is given by

$$V(t) = V(4, t) + V(4, t - 7.5)\,|_{t>7.5} + V(4, t - 15)\,|_{t>15} + V(4, t - 22.5)\,|_{t>22.5} + V(4, t - 30)\,|_{t>30}, \quad (5\text{-}22)$$

where $V(4, t)$ is given in Table 5-1.

Without modification, the functional notation is correct for 100-mV PSPs launched at all 16 peripheral terminals. The modifications for other input locations are illustrated below.

We next look at what happens if there is massive excitation: every peripheral terminal and every node simultaneously receives an AP, as in Fig. 5-4:

z value	Node or branch #	Time in ms	PSP peak in mV
2	01 to 16	0.0	100
2.5	01 to 08	0.0	100
3	01 to 04	0.0	100
3.5	01 & 02	0.0	100

Thirty simultaneous APs at strategic locations! This corresponds to the solid set of curves in Fig. 5-4. We can examine each level as follows:

$z = 2$: The curve is a 100-mV PSP, exactly the same as that of Figs. 5-1 and 5-3.

$z = 2.5$: The PSPs that originated at $z = 2$ are around 50 mV high at $z = 2.5$. Here they combine with the new 100-mV PSPs originating at $z = 2.5$. Since the $z = 2$ PSPs arrive wider and a bit later (more to the right), the sum of the DPs and PSPs is somewhat less than 150 mV; it is 144 mV.

Here there is a dilemma. Dendritic and axonal voltages are generated by the transmembrane migration of (mostly) sodium ions. It is not reasonable to expect a DP of much more than 100 mV. In the preceding case, where a simple summation yields 144 mV, we can expect the waveforms of Fig. 5-4 to be flattened or clipped. This clipping is ignored in what follows because the purpose of the chapter is to illustrate dendritic *summation*; the PSPs may be much less than 100 mV, in which event clipping may not occur.

z = 3: The PSPs and DPs from z = 2.5 are around 72 mV high at z = 3. Here they combine with the new 100-mV PSPs originating at z = 3. Since the z = 2.5 signals arrive wider and a bit later, the sum of the DPs and PSPs is somewhat less than 172 mV—it is 161 mV.

z = 3.5: The PSPs and DPs from z = 3 are around 81 mV high at z = 3.5. Here they combine with the new 100-mV PSPs originating at z = 3.5. Since the z = 3 signals arrive wider and a bit later, the sum of the DPs and PSPs is somewhat less than 181 mV—it is 168 mV.

z = 4: The PSPs and DPs from z = 3.5 arrive wider and a bit later. The approximately 2-to-1 attenuation between levels yields a peak height of 87 mV as shown.

The z = 4 waveform displays a slowly decreasing tail; that is, the leading edge is much steeper than the trailing edge. The reason for this is that the z = 4 signal includes DPs that have traveled a long distance. The contribution from z = 2 has descended four levels; the PSP from z = 2.5 has propagated through three levels; and so forth. The older DPs suffer greater time delay, and therefore appear farther to the right. They are responsible for the long tail in the z = 4 waveform.

Mathematically, the solid curve at z = 4 in Fig. 5-4 is given by (ms units)

$$V(t) = V(4, t) + V(3.5, t) + V(3, t) + V(2.5, t). \tag{5-23}$$

Here the functional notation can be confusing; $V(2.5, t)$ does *not* stand for a DP that originated at a z = 2.5 location; instead,

$$\underset{\text{notation}}{z_{\text{functional}}} = 2 + \underset{\text{location}}{z_{\text{measuring}}} - \underset{\text{location}}{z_{\text{input}}} \tag{5-24}$$

In Eq. (5-23) we are looking at the waveform at z = 4, so this is the *measuring location*. The *input location* is where the DP originates. Expanding Eq. (5-23), the terms become

$$V(4, t) = V(2 + 4 - 2, t) \qquad \text{(DP originates at } z = 2 \text{ location)}$$

$$V(3.5, t) = V(2 + 4 - 2.5, t) \qquad \text{(DP originates at } z = 2.5 \text{ location)}$$

$$V(3, t) = V(2 + 4 - 3, t) \qquad \text{(DP originates at } z = 3 \text{ location)}$$

$$V(2.5, t) = V(2 + 4 - 3.5, t) \qquad \text{(DP originates at } z = 3.5 \text{ location)}$$

[S.D. writes: "I once fell victim to massive excitation in which everything happens at once: My first job was that of an electric motor technician. One hot day ('twas in the days before air conditioning) I was sitting on a grounded pipe (an excellent conductor) and my forehead touched a brush cap which one of my colleagues neglected to insulate. Some 110 volts dc passed through my body in a flash. What is it like to be electrocuted? It felt as if somebody came down on my head full force with a board. My entire visual firmament filled up with stars! You really *do* see stars, as from a blow on the head, from which the expression actually arose. Fortunately for me, the shock threw my head back. I picked myself up from the floor, a bit shaken but without any apparent loss of short- or long-term memory."]

The next combination of messages tries a variation on "everything happens at once": If the APs arriving at the lower levels are delayed, they can arrive at approximately the same time as the upper-level APs. The dashed set of curves in Fig. 5-4 again corresponds to massive excitation, with 30 APs arriving at strategic locations, but they do not arrive simultaneously. Notice that the APs occur at time = 0, 1, 2.5, and 4 ms, respectively:

z value	Node or branch #	Time in ms	PSP peak in mV
2	01 to 16	0.0	100
2.5	01 to 08	1.0	100
3	01 to 04	2.5	100
3.5	01 & 02	4.0	100

Each level is examined as follows:

$z = 2$: This is the same for both solid and dashed sets of conditions.

$z = 2.5$: The PSPs that originated at $z = 2$ are around 50 mV high at $z = 2.5$. Here they combine with the new 100-mV PSPs originating, 1 ms later than before, at $z = 2.5$. The sum of the DPs and PSPs yields a net curve that is to the right of the previous curve. It is not exactly 1 ms to the right because much of the curve is contributed by the DPs from $z = 2$, which started out at time = 0.

$z = 3$: The PSPs and DPs from $z = 2.5$ combine with the new 100-mV PSPs originating, at time = 2.5 ms, at $z = 3$. The net curve appears to be approximately 2 ms to the right of the previous curve at $z = 3$.

$z = 3.5$: The PSPs and DPs from $z = 3$ combine with the new 100-mV PSPs originating, at time = 4 ms, at $z = 3.5$. The net curve appears to be approximately 3 ms to the right of the previous curve at $z = 3.5$.

$z = 4$: The PSPs and DPs from $z = 3.5$ arrive wider, a bit later, and approximately half as high.

The solid curve at $z = 4$ has a rapidly rising leading edge and a slowly falling trailing edge. For the new set of inputs, as a result of the AP time shifts, the dashed curve displays a leading edge that rises more slowly than before, and a trailing edge that falls more rapidly than before.

Mathematically, the dashed curve at $z = 4$ in Fig. 5-4 is given by (ms units)

$$V(t) = V(4, t) + V(3.5, t - 1)|_{t>1} + V(3, t - 2.5)|_{t>2.5} + V(2.5, t - 4)|_{t>4}. \quad (5\text{-}25)$$

In Fig. 5-5 an example is given in which inhibitory PSPs arrive at both nodes at $z = 3.5$, while the higher levels of the tree receive massive excitatory PSP stimulation:

z value	Node or branch #	Time in ms	PSP peak in mV
2	01 to 16	0.0	100
2.5	01 to 08	0.0	100
3	01 to 04	0.0	100
3.5	01 & 02	0.0	−92

The inhibitory PSPs at $z = 3.5$ are shown as negative inputs, but why 92 mV? Because this is the amplitude needed to approximately cancel the positive DPs. The purpose of the example is to show that 10-mV inhibitory PSPs at $z = 3.5$ are much too weak to do a good job of cancellation.

The waveforms at levels $z = 2$, 2.5, and 3 in Fig. 5-5 are exactly the same as the corresponding solid set of curves in Fig. 5-4. At $z = 3.5$ we add a 92-mV negative PSP to the positive DPs arriving from $z = 3$. The negative PSP is similar to the PSP at $z = 2$ except that it is slightly smaller, and the curve, of course, goes down instead of up. Its leading edge precedes that of the DPs arriving from $z = 3$, so the net voltage at $z = 3.5$ at first swings negative as shown. At the zero crossing the negative PSP exactly cancels the positive DPs. After that the positive DPs, which are shifted to the right because of the time delay in reaching $z = 3.5$, are stronger than the negative PSP; the result is that the net voltage curve swings positive.

As in our previous examples, the PSPs and DPs from $z = 3.5$ arrive at $z = 4$ somewhat wider, a bit later, and attenuated by a factor of around 2. The negative peak reaches −22 mV, the zero crossing occurs at time = 3.6 ms, and the positive peak reaches 15 mV.

Mathematically, the waveform at $z = 3.5$ in Fig. 5-5 is given by (ms units)

$$V(t) = V(3.5, t) + V(3, t) + V(2.5, t) - 0.92V(2, t) \quad (5\text{-}26)$$

while the waveform at $z = 4$ is given by

$$V(t) = V(4, t) + V(3.5, t) + V(3, t) - 0.92V(2.5, t). \quad (5\text{-}27)$$

REFERENCES

S. Deutsch, RCG cable analysis of a dendritic tree based on Rall's idealized model, *IEEE Trans. Syst., Man, Cybern.*, vol. SMC-13, pp. 1007–1010, Sept./Oct. 1983.

S. Deutsch and E. M-Tzanakou, *Neuroelectric Systems*. New York: New York Univ. Press, 1987.

M. Ito, *The Cerebellum and Neural Control*. New York: Raven Press, 1984.

R. E. Kalil, Synapse formation in the developing brain, *Sci. Am.*, vol. 261, pp. 76–85, Dec. 1989.

C. Koch, T. Poggio, and V. Torre, Retinal ganglion cells: A functional interpretation of dendritic morphology, *Phil. Trans. Roy. Soc. London*, vol. B298, pp. 227–264, 1982.

A. F. Kohn, Dendritic transformation of random synaptic inputs as measured from a neuron's spike train—Modeling and simulation, *IEEE Trans. Biomed. Eng.*, vol. 36, pp. 44–54, Jan. 1989.

R. H. Masland, The functional architecture of the retina, *Sci. Am.*, vol. 255, pp. 102–111, Dec. 1986.

H. D. Patton, Spinal reflexes and synaptic transmission, in *Neurophysiology*, 2d ed., ed. T. C. Ruch et al. Philadelphia: Saunders, pp. 153–180, 1965.

A. Pellionisz and R. Llinas, A computer model of cerebellar Purkinje cells, *Neuroscience*, vol. 2, pp. 37–48, 1977.

E. W. Pottala, T. R. Colburn, and D. R. Humphrey, A dendritic compartment model neuron, *IEEE Trans. Biomed. Eng.*, vol. BME-20, pp. 132–139, March 1973.

W. Rall, Theoretical significance of dendritic trees for neuronal input-output relations, in *Neural Theory and Modeling*, ed. R. F. Reiss. Stanford, Calif.: Stanford Univ. Press, pp. 73–97, 1964.

W. Rall, Time constants and electrotonic lengths of membrane cylinders and neurons, *Biophys. J.*, vol. 9, pp. 1483–1508, 1969.

W. Rall, Core conductor theory and cable properties of neurons, in *Handbook of Physiology*, vol. 1, ed. E. R. Kandel. Bethesda, Md.: American Physiological Society, pp. 39–98, 1977.

G. M. Shepherd, *The Synaptic Organization of the Brain*. Oxford: Oxford Univ. Press, 1974.

G. M. Shepherd, Microcircuits in the nervous system, *Sci. Am.*, vol. 238, pp. 93–103, Feb. 1978.

Problems

1. According to the "s-shift" theorem of the Laplace transformation, if $\mathcal{L}^{-1}f(s) = f(t)$, then $\mathcal{L}^{-1}(s + \alpha) = f(t)\epsilon^{-\alpha t}$. Apply this to Eq. (4-8) in order to derive Eq. (5-6).
2. Carry out the step-by-step derivation of Eq. (5-11).
3. Carry out the step-by-step derivation of Eq. (5-12).
4. Carry out the step-by-step derivation of Eq. (5-13).
5. Plot the $z = 2$ and $z = 4$ waveforms of Fig. 5-1 (and Table 5-1).
6. Verify all of the Table 5-2 values for the $z = 2$ to 2.5 branches given such basic information as: The diameter of the $z = 3.5$ to 4 branches is 10 μm; the characteristic resistance of the ensemble is constant; the length of each branch is half its length constant; and so forth.

7. For each branch of the dendritic tree of Fig. 5-1, find R, C, and G as a function of diameter d, and verify the values given in Table 5-2.
8. In the dendritic tree of Fig. 5-1, PSPs are launched in accordance with:

z value	Node or branch #	Time in ms	PSP peak in mV
2.35	01 to 16	1.5	50

Write the DP equation at (a) $z = 2.5$; (b) $z = 3$; (c) $z = 3.5$; (d) $z = 4$. (e) For the $z = 4$ waveform, find the time of the peak (t_p), its height (V_p), and the nominal curve width (t_w). Also, plot the curve. [Ans.: (d) (mV, ms units)

$$V(3.65, t - 1.5) = \frac{3144}{(t - 1.5)^{3/2}} \exp\left[-\left(\frac{13.3225}{t - 1.5} + \frac{t - 1.5}{4} \right) \right] \Bigg|_{t>1.5};$$

(e) $t_p = 6.392$ ms, $V_p = 5.616$ mV, $t_w = 7.067$ ms.]
9. Plot the $z = 2$ and $z = 4$ waveforms of Fig. 5-3.
10. Plot the solid and dashed waveforms for $z = 4$ of Fig. 5-4.
11. Given the conditions of Fig. 3-2(a) and (b)—the figure is modified because the neuron receives an inhibitory input that raises its threshold level from -82 mV to -81 mV—find the resulting AP frequencies. [Ans.: (a) $f = 7$ Hz; (b) $f = 43.4$ Hz.]
12. Given the conditions of Fig. 3-2(a)—the figure is modified because the neuron receives an inhibitory input that lowers the AP frequency from 10 Hz to 1 Hz—find the new threshold level. [Ans.: $V_F = -80.000\ 001$ mV.]
13. Plot the $z = 3.5$ and $z = 4$ waveforms of Fig. 5-5.
14. Expand Eqs. (5-26) and (5-27) to show $z_{\text{measuring location}}$ and $z_{\text{input location}}$ values.
15. Using functional notation, write $V(t)$ for Fig. 5-5 for (a) $z = 2.5$ location; (b) $z = 3$ location. [Ans.: (a) $V(t) = V(2.5, t) + V(2, t)$; (b) $V(t) = V(3, t) + V(2.5, t) + V(2, t)$.]

6

Lateral Inhibition

ABSTRACT. The nervous system uses lateral inhibition to improve spatial resolution and contrast. Suppose, for example, that we have a stimulus distribution shaped like a bell. In lateral inhibition the stimulus distribution is shifted laterally (left and right in this case, say) by lateral branches of afferent axons, and subtracted (hence the designation *inhibition*) from the original stimulus curve. This yields a narrower curve (the sides of the "bell" are steeper).

Suppose that we have two bell-shaped stimulus curves, so close together that they partially merge to yield a single stimulus peak. Lateral inhibition may nevertheless be able to reveal that two stimuli are actually present.

A hypothetical three-stage model is examined in which the primary stimulus is a blunt "compass" point pressing against the hand, and lateral inhibition is applied in the spinal cord, thalamus, and somatosensory cortex.

Several two-dimensional models are also examined. A special case known as *zero-sum lateral inhibition* is especially important because it can extract the edges of the spatial stimulus curve.

6-1 A Single Stage of Lateral Inhibition

How does the brain process the huge amount of information coming in from the countless number of sensory receptors? We are a long way from fully answering this question, but we do know some of the tricks that are used. This chapter is about *lateral inhibition*, a scheme for the improvement of spatial resolution and contrast. We know definitely that the retina uses lateral inhibition. Beyond that the chapter is *entirely hypothetical* because it is assumed that, since lateral inhibition is so "cost effective," it is also employed by other senses such as touch, hearing, taste, and olfaction.

Consider first spatial resolution. Two pinpoints of light on a train, say, are moving away from you. Eventually they are so far away that it appears to you as if the light originates from a single pinpoint. You are no longer able to resolve the light into its dual sources. Obviously, any improvement in this spatial resolution corresponds to sharper vision, an important concomitant to survival. In studying the eyes of the horseshoe crab, H. Keffer Hartline found that improved spatial resolution and contrast are bestowed by lateral inhibitory neuronal circuits. For this he was awarded a Nobel Prize in physiology and medicine in 1967 (shared with R. Granit and G. Wald). [F. Ratliff, 1965; F. Ratliff, H. K. Hartline & W. H. Miller, 1963; G. von Bekesy, 1967.]

It is easier to illustrate lateral inhibition using the sense of touch rather than vision. Ask somebody (preferably a friend) to apply simultaneously two blunt compass points, about 8 cm (3 in.) apart, to your back. (A compass-type instrument is actually used because the spacing between points is easily adjusted.) You should be able to report that the stimulus is due to two separate points of pressure. (In this chapter it is assumed that such "reports" are solicited after adaptation has run its course; that is, after the initial shock of being jabbed in the back has subsided.) But when the compass points are only 6.7 cm apart, the average person cannot tell the difference between one- or two-point stimuli. The 6.7-cm distance is known as the *two-point threshold* for the back. Here the evidence for lateral inhibition is not at all clear-cut as it is for the horseshoe crab (or mammalian) eye; nevertheless, it is assumed here that the sense of touch also uses lateral inhibition. According to calculations given below, *without* lateral inhibition, two compass points on the back that are a bit less than 8.4 cm apart would be judged to be a single point. *With* lateral inhibition they are perceived easily as being two separate points, and the two-point threshold for the back decreases to 6.7 cm.

Incidentally, the two-point threshold is one of the standard neurological determinations. A neurologist may actually probe the patient's responses with the equivalent of two blunt compass points. The threshold for a normal individual varies widely, from 6.7 cm for the back, thigh, and upper arm to 0.2 cm for the fingertips [T. C. Ruch, 1965]. Obviously, where high spatial resolution for the analysis of texture is required, as for the fingertips, the two-point threshold is correspondingly small. The threshold for the tongue is even smaller than for the fingertips. In a neurological examination, corresponding areas on both sides of the body are compared since, in many of the pathological assaults upon the nervous system, only one side of the body is affected.

The palm of the hand, as depicted in Fig. 6-1(a), is a convenient although very approximate model with which to illustrate lateral inhibition. Its two-point threshold is around 1.125 cm (almost ½ in.). Curve (b) in Fig. 6-1 represents the outcome if the blunt compass point of Fig. 6-1(a) is pressed

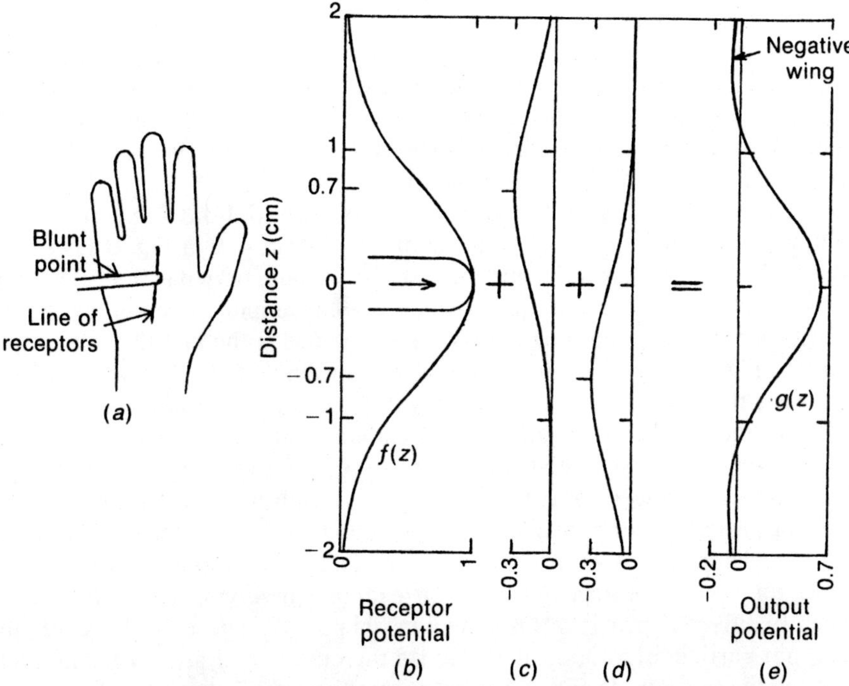

Fig. 6-1 Hypothetical curves that illustrate how lateral inhibition improves spatial resolution. (*a*) Inset depicting blunt point pressed against a line in the palm of the hand. (*b*) Response of touch receptors along the line. (*c*) Curve (*b*) multiplied by −0.3 and shifted laterally up. (*d*) Curve (*b*) multiplied by −0.3 and shifted laterally down. (*e*) Sum of curves (*b*), (*c*), and (*d*), which is substantially narrower than the initial response, curve (*b*).

toward the right against the palm of the right hand. The indentation of the palm along one of its lines, say, is depicted by the curve running from 2 cm above to 2 cm below the compass point. Better still, suppose that the curve represents voltage generated by the sensory receptors that transduce pressure into receptor potential along the line in question. Underneath the blunt point the receptor potential has a relative value 1; at the +1- and −1-cm distances it is 0.37, and so forth.

The lateral inhibitory neuronal circuits sharpen the receptor potential curve by a method so simple that one can only marvel, as usual, at the trial-and-error inventiveness of evolution, if it is given sufficient time. The method is to shift laterally (just how this is done is explained later on) smaller versions of the original curve and to subtract the shifted curves from the original. In Fig. 6-1(*c*) the curve is the original shifted up 0.7 cm and multiplied by 0.3; the curve of Fig. 6-1(*d*) is the original shifted down 0.7 cm and multiplied by 0.3. Subtraction is indicated by the fact that the shifted curves are negative. When the shifted curves are algebraically added to the original, we get the curve of Fig. 6-1(e). As you can see, the output curve covers a substantially narrower length than the original curve; that is, spatial resolution is improved by this simple technique. If the original consists of more than one stimulus, it turns out that this improvement in resolution is equivalent to an enhancement of contrast between the stimuli [W. Reichardt and G. MacGinitie, 1962; H. R. Wilson and J. D. Cowan, 1972].

There is a price paid for improved resolution. Subtraction results in an output curve that is weaker than the original. At $z = 0$ in Fig. 6-1, each of the shifted curves has the value 0.18, so the output curve has the peak value 0.64. The loss is not serious because neurons can amplify receptor potentials, if need be, to make up for the loss. More serious is the fact that the output curve, Fig. 6-1(e), is a distortion of the original; it has negative wings, whereas the original never swings negative. For the simple one-point stimulus of Fig. 6-1, large negative wings cannot result in serious distortion. If the stimulus is complicated, however, large negative wings (which are inhibitory) can cancel large positive peaks (which are excitatory), thereby erasing part of the stimulus; usually, this represents unacceptable distortion.

In vision, the distortion that accompanies lateral inhibition is responsible for certain optical illusions; an example involving parallel strips of slightly different gray levels is shown in Chapter 10. These illusions are mild and not particularly noticeable because the curves that are shifted laterally are relatively weak. In Fig. 6-1, the lateral shifts of 0.7 cm and multiplication by 0.3 yield negative wings that are about 10% as high as the positive peak, and this is a reasonable compromise that is incapable of serious distortion even if the spatial stimulus is complicated.

In lateral inhibition, in general, using functional notation, we have [S. Deutsch, 1977; S. Deutsch and E. M-Tzanakou, 1987]

$$g(z) = A\{f(z) - W[f(z - \delta) + f(z + \delta)]\}, \tag{6-1}$$

where A = neuronal amplification factor,
W = inhibitory weighting coefficient,
δ = lateral shift distance in the z direction.

In Fig. 6-1, $A = 1$, $W = 0.3$, and $\delta = 0.7$ cm.

The equation used in plotting Fig. 6-1 is the "Gaussian" function,

$$f(z) = \epsilon^{-z^2}. \tag{6-2}$$

For this simple curve, it is relatively easy to calculate the output in Fig. 6-1(e). Substituting into Eq. (6-1),

$$g(z) = \epsilon^{-z^2} - 0.3[\epsilon^{-(z-0.7)^2} + \epsilon^{-(z+0.7)^2}]. \tag{6-3}$$

In general, however, $f(z)$ may be a complicated analytic function, or a nonanalytic function containing straight lines and/or step discontinuities, resulting in an unwieldy $g(z)$. In that event it may be more sensible to draw $f(z)$, shift laterally, and add algebraically to get $g(z)$ as in Fig. 6-1. Alter-

TABLE 6-1 Tabular Method of Analysis[a]

z	0	0.35	0.7	1.05	1.4	1.75	2.1	2.45	2.8
$f(z)$	1	0.885	0.613	0.332	0.141	0.047	0.012	0.002	0
$f(z-0.7)$	0.613	0.885	1	0.885	0.613	0.332	0.141	0.047	0.012
$f(z+0.7)$	0.613	0.332	0.141	0.047	0.012	0.002	0		
$g(z)$	0.632	0.520	0.270	0.053	−.046	−.054	−.030		
$g_n(z)$	1	0.822	0.428	0.083	−.074	−.085	−.048		

[a] This method is applied to $f(z) = \exp(-z^2)$, $g(z) = f(z) - 0.3[f(z-0.7) + f(z+0.7)]$, g_n normalized so that $g_n = 1$ at $z = 0$. This corresponds to the curves of Figs. 6-1 and 6-3.

natively, one may use a tabular method, as is explained in Appendix A6-1 and illustrated in Table 6-1 for Fig. 6-1.

Also considered in Appendix A6-1 is the Taylor's series method of analysis, which reveals that *lateral inhibition tends to give the negative of the second spatial derivative of the input stimulus.*

Not surprisingly, the neuronal circuits responsible for the lateral shifts are relatively simple. Even so, a true representation showing synaptic junctions on a dendritic tree can become very confusing. In Chapter 5 we carefully followed dendritic potentials as they propagated toward the neuron body, rapidly attenuating, becoming wider, and suffering an increasing time delay. In the present chapter it is wise to ignore the dendritic potential waveform and to concentrate only on its attenuation. This modification is shown in Fig. 6-2(*a*). Here the neuron receives only three inputs: the central dendrite attenuates its signal by a factor of 10 to 1, as indicated by $\times(0.1)$, while each of the outer dendrites supplies an *inhibitory* (negative) signal, attenuated by a factor of 33 to 1, as indicated by $\times(-0.03)$. The neuron amplifies the net signal reaching its body by a factor of 2, as indicated by $\times 2$.

How can a neuron "amplify"? A good example is provided by the right-

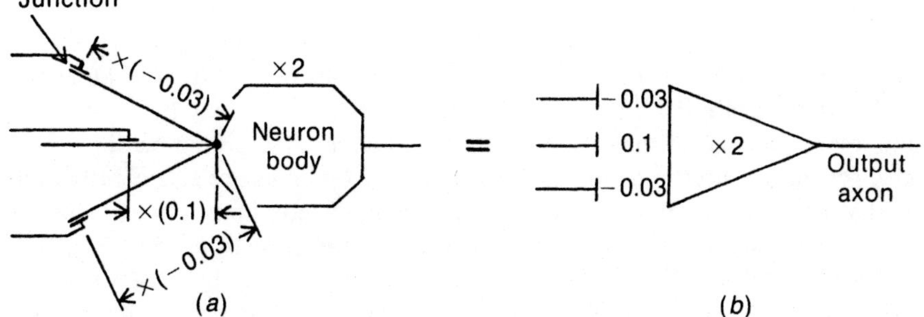

Fig. 6-2 Idealized neuron models. (*a*) Attenuation between junction input and neuron body is represented by $\times(0.1)$ (times 0.1) and $\times(-0.03)$ (times 0.03, inhibitory). The neuron amplifies the net signal reaching its body by a factor of 2 as indicated by $\times 2$ (times 2). (*b*) Simplified representation of part (*a*).

hand scale of Fig. 5-3. Here the AP input frequency is 133 spikes/s, which results in an approximately steady overpowering of the threshold level. The neuron thereupon fires at its own rate as determined by its own characteristics. If it fires at a frequency of 266 spikes/s, one can say that it has amplified the input signal by a factor of 2.

Similarly, if the input to a neuron is a graded potential, and its output is also a graded potential, the neuron can amplify the *changes* in potential. For example, if the output increases from 5 to 25 mV as the input increases from 12 to 22 mV, the output change of 20 mV corresponds to an input change of 10 mV and we can again say that amplification by a factor of 2 has occurred.

If the input to a neuron consists of action potentials (APs) while it generates a graded potential, or if the input consists of a graded potential while it generates APs, amplification is meaningless. We will try to avoid these situations in our numerical models.

Returning to Fig. 6-2, further simplification is possible and desirable. In Fig. 6-2(*b*) we dispense completely with the dendrite branches. Only the junctions remain, and the attenuation that each dendritic potential receives is indicated as shown. The multiplication sign is omitted; it is understood that these are multiplicative weighting coefficients or weighting factors or, simply, *weights*. The neuron body is shown as a triangle because, in electronic circuits, this is the standard symbol for an amplifier. The degree of amplification, $\times 2$, is entered as shown.

From here on in this book, the simplified neuronal notation of Fig. 6-2(*b*) is used: A neuron is a triangle feeding an output axon; each input junction is shown as a T, sometimes along with its weighting factor. If the junction is excitatory, the weight is positive or E; if inhibitory it is negative or I.

Using the preceding model for a neuron, the lateral inhibitory neuronal circuits are shown in Fig. 6-3. In Fig. 6-3(*a*) the input curve of Fig. 6-1(*b*) is repeated. The curve represents the idealized receptor potential response if a blunt point is pressed against one of the lines in the palm of the hand. There are many pressure sensory receptors along this line, but space limitations in the drawing allow only five afferent axons to be shown. (The axons in the drawing are 0.7 cm apart, whereas in reality they may be only 1 or 2 mm apart. In general, neuron density varies depending on the need for sensory resolution. In the palm, however, neuron densities are sufficiently high so that it is reasonable to assume a homogeneous continuum, with neurons everywhere, with the output independent of neuron distribution.) The 0.7-cm lateral shifts of Fig. 6-1(*c*) and (*d*) are effected by branches running laterally to the neurons that are 0.7 cm away on each side; the lateral branches terminate on inhibitory junctions whose weights are -0.3, whereas the direct branches terminate on junctions whose weights are $+1$. In Fig. 6-1 lateral inhibition causes a loss of output in exchange for a narrower response: a receptor potential of 1 at $z = 0$ results in a graded potential output of 0.632, according to Table 6-1. In Fig. 6-3(*b*) each neuron instead amplifies by a factor of 1.58, as indicated, so the 1 output at $z = 0$ is restored.

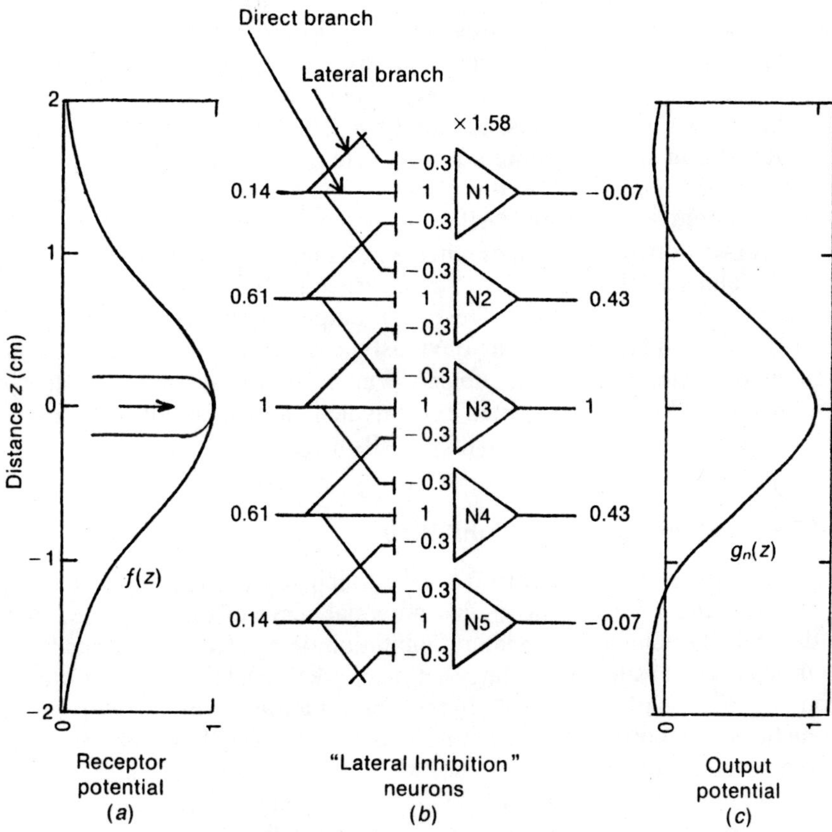

Fig. 6-3 Hypothetical lateral inhibitory neuronal circuit for the case of a blunt instrument pressed against one of the lines in the palm of the hand. (*a*) Input curve, $f(z)$. (*b*) Highly simplified representation of the circuit, showing five afferent axons from five pressure sensory receptors. The conditions of Fig. 6-1 are duplicated except that each neuron amplifies by a factor of 1.58, thus yielding a peak output amplitude of 1. The values 0.14, 0.61, . . . correspond to receptor potentials at that location, while −0.07, 0.43, . . . correspond to output graded potentials. (*c*) Output curve, $g_n(z)$.

The graded potentials carried by each output axon have the values −0.074, 0.428, 1, 0.428, and −0.074, respectively, as shown in the $g_n(z)$ row of Table 6-1. The output curve is obviously much sharper, with a higher resolution, than the input curve. The subjective impression would be that the stimulus is "sharper" than without lateral inhibition. Normalization is very convenient in the table because it shows, at a glance, how much narrower $g_n(z)$ is than $f(z)$.

Figure 6-3 is highly oversimplified. The row of five neurons is located physically in or near the spinal cord and not in the palm of the hand. Neurons associated with the sensory receptors in the palm generate APs for the one-meter-long journey through a nerve trunk in the arm to the spinal cord; here is located the first of several relay stations. The synaptic junctions of Fig. 6-3(b) relay the pressure stimuli toward the brain. Synaptic junctions are also made with interneurons as part of a reflex system to achieve rapid withdrawal from noxious stimuli. It is a world dominated by AP spikes. Most of the "graded potentials" in this chapter are a convenient fiction, since the neurons should be generating AP frequencies. For example, in Fig. 6-3(b), N1 to N5 can operate at a spontaneous, unstimulated rate of 100 Hz, say, and the values -0.07, 0.43, and 1 can, respectively, correspond to AP frequencies -7, 43, and 100 Hz. The actual output frequencies would then be 93, 143, and 200 Hz. All of this obviously adds many unnecessary complications to an already difficult subject. We will simply acknowledge that, in this chapter, APs are represented by equivalent graded potentials, confident that we can always transform back to APs if need be.

6-2 Three Stages of Lateral Inhibition

How does the stimulus pattern from the touch sensory receptors reach the brain? The equivalent of the graded potential curve of Fig. 6-3(c), with hypothetical sharpening due to lateral inhibition, is located in the spinal cord. Neuroanatomical studies indicate that there are two additional relay stations along the main path. First, the afferent output axons of Fig. 6-3(b) proceed along the spinal cord to the thalamus. This is a relatively large approximately spherical mass of neurons at the base of the brain. (It *has* to be large because practically all sensory receptors feed to the thalamus.) From here the stimulus pattern projects to various parts of the brain; the projection to the cortex terminates upon synaptic junctions and a final set of neurons (known as *somatosensory neurons* if the sense of touch is involved). The cortex is the outer layer of the brain, which is so wrinkled and convoluted in man that it is somewhat misleading to call it a "layer."

It turns out that two stages of lateral inhibition are better than a single stage, and three stages are still better. Therefore, since the "cost" (some loss of signal amplitude and the growth of lateral inhibitory branches) is small compared to the benefits (enhanced spatial resolution and contrast), we will assume that lateral inhibition is incorporated into each of the relay stations mentioned earlier. It is next to impossible, however, to verify this conjecture, so what follows is *hypothetical.*

For a blunt point pressed against one of the lines in the palm of the human hand, the complete three-stage narrative is summarized in Fig. 6-4. The curves in this figure depict the following:

Curve (a): $f(z)$, the output of the pressure receptors, identical to the curves of Figs. 6-1(b) and 6-3(a);

Curve (b): $g_n(z)$, the distribution of graded potentials along the spinal cord;

Curve (c): $h_n(z)$, the pattern between the thalamus and cortex;

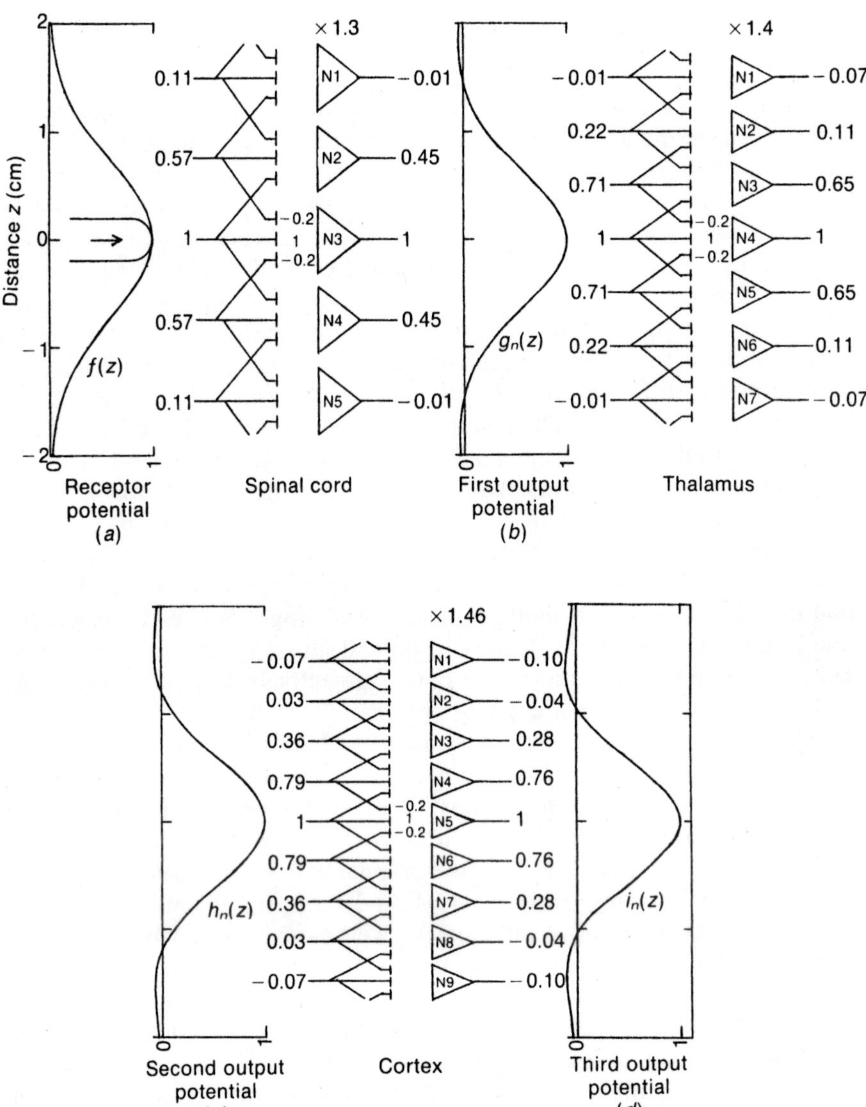

Fig. 6-4 Hypothetical three stages of lateral inhibition. Weight of −0.2 is assigned to *each* of the lateral synaptic junctions. (*a*) Response of touch receptors, along a line in the palm of the hand, to a blunt compass point, $f(z)$. (*b*) Between spinal cord and thalamus after lateral shift distance of 0.75 cm and amplification by factor of 1.3 is applied, $g_n(z)$. (Shift distances are *relative to the palm.*) (*c*) Between thalamus and cortex after lateral shift distance of 0.5 cm and amplification by factor of 1.4 is applied, $h_n(z)$. (*d*) At the cortex after lateral shift distance of 0.375 cm and amplification by factor of 1.46 is applied, $i_n(z)$.

Curve (d): $i_n(z)$, the final spatial distribution in the somatosensory cortex.

(All of the z distance scales in Fig. 6-4 are, of course, *relative to the palm*. Because the axon bundles converge and diverge, the actual distance may be completely different from the centimeter value shown in Fig. 6-4. At the thalamus in Fig. 6-4, seven neurons are shown compared to five at the spinal cord. Unfortunately, this can create the impression that thalamus neurons are smaller than those of the spinal cord; this may or may not be true, but neuron size in Fig. 6-4 bears no relation to that of actual neurons. At the cortex nine small neurons are shown on the drawing, again without regard to the actual sizes of cortical neurons.)

Curve (*d*) in Fig. 6-4 is correctly shown as a graded potential because the signal is processed locally and the distances are small. The other curves, however, should be shown as AP frequencies but, as explained in the previous section, are shown instead as receptor and graded potentials.

The "spinal cord" neuronal circuit of Fig. 6-4 [between curve (*a*) and curve (*b*)] is not the same as that of Fig. 6-3(*b*). The reason is that, if we simply repeat the circuit of Fig. 6-3(*b*) three times, the result is a final output that displays excessive negative wings. These wings could cause erroneous touch illusions that are analogous to optical illusions. In fingering a complicated texture, for example, one could subjectively feel additional ridges and/or valleys that are not actually present.

We have employed two rules in designing Figs. 6-3 and 6-4. (1) The final output curve should have negative wings that are about 10% as high as the positive peak. In Fig. 6-3 this is satisfied by lateral weights of -0.3. (2) The lateral shift distance should be optimum in the sense that the final output curve is appreciably sharper than the original without requiring a great deal of neuron amplification. In Fig. 6-3(*b*) the lateral shift distance is 0.7 cm.

The effects of using insufficient as well as excessive lateral shift distances are illustrated in Fig. 6-5. In Fig. 6-5(*b*), (*c*), and (*d*) the solid curves belong to a shift distance of 0.35 cm. The output, in Fig. 6-5(*d*), displays an excessive loss of amplitude: at $z = 0$ the output is 0.47 compared to 0.63 for a shift distance of 0.7 cm. The reason is obvious: with a lateral shift of only 0.35 cm, the inhibitory branches are gnawing away at the peak of the input stimulus rather than its sides. Incidentally, curve (*d*) in Fig. 6-5 is appreciably narrower than curve (*a*), and it does have negative wings, but they are too small to show up on the drawing.

The dashed curves of Fig. 6-5(*b*), (*c*), and (*d*) belong to a shift distance of 1.4 cm. Now it is obvious that the negative wings are excessive. The amplitude at $z = 0$ is good—0.92—but the wings dip down to -0.22 (at $|z| = 1.73$ cm). The negative wings are 24% as high as the positive peak, and the output is not much narrower than the input.

Applied to Fig. 6-4, rule (1) with regard to negative wings is satisfied by a weight of -0.2 assigned to *each* of the lateral inhibitory synaptic junctions. Curve (*b*) of Fig. 6-4 has negative wings of 3.5%; for curve (*c*) the

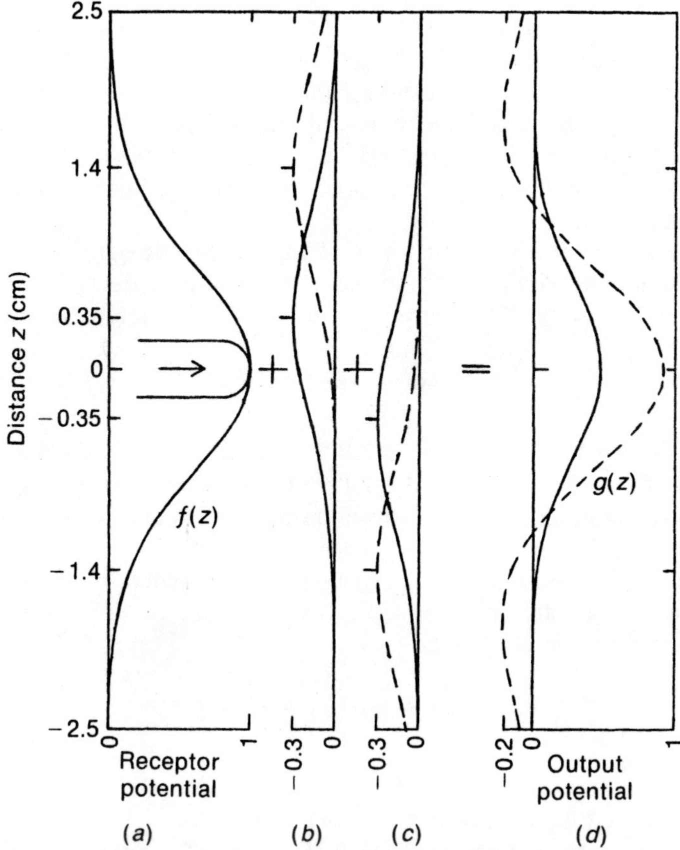

Fig. 6-5 Figure 6-1 repeated with insufficient lateral shift distance (0.35 cm, solid curves) and excessive shift distance (1.4 cm, dashed curves). (*a*) Response of touch receptors for which the optimum shift distance is 0.7 cm. (*b*) Lateral up shift. (*c*) Lateral down shift. (*d*) Sum of parts (*a*), (*b*), and (*c*).

value is 7.1%; for curve (*d*), as desired, the negative wings are 10% as high as the positive peak.

What is the optimum lateral shift distance? Given a symmetrical curve, trial and error shows that δ_{opt} is the distance from $z = 0$ to the inflection point, where d^2f/dz^2 is zero. For $f(z) = \exp(-z^2)$,

$$\frac{df}{dz} = -2z\epsilon^{-z^2} \tag{6-4}$$

and

$$\frac{d^2f}{dz^2} = 2(2z^2 - 1)\epsilon^{-z^2}. \tag{6-5}$$

From Eq. (6-5), $d^2f/dz^2 = 0$ at $|z| = \sqrt{1/2} = 0.7071$ cm. It is of course more convenient to use $\delta = 0.7$ than 0.7071 cm in drawing Figs. 6-1 and 6-3. The value $\delta_1 = 0.75$ cm is used for the spinal cord network in Fig. 6-4.

For the hypothetical three stages of lateral inhibition of Fig. 6-4, δ_{opt} decreases as the curves become narrower and the distance from $z = 0$ to the inflection point decreases. One should therefore calculate a different δ_{opt} for each network.

According to the Taylor's series of Eq. (6-15), the $g_n(z)$ curve of Fig. 6-4(b) is dominated by d^2f/dz^2, so the location of its inflection point is approximately given by $d^4f/dz^4 = 0$. From Eq. (6-5), differentiating twice,

$$\frac{d^4f}{dz^4} = 4(3 - 12z^2 + 4z^4)\epsilon^{-z^2}. \tag{6-6}$$

The first zero crossing of d^4f/dz^4 occurs at $|z| = \sqrt{1.5 - \sqrt{1.5}} = 0.5246$. The value $\delta_2 = 0.5$ cm is used for the thalamus network in Fig. 6-4. (A smaller value for δ gives a narrower output, but requires more neuronal amplification.)

Similarly, the $h_n(z)$ curve of Fig. 6-4(c) is dominated by d^4f/dz^4, so the location of its inflection point is approximately given by $d^6f/dz^6 = 0$. From Eq. (6-6), differentiating twice,

$$\frac{d^6f}{dz^6} = -8(15 - 90z^2 + 60z^4 - 8z^6)\epsilon^{-z^2}. \tag{6-7}$$

The first zero crossing of d^6f/dz^6 occurs at $|z| = 0.4361$. The value $\delta_3 = 0.375$ cm is used for the cortical network in Fig. 6-4.

The reason for employing $\delta_1 = 0.75$, $\delta_2 = 0.5$, and $\delta_3 = 0.375$ cm is that the tabular method of deriving $i(z)$ requires simple integer relationships between the δs. We have $0.75/6 = 0.5/4 = 0.375/3 = 0.125$. Since Fig. 6-4 extends to $z = 2$, the corresponding table should extend to $z = 2 + 0.75 + 0.5 + 0.375 = 3.625$. Because of space limitations, Table 6-2 cuts off at $z = 2$, but the original work sheet extended to $z = 3.625$.

The curves of Fig. 6-4 are normalized so that the output peaks at 1 in each case:

For $g_n(z)$ we require $A_1 = 1.295$.

For $h_n(z)$ we require $A_2 = 1.397$.

For $i_n(z)$ we require $A_3 = 1.462$.

There is nothing especially significant concerning the values of these neuronal amplification factors. Because of normalization, we can see by comparing f, g_n, h_n, and i_n values in Table 6-2 how lateral inhibition yields an ever narrower curve. This is also shown by the zero crossings in Fig. 6-4. Curve (d) of Fig. 6-4 crosses at $|z| = 1.06$ cm, while curve (c) of Fig. 6-3 crosses at $|z| = 1.18$ cm. Admittedly the improvement is small, but evo-

TABLE 6-2 Tabular Method of Analysis[a]

z	0	0.125	0.25	0.375	0.5	0.625	0.75	0.875	1	1.125	1.25	1.375	1.5	1.625	1.75	1.875	2
$f(z)$	1	0.984	0.939	0.869	0.779	0.677	0.570	0.465	0.368	0.282	0.210	0.151	0.105	0.071	0.047	0.030	0.018
$f(z - 0.75)$	0.570	0.677	0.779	0.869	0.939	0.984	1	0.984	0.939	0.869	0.779	0.677	0.570	0.465	0.368	0.282	0.210
$f(z + 0.75)$	0.570	0.465	0.368	0.282	0.210	0.151	0.105	0.071	0.047	0.030	0.018	0.011	0.006	0.004	0.002	0.001	0.001
$g_n(z)$	1	0.979	0.919	0.827	0.711	0.583	0.452	0.329	0.221	0.132	0.066	0.017	−.013	−.030	−.035	−.034	−.031
$g_n(z - 0.5)$	0.711	0.827	0.919	0.979	1	0.979	0.919	0.827	0.711	0.583	0.452	0.329	0.221	0.132	0.066	0.017	−.013
$g_n(z + 0.5)$	0.711	0.583	0.452	0.329	0.221	0.132	0.066	0.017	−.013	−.030	−.035	−.034	−.031	−.025	−.019	−.013	−.010
$h_n(z)$	1	0.974	0.901	0.790	0.652	0.504	0.356	0.224	0.114	0.030	−.024	−.059	−.071	−.072	−.062	−.049	−.037
$h_n(z - 0.375)$	0.790	0.901	0.974	1	0.974	0.901	0.790	0.652	0.504	0.356	0.224	0.114	0.030	−.024	−.059	−.071	−.072
$h_n(z + 0.375)$	0.790	0.652	0.504	0.356	0.224	0.114	0.030	−.024	−.059	−.071	−.072	−.062	−.049	−.037	−.025	−.016	−.008
$i_n(z)$	1	0.970	0.885	0.758	0.603	0.440	0.281	0.144	0.037	−.039	−.080	−.101	−.098	−.087	−.066	−.046	−.031

[a] This method is applied to $f(z) = \exp(-z^2)$, $g(z) = f(z) - 0.2[f(z - 0.75) + f(z + 0.75)]$, $h(z) = g_n(z) - 0.2[g_n(z - 0.5) + g_n(z + 0.5)]$, $i(z) = h_n(z) - 0.2[h_n(z - 0.375) + h_n(z + 0.375)]$. This corresponds to the curves of Fig. 6-4.

lution is full of examples in which small improvements appreciably enhance survivability.

Table 6-2 also shows that the final output, $i_n(z)$, has negative wings of relative amplitude 10%.

6-3 Two-Point Examples

Just how good is three-stage lateral inhibition? To answer the question two two-stimulus-point examples are considered in this section.

Again take the palm of the right hand and hold it vertically. Along one of its lines, at a location conveniently called $z = 0$, press to the right with a blunt compass point [solid curve in Fig. 6-6(*a*)]. At a location 1.375 cm away, press to the right with the other (blunt) compass point, but with a force only 60% as strong as at $z = 0$ [dashed curve in Fig. 6-6(*a*)]. The arrow representing the second point, at $z = 1.375$ cm, is only 60% as long as at

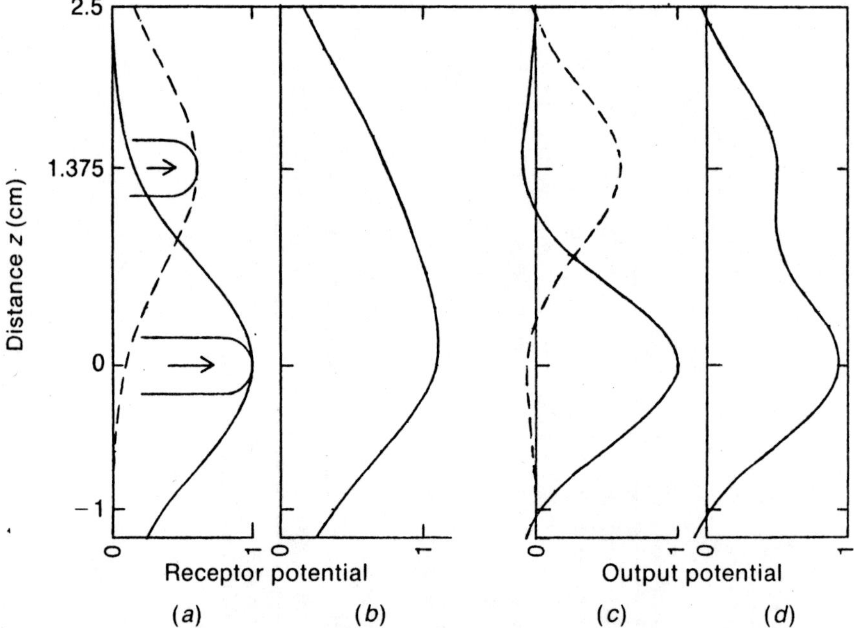

Fig. 6-6 Hypothetical three stages of lateral inhibition applied to two-stimulus-point example. Along a line in the palm of the hand, we have blunt compass points applied at distance = 0 (solid curves) and at distance = 1.375 cm, but with 60% of the force of the first stimulus (dashed curves). (*a*) Response of touch receptors to each stimulus alone (solid line and dashed line). (*b*) Combined response. The curve has a single peak at 0.18 cm. (*c*) Cortical output corresponding to each stimulus alone. (*d*) Combined output. The cortex is able to resolve the two-point stimulus, as shown, because two peaks are extracted.

the first point. Curve (*b*) of Fig. 6-6 shows the deflection of the skin or, equivalently, the receptor potential output of the sensory receptors that lie along the line. The stimulus pattern shows a small bump at around the z = 1.375-cm location.

Despite the vagueness of the bump, the three-stage lateral inhibition output curve is able to resolve the two-point stimulus. The two components of the output are shown in Fig. 6-6(*c*). The solid curve is a repeat of Fig. 6-4(*d*), while the dotted curve is Fig. 6-4(*d*) shifted up by the equivalent of 1.375 cm and multiplied by 0.6. Not only is the somatosensory cortex curve of Fig. 6-6(*d*) able to "see" two points, it is able to locate accurately the point at $z = 0$ despite the fact that the peak in the receptor potential curve occurs at $z = 0.18$ cm.

Using functional notation, the input stimulus in Fig. 6-6 is

$$f_t(z) = f(z) + 0.6f(z - 1.375), \tag{6-8}$$

where $f(z) = \exp(-z^2)$. Because it is assumed that this is a linear system, the output graded potential is described by

$$i_t(z) = i_n(z) + 0.6i_n(z - 1.375). \tag{6-9}$$

Given the $i_n(z)$ values of Table 6-2, it is easy to use the tabular method to find $i_t(z)$, curve (*d*) in Fig. 6-6.

Although we are using the palm of the hand as a convenient vehicle for discussion, the curves of Fig. 6-6 are approximately correct for any sensory system that is modified by lateral inhibition. Suppose, for example, that the two-point stimuli represent two pinpoints of light, so the receptor potential curve belongs to photoreceptors in the eye. The photoreceptors see a single diffuse light source at $z = 0.18$ cm and the equivalent of a small bump at $z = 1.375$ cm. The visual cortex extracts a pinpoint of light at $z = 0$ and a weaker, diffuse source at around $z = 1.3$ cm. The improvement in spatial resolution is accompanied by an improvement in contrast; the two effects are inseparable.

The second example illustrates the two-point threshold along the line of the palm. As the curves in Fig. 6-7(*a*) indicate, we again have two blunt compass points, but now they are only 1.125 cm apart. The pressures are equal so the curves are symmetrical. The resulting stimulus pattern in curve (*b*) of Fig. 6-7 has a single peak at $z = 0.5625$ cm.

The two components of the output are shown in Fig. 6-7(*c*): they are the curve of Fig. 6-4(*d*) centered on $z = 0$ and the same curve centered on $z = 1.125$ cm. Why does the output pattern of Fig. 6-7(*d*) illustrate the two-point threshold? Because it is a flat-topped curve. If the compass points are moved closer together, a peak will develop at $z = 0.5625$ cm; if the points are moved farther apart, a valley will develop and the somatosensory cortex, which is your subjective feeling in the matter, will "see" two points.

In Fig. 6-7, using functional notation, the input stimulus is

$$f_t(z) = f(z) + f(z - 1.125) \tag{6-10}$$

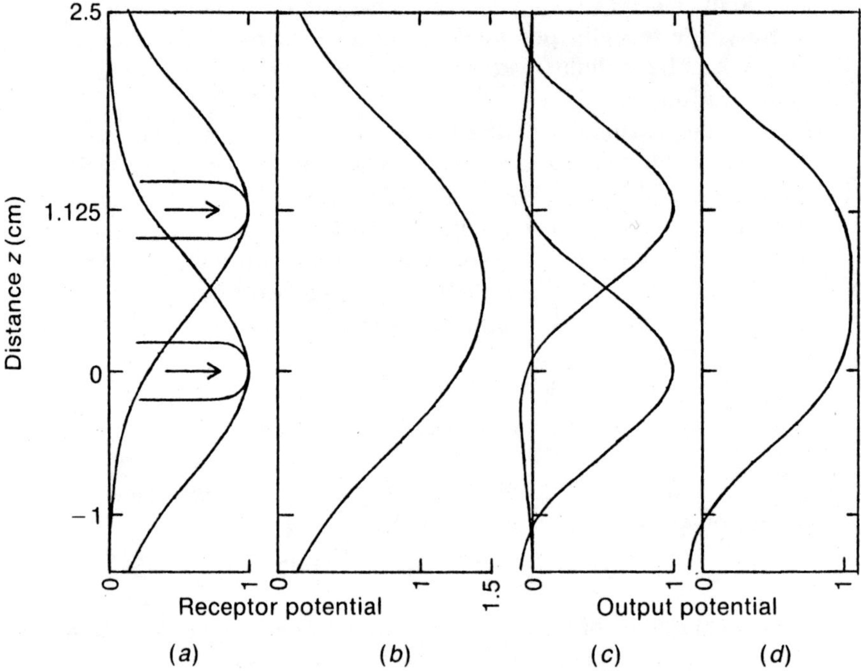

Fig. 6-7 Hypothetical three-stage lateral inhibition applied to determination of the two-point threshold along a line in the palm of the hand. We have blunt compass points applied at distance = 0 and at distance = 1.125 cm. (*a*) Response of touch receptors to each stimulus alone. (*b*) Combined response. The curve has a single peak at 0.5625 cm. (*c*) Cortical output to each stimulus alone. (*d*) Combined output, which has a flat top.

while the output graded potential appears as

$$i_t(z) = i_n(z) + i_n(z - 1.125). \tag{6-11}$$

What would be the two-point threshold without lateral inhibition? For curve (*a*) of Fig. 6-4 it is 1.41 cm, 25% larger than the 1.125-cm value of Fig. 6-7.

If lateral inhibition can improve the spatial resolution and contrast of visual images, why is it not used commercially to sharpen up photographs? It is *exactly* what is used, except that it is not called "lateral inhibition." Remember, also, that it can be overdone: too much lateral inhibition generates excessive negative wings that can introduce false ridges and valleys. A photograph, which is of course two-dimensional, first has to be broken up into small picture elements. The gray levels are digitized. Each small element of the picture is then processed by taking into account the surrounding picture elements. Because a photograph can easily contain one million picture elements, the processing has to be done automatically with the aid of a high-speed computer (it is a good application for a parallel processor). Military people routinely sharpen up spy photographs; in general, all images taken from satellites can benefit from a "lateral inhibition" type of processing [R. N. Bracewell, 1955; A. Rosenfeld, 1978]. This is, in fact, the topic considered in the remainder of this chapter, except that we will examine "photographs" that contain, at the most, 169 elements. The processing can then be done with a slow-speed computer, the human brain.

6-4 Two-Dimensional Lateral Inhibition

A line in the palm of the hand, which has been the source of sensory receptors in the preceding sections of this chapter, is a very special case of a one-dimensional distribution, because a line has only one dimension—length. The auditory sensory receptors in each inner ear also form a one-dimensional distribution, and lateral inhibition is probably important in sharpening the perception of input sound [W. H. Huggins and J. C. R. Licklider, 1951]. But all of the other major sensory systems are two-dimensional; that is, they entail two-dimensional areas. The sense of touch, removed from the special case of a line in the palm, is of course two-dimensional. The sensory receptors for cold, warmth, pain, taste, olfaction, and vision all form two-dimensional arrays, like a map. In the case of vision, at least, we definitely know that the input stimuli are processed with the aid of two-dimensional lateral inhibition.

It is difficult to draw a two-dimensional ensemble of neurons and axons. We have a sheet of sensory receptors in the skin and, in effect, their neurons project to a sheet of neural tissue in the spinal cord. Suppose that each sensory receptor–neuron axon forms a synaptic junction with a spinal-cord neuron, and also sends out lateral inhibitory branches to surrounding spinal-cord neurons. It may be easy for the nervous system to grow all of these axons and lateral branches, but if we try to make a diagram of this neuronal network we get a confusing hodgepodge. Even a single axon, as in Fig. 6-8(a), is disconcerting. In the one-dimensional example of Fig. 6-3(b), on the other hand, we cannot only draw the neuronal network but also show input and output potentials.

The situation for two-dimensional lateral inhibition is not hopeless, but we have to abandon the luxury of a schematic diagram, and use a different system to represent receptor and graded potentials. A numerical example is given in Fig. 6-8. In Fig. 6-8(b) we are looking at the cross section of a 4×5 bundle of sensory receptor–neuron axons. In this highly idealized example the axons have a square cross section. The number in each square is the graded-potential output of each axon, in millivolts. *Each* axon synapses with a sheet of spinal-cord neurons in accordance with the 3×3 array in Fig. 6-8(c). The numbers here represent multiplicative weights. The graded-potential value coming from the central axon is multiplied by 1. There are four lateral branches—left, right, up, and down—and the graded-potential value at the synaptic junction of each lateral branch is multiplied by -0.2.

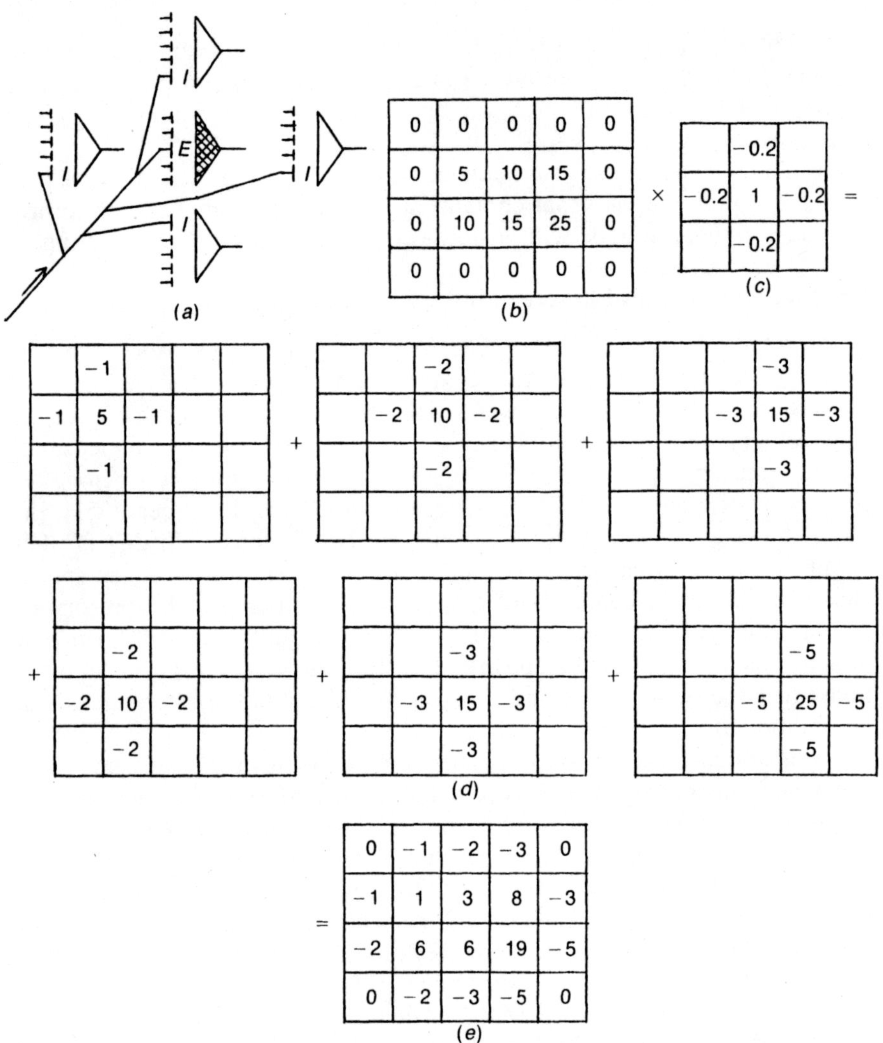

Fig. 6-8 Hypothetical example of two-dimensional lateral inhibition. (*a*) View of a sensory receptor–neuron axon that synapses with spinal-cord neurons. The axon terminates in the excitatory junction (*E*) of a direct branch (cross-hatched neuron) surrounded by four inhibitory junctions (*I*) of lateral branches. (*b*) Cross section of a 4 × 5 bundle of these sensory receptor–neuron axons. The number in each square is the graded-potential output of each axon, in millivolts. (*c*) Multiplicative weights $E = 1$ applied to the central axon and $I = -0.2$ to the lateral branches. (*d*) Cross sections of the 4 × 5 bundle of spinal-cord neuron axons. (*e*) Net summations of the individual contributions of part (*d*). The numbers in part (*d*) are the graded-potential contributions if *each* axon of part (*b*) synapses with part (*d*) in accordance with part (*c*).

To calculate the output for this example of two-dimensional lateral inhibition, we combine the contributions from each sensory receptor neuron as shown in the second and third rows of 4 × 5 arrays, in Fig. 6-8(*d*). For instance, the 5 surrounded by four −1s is the contribution from the 5-mV cell near the upper-left corner of Fig. 6-8(*b*). When all of the contributions are added we get the neuronal sheet at the bottom of Fig. 6-8(*e*). Two of the spinal-cord neurons have a net output of −5 mV; the maximum positive output is +19 mV, and so forth.

There is a simple parity check you can apply to the output array. The sum of input values in Fig. 6-8(*b*) multiplied by the sum of weight values in Fig. 6-8(*c*) has to equal the sum of output values in Fig. 6-8(*e*). In the figure the sum of input values is 5 + 10 + · · · + 25 = 80, the sum of weight values is 1 − (4)(0.2) = 0.2, so the sum of output values has to be 80 × 0.2 = 16. You will find that the numbers in the output array add up to 16 as required. (Sometimes, during an examination, a student makes a mistake in calculating the *parity* check, and wastes a lot of time going over previous work that happens to be correct!)

The example of Fig. 6-8 is of no special significance, and is only presented to illustrate how one can calculate the output of a two-dimensional array. More realistic is the array of Fig. 6-9, where the stimulus in part (*a*) is a single blunt compass point applied somewhere to the skin. The smooth curve of Fig. 6-3(*a*) is approximated, in Fig. 6-9(*b*), by a series of steps.

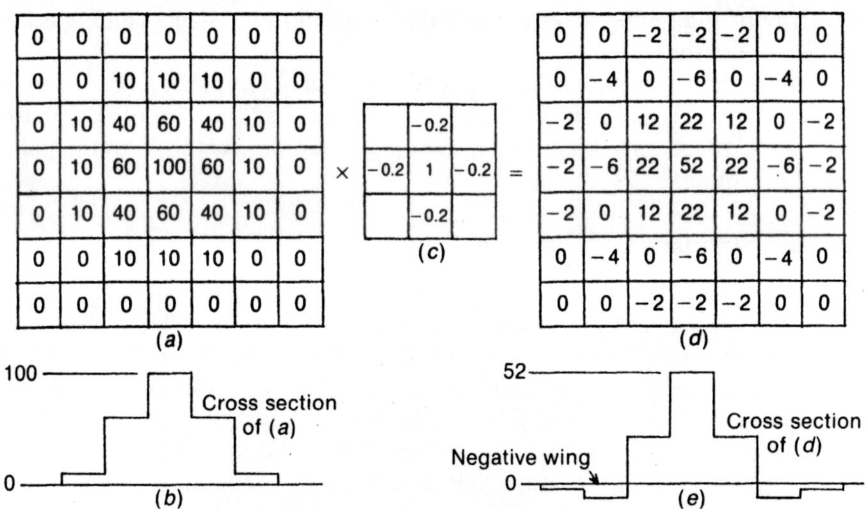

Fig. 6-9 Hypothetical example of two-dimensional lateral inhibition. (*a*) Assumed graded-potential outputs of touch receptors in response to a blunt compass point applied to the skin. (*b*) Cross section through center of part (*a*). (*c*) Multiplicative weights. (*d*) Graded-potential outputs of spinal-cord neurons if *each* axon of part (*a*) synapses with part (*d*) in accordance with part (*c*). (*e*) Cross section through center of part (*d*).

Whereas the optimum lateral shift distance in Fig. 6-3 is 0.7 cm, in Fig. 6-9 it is 1 box. The result of applying the two-dimensional lateral inhibitory weights of Fig. 6-9(*c*) to (*a*) is shown in (*d*). (Most people enjoy working out these exercises, somewhat like doing the Sunday puzzles in a newspaper. Since Fig. 6-9 is symmetrical, you only have to calculate the output for one quadrant.) The parity check appears as follows: The sum of input values in Fig. 6-9(*a*) is 620, the sum of weight values in Fig. 6-9(*c*) is 0.2, so the sum of output values in Fig. 6-9(*d*) should be $620 \times 0.2 = 124$. This of course turns out to be the case.

The output in Fig. 6-9(*e*) displays the sharpening that we expect from lateral inhibition. It also displays the disadvantages: A loss of amplitude, from 100 to 52, and the development of a negative valley instead of negative wings, with its attendant distortion, surrounding the positive peak.

Stimulation with two blunt compass points, the equivalent of two boxes apart, is depicted in Fig. 6-10. The stimulus points are indicated by two arrows in Fig. 6-10(*b*). The graded-potential outputs of Fig. 6-10(*a*) are obtained from Fig. 6-9(*a*) as follows: Center the array of Fig. 6-9(*a*) over each of the arrows in Fig. 6-10(*b*) and add the two sets of values. The series of steps in Fig. 6-10(*b*) outline approximately the net stimulus curve, which has a single peak. After processing by the two-dimensional lateral inhibitory weights of Fig. 6-10(*c*), however, the output in Fig. 6-10(*d*) shows that the two input points are resolved. An easy way to derive Fig. 6-10(*d*) is to center the array of Fig. 6-9(*d*) over each of the arrows in Fig. 6-10(*e*) and add the two sets of values.

It is well to remember that Fig. 6-10 could also represent a set of visual stimuli, such as two pinpoints of light barely resolved, or a set of stimuli for any of the other two-dimensional sensory receptors (cold, warmth, pain, taste, and smell).

6-5 Zero-Sum Lateral Inhibition

The chapter ends with the remarkable properties of a zero-sum lateral inhibitory network. The zero-sum network is shown in Fig. 6-11(*c*). Here the multiplicative weight of each lateral branch is -0.25 compared to -0.2 of Figs. 6-8, 6-9, and 6-10. The sum of the weights is obviously zero. If this happens to be the case, the lateral inhibitory processor becomes an edge extractor; that is, it can reveal the edges of an input stimulus.

In Fig. 6-11(*a*) the stimulus has the shape of a miniature camping tent rather than a blunt compass point; that is, the point has the shape of a pyramid (four sides coming to a point), but the pyramidal angle changes (becomes steeper) a short distance away from the point. When this complex object is pressed into the skin, the touch receptor neuron at the center, at the point, generates a potential of 36 mV. The potential decreases at the rate of 4 mV per box until 24 mV is reached; the values are 36, 32, 28, and 24 mV. Thereafter the potential decreases at the rate of 8 mV per box; the values are 24, 16, 8, and 0 mV.

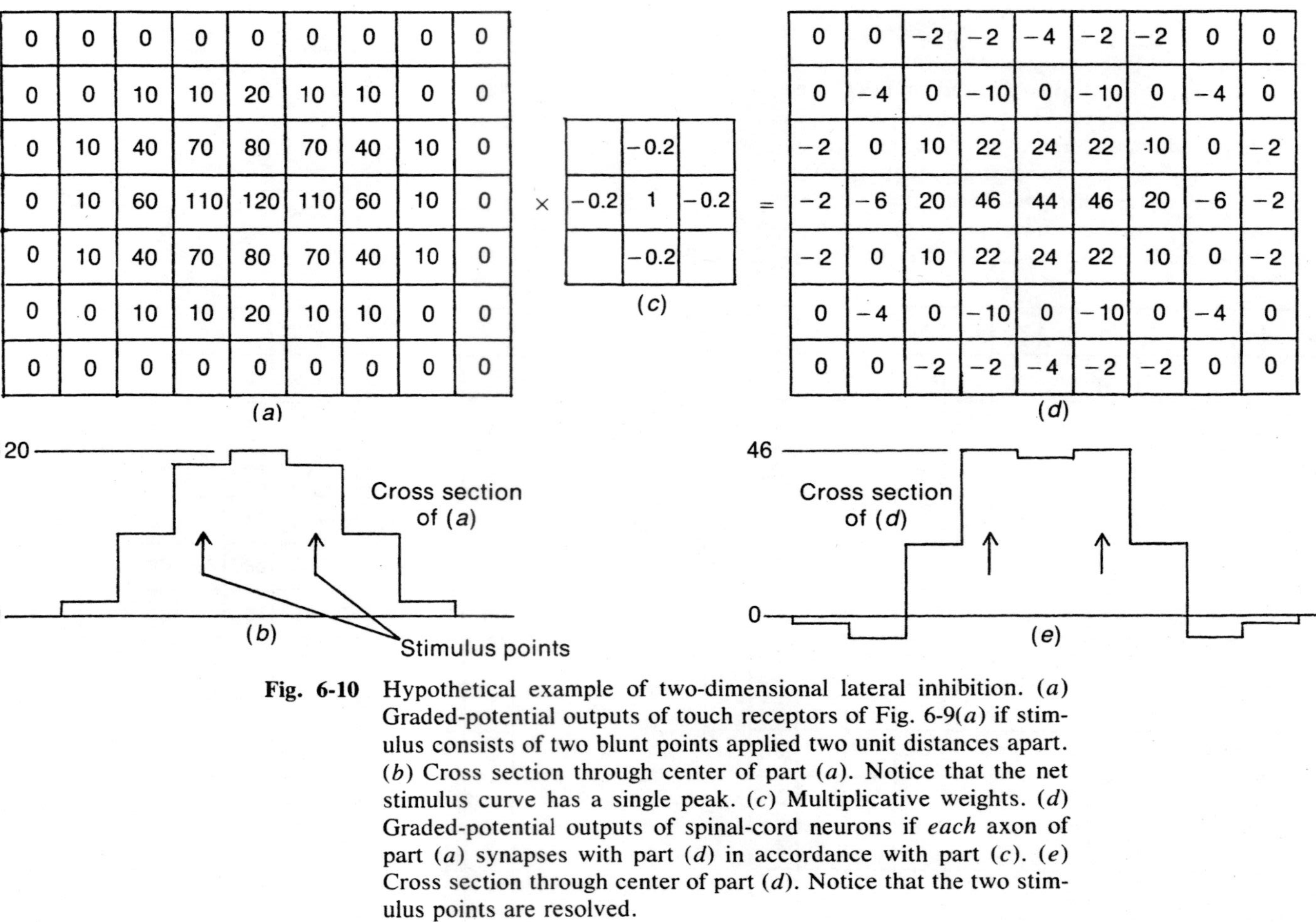

Fig. 6-10 Hypothetical example of two-dimensional lateral inhibition. (*a*) Graded-potential outputs of touch receptors of Fig. 6-9(*a*) if stimulus consists of two blunt points applied two unit distances apart. (*b*) Cross section through center of part (*a*). Notice that the net stimulus curve has a single peak. (*c*) Multiplicative weights. (*d*) Graded-potential outputs of spinal-cord neurons if *each* axon of part (*a*) synapses with part (*d*) in accordance with part (*c*). (*e*) Cross section through center of part (*d*). Notice that the two stimulus points are resolved.

0	0	0	0	0	0	0	0	0	0	0	0	0
0	8	8	8	8	8	8	8	8	8	8	8	0
0	8	16	16	16	16	16	16	16	16	16	8	0
0	8	16	24	24	24	24	24	24	24	16	8	0
0	8	16	24	28	28	28	28	28	24	16	8	0
0	8	16	24	28	32	32	32	28	24	16	8	0
0	8	16	24	28	32	36	32	28	24	16	8	0
0	8	16	24	28	32	32	32	28	24	16	8	0
0	8	16	24	28	28	28	28	28	24	16	8	0
0	8	16	24	24	24	24	24	24	24	16	8	0
0	8	16	16	16	16	16	16	16	16	16	8	0
0	8	8	8	8	8	8	8	8	8	8	8	0
0	0	0	0	0	0	0	0	0	0	0	0	0

(b)

×

	$-\frac{1}{4}$	
$-\frac{1}{4}$	1	$-\frac{1}{4}$
	$-\frac{1}{4}$	

(c)

=

	−2	−2	−2	−2	−2	−2	−2	−2	−2	−2	−2	
−2	4										4	−2
−2		4								4		−2
−2			4	1	1	1	1	1	4			−2
−2			1	2				2	1			−2
−2			1		2		2		1			−2
−2			1			4			1			−2
−2			1		2		2		1			−2
−2			1	2				2	1			−2
−2			4	1	1	1	1	1	4			−2
−2		4								4		−2
−2	4										4	−2
	−2	−2	−2	−2	−2	−2	−2	−2	−2	−2	−2	

(d)

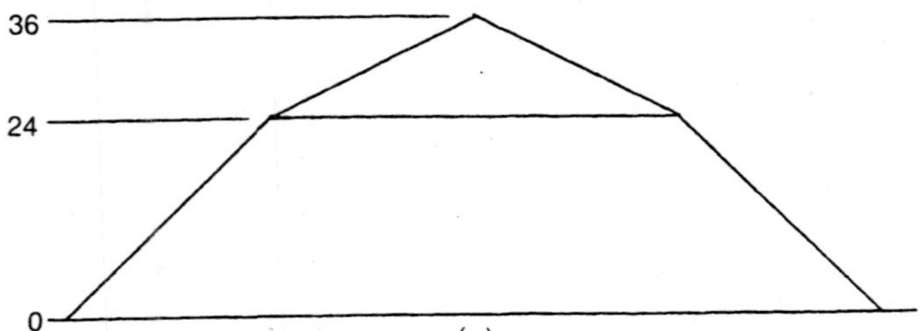

(a)

Fig. 6-11 Hypothetical example of zero-sum two-dimensional lateral inhibition. (*a*) An object that has the shape of a "miniature camping tent" is pressed against the skin. (*b*) Resulting graded-potential outputs of touch receptors. (*c*) Zero-sum multiplicative weights. (*d*) Graded-potential outputs of spinal-cord neurons if *each* axon of part (*b*) synapses with part (*d*) in accordance with part (*c*). The potential value in blank boxes is zero. Notice that all of the edges of part (*a*) are extracted.

If your nervous system operates on Fig. 6-11(*b*) with the zero-sum weights of Fig. 6-11(*c*), you would subjectively feel only the edges, as in Fig. 6-11(*d*). You would not feel the blank areas of Fig. 6-11(*d*). All planar surfaces of the stimulus in Fig. 6-11(*b*) yield zeros in Fig. 6-11(*d*), but these "zero" boxes have been left blank in order to emphasize the nonzero edges. Not only is each edge extracted, but its numerical value is a rough indication of how sharp the edge is. The sharp point at the center has the value 4 mV. The four edges radiating away from center are at 2 mV. The change between the blunt and sharp pyramidal shapes, which is a weak edge, is represented by only 1 mV. Edges of the sharp pyramid yield 4 mV. The edge at the base of the miniature camping tent, where it becomes horizontal, is reasonably sharp; here the value is -2 mV. The value is negative if the edge is concave (as seen by standing above and looking down on the "tent"), and positive if it is convex.

The parity check for a zero-sum processor has to be zero. The positive values in Fig. 6-11(*d*) add up to 88; there are 44 boxes with the value -2, so the net sum is indeed zero.

Why does a plane surface in Fig. 6-11(*b*) yield zero in Fig. 6-11(*d*)? Because, if the surface is at least three boxes wide, it spans the full width of Fig. 6-11(*c*), and multiplication by the weights of Fig. 6-11(*c*) then yields positive and negative contributions that sum to zero.

Although this may be the stuff out of which Sunday newspaper puzzles are made, it is, literally, vitally important. No claim is made that, if the pyramidal point of Fig. 6-11(*a*) is pressed into your skin, your spinal-cord neurons will extract the edge information of Fig. 6-11(*d*). But the brain *does* seem to extract edges since these are the salient features by which we recognize an object, and zero-sum lateral inhibition is compatible with known neuronal circuits.

It is estimated that the brain contains some 10^{12} neurons distributed over 50 to 100 subareas. Most of these subareas are engaged in specialized processing, of which we are not aware consciously. Some of the subareas may extract edge information. For example, what goes on when you see a new face? The salient features, such as edges, are extracted along with many other characteristics, such as skin color. The caricature is somehow stored in visual memory, along with association fibers to the person's name, voice spectrum, and so forth. A dog also runs association fibers to the person's odor spectrum (and you probably do the same, if there is anything noteworthy to record in olfactory memory). A lifetime later—10 years—you again see the person. The salient features are extracted and the caricature is again stored in visual memory. But this time, when the caricature is routed through visual memory, you almost instantly recognize that these features are already on file, along with a name and voice spectrum (and, perhaps, an odor spectrum). (Note, however, that the edges of a photographic negative are the same as those of a positive, but it is generally impossible to recognize a face from the negative. There is more to face recognition than edge detection.)

The military has urgent need for the automatic recognition of certain objects in spy photographs, and the post office would like to automatically read Zip codes; here an important step entails the extraction of edges of objects and handwritten numbers. A tremendous amount of research money has been and will be devoted to these problems. The human brain is, of course, able to almost instantly recognize complex objects, such as a face seen 10 years ago, despite the often minuscule differences between one face and the next. Edge extraction via zero-sum lateral inhibition may play an important role in this process.

A6-1 TABULAR AND TAYLOR'S SERIES METHODS

The tabular method gives discrete points on the $g(z)$ curve, and one must fill in-between regions as is customary in plotting any curve. The z columns in Table 6-1 are based on the lateral shift distance of Fig. 6-1, 0.7 cm. The column spacing must therefore be 0.7 divided by an integer such as 2 (which yields a spacing of 0.35 cm), or 3 (spacing 0.2333 cm), or 4 (spacing 0.175 cm), and so forth. A spacing of 0.35 cm is employed in Table 6-1, but a finer spacing such as 0.175 cm will, of course, result in a smoother curve at the cost of increased computation effort.

Figure 6-1 extends to $z = 2$, so the corresponding table should extend to $z = 2 + 0.7$; since the last column in Table 6-1 belongs to $z = 2.8$, the requirement is satisfied. Since we are applying lateral inhibition to a symmetrical curve, it is not necessary to include negative-z columns in the table. The $f(z)$ row is, of course, $\exp(-z^2)$. The $f(z - 0.7)$ row is the $f(z)$ row shifted 0.7 cm to the *right*, as shown. Similarly, the $f(z + 0.7)$ row is the $f(z)$ row shifted 0.7 cm to the *left*. The $g(z)$ row is given, in functional notation, by

$$g(z) = f(z) - 0.3[f(z - 0.7) + f(z + 0.7)]. \qquad (6\text{-}12)$$

If we apply Taylor's series expansion to lateral inhibition, we can express the output $g(z)$ in terms of familiar and easily visualized concepts such as the spatial derivatives of the input $f(z)$. Taylor's series is appropriate if $f(z)$ is analytic. We get

$$f(z - \delta) = f(z) - \delta \frac{df(z)}{dz} + \frac{\delta^2}{2!}\frac{d^2f(z)}{dz^2} - \frac{\delta^3}{3!}\frac{d^3f(z)}{dz^3} + \cdots. \qquad (6\text{-}13)$$

Applying Eq. (6-13) to the right-hand portion of Eq. (6-1),

$$f(z - \delta) + f(z + \delta) = 2\left[f(z) + \frac{\delta^2}{2!}\frac{d^2f(z)}{dz^2} + \frac{\delta^4}{4!}\frac{d^4f(z)}{dz^4} + \cdots\right] \qquad (6\text{-}14)$$

because all odd-ordered derivatives cancel. For the *special case* $W = 0.5$, substituting Eq. (6-14) into Eq. (6-1), we get

$$g(z) = -A\left[\frac{\delta^2}{2!}\frac{d^2f(z)}{dz^2} + \frac{\delta^4}{4!}\frac{d^4f(z)}{dz^4} + \cdots\right]. \qquad (6\text{-}15)$$

If typical numerical values are applied to Eq. (6-15), it will be found that the $d^2f(z)/dz^2$ term dominates. In other words, *lateral inhibition tends to give the negative of the second spatial derivative of the input stimulus.* This is very helpful if one wishes to make a rough sketch of $g(z)$ given $f(z)$, even if $f(z)$ is not analytic.

REFERENCES

R. N. Bracewell, Simple graphical method of correcting for instrumental broadening, *J. Opt. Soc. Am.*, vol. 45, pp. 873–876, 1955.

S. Deutsch, Two-stage lateral inhibition for auditory selectivity, *Bull. Math. Biol.*, vol. 39, pp. 259–266, 1977.

S. Deutsch and E. M-Tzanakou, *Neuroelectric Systems*. New York: New York Univ. Press, 1987.

W. H. Huggins and J. C. R. Licklider, Place mechanisms of auditory frequency analysis, *J. Acoust. Soc. Am.*, vol. 23, pp. 290–299, 1951.

F. Ratliff, *Mach Bands: Quantitative Studies on Neural Networks in the Retina.* San Francisco: Holden-Day, 1965.

F. Ratliff, H. K. Hartline, and W. H. Miller, Spatial and temporal aspects of retinal inhibitory interactions, *J. Opt. Soc. Am.*, vol. 53, pp. 110–120, 1963.

W. Reichardt and G. MacGinitie, On theory of lateral inhibition, *Kybernetik*, vol. 1, pp. 155–165, 1962.

A. Rosenfeld, Iterative methods in image analysis, *Pattern Recognition*, vol. 10, pp. 181–187, 1978.

T. C. Ruch, Somatic sensation, and T. C. Ruch, Neural basis of somatic sensation, both in *Neurophysiology*, 2d ed., eds. T. C. Ruch et al. Philadelphia: Saunders, pp. 302–344, 1965.

G. von Bekesy, *Sensory Inhibition.* Princeton, N.J.: Princeton Univ. Press, 1967.

H. R. Wilson and J. D. Cowan, Excitatory and inhibitory interactions in localized populations of model neurons, *Biophys. J.*, vol. 12, pp. 1–24, 1972.

Problems

1. Given the stimulus $f(z) = 100{,}000 \exp(-z^2)$, (cm units), which is processed by a lateral inhibitory circuit whose lateral shift distance is 0.75 cm, with inhibitory weights 0.2 and unity gain: (a) write the equation for the output, $g(z)$, using functional notation; (b) substitute for $f(z)$, then calculate and plot $f_n(z)$ and $g_n(z)$ for $-2.25 < z < +2.25$; (c) write the Taylor's series for $g(z)$ up to and including the second derivative term. Compare the Taylor's series approximation (TSA) versus the accurate value at $z = 0, 0.25, 0.5$, and 0.75 cm. [Ans.: (a) $g(z) = f(z) - 0.2[f(z - 0.75) + f(z + 0.75)]$; (b) $g(z) = 100{,}000 (\exp(-z^2) - 0.2\{\exp[-(z - 0.75)^2] + \exp[-(z + 0.75)^2]\})$; (c) $g(z) = 100{,}000 \exp(-z^2)[0.6 - 0.225(2z^2 - 1)]$; $g(0) = 77209$, TSA = 82500; $g(0.25) = 71008$, TSA = 74859; $g(0.5) = 54900$, TSA = 55490; $g(0.75) = 34870$, TSA = 32584.]
2. Given the stimulus $f(z) = 1 + \cos z$ in the range $-180° < z < +180°$, and $f(z) = 0$ elsewhere, which is processed by a unity-gain lateral inhibitory circuit: (a) assume that the optimum lateral shift distance is given by the inflection point, at $d^2f/dz^2 = 0$, find δ_{opt}. (b) Assuming that the inhibitory weights W are reasonably small, find $z(W)$ at which the negative wings peak and the corresponding $g(W)$ at the negative peak. (c) Find W so that the negative wing peak height is 10% of

the positive peak height. Also find these peak heights. (d) Using δ_{opt} and the W of part (c), apply the tabular method, with columns spaced $z = 22.5°$ apart through 270°, to find $g(z)$. Plot the $f_n(z)$ and $g_n(z)$ curves. [Ans.: (a) $\delta_{opt} = 90° = \pi/2$; (b) $|z| = \arctan(-W)$, $g = 1 - W - \sqrt{1 + W^2}$; (c) $W = 0.1565$, $g_{negpeak} = -0.1687$, $g_{pospeak} = 1.687$.]

3. Carry out the step-by-step derivation of Eqs. (6-5), (6-6), and (6-7). Show that $\delta_{1opt} = 0.7071$, $\delta_{2opt} = 0.5246$, and $\delta_{3opt} = 0.4361$, respectively.
4. Given the stimulus $f(z) = 100{,}000 \exp(-z^2)$, (cm units), which is processed by three stages of lateral inhibition, with lateral shift distances 0.75, 0.5, and 0.375 cm, respectively, each stage has inhibitory weight 0.2 and unity gain. (a) Using the tabular method, find the final output $i(z)$ at $z = 0, 0.125, 0.25, \ldots, 1.125$. (b) Find $i_n(z)$. Plot $f_n(z)$ and $i_n(z)$ for $-1.125 < z < +1.125$ cm. [Ans.: (a) $i(z) = 37784, 36679, 33505, \ldots, -1457$; (b) $i_n(z) = 1, 0.971, 0.887, \ldots, -0.039$; $f_n(z) = 1, 0.984, 0.939, \ldots, 0.282$.]
5. Given the stimulus $f(z) = 100{,}000 \exp(-z^2)$, (cm units), which is processed by a lateral inhibitory circuit whose lateral shift distance is 0.75 cm, with inhibitory weights 0.2 and unity gain: (a) calculate the Taylor's series value for $g(z)$, up to and including the second derivative term, at $z = 0, 0.25, 0.5, \ldots, 2$. (b) Add the values given by the fourth derivative term to the calculations of part (a). (c) Add the values given by the sixth derivative term to the calculations of part (b). [Ans.: (a) $g(z) = 82500, 74859, 55490, \ldots, -1786$; (b) $g(z) = 76172, 70370, 55079, \ldots, -2520$; (c) $g(z) = 77358, 71084, 54840, \ldots, -2371$.]
6. Given the stimulus $f(z) = 0.1372 - 0.5918z^2 + 0.945z^4 - 1.12z^6 + z^8$ in the range $-0.7009 < z < +0.7009$, and $f(z) = 0$ elsewhere, which is processed by three stages of lateral inhibition, assume that the optimum lateral shift distances are given, respectively, by $d^2f/dz^2 = 0$, $d^4f/dz^4 = 0$, and $d^6f/dz^6 = 0$. (a) Plot the stimulus curve; (b) find the value of $|z|$ at which $f(z)$ becomes zero; (c) find the value of $|z|$ at which df/dz becomes zero; (d) find δ_{1opt}; (e) find δ_{2opt}; (f) find δ_{3opt}. [Ans.: (b) $|z| = 0.7009$; (c) $|z| = 0.7009$; (d) $\delta_{1opt} = 0.4$; (e) $\delta_{2opt} = 0.3$; (f) $\delta_{3opt} = 0.2$.]
7. Given the stimulus $f(z) = 0.1372 - 0.5918z^2 + 0.945z^4 - 1.12z^6 + z^8$ in the range $-0.7009 < z < +0.7009$, and $f(z) = 0$ elsewhere, which is processed by a lateral inhibitory circuit whose lateral shift distance is 0.4, with inhibitory weights 0.2 and unity gain: (a) use the tabular method, with columns spaced $z = 0.1$ apart through $z = 1.1$, to find $g(z)$. (b) A second stimulus is applied Δ away from the first. Find the two-point threshold [i.e., Δ at which the total $g(z)$ curve becomes flattopped]. Draw the total $g(z)$ curve. (c) Repeat part (b) in the absence of lateral inhibition. Find the two-point threshold [i.e., Δ at which the total $f(z)$ curve becomes flattopped]. Draw the total $f(z)$ curve. [Ans.: (b) $\Delta \cong 0.7$; (c) $\Delta \cong 0.8$.]
8. Given the cross section of a 5 × 5 bundle of axons in part (a) below; the number in each "box" is the graded-potential output of each axon, in millivolts. Each axon is multiplied by the weights of part (b). Find the net output of part (c), and write the parity check equation.

$$\begin{matrix} 0 & 0 & 0 & 0 & 0 \\ 0 & 5 & 10 & 15 & 0 \\ 0 & 10 & 15 & 20 & 0 \\ 0 & 20 & 25 & 35 & 0 \\ 0 & 0 & 0 & 0 & 0 \end{matrix} \times \begin{matrix} 0 & -0.2 & 0 \\ -0.2 & 1 & -0.2 \\ 0 & -0.2 & 0 \end{matrix} = \text{(c)}.$$

(a) (b)

$$\text{[Ans.: (c)} = \begin{matrix} 0 & -1 & -2 & -3 & 0 \\ -1 & 1 & 3 & 9 & -3 \\ -2 & 2 & 2 & 7 & -4, \\ -4 & 13 & 11 & 26 & -7 \\ 0 & -4 & -5 & -7 & 0 \end{matrix} \qquad 155 \times 0.2 = 31.]$$

9. Given the cross section of a 5 × 5 bundle of axons in part (a) below—the number in each "box" is the graded-potential output of each axon, in millivolts—each axon is multiplied by the zero-sum weights of part (b). Find the net output of part (c), and write the parity check equation.

$$\underset{\text{(a)}}{\begin{matrix} 0 & 0 & 0 & 0 & 0 \\ 0 & 5 & 10 & 15 & 0 \\ 0 & 10 & 15 & 20 & 0 \\ 0 & 20 & 25 & 35 & 0 \\ 0 & 0 & 0 & 0 & 0 \end{matrix}} \times \underset{\text{(b)}}{\begin{matrix} 0 & -0.2 & 0 \\ -0.2 & 0.8 & -0.2 \\ 0 & -0.2 & 0 \end{matrix}} = \text{(c)}.$$

$$\text{[Ans.: (c)} = \begin{matrix} 0 & -1 & -2 & -3 & 0 \\ -1 & 0 & 1 & 6 & -3 \\ -2 & 0 & -1 & 3 & -4, \\ -4 & 9 & 6 & 19 & -7 \\ 0 & -4 & -5 & -7 & 0 \end{matrix} \qquad 155 \times 0 = 0.]$$

7

Simple Neuronal Systems

ABSTRACT. Two circuits, each having three interacting neurons, are analyzed.

The first configuration is a "start-and-stop" circuit: When a sensory stimulus starts, as signaled by a sudden change of graded potential or a burst of action potentials (APs), the circuit generates a "start" pulse. When the stimulus stops, the circuit generates a "stop" pulse.

The second arrangement includes feedback from some output points to input points. With feedback, depending on synaptic junction weighting factors and on whether junctions are excitatory or inhibitory, one can get a rich variety of different characteristics from what appears to be the same circuit.

The following behaviors are elicited in response to the onset of a steady stream of APs:

(a) An unstable circuit in which the output rapidly increases without limit;
(b) Outputs that approach a finite limit when the weighting factors are reduced from 1 to 0.5;
(c) With one or more inhibitory junctions, an unstable circuit in which oscillations rapidly build up without limit;
(d) A circuit that generates rising voltage steps;
(e) A different arrangement that also generates rising voltage steps;
(f) A circuit that generates peculiar repeating patterns;
(g) A circuit that generates square voltage pulses;
(h) A circuit that generates rectangular pulses, 60% on and 40% off.

Some of the oscillatory circuits may be applicable to "oscillatory" activities such as walking and breathing.

7-1 A "Start-and-Stop" Circuit

In the previous chapter we stated that the brain contains some 10^{12} neurons (which, written as one trillion, "sounds" a lot less than it really is, as every politician knows). In the distant future, humans, in our quest for knowledge, will decipher how this mass of cells interacts to produce human thought, as well as coordinate movement and other activities. For now, we have to remain with small groups of interacting neurons, such as the lateral inhibitory arrays of Chapter 6.

On the other hand, complex behavior does not require tens of control knobs and hundreds of transistors. If you are an engineer, you probably have never looked at a living cell under a microscope. It can be a fascinating experience. Get a vase in which flower stems have been standing in water for a week, say. Place a drop of this homemade "pond" water on a glass slide, and look through a low-power microscope. You will see cells swimming across the drop in every conceivable direction. Their swim propulsion systems are not visible; instead, they look like workers in Times Square during a lunch break, each scurrying about as if it has a destination and a limited time to get there.

Biologists inform us that the direction in which a one-celled *animal* swims is determined by only two "sensory" receptors: a chemical receptor that tells it to head for food and away from noxious ions, and a thermal receptor that tells it to head for an appropriate temperature so as to avoid an environment that is too hot or too cold. (A well-fed cell will stop to divide in two every 24 hours or so, and it will also stop swimming if its energy source is depleted.)

The present chapter is devoted to a modest consortium of cells, such as three interacting neurons. The reason for choosing three is that the repertoire of one or two neurons is too limited; three already becomes so complicated that one must oversimplify to keep the discussion within reasonable limits. Nevertheless, the following treatment can give us a theoretical basis for ultimately understanding the complex functioning of neurons in the brain [M. Abeles, 1982; E. R. Caianiello, 1968; S. Deutsch, 1967; A. C. Guyton, 1986; L. D. Harmon, 1961; E. R. Kandel, 1979; S. Ochs, 1965].

Our oversimplified neurons follow the model of Fig. 6-2(*b*). Even this is too complex; in the present chapter we assume that the neurons do not amplify. One could write "$\times 1$" inside the triangle representing each neuron, but this becomes redundant. Each neuron is identified instead as N1, N2, or N3 inside the triangle.

What makes our ensembles of three neurons so complicated is that the element of time, which was omitted in the previous chapter, is now reintroduced. In Chapter 4 the time it takes for an action potential (AP) to travel from one part of the body to another is emphasized via the *velocity of prop-*

agation. In Chapter 5 the time delay introduced by a synaptic junction and propagation along a dendritic tree is very much in evidence. The net time delay is a function of axon type, diameter, and length, and the particular path followed along dendritic branches. In the present chapter it is assumed that the delay introduced by a synaptic junction and its dendritic branch is the same for every signal path: namely, *one unit of time*. As a matter of convenience, this unit of time delay is shown as if it were concentrated in the synaptic junction. It is further assumed that the propagation time along axons can be neglected, or can be included in the "one unit of time" delay. The assumption that all time delays are equal allows for a reasonably simple general discussion.

In this chapter the voltages (Vs) are shown as graded potentials because, as usual, it is easier to draw a rectangle than a series of spikes. However, the Vs may actually be trains of action potentials.

Our first example of a useful arrangement of three interacting neurons—a vitally important function, in fact—is a *start-and-stop* circuit that provides a pulse when an input signal starts and another pulse when it stops. It does not take a great deal of introspection to realize that the nervous system is challenged by sudden changes of this type. A new stimulus may represent danger to be avoided; the sudden loss of a steady-state stimulus may also signify a danger to be avoided. This perpetual state of alertness is due to subconscious processing carried on in one or more of the subareas of the brain: A fly lands on your foot and you immediately become aware of it. Somebody sneaks up behind you, but his body shields and/or absorbs environmental sound; the sudden change is detected by the auditory cortex. You are chewing away while engrossed in a suspense thriller, but stop at something that tastes "funny." You suddenly smell cooking gas because somebody left an unlit burner open. Your peripheral vision is constantly on the lookout for motion; in fact, an entire layer of cells in the retina is devoted to the detection of moving "bugs."

The start-and-stop circuit is shown in Fig. 7-1, along with its waveforms. Some of the techniques displayed here are borrowed from electrical circuit drawings. We have axons in place of wires; neurons N1, N2, and N3 in place of amplifiers; and junctions that introduce appreciable time delay in place of transistor junctions. In Fig. 7-1 the multiplicative synaptic junction weights are either excitatory $(+1)$ or inhibitory (-1). The input signal, V_0, starts at zero time and stops some time later, while V_1, V_2, and V_3 represent output signals. Their graded-potential waveforms are given, in relation to each other, on the right.

Neuron N1 does not alter the shape of V_0, but only delays it as shown by the V_1 versus V_0 waveforms. In N2, V_0 enters an excitatory junction and V_1 an inhibitory junction; accordingly, V_1 has to be subtracted from V_0. This maneuver results in a positive "start" pulse, which is delayed one unit of time as shown by the V_2 waveform. The same process, when V_0 ends, generates a negative "stop" pulse.

Only two neurons, N1 and N2, are sufficient thus far. However, the

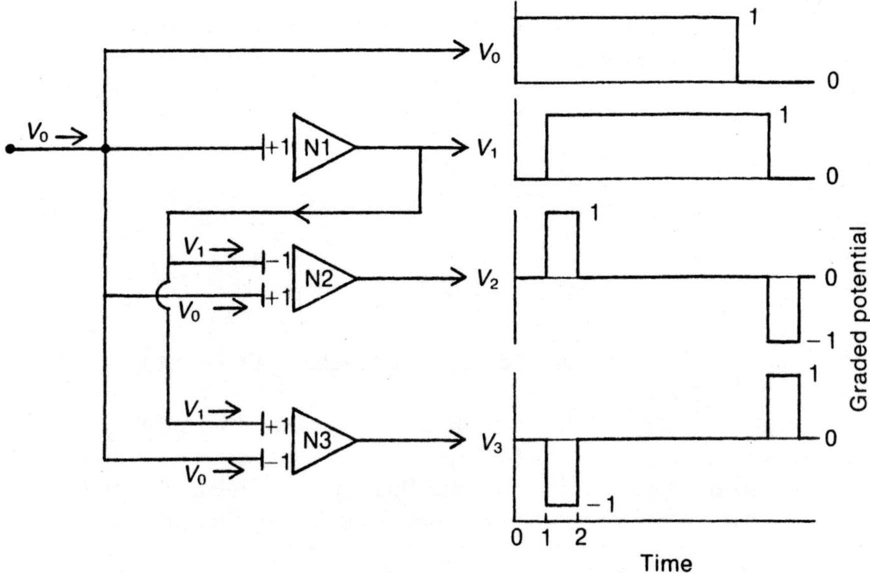

Fig. 7-1 Hypothetical start-and-stop circuit. Throughout this chapter, the synaptic junction weighting factors are multiplicative, each synaptic junction introduces one unit of time delay, and each neuron multiplies its net input by one. The waveforms on the right show relative values of graded potential versus relative values of time.

nervous system may ignore the stop pulse because it is negative; therefore, in N3, V_1 enters an excitatory junction and V_0 an inhibitory junction so that V_0 is subtracted from V_1. This provides a positive stop pulse, as shown.

In functional notation, the circuit of Fig. 7-1 is described by

$$\begin{aligned} V_1 &= V_0(t - \tau), \\ V_2 &= (V_0 - V_1)(t - \tau), \\ V_3 &= (V_1 - V_0)(t - \tau), \end{aligned} \tag{7-1}$$

where each synaptic junction introduces τ units of time delay. It is assumed that the system is linear; therefore, these equations can be expressed in closed form by writing their Laplace transforms, since a delay of τ units transforms into multiplication by $\epsilon^{-s\tau}$:

$$\begin{aligned} V_1(s) &= V_0(s)\epsilon^{-s\tau}, \\ V_2(s) &= [V_0(s) - V_1(s)]\epsilon^{-s\tau}, \\ V_3(s) &= [V_1(s) - V_0(s)]\epsilon^{-s\tau}. \end{aligned} \tag{7-2}$$

[In the interest of brevity, we will omit the "function of s" reminder, (s), in what follows since it is obvious from the other variables whether we have a time function, $V(t)$, or its transform, $V(s)$.]

As an example, suppose that V_0 is a unit step, as in the left-hand portion of the Fig. 7-1 waveforms. Substituting $1/s$ for $V_0(s)$ in Eq. (7-2),

$$V_1 = \frac{1}{s}\epsilon^{-s\tau},$$

$$V_2 = \frac{1}{s}(\epsilon^{-s\tau} - \epsilon^{-2s\tau}), \qquad (7\text{-}3)$$

$$V_3 = -\frac{1}{s}(\epsilon^{-s\tau} - \epsilon^{-2s\tau}).$$

Visual inspection shows that these equations agree with the left-hand portion of the waveforms.

Neuronal start-and-stop circuits certainly exist, although they may be much more intricate than that of Fig. 7-1. The nervous system uses many neurons to do the same job because this improves reliability; the actual network may therefore be no more complicated, basically, than that of Fig. 7-1.

7-2 A Circuit That Can "Blow Up"

Our second three-neuron circuit is that of Fig. 7-2. Whereas the network of Fig. 7-1 has been designed for a specific purpose, Fig. 7-2 is something of a mystery. It was published way back in 1938 by Raphael Lorente de No (1902–1990), based on his neuroanatomical studies using dyes to outline various details of the nervous system. He drew round neurons, and the synaptic junctions did not show weighting factors, but the circuits are equivalent. Lorente de No called this circuit a *closed chain* type because it includes feedback; that is, part of the output, V_2 and V_3 in this case, is fed back to the input N1 [T. J. Carew et al., 1979; E. R. Lewis, T. E. Everhart & Y. Y. Zeevi, 1969].

One cannot find Lorente de No's circuit by looking at tissue under a microscope. The microscope shows bits and pieces of neuron bodies, axons, and synaptic junctions; the whole must be pieced together painstakingly by examining many consecutive cross sections. There is no decisive visible difference, under a microscope, between excitatory and inhibitory synaptic junctions. The equivalent of weighting factors is, of course, completely unrevealed. What Lorente de No gave us, in other words, was a circuit that exists in a biological system without the all-important weights that tell us whether a junction is excitatory or inhibitory, and to what degree.

[S.D. writes: "When I found Lorente de No's 'closed chain' a few years ago, I decided to analyze it by assuming various combinations of excitatory (positive) and inhibitory (negative) junctions. The first (upper) junction of N1 has to be excitatory or the entire circuit will forever remain dormant. Similarly, the N2 junction has to be excitatory or N2 will never fire. At least one of the N3 junctions has to be positive, and so forth. When I analyzed

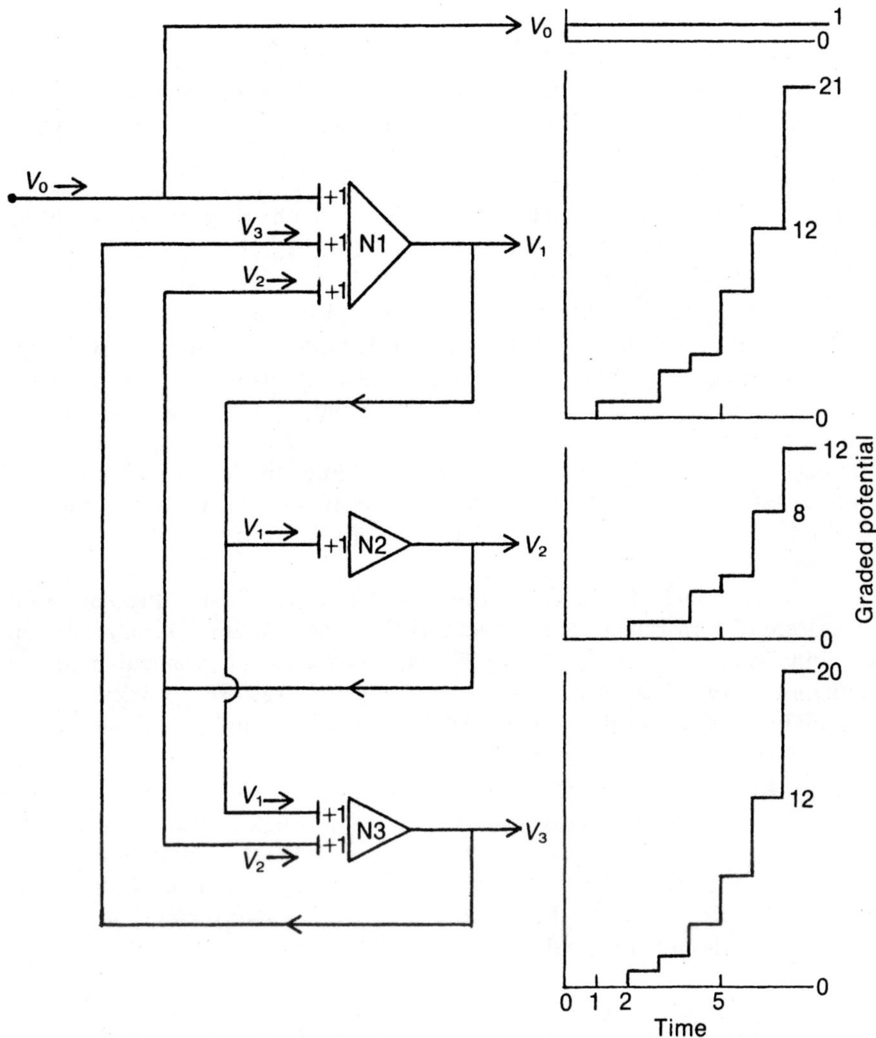

Fig. 7-2 Idealized Raphael Lorente de No closed chain (feedback) type of circuit with all synaptic junctions excitatory. With weighting factors of +1, the output potentials "blow up."

all of the possible synaptic junction combinations, I was surprised to find seven viable combinations—viable in the sense that all three neurons are active and generate nontrivial waveforms."]

These seven combinations are reviewed in the remainder of this chapter.

Figure 7-2 is the "all positive" version in which every junction is excitatory. We will first look at this circuit with all weights equal to 1, and later with the weights reduced to 0.5.

We all have heard of biofeedback—it has become a fashionable household expression. Figure 7-2 is a good illustration of biofeedback. To qualify for the biofeedback category, a biological system must have (a) a forward path between an input point to an output point: in Fig. 7-2 this is neuron N1, which bridges the path between an input on the left (V_0) and an output on the right (V_1); and (b) a feedback path between the output and input points: in Fig. 7-2, V_1 feeds to N2 whose output, V_2, feeds back to an input junction on N1. There is also a second feedback path: V_1 feeds to N3 whose output, V_3, feeds back to another input junction on N1.

Incidentally, much of our everyday life is governed by feedback. A classic example is the thermostatic control of a home heating system. The forward path goes from the on–off switch (the input point) to the heat source to the warm air it produces (the output point). The feedback path is traced out from the warm air to the thermostat to the on–off switch. When the temperature rises to 20°C, say, the thermostat opens the on–off switch and the heat source stops. After a fall of a few degrees the feedback causes the thermostat to close the switch, turning the heat source on again. A similar feedback system controls the cooling cycle of your refrigerator and air conditioner.

With feedback, the simple arrangement of Fig. 7-2 goes through a rich repertoire of performances as a function of excitatory and inhibitory junction combinations. Is Lorente de No's closed chain circuit applicable to the human condition? As far as we know, epileptic seizure convulsions are caused by feedback of the type described by the circuit in Fig. 7-2. However, it is virtually impossible to look at the circuit and predict its behavior, especially since it may be nonlinear. If it is nonlinear, one cannot take advantage of the simplifications introduced by the Laplace transform in analyzing the network. However, the following simple step-by-step procedure is always valid, regardless of whether the system is linear or nonlinear. If you do not wish to derive the waveforms there is, nevertheless, something you can and should do: check our answers! Not because there is the slightest possibility that we ever make a mistake, but because you may enjoy doing the simple "check" exercises, and develop an appreciation for some of the nuances of feedback and time delay in a neuronal circuit.

In Fig. 7-2, at $t = 0$, the V_0 stimulus enters the system with a constant amplitude of 1. The resulting V_1, V_2, and V_3 excursions are given underneath the V_0 waveform. In order for you to check our answers, notice from the diagram that the neuron outputs are described by

$$\begin{aligned} &\text{(a)}\ V_1 = (V_0 + V_2 + V_3) && \text{delayed 1 unit of time,} \\ &\text{(b)}\ V_2 = V_1 && \text{delayed 1 unit of time,} \qquad \text{(7-4)} \\ &\text{(c)}\ V_3 = (V_1 + V_2) && \text{delayed 1 unit of time.} \end{aligned}$$

The waveforms are derived by a three-step procedure. Changes take place only at $t = 0, 1, 2, \ldots$. From the preceding equations, the three-step iterative (repeating) agenda follows:

(a) Calculate $V_1 = V_0 + V_2 + V_3$. Enter the value in the *next* time slot.

(b) The value for V_2 is the same as V_1, but entered in the *next* time slot.

(c) Calculate $V_3 = V_1 + V_2$. Enter the value in the *next* time slot.

Time slots are listed from left to right. For Fig. 7-2 the initial step is given by

$$\begin{array}{rlll} t = 0 & 1 & 2 & \cdots \\ V_0 = 1 & 1 & 1 & \cdots \\ V_1 = 0 & & & \\ V_2 = 0 & & & \\ V_3 = 0 & & & \end{array}$$

because, at $t = 0$, V_0 comes on with the value 1, but V_1, V_2, and V_3 have not had time to change. From (a) in the previous list we get (new entries are underlined),

$$\begin{array}{rlll} t = 0 & 1 & 2 & \cdots \\ V_0 = 1 & 1 & 1 & \cdots \\ V_1 = 0 & \underline{1} & & \\ V_2 = 0 & & & \\ V_3 = 0 & & & \end{array}$$

From (b),

$$\begin{array}{rlll} t = 0 & 1 & 2 & \cdots \\ V_0 = 1 & 1 & 1 & \cdots \\ V_1 = 0 & 1 & & \\ V_2 = 0 & \underline{0} & \underline{1} & \\ V_3 = 0 & & & \end{array}$$

Here we have filled in two time slots. It turns out that one can fill in two values "simultaneously" from here on. The three-step procedure is completed, from (c), with

$$\begin{array}{rlll} t = 0 & 1 & 2 & \cdots \\ V_0 = 1 & 1 & 1 & \cdots \\ V_1 = 0 & 1 & & \\ V_2 = 0 & 0 & 1 & \\ V_3 = 0 & \underline{0} & \underline{1} & \end{array}$$

In the next round, from (a),

$t = 0$	1	2	3	...
$V_0 = 1$	1	1	1	...
$V_1 = 0$	1	$\underline{1}$	$\underline{3}$	
$V_2 = 0$	0	1		
$V_3 = 0$	0	1		

From (b),

$t = 0$	1	2	3	4	...
$V_0 = 1$	1	1	1	1	...
$V_1 = 0$	1	1	3		
$V_2 = 0$	0	1	$\underline{1}$	$\underline{3}$	
$V_3 = 0$	0	1			

From (c),

$t = 0$	1	2	3	4	...
$V_0 = 1$	1	1	1	1	...
$V_1 = 0$	1	1	3		
$V_2 = 0$	0	1	1	3	
$V_3 = 0$	0	1	$\underline{2}$	$\underline{4}$	

and so forth. It is easy to make a mistake, and it "louses up" everything beyond that point. In grading an examination paper with this type of problem, we only take credit off for the initial error. It is necessary to check your entries by applying (a), (b), and (c) of Eq. (7-4) to each column in sequence. The complete listing for the waveforms of Fig. 7-2, which ends after time slot 7, follows:

$t = 0$	1	2	3	4	5	6	7	...
$V_0 = 1$	1	1	1	1	1	1	1	...
$V_1 = 0$	1	1	3	4	8	12	21	...
$V_2 = 0$	0	1	1	3	4	8	12	...
$V_3 = 0$	0	1	2	4	7	12	20	...

It is obvious that the output potentials of Fig. 7-2 are rapidly increasing without limit; an engineer would say that the system is "unstable," that it is "blowing up."

It will be shown below that the network of Fig. 7-2 becomes stable if each of the weights is reduced to $(\sqrt{5} - 1)/2$, or 0.6180. According to this, Fig. 7-3, which is characterized by weight values of 0.5, should be stable. Here the neuron equations become

$$\begin{aligned}
&\text{(a)}\ V_1 = (V_0 + V_2 + V_3)/2 && \text{delayed 1 unit of time,} \\
&\text{(b)}\ V_2 = V_1/2 && \text{delayed 1 unit of time,} \qquad (7\text{-}5) \\
&\text{(c)}\ V_3 = (V_1 + V_2)/2 && \text{delayed 1 unit of time.}
\end{aligned}$$

The potentials shown in Fig. 7-3 are given by

$t = 0$	1	2	3	4	5	6	7	. . .	∞
$V_0 = 1$	1	1	1	1	1	1	1	. . .	1
$V_1 = 0$	0.5	0.5	0.75	0.81	0.94	1	1.07	. . .	1.33
$V_2 = 0$	0	0.25	0.25	0.38	0.41	0.47	0.5	. . .	0.67
$V_3 = 0$	0	0.25	0.38	0.5	0.59	0.67	0.73	. . .	1

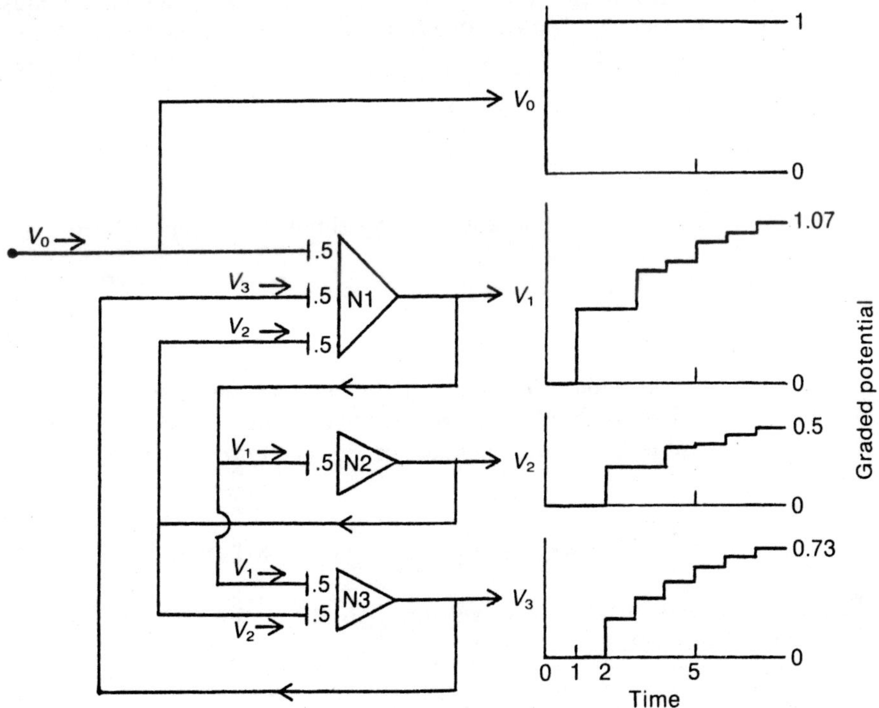

Fig. 7-3 The feedback circuit of Fig. 7-2 if all weighting factors are reduced to ½. The system is now stable. At time = infinity, the potentials approach $V_1 = 1.33$, $V_2 = 0.67$, and $V_3 = 1$.

The last column at the right gives the values if we let time approach infinity: $V_1 \to 4/3$, $V_2 \to 2/3$, and $V_3 \to 1$. The system is indeed stable. Notice the tremendous change that is achieved by only a 2-to-1 reduction in weight values. Similarly, psychoactive drug therapy may be successful, in part, because a small reduction in excitability goes a long way toward calming the patient.

A mathematical analysis of Figs. 7-2 and 7-3 follows. It is restricted to the special case in which each of the weighting factors has the same value, W. We can find W for which the circuit becomes unstable in the classic manner: in the denominator of the transfer function, solve for s at which the denominator becomes zero. First, in functional notation, the circuit of Figs. 7-2 and 7-3 appears as

$$\begin{aligned} V_1 &= W(V_0 + V_2 + V_3)(t - \tau), \\ V_2 &= WV_1(t - \tau), \\ V_3 &= W(V_1 + V_2)(t - \tau). \end{aligned} \tag{7-6}$$

In the step-by-step method for finding the waveforms of Figs. 7-2 and 7-3, all of the operations are linear. Because all of the junction weights are positive, none of the graded potentials ever becomes negative, and it is obvious that the system is linear. We can therefore write the Laplace transforms

$$\begin{aligned} V_1 &= W(V_0 + V_2 + V_3)\epsilon^{-s\tau}, \\ V_2 &= WV_1\epsilon^{-s\tau}, \\ V_3 &= W(V_1 + V_2)\epsilon^{-s\tau}. \end{aligned} \tag{7-7}$$

Now solve for the transfer function V_1/V_0 by eliminating V_2 and V_3 from the simultaneous equations. We get

$$\frac{V_1}{V_0} = \frac{W\epsilon^{-s\tau}}{1 - 2W^2\epsilon^{-2s\tau} - W^3\epsilon^{-3s\tau}}. \tag{7-8}$$

Next, solve for s at which the denominator becomes zero. It is shown in Appendix A7-1 that the borderline case between stability and instability is given by $W = 0.6180$. The denominator of Eq. (7-8) then becomes zero at $s = 0$ and at

$$s = \frac{1}{\tau}(-0.9624 \pm j\pi), \qquad s = \frac{1}{\tau}(-0.4812 \pm j\pi). \tag{7-9}$$

We can also derive the first few waveform steps of V_1 in Fig. 7-2. Substituting $W = 1$ into Eq. (7-8),

$$V_1 = \frac{V_0\epsilon^{-s\tau}}{1 - 2\epsilon^{-2s\tau} - \epsilon^{-3s\tau}}. \tag{7-10}$$

Long division yields

$$V_1 = V_0\epsilon^{-s\tau} + 2V_0\epsilon^{-3s\tau} + V_0\epsilon^{-4s\tau} + 4V_0\epsilon^{-5s\tau} + \cdots. \tag{7-11}$$

The inverse transform results in

$$V_1 = V_0(t - \tau) + 2V_0(t - 3\tau) + V_0(t - 4\tau) + 4V_0(t - 5\tau) + \cdots, \tag{7-12}$$

which, as expected, describes $V_1(t)$ of Fig. 7-2.

The conditions of Fig. 7-3 are reproduced with $W = 0.5$ in Eq. (7-8):

$$\frac{V_1}{V_0} = \frac{0.5\epsilon^{-s\tau}}{1 - 0.5\epsilon^{-2s\tau} - 0.125\epsilon^{-3s\tau}}. \tag{7-13}$$

The limit of the unit step response, as $t \to \infty$, is given by letting $s \to 0$ in Eq. (7-13). This yields $V_1(t) \to 1.333$.

During an epileptic seizure, the voltages picked up by electroencephalographic (EEG) electrodes are much higher (50 μV) than during normal brain activity (10 μV). Each neuron "saturates" at some relatively high potential; that is, its output levels off. Thereafter, the biofeedback causes the potentials to rapidly decrease (i.e., increase in a *negative* direction) until each neuron saturates at whatever its negative limit happens to be. The process then repeats, and we have uncontrollable convulsive oscillations. It is a frightening experience to witness these oscillations, let alone undergo them. Fortunately, for epileptic seizure patients there are modern drugs that can usually reduce the weighting factors sufficiently so as to stabilize the system [M. R. Guevara et al., 1983; C. Hayashi, 1964; A. I. Selverston, J. P. Miller & M. Wadepuhl, 1983; R. Thom, 1975].

7-3 Circuits with Inhibition, Unstable with $W = 1$

What changes take place if one or more of the synaptic junctions in Fig. 7-2 is inhibitory? Because of inhibition, some of the potentials may become negative. Here one must make a decision—are negative values to be allowed? In the remainder of this chapter it is assumed that the potentials correspond to *action potential* frequencies. In other words, V_0 is actually a train of APs that starts at $t = 0$. If $V_0 = 1$, the value $V_1 = 20$ actually corresponds to an AP frequency that is 20 times that of V_0, and so forth. Since a negative AP frequency is meaningless, negative values are *not* allowed in the remainder of this chapter. (Alternatively, one can assume that the potentials are graded, as shown, and that the voltages never go below the zero [resting potential] level.)

In the present section, we consider a few circuits that are unstable with $W = 1$ ($W = -1$ for inhibitory junctions). There do not appear to be any examples of these circuits functioning in this unstable mode in "real life." As we will show, this is not normally a viable state for a biological system. We are dealing here with strange creatures. But as with most theories, maybe we just have not come across the set of data to fit this model. Besides, one can regard these nonviable arrangements as possible examples of artificial neural networks (ANNs) since the circuits become stable if the amplification is reduced sufficiently.

Our first example, in which the lower synaptic junction of N3 in Fig. 7-2 is inhibitory, is illustrated in Fig. 7-4. The neuron equations of Fig. 7-4 are written as

$$\begin{aligned} V_1 &= (V_0 + V_2 + V_3) && \text{delayed 1 unit of time,} \\ V_2 &= V_1 && \text{delayed 1 unit of time,} \\ V_3 &= (V_1 - V_2) && \text{if positive, otherwise zero; delayed 1 unit of time.} \end{aligned} \tag{7-14}$$

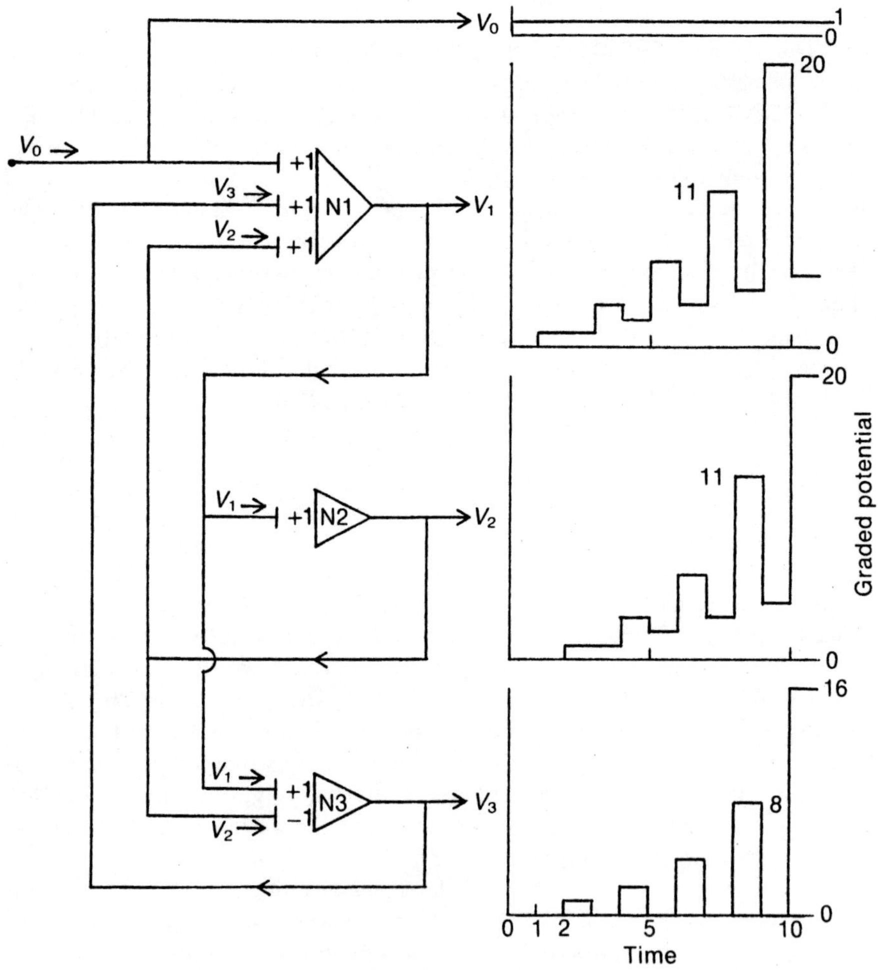

Fig. 7-4 The feedback circuit of Fig. 7-2 if the lower synaptic junction of N3 is inhibitory. The system is unstable. In this and the remaining circuits of this chapter, negative potentials are not allowed and are clipped (set equal to zero) instead.

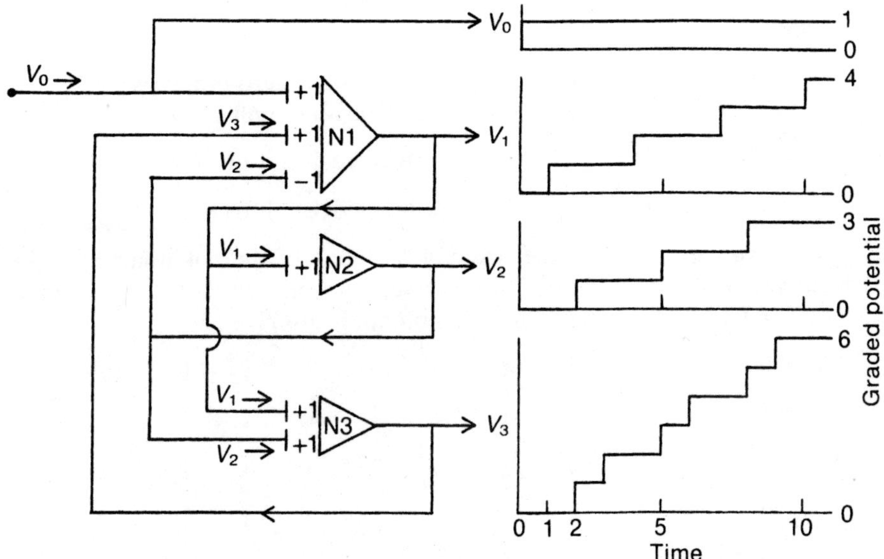

Fig. 7-5 The feedback circuit of Fig. 7-2 if the lower synaptic junction of N1 is inhibitory. The system is on the borderline between stability and instability.

(In the preceding group, V_1 and V_2 cannot become negative because they are only made up of positive components.)

The potentials are given by

t = 0	1	2	3	4	5	6	7	8	9	10	...
V_0 = 1	1	1	1	1	1	1	1	1	1	1	...
V_1 = 0	1	1	3	2	6	3	11	4	20	5	...
V_2 = 0	0	1	1	3	2	6	3	11	4	20	...
V_3 = 0	0	1	0	2	$\underline{\underline{0}}$	4	$\underline{\underline{0}}$	8	$\underline{\underline{0}}$	16	...

If you are checking these answers, you will notice that wherever $V_1 - V_2$ is negative, V_3 is replaced by zero (and shifted one time unit to the right). The negative excursions clipped to zero are indicated by double underlines.

Since the potentials of Fig. 7-4 rise rapidly, the system is unstable. The values are more regular, however, than those of Fig. 7-2. In Fig. 7-4, the V_1 and V_2 values alternate between two series: 1, 2, 3, 4, . . . and 1, 3, 6, 11, 20, . . . (the latter series is given by $2^n + n$, where n is an integer). The V_3 values alternate between 0 and 1, 2, 4, 8, 16, . . . (the latter series is obviously given by 2^n).

Another circuit that does not occur in an unstable state in a biological system, as far as we know, is that of Fig. 7-5, where the lower synaptic junction of N1 is inhibitory. The neuron equations are

$$V_1 = (V_0 + V_3 - V_2) \quad \text{if positive, zero otherwise; delayed 1 unit of time,}$$

$$V_2 = V_1 \quad \text{delayed 1 unit of time,} \tag{7-15}$$

$$V_3 = (V_1 + V_2) \quad \text{delayed 1 unit of time.}$$

As we can see from the waveforms, none of the potentials ever goes negative, so the admonition "if positive, zero otherwise" for V_1 is unnecessary if V_0 is a step stimulus. The potentials are given by

$t = 0$	1	2	3	4	5	6	7	8	9	10	...
$V_0 = 1$	1	1	1	1	1	1	1	1	1	1	...
$V_1 = 0$	1	1	1	2	2	2	3	3	3	4	...
$V_2 = 0$	0	1	1	1	2	2	2	3	3	3	...
$V_3 = 0$	0	1	2	2	3	4	4	5	6	6	...

The potentials consist of a series of steps. Since the outputs increase without limit, a casual assessment would be that Fig. 7-5 is an unstable system. There is an important difference, however, between the outputs of Figs. 7-2 and 7-5. In Fig. 7-2 the rate of increase keeps increasing, justifying the classification of a system that is "blowing up." In Fig. 7-5 the rate of increase is constant and is perhaps due to the positive stimulus, V_0, which is "fueling" the increase. If in fact V_0 returns to zero, removing the external stimulus, the potentials of Fig. 7-5 stop increasing. This is shown as follows (the last three columns of the preceding listing are repeated, followed by $V_0 = 0$):

$t = 8$	9	10	11	12	13	14	15	16	17	...
$V_0 = 1$	1	1	0	0	0	0	0	0	0	...
$V_1 = 3$	3	4	4	3	4	4	3	4	4	...
$V_2 = 3$	3	3	4	4	3	4	4	3	4	...
$V_3 = 5$	6	6	7	8	7	7	8	7	7	...

Figure 7-2 is genuinely unstable because the outputs will continue to increase even if the V_0 stimulus ends. According to the previous tabulation, on the other hand, if the V_0 stimulus terminates after the tenth time slot, V_1 and V_2 follow the series 3, 4, 4, 3, 4, 4, . . . , while V_3 follows the series 7, 7, 8, 7, 7, 8, ... The system of Fig. 7-5 is therefore on the borderline between stability and instability; a slight increase in the positive weighting factors will push it over the border. One would expect that, if V_0 decreases to zero, the output potentials should, within a few time units, also decrease to zero. The preceding sequences show that it is possible for potential to become "trapped" so that, when V_0 decreases to zero, voltage is forever passed along from one neuron to the next by means of feedback and feedforward.

Since none of the waveforms is clipped, the system is linear *for the given* V_0 *waveforms*, and analysis via the Laplace transform is permitted. In general, however, inhibitory systems where negative clipping occurs, or excitatory systems where saturation takes place, are nonlinear systems. Here one should employ a step-by-step analysis in the time domain.

This section ends with another "borderline" case, that of Fig. 7-6. Here the middle junction of N1 and the lower junction of N3 are inhibitory. The neuron equations are

$$
\begin{aligned}
V_1 &= (V_0 + V_2 - V_3) && \text{if positive, zero otherwise; delayed 1 unit of time,} \\
V_2 &= V_1 && \text{delayed 1 unit of time,} \\
V_3 &= (V_1 - V_2) && \text{if positive, zero otherwise; delayed 1 unit of time.}
\end{aligned}
\tag{7-16}
$$

The potentials are given by

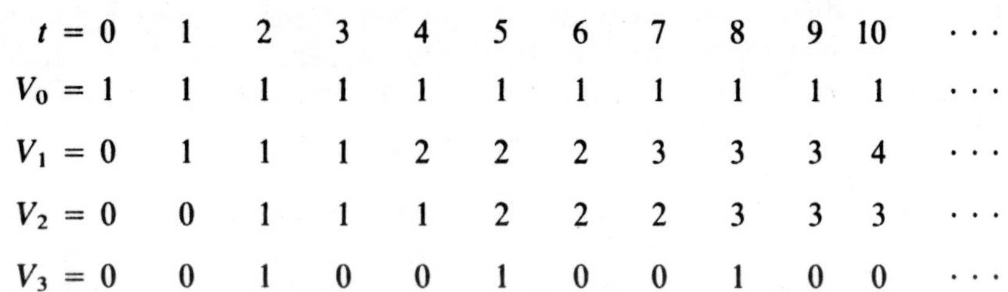

$t = 0$	1	2	3	4	5	6	7	8	9	10	...
$V_0 = 1$	1	1	1	1	1	1	1	1	1	1	...
$V_1 = 0$	1	1	1	2	2	2	3	3	3	4	...
$V_2 = 0$	0	1	1	1	2	2	2	3	3	3	...
$V_3 = 0$	0	1	0	0	1	0	0	1	0	0	...

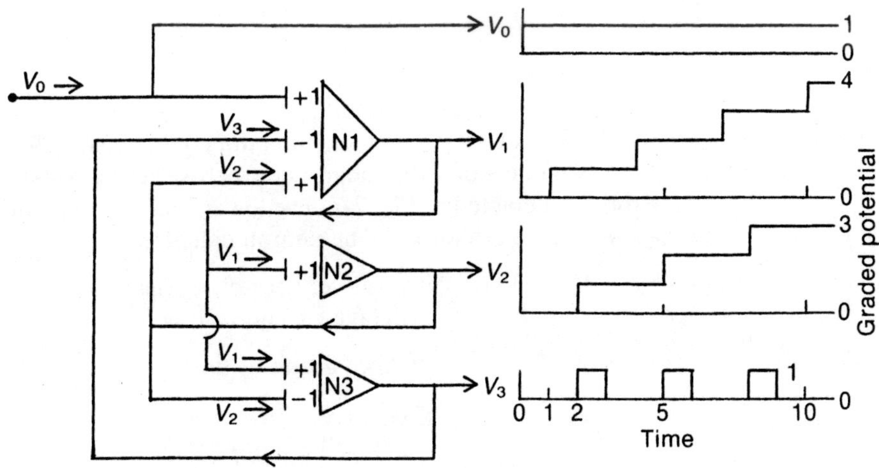

Fig. 7-6 The feedback circuit of Fig. 7-2 if the middle synaptic junction of N1 and the lower junction of N3 are inhibitory. The system is on the borderline between stability and instability.

Looking at the waveforms, one would guess that V_3 repeatedly swings below zero, and that it is clipped off at zero because of the "if positive, zero otherwise" rule. A glance at the V_1 and V_2 rows in the preceding display shows that this guess is wrong: $V_3 = V_1 - V_2$ never goes negative and, in fact, the "if positive, zero otherwise" admonition is unnecessary for V_1 and V_3 if V_0 is a step stimulus.

Comparing Figs. 7-5 and 7-6, we see that their inhibitory junctions are not the same, yet output potentials V_1 and V_2 are identical. This is only a coincidence; their V_3 potentials are completely different.

Is the circuit of Fig. 7-6 stable? Let us see what happens if V_0 returns to zero (after the last three columns of the previous listing are repeated):

$t =$	8	9	10	11	12	13	14	15	16	17	...
$V_0 =$	1	1	1	0	0	0	0	0	0	0	...
$V_1 =$	3	3	4	4	3	4	3	3	3	3	...
$V_2 =$	3	3	3	4	4	3	4	3	3	3	...
$V_3 =$	1	0	0	1	0	$\underline{\underline{0}}$	1	$\underline{\underline{0}}$	0	0	...

The result is that V_3 tries to go negative in two places, as indicated by the double underlines. After the second time this happens, on the fifteenth time slot, the output settles down to the steady values $V_1 = V_2 = 3$, while $V_3 = 0$. This outcome is rather unexpected; we again have an example of a "trapped" potential.

The system of Fig. 7-6 is evidently on the borderline between stability and instability; a slight increase in the positive weighting factors will push it into instability. The system is linear if V_0 is a step, but not if V_0 is a rectangular pulse.

7-4 Circuits with Inhibition, Stable with $W = 1$

Our last three variations of the feedback circuit of Fig. 7-2, discussed in this section, are stable with $W = 1$ ($W = -1$ for inhibitory junctions). They can conceivably serve as models of living neuronal processes (or ANNs).

The first of the three is depicted in Fig. 7-7. The lower synaptic junction of N1 and the upper of N3 are inhibitory. The neuron equations are

$$\begin{aligned} V_1 &= (V_0 + V_3 - V_2) && \text{if positive, zero otherwise; delayed 1 unit of time,} \\ V_2 &= V_1 && \text{delayed 1 unit of time,} \\ V_3 &= (V_2 - V_1) && \text{if positive, zero otherwise; delayed 1 unit of time.} \end{aligned} \tag{7-17}$$

There is nothing unusual here, but the resulting waveforms are strange indeed. It seems as if they are trying to generate some kind of abstract design. Upon closer inspection, however, there is some method to the madness: V_1

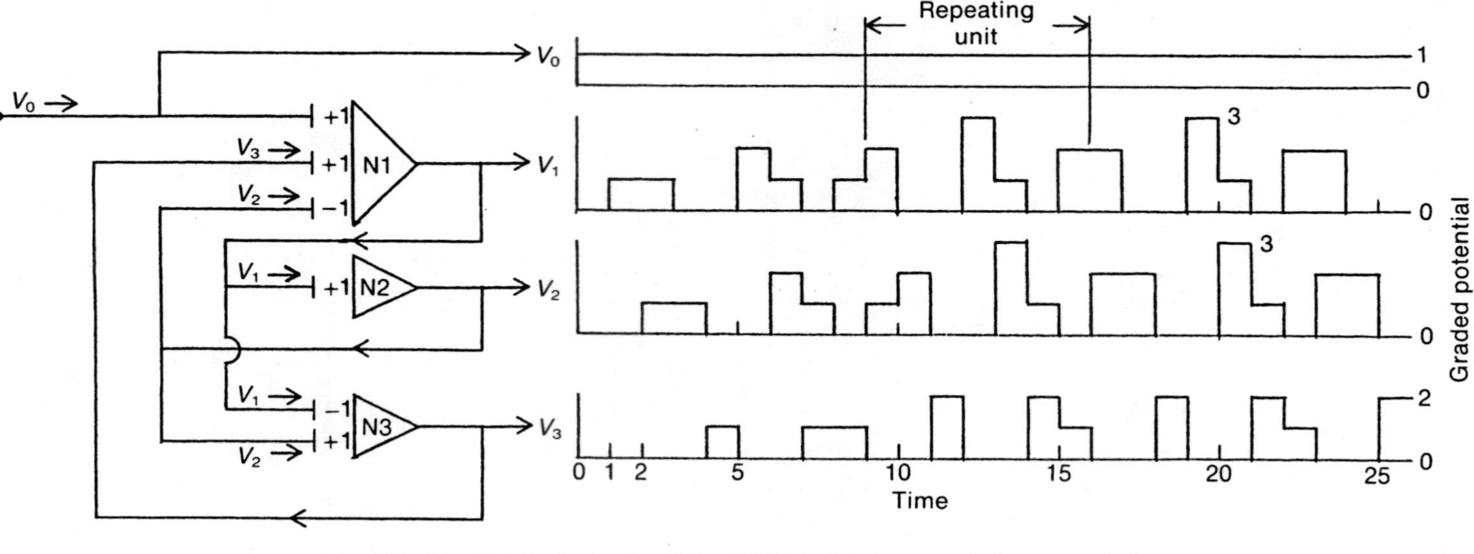

Fig. 7-7 The feedback circuit of Fig. 7-2 if the lower synaptic junction of N1 and upper junction of N3 are inhibitory. The system is stable. After an initial transient period, each of the output potentials repeats over and over again a pattern seven time units wide.

goes through a hop-skip-and-jump routine until $t = 9$. After that it repeats, over and over again, a pattern seven time units wide, as indicated by "repeating unit." Of course, V_2 does the same, but delayed by one unit of time. Also, V_3 starts to repeat over and over again, at $t = 8$, a pattern seven time units wide. Here are complicated outputs, then, generated by only three interconnected neurons. Perhaps this is a good example to show the complexity that can be achieved by only three neurons.

The potentials are given by

$$
\begin{array}{rcl}
t & = & \begin{matrix} 0&0&0&0&0&0&0&0&0&0&1&1&1&1&1&1&1&1&1&1&2&2&2&2&2&2 \\ 0&1&2&3&4&5&6&7&8&9&0&1&2&3&4&5&6&7&8&9&0&1&2&3&4&5 \end{matrix} \cdots \\
V_0 & = & 1\ 1\ 1\ 1\ 1\ 1\ 1\ 1\ 1\ 1\ 1\ 1\ 1\ 1\ 1\ 1\ 1\ 1\ 1\ 1\ 1\ 1\ 1\ 1\ 1\ 1 \cdots \\
V_1 & = & 0\ 1\ 1\ 0\ 0\ 2\ 1\ \underline{\underline{0}}\ 1\ 2\ 0\ \underline{\underline{0}}\ 3\ 1\ \underline{\underline{0}}\ 2\ 2\ \underline{\underline{0}}\ \underline{\underline{0}}\ 3\ 1\ \underline{\underline{0}}\ 2\ 2\ \underline{\underline{0}}\ \underline{\underline{0}} \cdots \\
V_2 & = & 0\ 0\ 1\ 1\ 0\ 0\ 2\ 1\ 0\ 1\ 2\ 0\ 0\ 3\ 1\ 0\ 2\ 2\ 0\ 0\ 3\ 1\ 0\ 2\ 2\ 0 \cdots \\
V_3 & = & 0\ 0\ \underline{\underline{0}}\ 0\ 1\ 0\ \underline{\underline{0}}\ 1\ 1\ \underline{\underline{0}}\ \underline{\underline{0}}\ 2\ 0\ \underline{\underline{0}}\ 2\ 1\ \underline{\underline{0}}\ 0\ 2\ 0\ \underline{\underline{0}}\ 2\ 1\ \underline{\underline{0}}\ 0\ 2 \cdots
\end{array}
$$

The V_1 and V_3 potentials frequently "try" to become negative, as indicated by the many double underlines. If the double underlines are taken into account, in fact, V_1 and V_3 do not start repeating until $t = 11$. The circuit is certainly stable because the potentials oscillate between 0 and 2 or 0 and 3, never more, if the stimulus V_0 is one unit high.

Although the waveforms of Fig. 7-7 are stable, they are too erratic, too unnatural to act as models for a living system. Finally—we are saving the best examples for last—our next circuit, that of Fig. 7-8, generates perfectly regular and symmetrical oscillations that could be used in a biological sys-

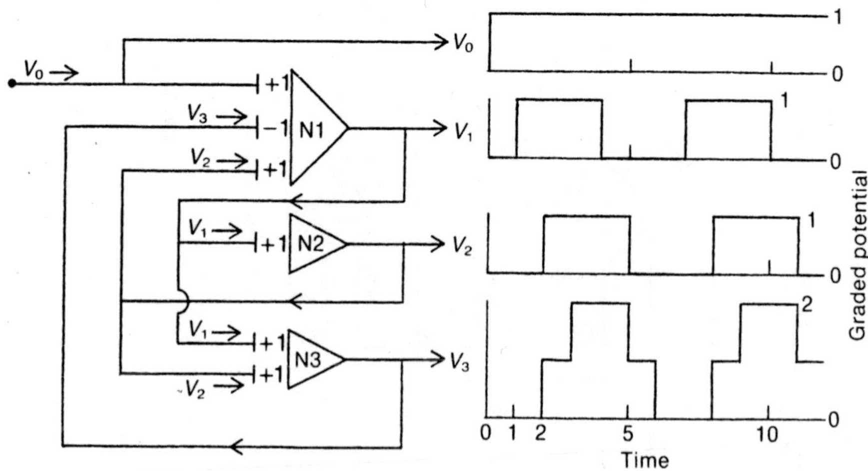

Fig. 7-8 The feedback circuit of Fig. 7-2 if the middle synaptic junction of N1 is inhibitory. The system is stable.

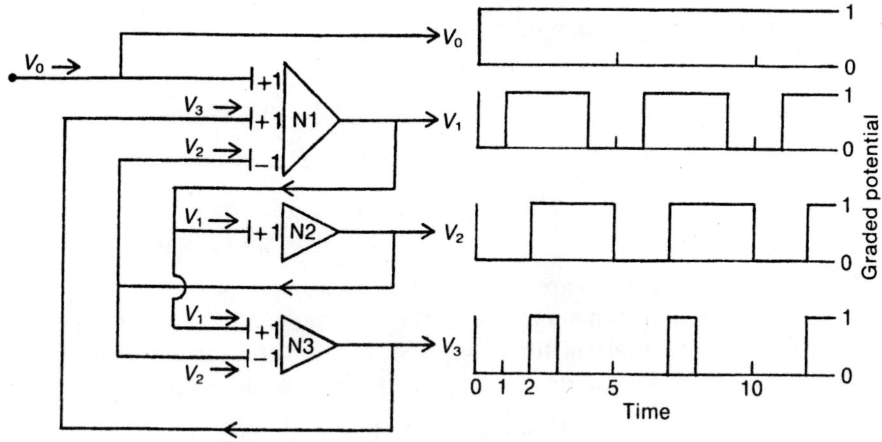

Fig. 7-9 The feedback circuit of Fig. 7-2 if the lower synaptic junctions of N1 and N3 are inhibitory. The system is stable.

tem. Here the middle junction of N1 is inhibitory. The neuron equations are

$$
\begin{aligned}
V_1 &= (V_0 + V_2 - V_3) && \text{if positive, zero otherwise; delayed 1 unit of time,} \\
V_2 &= V_1 && \text{delayed 1 unit of time,} \qquad (7\text{-}18) \\
V_3 &= (V_1 + V_2) && \text{delayed 1 unit of time.}
\end{aligned}
$$

The V_1 and V_2 potentials oscillate between +1 and 0, with a cycle that is six units of time wide. The V_3 potential somewhat more gently oscillates between +2 and 0. These are the neuronal patterns that are reminiscent of walking or respiration.

The potentials are given by

$t =$	0	1	2	3	4	5	6	7	8	9	10	11	⋯
$V_0 =$	1	1	1	1	1	1	1	1	1	1	1	1	⋯
$V_1 =$	0	1	1	1	0	0	0	1	1	1	0	0	⋯
$V_2 =$	0	0	1	1	1	0	0	0	1	1	1	0	⋯
$V_3 =$	0	0	1	2	2	1	0	0	1	2	2	1	⋯

None of the potentials ever swings negative, so the "if positive, zero otherwise" admonition for V_1 is unnecessary if V_0 is a step stimulus. In other words, the circuit is well behaved.

If V_0 returns from 1 to 0, the output potentials soon become zero. The system is linear for the given V_0 waveforms.

Our last circuit is that of Fig. 7-9, where the lower junctions of N1 and

N3 are inhibitory. The neuron equations are

$$
\begin{aligned}
V_1 &= (V_0 + V_3 - V_2) && \text{if positive, zero otherwise; delayed 1 unit of time,} \\
V_2 &= V_1 && \text{delayed 1 unit of time,} \\
V_3 &= (V_1 - V_2) && \text{if positive, zero otherwise; delayed 1 unit of time.}
\end{aligned} \tag{7-19}
$$

The potential waveforms are rectangular, with a five-unit-of-time cycle. The V_1 and V_2 potentials remain at level 1 for three units of time and then return to level 0 for two units of time, while V_3 is at level 1 for one unit of time and at level 0 for four units of time. The potentials are given by

$t =$	0	1	2	3	4	5	6	7	8	9	10	11	12	...
$V_0 =$	1	1	1	1	1	1	1	1	1	1	1	1	1	...
$V_1 =$	0	1	1	1	0	0	1	1	1	0	0	1	1	...
$V_2 =$	0	0	1	1	1	0	0	1	1	1	0	0	1	...
$V_3 =$	0	0	1	0	0	$\underline{\underline{0}}$	0	1	0	0	$\underline{\underline{0}}$	0	1	...

We see that V_1 never tries to go negative, but V_3 does, as indicated by the double underlines, and is clipped to zero.

The stable and repetitive waveforms of Figs. 7-8 and 7-9 are the types of waveforms that are picked up by electrodes monitoring the leg muscles in walking, or the diaphragm muscle in breathing, or the heart muscles, or the numerous muscles engaged in peristaltic contractions. In the case of a symmetrical activity such as walking, neurons on one side of the body inhibit their counterparts on the other side and vice versa, so that contraction of a particular muscle is accompanied by relaxation of antagonistic muscles. You *can* hop, if you must, but it is more natural to walk.

A7-1 DENOMINATOR ZEROS OF A LINEAR SYSTEM

We solve for s at which the denominator of Eq. (7-8) becomes zero:

$$1 - 2W^2\epsilon^{-2s\tau} - W^3\epsilon^{-3s\tau} = 0. \tag{7-20}$$

One can solve for the three roots of this cubic equation, but there is an easier way. We can guess that two of the solutions constitute a complex conjugate pair, while the third solution is given by $s = 0$. If $s = 0$, then $\epsilon^{-s\tau} = 1$ and Eq. (7-20) simplifies to

$$1 - 2W^2 - W^3 = 0. \tag{7-21}$$

Now we have a cubic equation that is easily solved. By inspection, $W = -1$ is a solution. Dividing by $1 + W$ to remove this root, there remains

$$1 - W - W^2 = 0. \tag{7-22}$$

The solutions to this quadratic equation are $W = 0.6180$ and -1.6180. The only meaningful solution is the positive value, $W = (\sqrt{5} - 1)/2 = 0.6180$.

Substituting $W = 0.6180$ into Eq. (7-20),

$$1 - 0.7639\epsilon^{-2s\tau} - 0.2361\epsilon^{-3s\tau} = 0. \tag{7-23}$$

However, we started out with one of the solutions, $\epsilon^{-s\tau} = 1$. Dividing by $1 - \epsilon^{-s\tau}$ to remove this root, there remains

$$1 + \epsilon^{-s\tau} + 0.2361\epsilon^{-2s\tau} = 0. \tag{7-24}$$

The solutions to this quadratic equation are $\epsilon^{s\tau} = -0.3820$ and -0.6180, or

$$s = \frac{1}{\tau}(-0.9624 \pm j\pi), \qquad s = \frac{1}{\tau}(-0.4812 \pm j\pi). \tag{7-25}$$

($\text{Log}(a + jb) = \log M + j\theta$, where $M = \sqrt{a^2 + b^2}$ and $\theta = \arctan(b/a)$. One should draw a vector representing $a + jb$ to place θ in the correct quadrant. $\text{Arctan}(b/a)$ is multivalued; only the principal values, $-\pi < \theta < \pi$, are considered in this chapter.) Since all of the zeros of the denominator of Eq. (7-8) are at the origin or in the left half of the s plane, $W = 0.6180$ is the correct solution. For $W < 0.6180$, the circuit of Figs. 7-2 and 7-3 is stable.

REFERENCES

M. Abeles, *Studies of Brain Function: Local Cortical Circuits*. New York: Springer-Verlag, 1982.

E. R. Caianiello, *Neural Networks*. New York: Springer-Verlag, 1968.

T. J. Carew, V. F. Castellucci, J. H. Byrne, and E. R. Kandel, Quantitative analysis of relative contribution of central and peripheral neurons to gill-withdrawal reflex in Aplysia Californica, *J. Neurophysiol.*, vol. 42, pp. 497–509, 1979.

S. Deutsch, *Models of the Nervous System*. New York: Wiley, 1967.

M. R. Guevara, L. Glass, M. C. Mackey, and A. Shrier, Chaos in neurobiology, *IEEE Trans. Syst., Man, Cybern.*, vol. SMC-13, pp. 790–798, Sept./Oct. 1983.

A. C. Guyton, *Textbook of Medical Physiology*. Philadelphia: Saunders, 1986.

L. D. Harmon, Studies with artificial neurons. Properties and functions of an artificial neuron, *Kybernetik*, vol. 1, pp. 89–101, 1961.

C. Hayashi, *Nonlinear Oscillations in Physical Systems*. New York: McGraw-Hill, 1964.

E. R. Kandel, Small systems of neurons, *Sci. Am.*, vol. 241, pp. 66–76, Sept. 1979.

E. R. Lewis, T. E. Everhart, and Y. Y. Zeevi, Studying neural organization in Aplysia with the scanning electron microscope, *Science*, vol. 165, pp. 1140–1143, Sept. 1969.

R. Lorente de No, Analysis of the activity of the chains of internuncial neurons, *J. Neurophysiol.*, vol. 1, pp. 207–244, 1938.

S. Ochs, *Elements of Neurophysiology*. New York: Wiley, 1965.

A. I. Selverston, J. P. Miller, and M. Wadepuhl, Neural mechanisms for the production of cyclic motor patterns, *IEEE Trans. Syst., Man, Cybern.*, vol. SMC-13, pp. 749–757, Sept./Oct. 1983.

R. Thom, *Structural Stability and Morphogenesis*. Reading, Mass.: Benjamin, 1975.

Problems

1. Given the circuit of Fig. 7-1, with all synaptic junction weights $W = +1$: (a) if V_0 is a unit step, find $V_2(s)$; (b) if V_0 is a one-unit-high pulse as in Fig. 7-1, draw the V_2 waveform. [Ans.: (a) $V_2 = 1/s(\epsilon^{-s\tau} + \epsilon^{-2s\tau})$.]
2. Carry out the step-by-step derivation of Eqs. (7-20) to (7-25).
3. Carry out the step-by-step derivation of Eqs. (7-10) to (7-12).
4. Given the circuit of Fig. 7-2, each of the synaptic junctions of N1 has weight W, while the junctions of N2 and N3 have weight $W = 2$. (a) Find the transfer function, V_1/V_0. (b) Find W and the zeros of the denominator if the network is on the borderline between stability and instability. If V_0 is a unit step, (c) find $V_1(W)$ up to and including the $t - 6\tau$ term; (d) plot V_1 in the range $0 < t < 6.5$ for $W = 0.1$. What does V_1 approach as $t \to \infty$? (e) Find V_1 at $t = 6.5$ for $W = 0.2$. [Ans.: (a) $V_1/V_0 = W\epsilon^{-s\tau}/(1 - 4W\epsilon^{-2s\tau} - 4W\epsilon^{-3s\tau})$; (b) $W = 0.125$; $s = 0, (-0.3466 \pm j2.356)/\tau$; (c) $V_1 = W(t - \tau) + 4W^2(t - 3\tau) + 4W^2(t - 4\tau) + 16W^3(t - 5\tau) + 32W^3(t - 6\tau) + \cdots$; (d) $V_1 \to 0.5$ as $t \to \infty$; (e) $V_1 = 0.904$.]
5. Check the step-by-step listing for Fig. 7-4, which follows Eq. (7-14).
6. Check the step-by-step listing for Fig. 7-5, which follows Eq. (7-15).
7. Given the circuit of Fig. 7-5, with V_0 a unit step, the system is linear. (a) Find $V_1(s)$ and the corresponding $V_1(t)$ up to and including the $t - 8\tau$ term. (b) Repeat for V_3. (c) Find the zeros of the denominator. [Ans.: (a) $V_1 = \epsilon^{-s\tau}/(1 - \epsilon^{-3s\tau})$; $V_1 = 1(t - \tau) + 1(t - 4\tau) + 1(t - 7\tau) + \cdots$; (b) $V_3 = (\epsilon^{-2s\tau} + \epsilon^{-3s\tau})/(1 - \epsilon^{-3s\tau})$; $V_3 = 1(t - 2\tau) + 1(t - 3\tau) + 1(t - 5\tau) + 1(t - 6\tau) + 1(t - 8\tau) + \cdots$; (c) $s = 0, \pm j2.094/\tau$.]
8. Check the step-by-step listing for Fig. 7-6, which follows Eq. (7-16).
9. Given the circuit of Fig. 7-6, with V_0 a unit step, the system is linear. (a) Find $V_1(s)$ and the corresponding $V_1(t)$ up to and including the $t - 8\tau$ term. (b) Repeat for V_3. (c) Find the zeros of the denominator. [Ans.: (a) $V_1 = \epsilon^{-s\tau}/(1 - \epsilon^{-3s\tau})$; $V_1 = 1(t - \tau) + 1(t - 4\tau) + 1(t - 7\tau) + \cdots$; (b) $V_3 = (\epsilon^{-2s\tau} - \epsilon^{-3s\tau})/(1 - \epsilon^{-3s\tau})$; $V_3 = 1(t - 2\tau) - 1(t - 3\tau) + 1(t - 5\tau) - 1(t - 6\tau) + 1(t - 8\tau) + \cdots$; (c) $s = 0, \pm j2.094/\tau$.]
10. Check the step-by-step listing for Fig. 7-7, which follows Eq. (7-17).
11. Given the circuit of Fig. 7-7, V_0 is a rectangular pulse that starts at $t = 0$. Show the step-by-step listing for V_0, V_1, V_2, and V_3, from $t = 17$ to 31, if (a) $V_0 = 1$ up to and including $t = 17$, and 0 afterwards; (b) $V_0 = 1$ up to and including $t = 18$, and 0 afterwards; (c) $V_0 = 1$ up to and including $t = 19$, and 0 afterwards. [Ans.:

	$t =$	17	18	19	20	21	22	23	24	25	26	27	28	29	30	31
(a)	$V_0 =$	1	0	0	0	0	0	0	0	0	0	0	0	0	0	0
	$V_1 =$	$\underline{\underline{0}}$	$\underline{\underline{0}}$	2	0	$\underline{\underline{0}}$	2	0	$\underline{\underline{0}}$	2	0	$\underline{\underline{0}}$	2	0	$\underline{\underline{0}}$	2
	$V_2 =$	2	0	0	2	0	0	2	0	0	2	0	0	2	0	0
	$V_3 =$	0	2	0	$\underline{\underline{0}}$	2	0	$\underline{\underline{0}}$	2	0	$\underline{\underline{0}}$	2	0	$\underline{\underline{0}}$	2	0

$$
\text{(b)}\begin{cases}
V_0 = & 1 & 1 & 0 & 0 & 0 & 0 & 0 & 0 & 0 & 0 & 0 & 0 & 0 & 0 & 0\\
V_1 = & \underline{\underline{0}} & \underline{\underline{0}} & 3 & 0 & \underline{\underline{0}} & 3 & 0 & \underline{\underline{0}} & 3 & 0 & \underline{\underline{0}} & 3 & 0 & \underline{\underline{0}} & 3\\
V_2 = & 2 & 0 & 0 & 3 & 0 & 0 & 3 & 0 & 0 & 3 & 0 & 0 & 3 & 0 & 0\\
V_3 = & 0 & 2 & 0 & \underline{\underline{0}} & 3 & 0 & \underline{\underline{0}} & 3 & 0 & \underline{\underline{0}} & 3 & 0 & \underline{\underline{0}} & 3 & 0
\end{cases}
$$

$$
\text{(c)}\begin{cases}
V_0 = & 1 & 1 & 1 & 0 & 0 & 0 & 0 & 0 & 0 & 0 & 0 & 0 & 0 & 0 & 0\\
V_1 = & \underline{\underline{0}} & \underline{\underline{0}} & 3 & 1 & \underline{\underline{0}} & 1 & 1 & \underline{\underline{0}} & \underline{\underline{0}} & 1 & 0 & \underline{\underline{0}} & 1 & 0 & \underline{\underline{0}}\\
V_2 = & 2 & 0 & 0 & 3 & 1 & 0 & 1 & 1 & 0 & 0 & 1 & 0 & 0 & 1 & 0\\
V_3 = & 0 & 2 & 0 & \underline{\underline{0}} & 2 & 1 & \underline{\underline{0}} & 0 & 1 & 0 & \underline{\underline{0}} & 1 & 0 & \underline{\underline{0}} & 1
\end{cases}
$$

12. Check the step-by-step listing for Fig. 7-8, which follows Eq. (7-18).
13. Given the circuit of Fig. 7-8, with $V_0 = a$ unit step, the system is linear. (a) Find $V_1(s)$ and the corresponding $V_1(t)$ up to and including the $t - 10\tau$ term. (b) Repeat for V_3. (c) Find the zeros of the denominator. [Ans.: (a) $V_1 = \epsilon^{-s\tau}/(1 + \epsilon^{-3s\tau})$; $V_1 = 1(t - \tau) - 1(t - 4\tau) + 1(t - 7\tau) - 1(t - 10\tau) + \cdots$; (b) $V_3 = (\epsilon^{-2s\tau} + \epsilon^{-3s\tau})/(1 + \epsilon^{-3s\tau})$; $V_3 = 1(t - 2\tau) + 1(t - 3\tau) - 1(t - 5\tau) - 1(t - 6\tau) + 1(t - 8\tau) + 1(t - 9\tau) - \cdots$; (c) $s = \pm j\pi/\tau, \pm j1.047/\tau$.]
14. Check the step-by-step listing for Fig. 7-9, which follows Eq. (7-19).

8

Skeletal Muscle Circuits

ABSTRACT. A skeletal muscle fiber receives action potentials (APs) from its motoneuron. Embedded in the skeletal muscles are (a) tendon organ receptors that monitor tendon stress, and (b) spindles that generate an error signal in accordance with the following equation: Error = reference (desired contraction) − distance (actual contraction), or $e(t) = r(t) - x(t)$.

Three feedback loops are involved in the control of muscle contraction as follows: (a) an annulospiral receptor in the spindle feeds the error signal back to the motoneuron, exciting it until the error is zero; (b) a flower-spray receptor in the spindle also feeds the error signal back if the error is relatively large, thereby further increasing motoneuron excitation; and (c) the tendon organ feeds an *inhibitory* signal to the motoneuron if the tendon stress is excessive.

A typical motoneuron receives excitatory signals from synergistic muscles and inhibitory signals from ipsilateral antagonistic muscles.

Clonus can occur if excessive time is taken for the spindle signal to travel from a wrist muscle, say, to its motoneuron in the spinal cord and from the latter back to the skeletal muscle. If the motoneuron synaptic junction weighting factor is too high, the system will oscillate.

8-1 Some Neuroanatomical Features

There is a tendency to regard the nervous system as consisting only of the brain and nerve trunks, and to omit muscles. Since muscles consist of excitable tissue, however, they are very much a part of the nervous system. The passage of an action potential (AP) along a skeletal muscle fiber is followed by contraction of the fiber. (This is explained in greater detail below.) Furthermore, much of the activity of the brain is directed toward the control of muscle contraction [H. H. Kornhuber, 1974; S. L. Lehman and L. Stark, 1983].

Approximately half of the body consists of muscle tissue. Physiologists recognize three kinds of muscle: cardiac, smooth, and skeletal. Cardiac muscle forms the heart, which is approximately a spherical structure with a hollow core. When it contracts (systole), blood is forced out of the center of the sphere into the major arteries of the body and lungs. When it relaxes (diastole), blood is gathered in from the major veins of the body and lungs. Smooth muscle is typified by muscle found in the walls of the gastrointestinal and genitourinary tracts and blood vessels, and by the piloerector muscles that make your hair stand up. It gets its name from the lack of cross striations. The contraction of smooth muscle is "involuntary," relatively slow, and is directed by axons of the autonomic nervous system.

This chapter is devoted to certain features of skeletal muscle. It is muscle that tries to bring together two relatively rigid members, such as skeletal bones, when it contracts. There is of course a large variation in muscle length and diameter: the shortest muscles are only 0.5 cm (0.2 in.) long, while the longest, the sartorius muscle (between hip and knee), typically measures 40 cm (16 in.) long [J. W. Woodbury, A. M. Gordon & J. T. Conrad, 1965].

A magnified longitudinal section taken through a small portion of a typical skeletal muscle is depicted in Fig. 8-1(*a*). Action potentials are brought in via "skeletal muscle nerve fibers," which have a diameter of 16 μm (this and the other dimensions of Fig. 8-1 are typical values for a human). The fibers terminate on small tabs that are called *end plates*. For an eye muscle that is devoted to fine, precise movements, a single motoneuron and its axon terminate on three end plates; at the other extreme, for a leg muscle devoted to coarse movements, a single motoneuron and its axon can terminate on 150 end plates. The end plates are similar to synaptic junctions; the APs spread out on both sides of each end plate as propagation proceeds in a manner analogous to that of an unmyelinated axon. All of the APs coming from the same skeletal muscle axon to a common set of end plates are synchronous because they have a common source, but the APs in muscle fibers controlled by different motoneurons have different frequencies, in general, and are not related to each other.

A muscle fiber has a diameter of 10 to 100 μm; a typical value is 50 μm. Its substructure consists of bundles of tubular fibrils. Each fibril is 0.5 μm in diameter and is filled with *actin* and *myosin* filaments, as shown in Fig. 8-1(*b*) and (*c*). These are protein molecules that form a repeating structure 2.2 μm wide in the relaxed state, as indicated by the *Z* lines in Fig. 8-1(*b*). Ionic movements associated with an AP (including those of calcium ions) stimulate the filaments so that they slide past each other until the repeating structure is 1.7 μm wide in the fully contracted state, as shown in Fig. 8-1(*c*). The attraction between filaments is due to electrostatic forces between positively and negatively charged molecules. Energy for the work

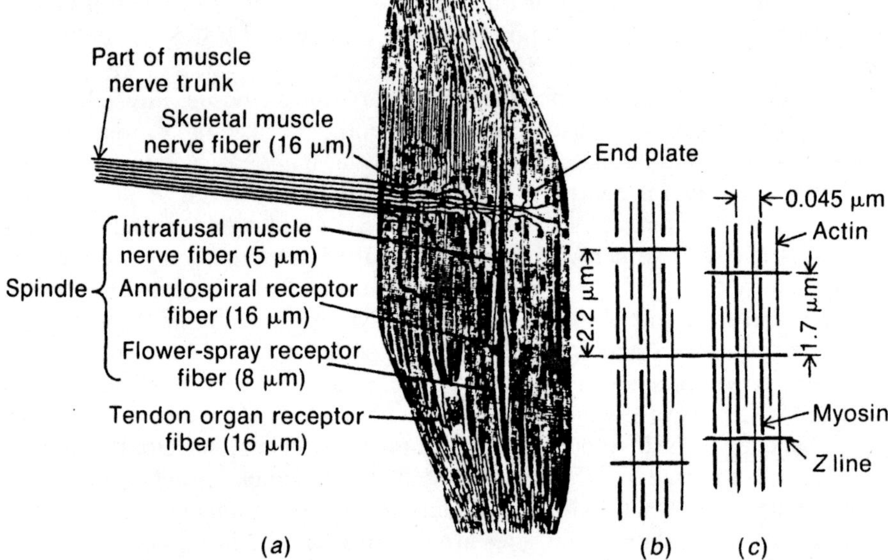

Fig. 8-1 Idealized diagram of a skeletal muscle and part of nerve trunk coming from motoneurons. (*a*) Longitudinal section taken through a small portion of a typical muscle. (Numbers in parentheses are axon diameters.) The tendon organ receptor monitors tendon stress. The spindle generates an error signal in accordance with Error = reference (desired contraction) minus distance (actual contraction). The intrafusal muscle, annulospiral receptor, and flower-spray receptor are parts of the spindle. (*b*) Highly magnified view of myosin and actin filaments in relaxed muscle. (*c*) Filaments in fully contracted muscle.

performed is of course derived from the food we eat [D. D. Vaccaro, G. C. Agarwal & G. L. Gottleib, 1988].

The muscle fiber AP is 10 ms wide compared to the 2-ms width of an unmyelinated fiber AP. The velocity of propagation is around 5 meters/second (m/s). (In a long muscle such as the sartorius, where the muscle ends may be 20 cm away from the end plates at the center, the muscle ends therefore contract 40 ms [20 cm divided by 5 m/s] after the center contracts.) A single contraction, called a *twitch*, rises to a peak in around 50 ms and more slowly tapers off to zero some 150 ms after the stimulating AP has ended.

A single twitch is inconsequential. Remember that only a small number of fibers are involved (according to the figures just given, a motor unit consists of somewhere between three and 150 fibers). It is the repeated, sustained overlapping, in space and time, of many twitches that yields significant force. Regarding space, at a given instant, many twitches occur simultaneously. Regarding time, during maximum contraction (called *tetanus*), the AP frequency may be 50 Hz, which corresponds to a period of 20 ms. Since a twitch is 150 ms long, there is a great deal of overlap if twitches start every 20 ms apart, and the net effect is a practically steady force of contraction. Only rarely, however, does a muscle go into tetanus. Normally, with lower AP frequencies, the force of contraction remains fairly steady because many twitches occur almost simultaneously. Nevertheless, we have all experienced how difficult it is to hold the hand steady when threading a needle; we become aware that the contraction of a muscle is quantized.

The values given previously are typical, but there are significant departures. For the internal rectus muscle of the eye, which is involved with very fast, fine rotation of the eyeball, tetanus requires an AP frequency of 350 Hz; for the slow soleus muscle (between knee and ankle), tetanus occurs at 30 Hz.

Returning to Fig. 8-1(*a*), we see that, embedded amongst the muscle fibers, there are *tendon organ* sensory receptors. The tendon organ is in series with part of the tendon connecting the muscle to its bone. It monitors the muscle stress (tension); when the latter exceeds some safety threshold, the tendon organ inhibits the motoneuron. In the absence of tendon organ inhibition, in a laboratory preparation, the muscle force of contraction is sometimes sufficient to tear the tendon loose from the bone. You will wonder why, then, people sometimes tear a tendon. Do they lack tendon organ receptors? Probably not. As Figs. 3-3 and 3-4 demonstrate, the system is noisy, so that mistakes are made. Some tendon organ circuits may give insufficient protection; this may not be too serious for a human, but a limping cheetah will probably starve to death. There is obviously a fine balance between too much and too little protection, since an otherwise healthy cheetah that is slowed by tendon organ inhibition will also miss out on a normal food supply.

We can think of one obvious technical reason why people sometimes tear a tendon: Although the tendon organ fiber is a high-speed conductor (96 m/s), it still takes appreciable time for the "take it easy" message to reach the muscle. It has to travel all the way from the tendon in your leg, say, eventually reaching the motoneuron in the spinal cord, and then back to the leg muscle, which seems like a poor way to design the equipment. Actually, there is a very good reason why the tendon alarm signal travels to the spinal cord—it goes to many other places besides its own motoneuron. It excites or inhibits other motoneurons, depending on where their muscles are located, so as to relieve the stress on its own muscle.

Also embedded amongst the muscle fibers are sensory receptors called *spindles*, which also play a vital role in muscle contraction. The *intrafusal* (within the spindle) muscle, *annulospiral* receptor, and *flower-spray* receptor are parts of the spindle. (Please do not get the impression from Fig. 8-1(*a*) that there is only one tendon organ receptor and one spindle. A typical skeletal muscle may contain hundreds of them; the drawing shows only one of each in a small part of the muscle.) When the motoneuron sends a burst of APs to the muscle, the degree of contraction depends on many other factors besides the discharge frequency. The muscle's energy sources may be partially depleted, and/or its load (bone, tissue, and external objects that resist the motion) may be heavier than expected, and so forth. The spindle, however, does not do any manual labor, so that it never gets tired normally. It is a white-collar worker. The same signal that goes to the motoneuron also goes to the spindle via its intrafusal muscle nerve fiber, so the spindle "knows" how much the muscle is supposed to contract. The spindle compares the actual movement with the desired movement; the difference is an *error* signal that the spindle feeds back to the motoneuron. This acts like a stick that prods a recalcitrant donkey: the spindle error signal encourages (i.e., excites) the motoneuron until actual and desired contractions are the same and the error signal is zero [H. Hemami and B. T. Stokes, 1983; J. C. Houk and L. Stark, 1962; G. F. Inbar and P. J. Joseph, 1976; H. D. Patton, 1965; R. E. Poppele and D. C. Quick, 1981].

The diameter of a fiber is a direct indication of how important conduction speed is for that fiber. All of the fibers listed in Fig. 8-1(*a*) are myelinated, so the velocity of propagation is 6×10^6 dia/s (this is discussed in Sect. 4-3). To get the speed in meters/second, therefore, we simply multiply the diameter in micrometers by 6. For the "skeletal muscle nerve fiber," as an example, we get $16\ \mu m \times 6 = 96$ m/s. This is a high-speed, 215-mph axon. It obviously *has* to be high speed to yield a fast response to noxious stimuli.

There is an interesting interplay of survival factors here. Each of the nerve fibers in Fig. 8-1(*a*) "tries" to increase its diameter since faster response is always beneficial. A larger nerve trunk, which may contain millions of axons, is deleterious because it uses precious body resources. Given a maximum permissible nerve trunk diameter, the individual fibers must compete amongst themselves for maximum diameter. Evolutionary survival of the fittest has determined that the skeletal muscle fiber should be high speed, while the intrafusal muscle nerve fiber can be relatively slow (30 m/s), and so forth for the other axons of Fig. 8-1(*a*).

All of the preceding seems like a terribly complicated system, but it gets the job done because it goes directly to the desired goal and keeps prodding the motoneuron until the desired goal is reached, regardless of how tired the muscle may be or how heavy the load may be.

8-2 With Feedback via Annulospiral Receptor Fiber

A highly simplified example is employed initially to illustrate the workings of a skeletal muscle. As depicted in Fig. 8-2, the tendon organ excessive-stress feedback is omitted. Only the spindle feedback, via its annulospiral receptor nerve fiber, is shown. The numerical values that follow have been calculated for this model.

Interneurons N1, N2, and N3, and the motoneuron MN, are located in or near the spinal cord. In a human, the muscle and its entourage—spindles, tendons, and load—can be as much as one meter away.

The desired motion or command input coming down from the brain,

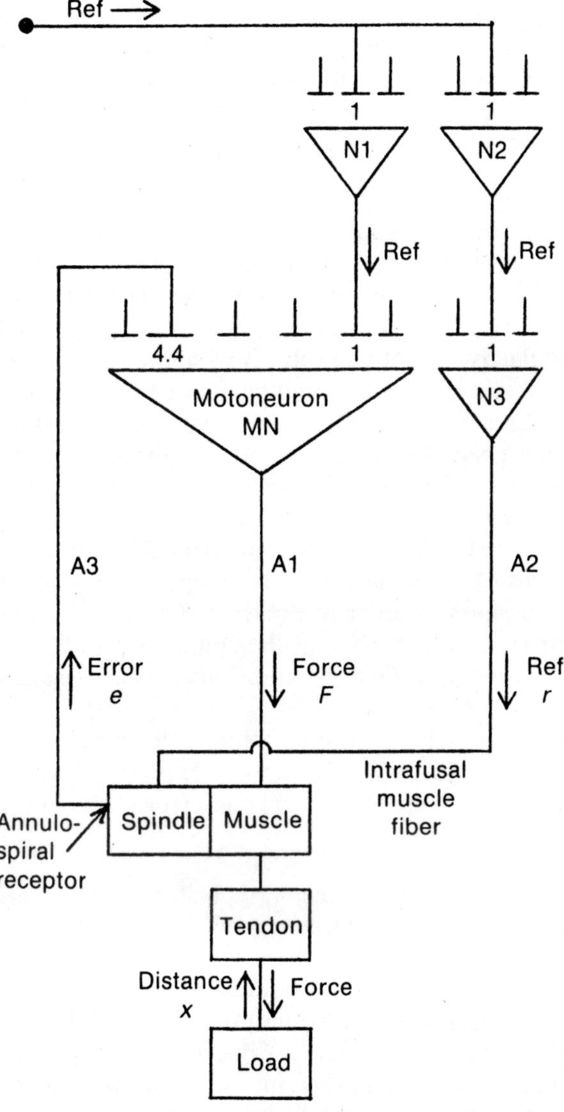

Fig. 8-2 Highly idealized and simplified schematic diagram of a skeletal muscle and one of its motoneurons, including spindle annulospiral receptor feedback. Ref is the reference (desired or command) contraction. Synaptic junction weighting factor values are taken from a numerical example that uses a bow-and-arrow type of load: Ref = 1; mass of arrow plus your arm = 2; frictional force = 3.3 times velocity; the bow pulls back with force = 1 when distance moved = 1.

say, is the reference signal, r [E. V. Evarts, 1980]. It reaches MN via N1; the synaptic junction weighting factors are 1. The spindle error signal, e, approaches MN via axon A3, and it enters MN at a synaptic junction that has a multiplicative weight of 4.445 (which is shown on the drawing as 4.4 because of space limitations). The force signal, F, is therefore given by

$$F = r + 4.445e. \tag{8-1}$$

The force signal reaches the muscle via axon A1. The amount that the muscle contracts, x, is symbolized by a "distance" arrow pointing upward. The "force" arrow points downward because the load is pulling down on the muscle.

We immediately run into a problem with units because here, as well as in Eq. (8-1), the units of force should be in newtons, say, while the units of distance should be in meters. Therefore, each equation should be accompanied by a conversion factor. Since this is inconvenient, and the numerical values are chosen on the basis of simplicity rather than to illustrate a typical human arm moving a load, *units will be omitted except for time in seconds*.

The annulospiral nerve fiber is another high-speed conductor (96 m/s). Because it is part of a feedback loop, it is important to minimize signal time delay around the loop in order to prevent instability or actual oscillations. The evolutionary adaptation is that the spindle stretch receptor and motoneuron form a *monosynaptic* reflex arc. This is discussed further in Section 8-6.

The reference signal reaches the spindle (the intrafusal muscle fiber) by way of interneurons N2 and N3 and axon A2. As the equation at the bottom of Fig. 8-2 states, the spindle generates an error signal in accordance with

$$e = r - x. \tag{8-2}$$

We are also reminded that "signals are positive; zero otherwise" because all of the signals consist of AP frequencies.

The intrafusal muscle nerve fiber has a conduction speed of only 30 m/s. It is not part of a feedback loop but, instead, is involved with "conscious" movements dictated by higher brain centers rather than rapid reflex withdrawal from a noxious stimulus. Conscious movements usually involve hundreds of milliseconds. The slow intrafusal response is rather like a telephone call halfway round the world, where the round-trip signal time is at least 135 ms, but this is scarcely noticed in comparison with the time taken to voice the answer to a question. The intrafusal muscle receives the same reference signal that the motoneuron receives, but its contraction is separate from that of the main muscle; as its name indicates, its contraction is confined to be within the spindle.

It is natural to show the load as if it were a weight being lifted, as in Fig. 8-2, but this is too specialized. We should work with a more general load, one that has mass (m), a frictional force proportional to velocity ($k_d v$), and the force of a spring ($k_r x$) that is trying to return the load to $x = 0$. Then

$$F = ma + k_d v + k_r x, \tag{8-3}$$

where a is the acceleration. The closest example of such a load from everyday life is one in which you are pulling back the arrow of a bow-and-arrow before it is released. (Perhaps a better example, nowadays, is one in which you are pulling up the hand brake of your automobile) [A. Ailon, G. Langholz & M. Arcan, 1984; P. E. Crago, J. T. Mortimer & P. H. Peckham, 1980].

Usually, in a classroom problem (but not in real life), the initial displacement $x(0)$ and initial velocity $v(0)$ are zero. In this chapter, however, some of the initial values differ from zero, so the Laplace transform of Eq. (8-3) has to include these elements:

$$F = ms^2X - msx(0) - mv(0) + k_dsX - k_dx(0) + k_rX, \tag{8-4}$$

where X is the transform of x. (Functional notation (t) or (s) is applied if it is not obvious as to whether we are dealing with $f(t)$ or $f(s)$.)

The numerical values of the example used are given as follows: the command reference calls for a unit step of motion distance ($r(t) = 1 \,|_{t>0}$); the mass of the arrow (plus that of your arm as it is pulling the arrow) is given by $m = 2$; the frictional force $= 3.3v$ (in other words, force is wasted in overcoming friction only when the arrow is moving); and the spring (the bow) pulls back with force $= 1$ when the distance moved $= 1$; that is,

$$m = 2; \qquad k_d = 3.3; \qquad k_r = 1. \tag{8-5}$$

Substituting the numerical values into Eq. (8-4), there results

$$F = 2s^2X - 2sx(0) - 2v(0) + 3.3sX - 3.3x(0) + X. \tag{8-6}$$

The curves of Fig. 8-3 describe the motion by means of distance x, error e, force F, and velocity v waveforms. The mathematical basis for these curves is given in Appendix A8-1.

It is worthwhile to examine some of the significant values in the curves of Fig. 8-3.

At $t = 0$: The command "Reference = 1" comes through at $t = 0$. Distance x and velocity v are, of course, zero. The error is 1. From Eq. (8-13), the force is 5.445.

At $t = 0.73$ s: You are pulling the arrow back with maximum velocity, 0.9 distance units/s. The motion has almost reached the halfway mark with $x = 0.45$, so the error is reduced to 0.55. From Eq. (8-13), the force is 3.4. Here, then, we can see how the spindle functions. Your spindle "knows" that x should be 1. Originally e was 1, and spindle feedback to the motoneuron was a major factor in getting a strong initial response, $F = 5.445$. Now, at $t = 0.73$ s, there is less error, so spindle feedback to the motoneuron results in a weaker response, $F = 3.4$.

At $t = 2.2$ s: There are some "strange goings on." The arrow has overshot, as your arm pulled it to $x = 1.16$. The error is therefore negative, $e = -0.16$. What does negative error mean? It implies that, for the *antag-*

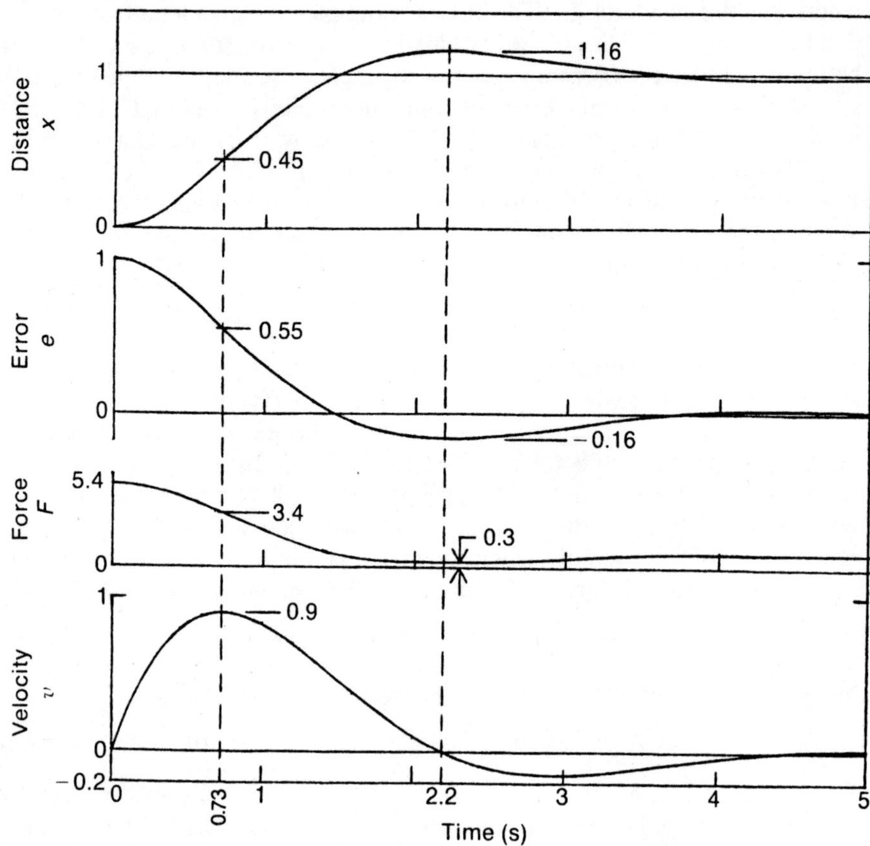

Fig. 8-3 Waveforms associated with Fig. 8-2 if Ref = 1 starting at $t = 0$. Velocity is maximum at $t = 0.73$ s, and there is a 16% overshoot at $t = 2.2$ s. The antagonistic muscle is active when the error is negative.

onistic muscle, the error is *positive*, so the antagonistic muscle will contract, pushing the arrow back toward the $x = 1$ line. From Eq. (8-13), $F = 0.3$. The velocity is zero when the arrow is about to start its journey back to the $x = 1$ line [D. K. Peterson and H. J. Chizeck, 1987].

After $t = 4$ s: Finally, the reference = 1 command is satisfied. The error is zero, the force is 1 (you are holding the arrow against the opposite pull of 1 exerted by the bow), and the velocity is zero.

8-3 With Feedback Added via Flower-Spray Receptor Fiber

When the command to load the bow-and-arrow came through it did not imply that you could take your time about it; quite the opposite is implied, that your life depends on reaching the distance = 1 mark *as soon as possible*. Initially, when you are far from the $x = 1$ line, it will save time if you can pull with a force much greater than 5.445. This is the purpose of the spindle flower-spray receptor.

As shown with axon A4 of Fig. 8-4, the flower-spray sensory receptor has a threshold level of 0.4. It is active when the error is greater than 0.4 (this is a numerical value made up for the bow-and-arrow example). When

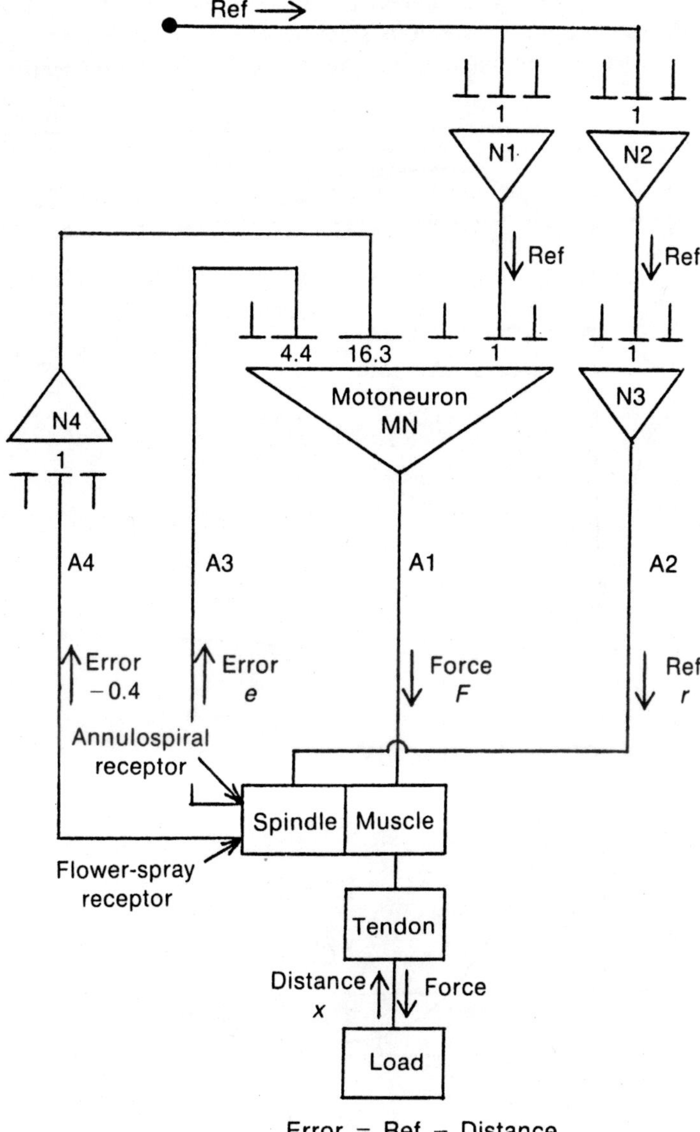

Fig. 8-4 Schematic diagram of Fig. 8-2 if feedback via spindle flower-spray receptor fiber is added. Numerical values are taken from the example that uses a bow-and-arrow type of load.

the error is less than 0.4, axon A4 becomes dormant and the circuit reverts to that of Fig. 8-2. The flower-spray receptor signal reaches MN via interneuron N4. It enters MN at a synaptic junction that has a multiplicative weight of 16.335 (which is shown on the drawing as 16.3 because of space limitations).

The flower-spray receptor fiber is a medium-speed conductor (48 m/s). This seems to contradict a previous statement, that it should be a high-speed fiber if it is part of a feedback loop. The reason it is medium speed probably

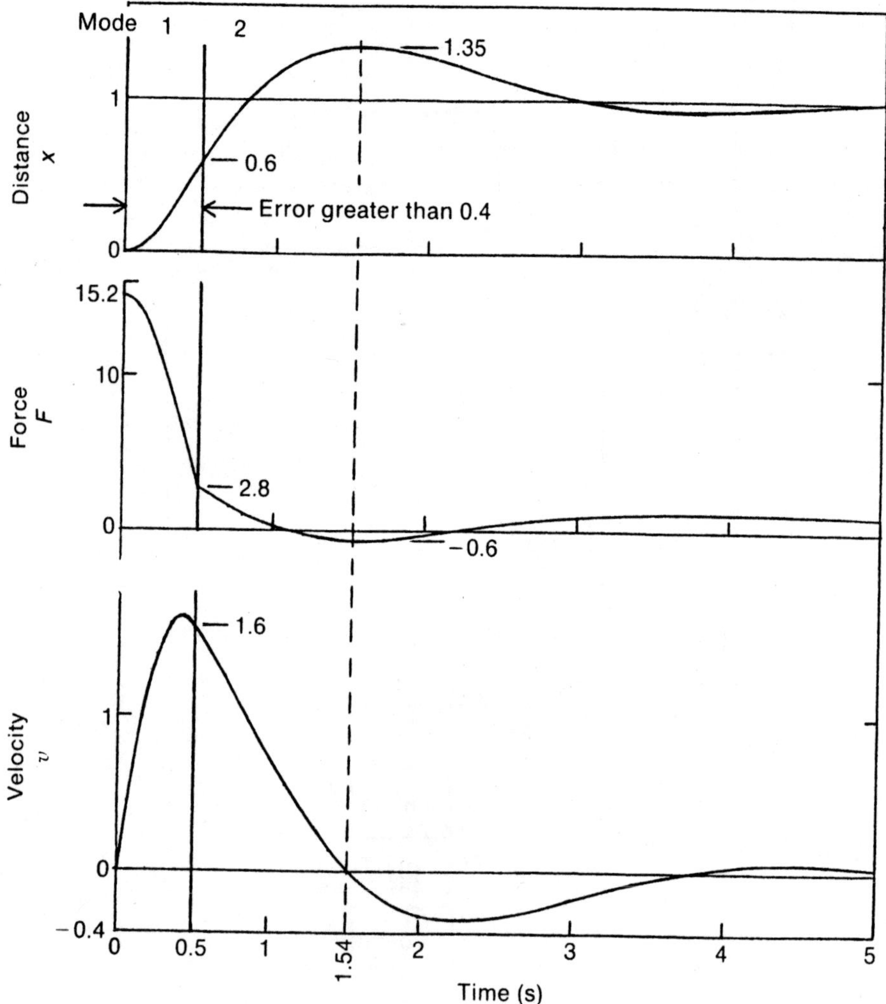

Fig. 8-5 Waveforms associated with Fig. 8-4 if Ref = 1 starting at $t = 0$. (The "error" waveform is the same as the "distance" curve turned upside down.) There is a 35% overshoot at $t = 1.54$ s. The antagonistic muscle is active when the force is negative.

is that the flower-spray fiber is in the loop only when the muscle contraction error is large; instability is more easily tolerated if it is associated with large error. The flower-spray nerve fiber is inactive when the muscle is "zeroing-in" and has almost achieved its objective.

The curves of Fig. 8-5 demonstrate the changes wrought by flower-spray receptor feedback. The mathematical basis for these curves is given in Appendix A8-2.

A glance at the distance curve shows that, alas, the high initial force is overdone. Your arm is moving so fast that it cannot stop in time; there is a large overshoot, to $x = 1.35$ at $t = 1.54$ s (compare with the distance curve in Fig. 8-3). Inertia is the culprit. The error and force are negative when the antagonistic muscle is trying to push the arrow back to the $x = 1$ line. After the large overshoot, it takes time to return to the command reference goal; nevertheless, comparing Figs. 8-3 and 8-5, it seems that the flower-spray receptor feedback *does* save time, and that is undoubtedly why it was conserved during evolution. It may seem as if the same result can be achieved if a weight of 16.3 is used, say, in place of 4.4 in Fig. 8-2, but this is not so; with a weighting factor of 16.3 the distance curve would overshoot and undershoot repeatedly, eventually settling down to $x = 1$ after an agonizingly long period of time. The strategy of initially using junction weights of 16.3 and 4.4 followed by a steady zeroing-in weight of 4.4 is much more effective.

8-4 With Feedback Added via Tendon Organ Receptor Fiber

The 35% overshoot of Fig. 8-5 is not in itself a serious problem, although we prefer the 16% overshoot of Fig. 8-3 because it is already close to the desired $x = 1$ value. The main problem with Fig. 8-5 is that the tendon stress is excessive. Accordingly, we add yet another, final feedback path, the tendon organ axon A5 of Fig. 8-6. The tendon organ has a threshold level of 8; that is, it is active when the stress is greater than 8 (this is a numerical value made up for the bow-and-arrow example). When the tendon stress is less than 8, axon A5 becomes dormant and the circuit reverts to that of Fig. 8-4 (or to Fig. 8-2 if the distance error is less than 0.4). The tendon organ receptor signal reaches MN via interneuron N5. It enters MN at an *inhibitory* synaptic junction that has a weight of -1 (remember that its purpose is to prevent the muscle from injuring the tendon during contraction).

Distance, force, and velocity waveforms are shown in Fig. 8-7 for the numerical example. The mathematical basis for these curves is given in Appendix A8-3.

At $t = 1.63$ s in Fig. 8-7 the arrow overshoots to $x = 1.32$. One would judge this to be an excessive overshoot, but somewhat less than the 1.35 value of Fig. 8-5, thanks to the tendon organ feedback for time less than 0.4 s.

The numerical example illustrated in Fig. 8-7 is, of course, highly oversimplified. When you load a bow-and-arrow in real life, many muscles are

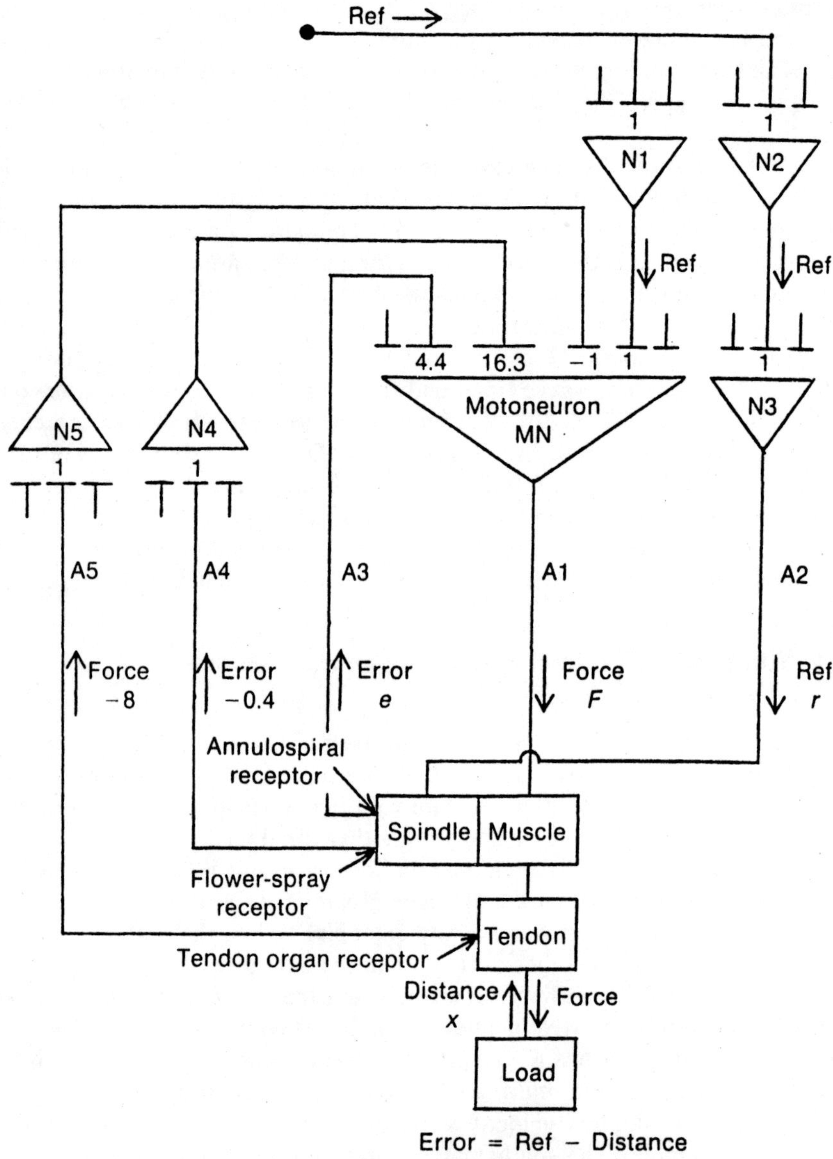

Fig. 8-6 Schematic diagram of Fig. 8-4 if feedback via tendon organ receptor fiber is added. Numerical values are taken from the example that uses a bow-and-arrow type of load.

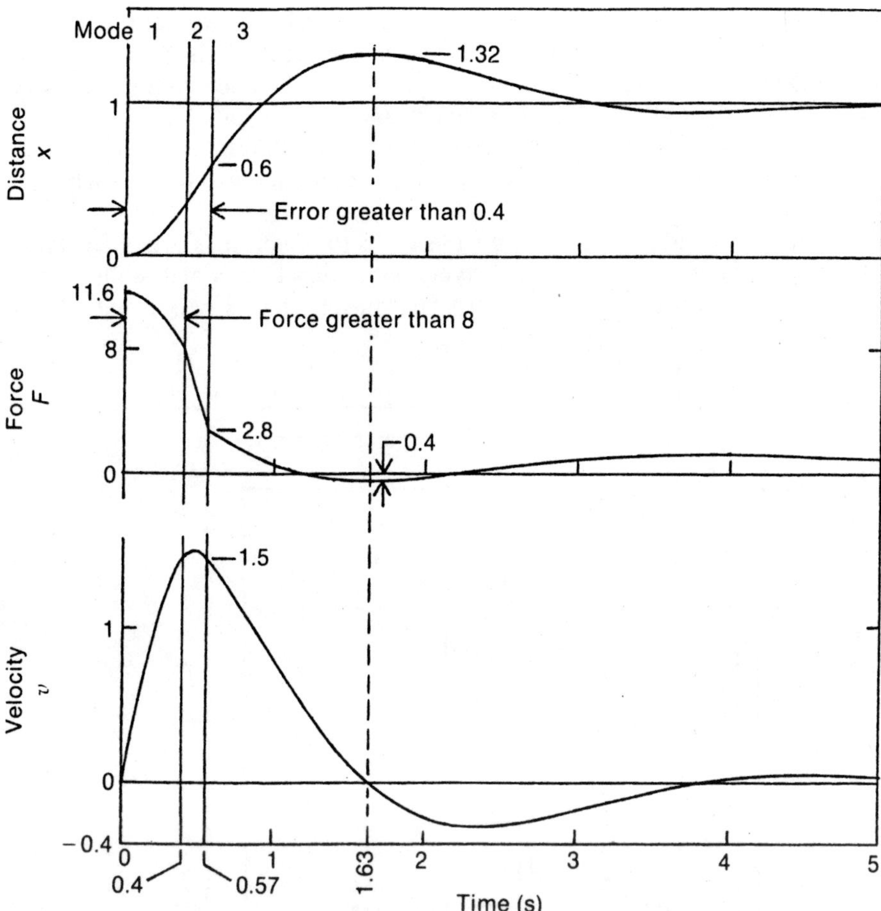

Fig. 8-7 Waveforms associated with Fig. 8-6 if Ref = 1 starting at $t = 0$. (The "error" waveform is the same as the "distance" curve turned upside down.) There is a 32% overshoot at $t = 1.63$ s.

involved rather than the single flexor muscle and (when the distance error becomes negative) its single antagonistic extensor muscle. Mammalian living systems are highly adaptive; you would soon notice that the overshoot is excessive, and modify the parameters of the numerical example by means of APs supplied by the higher brain centers, especially the cerebellum part of the brain (see Chap. 12).

8-5 Typical Motoneuron Circuit

Although the discussion in the previous sections concentrates on the local feedback loops associated with a spindle and tendon organ in the longitudinal section of Fig. 8-1(*a*), it is important to see how this local microcosm is integrated into the larger scheme of muscle control. The logical focal point of the next stage in the hierarchy is the motoneuron. A highly simplified representation of a typical motoneuron (MN) circuit is depicted in Fig. 8-8. Only a few of the thousands of synaptic junctions are shown.

The heavy line descending from MN is the efferent axon A1 of Fig. 8-6, the skeletal muscle nerve fiber, which excites the hatched muscle (M) in Fig. 8-8.

Muscle M pulls to the left on the bone to which it is attached. Also shown are three other muscles: a synergistic muscle that also pulls to the left; an ipsilateral (same side) antagonist muscle that is paired with M in

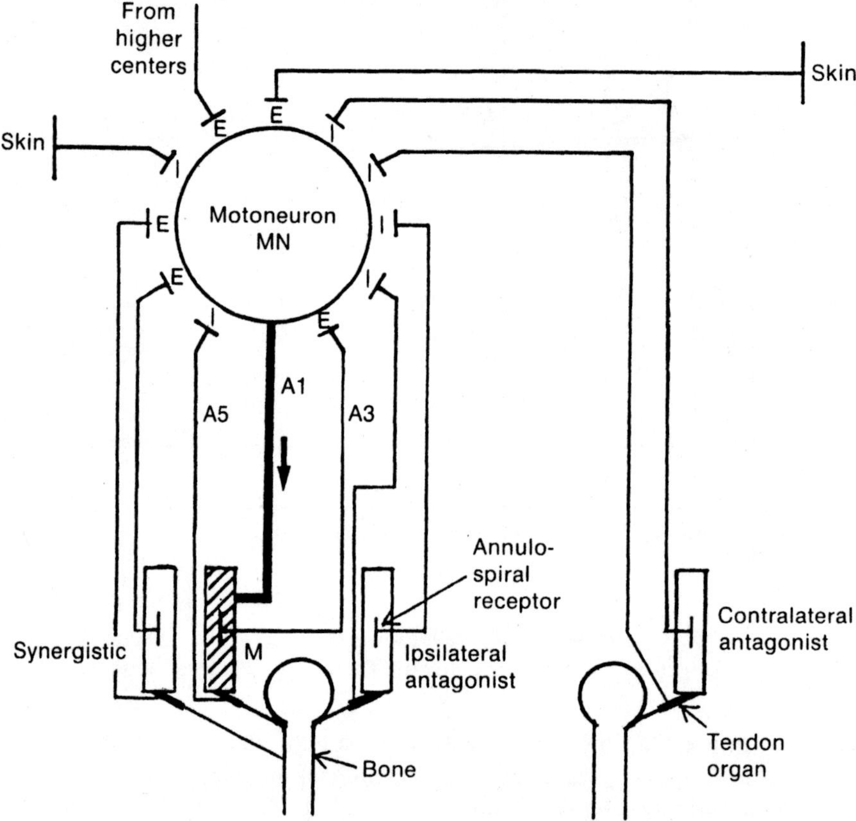

Fig. 8-8 Highly idealized and simplified representation of a typical motoneuron (MN) circuit. The heavy line descending from the motoneuron (A1) is the efferent axon, which excites the hatched muscle (M). Three related muscles are also included in the circuit. Feeding back to the motoneuron from each muscle are annulospiral receptor (A3) and tendon organ (A5) fibers. Intrafusal muscle and flower-spray receptor fibers, as well as interneurons, are omitted from the drawing. (From Deutsch and Tzanakou, *Neuroelectric Systems*, New York Univ. Press, New York, 1987.)

order to effect return motion by pulling to the right; and a contralateral (opposite side) antagonist muscle that also pulls to the right, but on the corresponding bone on the other side of the body.

Two fibers are shown feeding back from each of the four muscles: a tendon organ receptor fiber (axon A5 in Fig. 8-6) and a spindle annulospiral receptor fiber (axon A3 in Fig. 8-6). All of these fibers feed to the MN of muscle M (as well as to many other neurons). Intrafusal muscle and flower-spray receptor fibers are omitted from the drawing for the sake of simplicity; similarly, all interneurons are omitted.

Consider first the spindle error feedback (shown as originating at the center of each muscle). From M it arrives at MN via an excitatory (E) junction. Since the synergistic muscle is pulling in the same direction as M, its error signal also goes to an E junction on MN. Error in the ipsilateral antagonist tries to inhibit (I) MN since, if the pull of M to the left thereby decreases, the bone will move to the right so as to reduce the error of the ipsilateral antagonist muscle. The relationship of M to the contralateral antagonist is less clear-cut, but in the drawing it is assumed that its error decreases if M is inhibited.

The four tendon organ receptor fibers are shown as originating at the tendon at the bottom end of each muscle. In agreement with Fig. 8-6, M's tendon organ fiber arrives at MN via axon A5 followed by an I junction. If the synergistic muscle tendon is overstressed, it tries to excite MN since, if the contraction of M thereby increases, the bone will move to the left so as to remove some of the stress from the synergistic muscle. Similarly, if the ipsilateral antagonist tendon is overstressed, it tries to inhibit MN since, if the contraction of M decreases, the bone will move to the right so as to remove stress from the ipsilateral antagonist muscle. It is assumed that the same reasoning holds for tendon overstress of the contralateral antagonist muscle.

It is well known that, to relieve a cramp in M, you should try to contract the opposing muscle (the ipsilateral antagonist) because it will inhibit (relax) the MN of M.

The two "skin" inputs in Fig. 8-8 are meant to illustrate reflex action following noxious cutaneous stimulation. If the skin on the contralateral side (the right side in Fig. 8-8) touches a hot surface, MN is excited and M contracts in what is a built-in withdrawal movement. If the skin on the ipsilateral side touches a hot surface, however, it inhibits contraction on that side while causing contraction of the antagonist muscles via pathways to the antagonist motoneurons (not shown) [D. Adam, U. Windhorst & G. F. Inbar, 1978; G. F. Inbar, J. Madrid & P. Rudomin, 1979; K. Pearson, 1976; J. M. Winters and L. Stark, 1985].

One should appreciate the fact that most of the thousands of inputs to a typical motoneuron are part of built-in reflex pathways that enable us to execute normal contractions, such as those involved in walking or standing upright and withdrawal from noxious stimuli, with a minimum of command inputs from the higher brain centers.

8-6 Clonus

It is believed that the involuntary vibrations associated with Parkinson's disease are caused by an instability occurring in the brain. There is a similar pathological condition, known as spasticity or *clonus*, involving skeletal muscles. It is described as follows by Arthur C. Guyton in his *Textbook of Medical Physiology* (1986):

> Clonus ordinarily occurs only if the stretch reflex is highly sensitized by facilitatory impulses from the brain. For instance, in the decerebrate animal, in which the stretch reflexes are highly facilitated, clonus develops readily. Therefore, to determine the degree of facilitation of the spinal cord, neurologists test patients for clonus by suddenly stretching a muscle and keeping a steady stretching force applied to the muscle. If clonus occurs, the degree of facilitation is certain to be very high.

The equation "error = reference − distance," or $e = r - x$, which appears at the bottom of the models in this chapter, offers an interesting interpretation of what is happening when the neurologist tests patients for clonus by stretching a muscle. The patient is relaxed, so reference r is zero. When a muscle is voluntarily contracted, distance x is a positive quantity, but when the neurologist *stretches* a muscle, distance x is a *negative* quantity. In the equation, we get two negatives that yield a *positive* error. For example, $e = 0 - (-5) = +5$. In other words, during the neurological test, axon A3 in Fig. 8-6 becomes activated, and this in turn stimulates A1 to become active [L. Stark, 1986; S. Yurkovich, S. K. Hoffmann & H. Hemami, 1987].

As a matter of convenience, we will continue to use the numerical values of the bow-and-arrow example, except that the actual distance moved is very small because the command reference r is zero and the actual motion is an uncontrollable tremor. The patient's wrist is attached to the load of Eq. (8-5) (mass = 2 including the mass of the wrist, frictional force = 3.3 times the velocity, and a spring pulls back with force = 1 when distance moved = 1).

In the original model, axon A3 terminates on a synaptic junction that has a weight of 4.4. Clonus may occur, however, "if the stretch reflex is highly sensitized." This means, in effect, that the system may become unstable if the junction weighting factor is much higher than 4.4. In the present section we analyze the model when the junction weight is 20, presumably due to some kind of pathological condition.

We assume that the neurologist applies a reasonably gentle stretch to the patient's wrist, so that the error is low and the flower-spray receptors remain dormant. Similarly, the tendon stress is low so that the tendon organ receptors are inactive. The model, therefore, has to include only the spindle annulospiral receptor, as in Fig. 8-2.

With the numerical values we have been using, this system cannot possibly become unstable. In order to keep the discussion simple we left out a

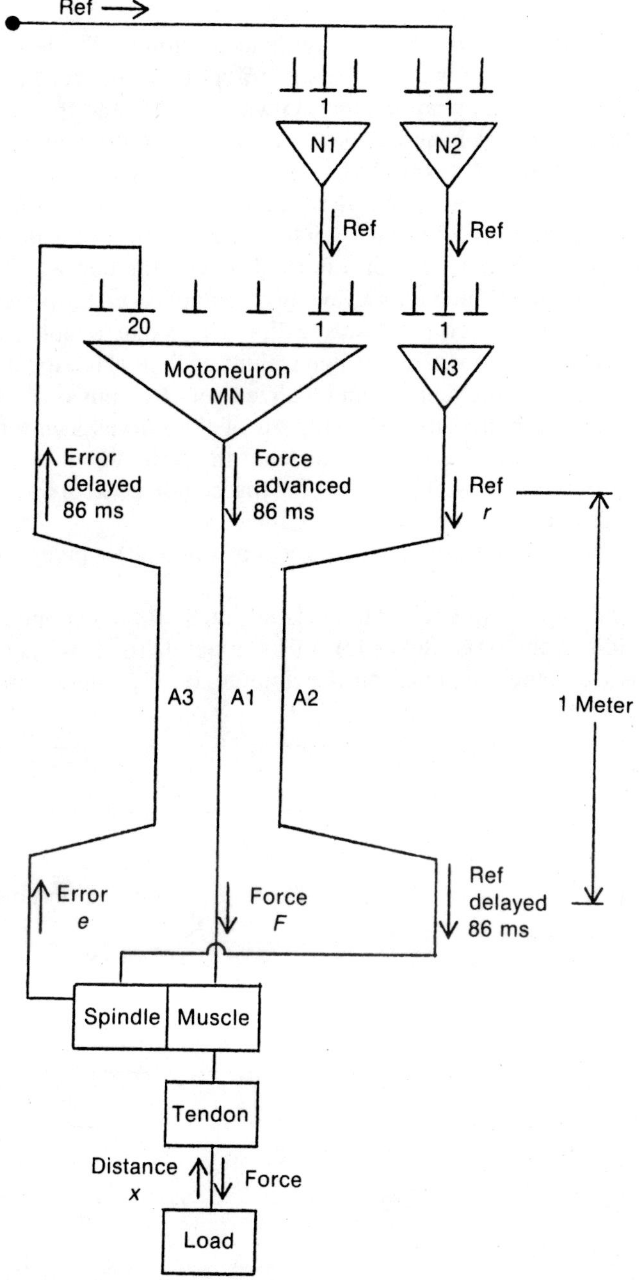

Fig. 8-9 Schematic diagram of Fig. 8-2 that features one particular set of conditions for which clonus occurs. Numerical values are taken from the example that uses a bow-and-arrow type of load. The time taken for APs to travel between motoneuron and muscle is 86 ms.

vital ingredient which, in real life, can lead to clonus. We left out the appreciable time required for AP signals to travel back and forth. It turns out that if the A3 feedback synaptic junction weight is 20, and the one-way time between the motoneuron and muscle is 86 ms, clonus will develop. The situation is summarized in Fig. 8-9.

The figure indicates a 1-m nerve trunk distance between the motoneuron in the spinal cord and the patient's ankle or wrist. The 1 m is not important; it is the 86 ms that is crucial. For a normal person the one-way time should be 20 ms (10-ms travel time plus a few junction delays). Our patient is therefore suffering from slow conduction speed, perhaps due to multiple sclerosis, in addition to "highly sensitized" (high weighting factor) feedback junctions.

For the numerical example, an idealized set of clonus oscillation waveforms results as depicted in Fig. 8-10. All of the curves swing negative as well as positive, so we have to assume that the wrist extensor muscle (and all of its motoneurons, etc.) accounts for the negative excursions, while the flexor muscle accounts for positive regions.

The mathematical basis for the curves of Fig. 8-10 is given in Appendix A8-4.

The oscillation frequency of 0.48 Hz in Fig. 8-10 corresponds to a period of 2.08 s; that is, the wrist flexes for 1.04 s, extends for 1.04 s, and so forth. The clonus frequency depends on the time delay, of course, but it also de-

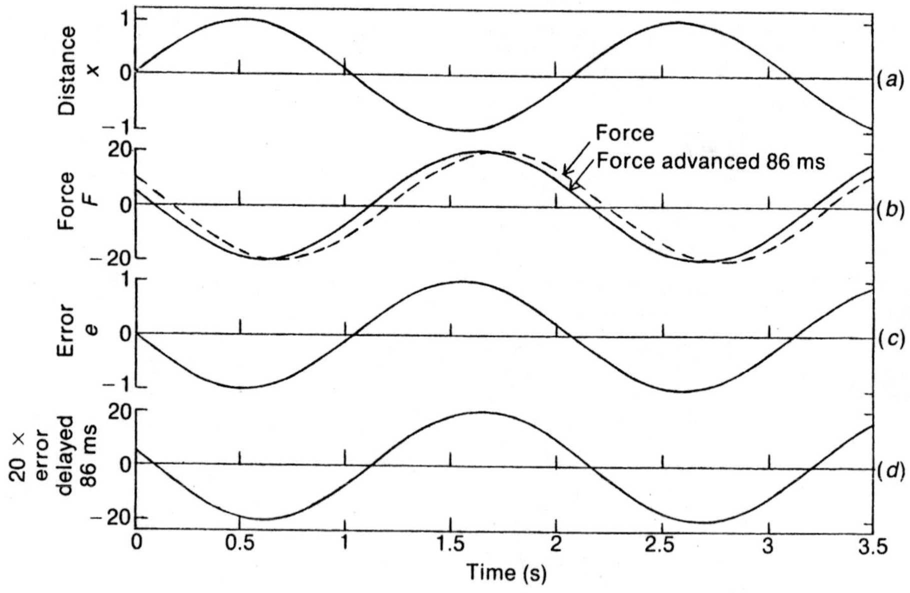

Fig. 8-10 Waveforms associated with Fig. 8-9 if the peak amplitude of clonus vibrations is 1. The frequency is 0.48 Hz. The antagonistic muscle accounts for negative regions of each curve.

pends on the synaptic junction weights and the load parameters. Without an external load, the clonus frequency can be as high as 10 vibrations/s.

Curve (*d*) in Fig. 8-10 is the feedback signal that reaches the motoneuron. It is derived from the error curve (*c*) in two steps: first, introduce a delay of 86 ms by shifting the curve 86 ms to the right; second, multiply by the synaptic junction weight of 20. In the highly idealized world of Fig. 8-10, the solid curve (*b*), labeled "Force advanced 86 ms," is the motoneuron output, and it is exactly equal to curve (*d*).

The force applied to the load is the "Force advanced 86 ms" curve shifted 86 ms to the right, thus introducing the delay suffered by the signal in traveling from the motoneuron to the muscle. The force curve is drawn dashed, next to its predecessor, to more clearly show this 86-ms time shift.

If the error feedback synaptic junction weight remains at 4.4, as in Fig. 8-2, clonus occurs when the time delay between motoneuron and muscle increases to 460 ms, and the oscillating frequency then decreases to 0.19 Hz. At a velocity of propagation of 100 m/s, the distance from the spinal cord to wrist or ankle would have to be 46 m (150 ft). Giraffes are in no danger of self-destruction due to clonus! (But we guess that for the largest animal, the blue whale, the velocity of propagation has to be much greater than 100 m/s.)

A8-1 CALCULATIONS RELATING TO ANNULOSPIRAL RECEPTOR FIBER

In Fig. 8-3 it happens that the initial values of x and v are zero, so that

$$F = 2s^2X + 3.3sX + X. \tag{8-7}$$

A second relation is given by Eq. (8-2). Since the curves are given for a unit step reference input signal, $r = 1/s$, we have

$$e = \frac{1}{s} - X. \tag{8-8}$$

A third relation is given by Eq. (8-1), which becomes

$$F = \frac{1}{s} + 4.445e. \tag{8-9}$$

Combining the preceding three equations so as to eliminate e and F,

$$X = \frac{2.7225}{s(s^2 + 1.65s + 2.7225)}. \tag{8-10}$$

The inverse transform is given by the Laplace pair

$$\begin{aligned} x(s) &= \frac{cs^2 + bs + \alpha^2 + \beta^2}{s(s^2 + 2s\alpha + \alpha^2 + \beta^2)}, \\ x(t) &= 1 + \epsilon^{-\alpha t}\left[\frac{b - \alpha(c+1)}{\beta}\sin\beta t + (c-1)\cos\beta t\right]. \end{aligned} \tag{8-11}$$

Applied to Eq. (8-10), this yields

$$x = 1 - \epsilon^{-0.825t}(0.5774 \sin 1.429t + \cos 1.429t). \tag{8-12}$$

The error curve is given by Eq. (8-8); for a unit step input, $e(t) = 1 - x$. The force curve is given by Eq. (8-9); combining with Eq. (8-8), we get

$$F(t) = 5.445 - 4.445x. \tag{8-13}$$

The velocity curve is of course given by dx/dt:

$$v = 1.905\epsilon^{-0.825t} \sin 1.429t. \tag{8-14}$$

A8-2 CALCULATIONS IF FLOWER-SPRAY RECEPTOR FIBER IS ADDED

For Fig. 8-5, see the following.

First Mode: At first, with an error greater than 0.4, the motoneuron of Fig. 8-4 specifies that

$$F = \frac{1}{s} + 4.445e + 16.335\left(e - \frac{0.4}{s}\right). \tag{8-15}$$

(Notice that the 0.4 threshold is introduced as a *step*, $0.4/s$.) Eliminating e and F from Eqs. (8-7), (8-8), and (8-15),

$$X = \frac{7.623}{s(s^2 + 1.65s + 10.89)}. \tag{8-16}$$

The Laplace transform pair Eq. (8-11) yields

$$x = 0.7[1 - \epsilon^{-0.825t}(0.2582 \sin 3.195t + \cos 3.195t)]. \tag{8-17}$$

The force curve is given by Eq. (8-15). Combining with Eq. (8-8), we get

$$F(t) = 15.246 - 20.78x. \tag{8-18}$$

The velocity curve is given by dx/dt:

$$v = 2.386\epsilon^{-0.825t} \sin 3.195t. \tag{8-19}$$

At $t = 0$, with $x = 0$, we get $F = 15.246$ as compared with the Fig. 8-3 value, 5.445. Now the system is really "hopping." The high force results in a much greater velocity than before.

The preceding equations are valid up to $t = 0.5046$ s. At this time the load has moved to $x = 0.6$, so the error is 0.4 and the flower-spray receptor branch switches off. The new conditions are described by Fig. 8-2, except that there is an initial displacement ($x = 0.6$) and an initial velocity ($v = 1.572$).

Second Mode: We continue with Fig. 8-2, but with a new time variable, t_1, where

$$t_1 = t - 0.5046. \tag{8-20}$$

Substituting the initial conditions into Eq. (8-6),

$$F = 2s^2X - 1.2s + 3.3sX - 5.124 + X. \tag{8-21}$$

Eliminating e and F from Eqs. (8-8), (8-9), and (8-21),

$$X = \frac{0.6s^2 + 2.562s + 2.7225}{s(s^2 + 1.65s + 2.7225)}. \tag{8-22}$$

The Laplace transform pair Eq. (8-11) yields

$$x = 1 + [\exp(-0.825t_1)][0.8691 \sin 1.429t_1 - 0.4 \cos 1.429t_1]. \tag{8-23}$$

The force curve is given by Eq. (8-13), and the velocity curve is given by dx/dt:

$$v = [\exp(-0.825t_1)][1.572 \cos 1.429t_1 - 0.1454 \sin 1.429t_1]. \tag{8-24}$$

One can verify that, at $t_1 = 0$ (or at $t = 0.5046$): $x = 0.6$, $F = 2.778$, and $v = 1.572$.

A8-3 CALCULATIONS IF TENDON ORGAN RECEPTOR FIBER IS ADDED

For Fig. 8-7, see the following.

First Mode: The distance error is greater than 0.4 and the force is greater than 8. The motoneuron of Fig. 8-6 specifies that

$$F = \frac{1}{s} + 4.445e + 16.335\left(e - \frac{0.4}{s}\right) - \left(F - \frac{8}{s}\right). \tag{8-25}$$

(Notice that the force threshold, 8, is introduced as a *step*, $8/s$.) Eliminating e and F from Eqs. (8-7), (8-8), and (8-25),

$$X = \frac{5.810}{s(s^2 + 1.65s + 5.695)}. \tag{8-26}$$

The Laplace transform pair Eq. (8-11) yields

$$x = 1.020[1 - \epsilon^{-0.825t}(0.3684 \sin 2.239t + \cos 2.239t)]. \tag{8-27}$$

The force curve is given by Eq. (8-25). Combining with Eq. (8-8), we get

$$F(t) = 11.62 - 10.39x. \tag{8-28}$$

The velocity curve is given by dx/dt:

$$v = 2.595\epsilon^{-0.825t} \sin 2.239t. \tag{8-29}$$

At $t = 0$, with $x = 0$, we get $F = 11.62$. The tendon organ feedback thus reduces the initial force from 15.2 in Fig. 8-5 to a much safer value in Fig. 8-7.

The preceding equations are valid up to $t = 0.3984$ s. At this time the force has decreased to 8, so the tendon organ receptor branch switches off.

The new conditions are described by Fig. 8-4, except that there is an initial displacement ($x = 0.3487$) and an initial velocity ($v = 1.454$).

Second Mode: We continue with Fig. 8-4, but with a new time variable, t_1, where

$$t_1 = t - 0.3984. \tag{8-30}$$

Substituting the initial conditions into Eq. (8-6),

$$F = 2s^2X - 0.6974s + 3.3sX - 4.059 + X. \tag{8-31}$$

Eliminating e and F from Eqs. (8-8), (8-15), and (8-31),

$$X = \frac{0.7(0.4981s^2 + 2.900s + 10.89)}{s(s^2 + 1.65s + 10.89)}. \tag{8-32}$$

The Laplace transform pair Eq. (8-11) yields

$$x = 0.7\{1 + [\exp(-0.825t_1)][0.5206 \sin 3.195t_1 - 0.5019 \cos 3.195t_1]\}. \tag{8-33}$$

The force curve is given by Eq. (8-18), and the velocity curve is given by dx/dt:

$$v = [\exp(-0.825t_1)][0.8218 \sin 3.195t_1 + 1.454 \cos 3.195t_1]. \tag{8-34}$$

One can therefore verify that at $t_1 = 0$ (or at $t = 0.3984$ s): $x = 0.3487$, $F = 8$, and $v = 1.454$.

These equations are valid up to $t = 0.5668$. At this time the load has reached $x = 0.6$, so the error is 0.4 and the flower-spray receptor branch switches off. The new conditions are described by Fig. 8-2, except that there is an initial displacement ($x = 0.6$) and an initial velocity ($v = 1.453$).

Third Mode: We continue with Fig. 8-2, but with a new time variable, t_2, where

$$t_2 = t - 0.5668. \tag{8-35}$$

Substituting the initial conditions into Eq. (8-6),

$$F = 2s^2X - 1.2s + 3.3sX - 4.886 + X. \tag{8-36}$$

Eliminating e and F from Eqs. (8-8), (8-9), and (8-36),

$$X = \frac{0.6s^2 + 2.443s + 2.7225}{s(s^2 + 1.65s + 2.7225)}. \tag{8-37}$$

The Laplace transform pair Eq. (8-11) yields

$$x = 1 + [\exp(-0.825t_2)][0.7862 \sin 1.429t_2 - 0.4 \cos 1.429t_2]. \tag{8-38}$$

The force curve is given by Eq. (8-13), and the velocity curve is given by dx/dt:

$$v = [\exp(-0.825t_2)][1.453 \cos 1.429t_2 - 0.077 \sin 1.429t_2]. \tag{8-39}$$

One can verify that at $t_2 = 0$ (or at $t = 0.5668$ s): $x = 0.6$, $F = 2.778$, and $v = 1.453$.

A8-4 CLONUS CALCULATIONS

For Fig. 8-10, the force versus distance the load moves is described by Eq. (8-7). The error versus distance relation, Eq. (8-8), is modified in two ways: first, instead of a unit step input, let the input be a generalized reference, $r(s)$ (this is the signal at the motoneuron); second, at the muscle, after a time delay of τ seconds between motoneuron and muscle, the signal is $r\epsilon^{-s\tau}$. Equation (8-8) becomes

$$e = r\epsilon^{-s\tau} - X. \tag{8-40}$$

Similarly, the motoneuron output, Eq. (8-1), has to be modified. The error signal is multiplied by 20 rather than 4.445, and time delays are introduced. The force at the muscle is then given by

$$F = r\epsilon^{-s\tau} + 20e\epsilon^{-2s\tau}. \tag{8-41}$$

Eliminating e and F from Eqs. (8-7), (8-40), and (8-41),

$$X = \frac{r\epsilon^{-s\tau} + 20r\epsilon^{-3s\tau}}{2s^2 + 3.3s + 1 + 20\epsilon^{-2s\tau}}. \tag{8-42}$$

The system becomes unstable when the denominator is zero. Substituting $j\omega$ for s, the denominator yields

$$-2\omega^2 + j3.3\omega + 1 + 20\cos 2\omega\tau - j20\sin 2\omega\tau = 0. \tag{8-43}$$

Separating the real and imaginary parts of the denominator,

$$-2\omega^2 + 1 + 20\cos 2\omega\tau = 3.3\omega - 20\sin 2\omega\tau = 0. \tag{8-44}$$

These simultaneous equations are easily solved, yielding $\tau = 86.38$ ms, and $\omega = 3.027$ radians per second (rps) (or $f = 0.4818$ Hz) as the frequency at which oscillations occur. (The reference input r is only used to derive Eq. (8-42); actually, it is zero.)

In Fig. 8-10, the waveforms are shown relative to displacement x. From Eq. (8-7),

$$\frac{F}{x} = -2\omega^2 + j3.3\omega + 1. \tag{8-45}$$

At the oscillating frequency, $\omega = 3.027$ rps, we get

$$\frac{F}{x} = 20 \arctan \frac{9.989}{-17.33} = 20\underline{/150^\circ}. \tag{8-46}$$

In other words, the F curve is 20 times as large as the x curve (the ratio would normally have units of newtons/meter; see the scale change in Fig. 8-10), and it leads x by 150°. At the relatively high frequency of the oscillation, almost all of the force goes into overcoming the inertia of the m =

2 mass. From Eq. (8-40), since r is actually zero, we get $e(t) = -x(t)$, so the error waveform is the negative of the distance waveform.

REFERENCES

D. Adam, U. Windhorst, and G. F. Inbar, The effects of recurrent inhibition on the cross-correlated firing patterns of motoneurons, *Biol. Cybern.*, vol. 29, pp. 229–235, 1978.

A. Ailon, G. Langholz, and M. Arcan, An approach to control laws for arm motion, *IEEE Trans. Biomed. Eng.*, vol. BME-31, pp. 605–610, Sept. 1984.

P. E. Crago, J. T. Mortimer, and P. H. Peckham, Closed-loop control of force during electrical stimulation of muscle, *IEEE Trans. Biomed. Eng.*, vol. BME-27, pp. 306–312, June 1980.

E. V. Evarts, Brain mechanisms in voluntary movement, in *Neural Mechanisms in Behavior*, ed. D. McFadden. New York: Springer-Verlag, 1980.

A. C. Guyton, *Textbook of Medical Physiology*, 7th ed. Philadelphia: Saunders, 1986.

H. Hemami and B. T. Stokes, A qualitative discussion of mechanisms of feedback and feedforward in the control of locomotion, *IEEE Trans. Biomed. Eng.*, vol. BME-30, pp. 681–689, Nov. 1983.

J. C. Houk and L. Stark, Analytical model of a muscle spindle receptor for simulation of motor coordination, *MIT-RLE Q. Progr. Rep.*, vol. 66, pp. 384–389, 1962.

G. F. Inbar and P. J. Joseph, Analysis of a model of the triceps surae muscle reflex control system, *IEEE Trans. Syst., Man, Cybern.*, vol. SMC-6, pp. 25–33, Jan. 1976.

G. F. Inbar, J. Madrid, and P. Rudomin, The influence of the gamma system on cross-correlated activity of Ia muscle spindles and its relation to information transmission, *Neurosci. Lett.*, vol. 13, pp. 73–78, 1979.

H. H. Kornhuber, Cerebral cortex, cerebellum, and basal ganglia: An introduction to their motor functions, in *The Neurosciences, Third Study Program*, ed. F. O. Schmitt and F. G. Worden. Cambridge, Mass.: MIT Press, 1974.

S. L. Lehman and L. Stark, Perturbation analysis applied to eye, head, and arm movement models, *IEEE Trans. Syst., Man, Cybern.*, vol. SMC-13, pp. 972–979, Sept./Oct. 1983.

H. D. Patton, Spinal reflexes and synaptic transmission, and H. D. Patton, Reflex regulation of movement and posture, both in *Neurophysiology*, 2d ed., eds. T. C. Ruch et al. Philadelphia: Saunders, 1965.

K. Pearson, The control of walking, *Sci. Am.*, vol. 235, pp. 72–86, Dec. 1976.

D. K. Peterson and H. J. Chizeck, Linear quadratic control of a loaded agonist-antagonist muscle pair, *IEEE Trans. Biomed. Eng.*, vol. BME-34, pp. 790–796, Oct. 1987; vol. 35, p. 282, Apr. 1988 (revised).

R. E. Poppele and D. C. Quick, Stretch-induced contraction of intrafusal muscle in cat muscle spindle, *J. Neurosci.*, vol. 1, pp. 1069–1074, Oct. 1981.

L. Stark, *Neurological Control Systems*. New York: Plenum, 1986.

D. D. Vaccaro, G. C. Agarwal, and G. L. Gottlieb, Nonlinear mechanical behavior in striated muscle and its relationship to underlying crossbridge activity, *IEEE Trans. Biomed. Eng.*, vol. 35, pp. 426–434, June 1988.

J. M. Winters and L. Stark, Analysis of fundamental human movement patterns through the use of in-depth antagonistic muscle models, *IEEE Trans. Biomed. Eng.*, vol. BME-32, pp. 826–839, Oct. 1985.

J. W. Woodbury, A. M. Gordon, and J. T. Conrad, Muscle, in *Neurophysiology*, 2d ed., eds. T. C. Ruch et al. Philadelphia: Saunders, 1965.

S. Yurkovich, S. K. Hoffmann, and H. Hemami, Stability and parameter studies of a stretch reflex loop model, *IEEE Trans. Biomed. Eng.*, vol. BME-34, pp. 547–553, July 1987.

Problems

1. Given the schematic diagram of Fig. 8-2—but feedback axon A3 is severed in an accident—the command reference is a unit step, and load conditions remain the same as before, as given by Eq. (8-5). Find (a) the distance waveform; (b) the velocity waveform. (c) Plot the curves and compare with those of Fig. 8-3. [Ans.: (a) $x = 1 - 1.471\epsilon^{-0.4t} + 0.471\epsilon^{-1.25t}$; (b) $v = 0.5882(\epsilon^{-0.4t} - \epsilon^{-1.25t})$.]
2. Apply the Laplace transforms of $\epsilon^{-\alpha t} \sin \beta t$ and $\epsilon^{-\alpha t} \cos \beta t$ to $x(t)$ of Eq. (8-11) to verify that $x(s)$ is obtained.
3. Carry out the step-by-step derivation of Eqs. (8-10), (8-12), (8-13), and (8-14). Plot the curves.
4. Given the schematic diagram of Fig. 8-2, with axon A3 exciting the motoneuron via a junction weight of 7, the command reference is a unit step, and the load is given by $m = 0.5$, $k_d = 2$, and $k_r = 1$. Find (a) distance; (b) force; (c) velocity waveforms. (d) Plot the waveforms. [Ans.: (a) $x = 1 - \epsilon^{-2t}(0.5774 \sin 3.464t + \cos 3.464t)$; (b) $F = 8 - 7x$; (c) $v = 4.619\epsilon^{-2t} \sin 3.464t$.]
5. Carry out the step-by-step derivation of Eqs. (8-16) through (8-19). Plot the curves.
6. Carry out the step-by-step derivation of Eqs. (8-22), (8-23), and (8-24). Plot the curves.
7. Given the schematic diagram of Fig. 8-4, with axon A3 exciting the motoneuron via a junction weight of 7, and A4 exciting via a weight of 10 when the error exceeds 0.6, the command reference is a unit step, and the load is given by $m = 0.5$, $k_d = 2$, and $k_r = 1$. For the first mode, find (a) distance; (b) force; (c) velocity waveforms; (d) t = start of second mode. For the second mode find (e) distance; (f) force; (g) velocity waveforms. (h) Plot the waveforms. {Ans.: (a) $x = 0.6667[1 - \epsilon^{-2t}(0.3536 \sin 5.657t + \cos 5.657t)]$; (b) $F = 12 - 17x$; (c) $v = 4.243\epsilon^{-2t} \sin 5.657t$; (d) $t = 0.2257$ s; (e) $x = 1 + [\exp(-2t_1)][0.4 \sin 3.464t_1 - 0.6 \cos 3.464t_1]$; (f) $F = 8 - 7x$; (g) $v = [\exp(-2t_1)][1.279 \sin 3.464t_1 + 2.585 \cos 3.464t_1]$.}
8. Carry out the step-by-step derivation of Eqs. (8-26) through (8-29). Plot the curves.
9. Carry out the step-by-step derivation of Eqs. (8-32), (8-33), and (8-34). Plot the curves.
10. Carry out the step-by-step derivation of Eqs. (8-37), (8-38), and (8-39). Plot the curves.
11. Given the schematic diagram of Fig. 8-6, with axon A3 exciting the motoneuron via a junction weight of 7, A4 exciting via a weight of 10 when the error exceeds 0.6, and A5 inhibiting via a weight of -1 when the force exceeds 4, the command reference is a unit step, and the load is given by $m = 0.5$, $k_d = 2$, and $k_r = 1$. For the first mode, find (a) distance; (b) force; (c) velocity waveforms; (d) t = start of second mode. For the second mode, find (e) distance; (f) force; (g) velocity waveforms; (h) t = start of third mode. For the third mode, find (i) distance; (j) force; (k) velocity waveforms. (l) Plot the waveforms. {Ans.: (a) x

$= 0.8421[1 - \epsilon^{-2t}(0.5164 \sin 3.873t + \cos 3.873t)]$; (b) $F = 8 - 8.5x$; (c) $v = 4.131\epsilon^{-2t} \sin 3.873t$; (d) $t = 0.2811$ s; (e) $x = 1.333\{1 + [\exp(-2t_1)][0.0736 \sin 2.236t_1 - 0.7 \cos 2.236t_1]\}$; (f) $F = 6 - 3.5x$; (g) $v = [\exp(-2t_1)][1.891 \sin 2.236t_1 + 2.086 \cos 2.236t_1]$; (h) $t = 0.3640$ s; (i) $x = 1 + [\exp(-2t_2)][0.3393 \sin 3.464t_2 - 0.4286 \cos 3.464t_2]$; (j) $F = 8 - 7x$; (k) $v = [\exp(-2t_2)][0.8061 \sin 3.464t_2 + 2.032 \cos 3.464t_2]$.}

12. Carry out the step-by-step derivation of Eq. (8-42), and derive $\tau = 86.38$ ms, $f = 0.4818$ Hz.
13. Given the schematic diagram of Fig. 8-2, with axon A3 exciting the motoneuron via a junction weight of 4.445, the command reference r is zero, but the load conditions are given by Eq. (8-5). The muscle is on the verge of clonus oscillations. Find (a) the oscillation frequency; (b) the signal time delay between motoneuron and muscle; (c) the magnitude and phase of F/x. [Ans.: (a) $f = 0.1929$ Hz; (b) $\tau = 0.4618$ s; (c) $F/x = 4.445\underline{/115.9^\circ}$.]
14. Given the schematic diagram of Fig. 8-2, with axon A3 exciting the motoneuron via a junction weight of W, the command reference r is zero, but the load conditions are given by Eq. (8-5). The muscle is on the verge of clonus oscillations with a frequency $f = 1/\pi = 0.3183$ Hz. Find (a) the signal time delay between motoneuron and muscle; (b) the weight W; (c) the magnitude and phase of F/x. [Ans.: (a) $\tau = 0.189$ s; (b) $W = 9.621$; (c) $F/x = 9.621\underline{/136.7^\circ}$.]

9

The Auditory System

ABSTRACT. Sound is funneled into the ear canal, setting the ear drum into vibration. The movement is transmitted through the bones of the middle ear and then through the oval window of the inner ear (the cochlea), thereby setting fluid and the basilar membrane (BM) into vibration. The latter separates the sound crudely into its frequency components: if a single tone is applied to the ear, the BM vibrates with maximum amplitude at a particular location along its 3.5-cm (1⅜-in.) length. The input tone is called the *characteristic frequency*, f_c, for that location. The f_c varies from 15,000 Hz at the base (proximal end of the cochlea) to 20 Hz at the apex (distal end).

A typical sound consists of many frequency components. In that event the movements of the BM are complex, with each section vibrating primarily in accordance with the component that agrees approximately with the characteristic frequencies of that section.

The BM vibrations stimulate 12,000 "hair cell" sensory receptors consisting of one column of inner hair cells (IHCs) and (according to views held until 1980) three columns of outer hair cells (OHCs). In the absence of stimulation, a hair-cell neuron generates random, spontaneous APs. As the loudness of a source increases, if the BM vibration frequency is low, the neuron generates an increasing proportion of periods that are synchronized to the BM period. As the loudness of the source increases, if the BM vibration frequency is so high that the neuron cannot follow, the neuron generates an increasing proportion of periods that are synchronized with a *multiple* of the BM period.

For example, if the sound is a 500-Hz whistle at the level of normal conversation, *every* neuron whose characteristic frequency is above 170 Hz will fully synchronize (saturate), with periods of 2, 4, 6, . . . ms. If the amplitude is increased to the maximum tolerable level, every neuron above 130 Hz will saturate. The primary effect of increasing amplitude is that more low-frequency hair-cell neurons are recruited into synchronization.

It is conjectured that sharp tuning occurs because outputs of the 3000 IHCs in each acoustic nerve are filtered by 3000 reverberatory circuits, each of which is tuned to the f_c of its input IHC. The frequency to which a reverberatory circuit is tuned is determined by the time delay around its feedback loop. The latter is formed by an afferent axon feeding a neuron in the cochlear nucleus, followed by an efferent axon that returns to the OHC which acts, in turn, as a muscle upon the BM.

It is also conjectured that a two-dimensional map is generated in the auditory cortex such that one dimension corresponds to frequency and the other to amplitude. For a complex sound, the map pinpoints each frequency component and its amplitude, thus creating a unique signature of the sound.

9-1 Anatomy of the Ear

In this chapter we consider the sensory receptors that are responsible for the reception of sound, and some of the preceding as well as subsequent processing steps. Like the touch receptors in the skin, the ear's receptors are sensitive to changes in pressure. For the ear, however, evolution has provided an unbelievably ingenious solution to the problem of transforming pressure changes in air into pressure changes in the fluid-filled inner ear. Accordingly, the first half of the chapter is devoted to this transformation apparatus. Although a narrow viewpoint would argue that the structures preceding the sensory receptors are not strictly part of the nervous system, we champion a broader viewpoint: that the mechanical structures responsible for excitation of the sensory receptors are a legitimate part, indeed they happen to be an especially interesting part, of the nervous system [J. M. Miller and A. L. Towe, 1979; T. F. Weiss, 1966].

The physical embodiment of sound is the mechanical vibration of molecules. It is called a *longitudinal* vibration because the molecules move back and forth in the same direction as the propagation of the sound. Each individual molecule undergoes a chaotic, random motion, in accordance with its temperature, as it is buffeted about by its neighbors; any additional motion due to sound is superimposed on the thermal motion. Sound propagates as a molecule moves, say, to the right; this compresses the material to the right, and is analogous to a spring being compressed. Expansion of the spring is resisted by the inertia (mass) of the material to the right. The rate at which the imaginary spring compresses and expands (the velocity of sound) is determined by the compressibility and density of the material. In an explosion, the shock front propagates at the speed of sound (in air at sea level at "room" temperature, around 340 meters/second (m/s) = 1100 ft/s = 750 mph = Mach 1).

A complex steady-state sound can be broken down into a few fundamentals and their *overtones* or *partials*. A *partial* is an approximate integer multiple of its fundamental—"approximate" because vibrating elements and air cavities have thickness as well as length. A *harmonic*, on the other hand, is an exact integer multiple of its fundamental. The *pitch* of a complex sound is the equivalent fundamental frequency as judged by a normal ear. The *timbre* of a complex sound is the quality that distinguishes it from other sounds of the same pitch because of the amplitudes and frequencies of its partials. *Tone* is often synonymous with timbre, but amongst engineers, and in this chapter, it is synonymous with fundamental frequency.

The anatomical elements involved in the mechanical processing of sound by the human ear are depicted in Fig. 9-1. External sound is funneled into the ear canal, to an appreciable degree, by the pinna. The vibrating air sets the ear drum into vibration; the mechanical resistance of the ear drum is similar to that of the ear canal, so it vibrates with approximately the same

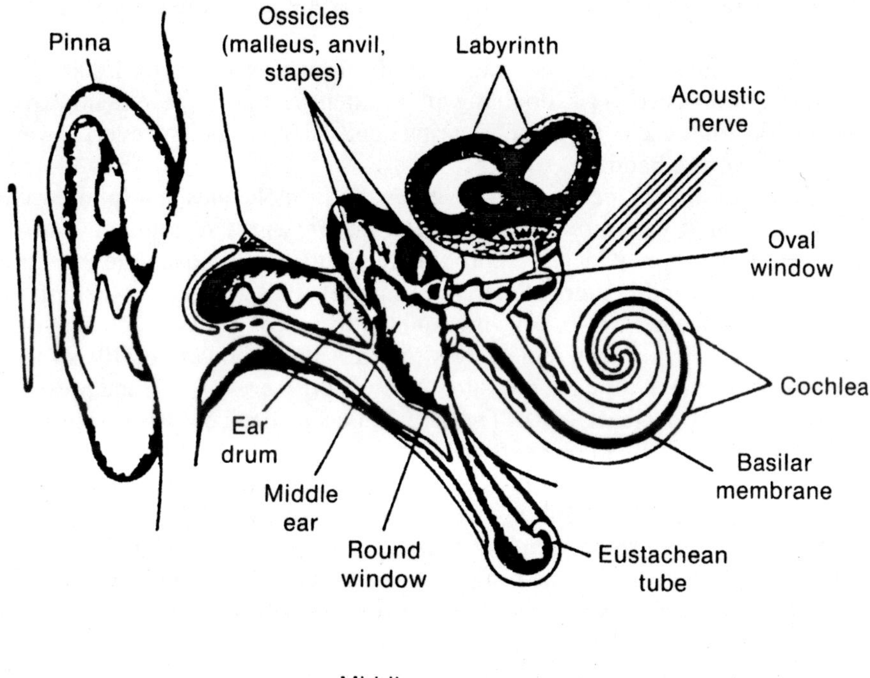

Fig. 9-1 A cross section showing main anatomical features of the human ear. Wiggles and small arrows indicate vibratory pathway: pinna, ear canal, ear drum, malleus, anvil, stapes, oval window, inner cochlear spiral, basilar membrane, outer cochlear spiral, and round window. The labyrinth is part of the balancing machinery of the body and is not involved with audition. (From Schroeder, Models of hearing, *Proc. IEEE*, vol. 63, p. 1333, 1975.)

magnitude as that of the air molecules [P. Dallos, 1973; M. R. Schroeder, 1975].

The magnitude can be unbelievably small. Indirect measurements show that a human with normal hearing ability can begin to detect sound when the peak displacement is 0.1 Å or 10^{-9} cm; that is, when the peak air movement is 0.1 Å to either side of the neutral position. By way of comparison, the diameter of a water molecule is 3 Å!

At the other end of the scale, measurements show that the sound amplitude can be increased by a factor of 10^7 before damage to the ear may result. In other words, the peak displacement of air molecules corresponding to the strongest tolerable audio signal is 0.1 mm. We evolved under natural conditions in which a peak displacement of 0.1 mm was rarely exceeded, but the modern military–industrial complex (and perhaps some of our so-called musical experiences) are full of loud blasts that subject nearby humans to damaging noise if they are not wearing ear plugs. Excessive environmental noise is responsible for a loss of hearing acuity among relatively young people.

The range of human hearing extends from around 20 to 15,000 Hz. Because one octave is a doubling of frequency, the range extends over log(15,000/20)/log 2 = 9.55 oct. Hearing acuity decreases, however, toward both ends of the band.

Because air has inertia, the movement of air molecules in a sound wave is accompanied by vibratory changes in air pressure. A common unit of pressure is the bar, which is approximately equal to atmospheric pressure at sea level (760 millimeters of mercury (mmHg) = 29.9 in.Hg = 14.7 lb/in.^{2}). A peak displacement of 0.1 Å corresponds to a peak pressure change of 3×10^{-10} bar. It is inconvenient to deal with these powers of 10, so it is customary to use a different measure of sound pressure—the sound pressure level (SPL) in decibels (dB). The SPL value in decibels is 20 times the logarithm, to base 10, of pressure value relative to that of the weakest audible sound. The weakest audible pressure change, 3×10^{-10} bar, therefore corresponds to 0 dB SPL. Multiplying 3×10^{-10} bar by 10^7 gives us the peak pressure change corresponding to the strongest tolerable audio signal, 3×10^{-3} bar or 140 dB SPL. The peak pressure change corresponding to normal human conversation is around 10^{-6} bar, or 70 dB SPL. The preceding can be summarized as follows:

$$\begin{aligned}\text{Weakest audible peak pressure change} &= 3 \times 10^{-10}\ \text{bar}\\ &= 0\ \text{dB SPL}\\ \text{Normal conversation peak pressure change} &= 1 \times 10^{-6}\ \text{bar}\\ &= 70\ \text{dB SPL}\\ \text{Strongest tolerable peak pressure change} &= 3 \times 10^{-3}\ \text{bar}\\ &= 140\ \text{dB SPL.}\end{aligned}$$

Returning to Fig. 9-1, vibrations of the ear drum are transmitted through the three ossicle bones to the *oval window*. The ossicles serve two purposes: first, they act as a lever, increasing the force by a factor of 1.3 (and decreasing displacement by a factor of 1.3); second, at high sound intensities, feedback causes muscles attached to the ossicles to attenuate the incoming vibrations. In this chapter the reasonable assumption is made that, starting with 100 dB SPL, the ossicular attenuation (which is called the *acoustic reflex*) amounts to 5 dB for every 10 dB of SPL input. An input of 140 dB SPL therefore receives 20 dB of acoustic reflex attenuation (sometimes called *compression*) [E. Borg and S. A. Counter, 1989].

The acoustic reflex is similar to the tendon organ feedback of Chapter 8, and it can fail for the same reason: it takes 10 ms for the ossicular attenuation to be activated. A sudden blast catches the ear "off guard," and can cause permanent damage. On the other hand, to prevent self-inflicted damage due to your own voice, the reflex is automatically activated *before* you speak.

Skin is the classic interface between air and liquid media; in the ear, the oval window of Fig. 9-1 is a thin skin through which vibrations of the stapes (one of the middle ear ossicles) on the left, in air, are transmitted to cochlear fluid on the right. The mechanical characteristics of the cochlear fluid are similar to those of water.

The cochlea is in the shape of a 2½-turn spiral, as shown. In humans, the total length of the spiral is 3.5 cm (1⅜ in.); the spiral shape is an example of packaging that yields a compact structure.

The cochlea is a crude mechanical filter that separates a complex incoming sound into its frequency components: high frequencies in the *base* of the spiral, near the oval window, and gradually into low frequencies in the *apex*, toward the end of the spiral. The path of the sound wave in Fig. 9-1 is depicted by means of wiggles. From the oval window the wiggles spiral counterclockwise inwardly, then unwind clockwise outwardly, ending at the *round window*. The wiggles do not quite accurately describe the sound wave, but they do bring out an important point: cochlear fluid, like water, is practically incompressible. When the oval window moves to the right, the round window is forced to move to the left. The round window is a thin skin that separates the cochlear fluid on the right, in Fig. 9-1, from air on the left. One would think that for the weakest sounds, where the peak movement of the oval window is less than one angstrom, the round window need not be flexible. Alas, a calcified, rigid round window is a relatively common cause for hearing deficiency. The cochlea is embedded in dense bone, and the round window is the only low-resistance outlet; besides, proper stimulation of the sensory receptors, called *hair cells*, requires that the sound-wave path return by way of the round window.

The vibrations are analyzed by 12,000 hair cells, in each ear, that are distributed along the 3.5 cm of cochlea length. These sensory hair cells are carried by the basilar membrane. They are too small to be shown in

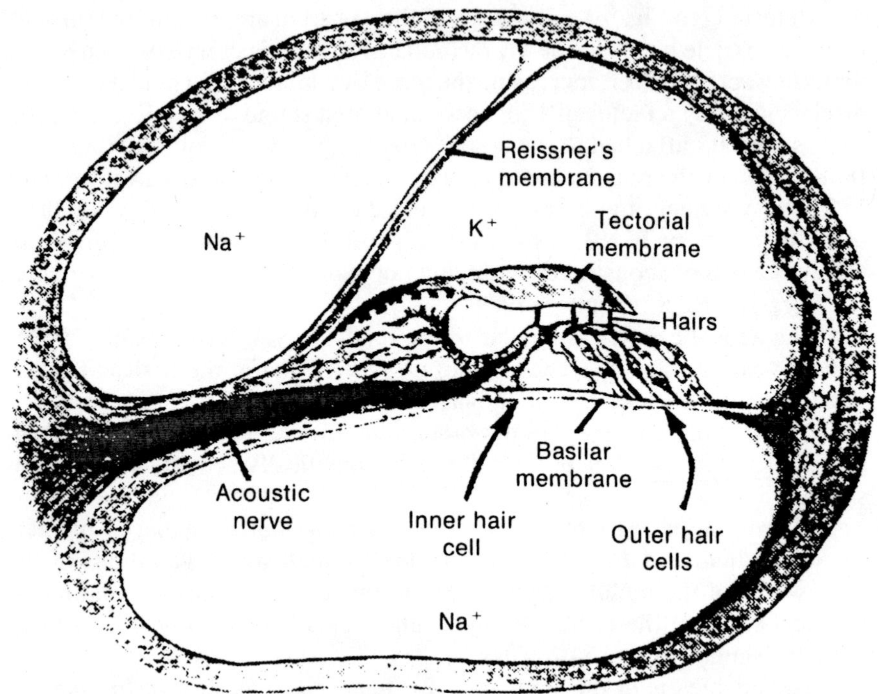

Fig. 9-2 Cross section taken through the cochlea. Its diameter is around 2 mm. Width of the basilar membrane varies from 0.1 mm at the high-frequency end of the cochlea to 0.5 mm at the low-frequency end. Dominant positive ion is indicated by Na^+ or K^+. (From Schroeder, Models of hearing, *Proc IEEE*, vol. 63, p. 1333, 1975.)

Fig. 9-1, but they are shown in a magnified cross section of the cochlea, Fig. 9-2. At the top of each cell are hairlike tufts that bridge the gap between the cell and tectorial membrane. The tectorial membrane also extends along the length of the cochlea. Like any of the other sensory receptors in the body, the hair cells of the ear have to operate in a liquid medium. They have to receive nutrients, and various ions have to diffuse from one location to another in order to generate graded and action potentials. Each hair cell is coupled to a neuron that generates action potentials (APs) that leave by way of the acoustic nerve.

There is one column of inner hair cells (IHCs) and three columns of outer hair cells (OHCs), as the cross section implies. The IHCs are on the side of the cochlear spiral toward the center, while the OHCs are on the outer side of the arc. Each column is 3000 hair cells long. Assuming that the rows are evenly spaced, the spacing between them is 3.5 cm/3000 rows $\cong$ 10 μm, which agrees with the anatomical measurement. The hair cells and their neuron axons are carried by, and form part of, the basilar membrane

(BM). In the absence of sound the discharge of a typical neuron is a spontaneous, relatively low-frequency sequence of random APs, such as those of Fig. 3-3(*a*). When the BM swings upward (toward the tectorial membrane) during a vibration cycle, the hairs are bent; this is the mechanical stimulus for generation of APs by the associated neuron. Presumably, hair bending causes changes in cell membrane permeability such that the resulting AP frequency momentarily increases. When the BM swings downward (away from the tectorial membrane), the hairs are unbent and the AP frequency momentarily decreases. More is said about these AP discharges later on in the chapter [A. Flock and S. Orman, 1983].

The labyrinth in Fig. 9-1 is part of the balancing machinery of the body and is not involved with audition.

The eustachean tube in Fig. 9-1 is an air connection to the pharynx; it is normally closed (undoubtedly to keep out infectious organisms), but it opens during swallowing and yawning. It is another example of an organ that has not kept pace with the demands of modern civilization. The eustachean tube is meant to supply or remove air from the middle ear (between the ear drum and cochlea) in response to *slow* changes in barometric pressure. It frequently cannot cope with rapid, large changes in pressure. When you ascend in an airplane, the pressure in the outer ear (and pharynx) decreases; the middle ear then forces excess air out through the eustachean tube. But when the airplane prepares to land, the pressure in the outer ear and the pharynx increases rapidly; if you do not swallow or yawn, a partial vacuum is created in the middle ear. In that event the ear drum in Fig. 9-1 bulges to the right; it can no longer vibrate freely, and considerable loss of hearing results. For many people, swallowing and/or yawning does not help because the eustachean tubes remain clogged, and bulging of the ear drums is accompanied by painful side effects.

[S.D. writes: "I am one of those who become partially deaf without pain. The condition is relieved instantly by pinching my nostrils closed and squeezing air into the nasal passages; the high pressure pries the eustachean tubes open. Whenever I travel from Jerusalem (half mile up) to Tel Aviv (sea level) I arrive quite deaf. It is fun to be immersed in this strange world in which the harsh noises of a large city are diluted. It wears off, of course, in an hour or two. The return trip to Jerusalem is punctuated by sudden changes in perceived sound as bubbles of air percolate back from the middle ear to the pharynx."]

In Fig. 9-2, the clockwise and counterclockwise wiggles of Fig. 9-1 are separated by the tectorial and basilar membranes. To a first approximation, the tectorial membrane is stationary while the BM vibrates up and down like a stretched rubber band. The dimensions are quite small: the diameter of the cochlea is around 2 mm, while the width of the BM varies from 0.1 mm at the high-frequency end (the oval and round windows) to 0.5 mm at the low-frequency end (the inner spiral).

Reissner's membrane is analogous to a cell membrane in that the region between it and the BM has a relatively high concentration of potassium ions, like the inside of a cell, as indicated. In Fig. 9-2, the regions to the left of Reissner's membrane and below the BM contain a relatively high concentration of sodium ions, like the outside of a cell. The ion separation is important for the generation of APs.

Reissner's membrane, as well as all of the other structures in the cochlea, determine its mechanical and hydrodynamic characteristics such that the BM responds to high-frequency components at the base end of the spiral, near the oval window, and to low-frequency components at the apex end.

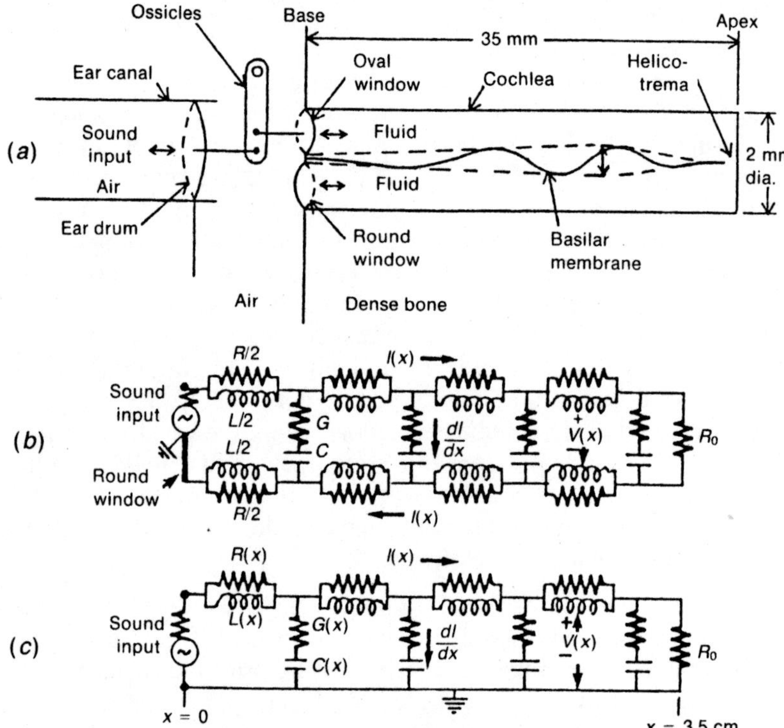

Fig. 9-3 Idealized salient mechanical features of the ear, and development of the electrical equivalent of the cochlea. (*a*) Mechanical model (not to scale), including ear canal and middle ear. The cochlear spiral is unwound and shown as a cylinder. Its actual length is 35 mm; cross section is shown in Fig. 9-2. Tectorial membrane and hair cells are omitted, but typical instantaneous up-and-down motion of the basilar membrane is shown (highly exaggerated); limits of the motion are indicated by a dashed envelope. (*b*) Electrical equivalent of part (*a*). (*c*) With lower half of part (*b*) swung into and combined with upper half so as to simplify the transmission-line model. (From Deutsch and M-Tzanakou, *Neuroelectric Systems*, New York Univ. Press, New York, 1987.)

The salient mechanical features of the ear are presented, in simplified form and not to scale, in Fig. 9-3(*a*). The ossicles appear as a lever rotating back and forth around a pivot, the lever arm lengths having a ratio of 1.3 as drawn. The cochlear spiral is unwound and shown as a cylinder, 2 mm in diameter and 35 mm long. Inside, in response to a typical sound input, the BM undulates up and down very much like water waves at the edge of a beach. At the very end of the cochlea there is a small hole, the *helicotrema*, which allows the pressure to slowly equalize between the upper and lower halves of the "cylinder."

The ear is primarily a mechanical device, with force broken up into three components in series: inertia (ma); a frictional force proportional to velocity ($k_d v$); and a spring that tries to return the load to $x = 0$ ($k_r x$). The relationship is given in Eq. (8-3) as $F = ma + k_d v + k_r x$, where this is applied to a muscle moving a load. In the present chapter, however, we are dealing mostly with a mechanical–hydrodynamic transmission line. It is easier to represent the line via an electrical circuit analogy in which inductance L, resistance R, and capacitance C are in series. With a trivial change (dv/dt in place of acceleration a), the analogy becomes:

$$\begin{aligned} F &= m\frac{dv}{dt} + k_d v + k_r x, \\ V &= L\frac{dI}{dt} + RI + \frac{q}{C}. \end{aligned} \tag{9-1}$$

The mechanical versus equivalent electrical parameters are summarized in Table 9-1. (Another analogy, based on $I = C\,dV/dt + GV + 1/L\int V\,dt$, is also possible, but is less convenient than the one given in Eq. (9-1) and Table 9-1 [W. Welkowitz, S. Deutsch & M. Akay, 1992].)

By way of further familiarization, consider next the excitation in the ear canal and cochlea when a single tone (frequency) at 1160 Hz at the level of normal conversation, 70 dB SPL, is applied. We are using 1160 Hz because the BM then vibrates with maximum amplitude at a convenient central location, 2 cm away from the base. The *peak* values are given in Table 9-2 as

TABLE 9-1 Mechanical versus Equivalent Electrical Parameters

Mechanical Parameter	Electrical Analogy
Mass, m	Inductance, L
Viscous resistance, k_d	Resistance, R
Compliance, $1/k_r$	Capacitance, C
Displacement, x	Charge, q
Velocity, v	Current, I
Force, F	Voltage, V
$F = m\frac{dv}{dt} + k_d v + k_r x$	$V = L\frac{dI}{dt} + RI + \frac{q}{C}$

TABLE 9-2 Peak Idealized Values[a] of Excitation in Ear Canal and Cochlea

1. Air pressure change	= 1 μbar
2. Characteristic R_0 of air	= 415 kg/s/m^2
3. Velocity of air change	= 0.021 cm/s
4. Displacement of air	= 0.03 μm
5. Characteristic R_0 of canal	= 0.012 Ω
6. Force at ear drum	= 2.5 μN
7. Force at oval window	= 3.3 μN
8. Velocity of oval window	= 0.017 cm/s
9. Characteristic R_0 of cochlea	= 0.02 Ω
10. Pressure at oval window	= 22 μbar
11. Velocity of BM at x = 2 cm	= 0.04 cm/s
12. Displacement of BM at x = 2 cm	= 0.06 μm

[a] These values occur when input is at normal conversation level (SPL = 70 dB) at 1160 Hz.

follows (most of the values in the table are rounded off to one- or two-place accuracy):

Row 1: At sea level, the static air pressure is around 1 bar, but propagation of the 1160-Hz sound wave is marked by increases and decreases around the static value. It has been experimentally determined that 0 dB SPL corresponds to a pressure change of 2×10^{-10} bar root mean square (rms). Then 70 dB SPL corresponds to a pressure change of 6.325×10^{-7} bar rms = 8.944×10^{-7} bar peak = 8.944×10^{-2} pascals peak. (The pascal (Pa), the MKS unit for pressure, is named after Blaise Pascal [1623–1662]. One bar = 100,000 Pa.)

Row 2: For a lossless electrical line, characteristic resistance is given by $R_0 = \sqrt{L/C} = Lv_p$ where v_p is the velocity of propagation. For a mechanical transmission line, the analogous expression is $R_0 = \rho_D v_s$, where ρ_D is density (analogous to L) and v_s is the velocity of sound. For air, the substitution of ρ_D and v_s values yields $R_0 = 415$ kg/s/m^2 at "room" temperature.

Row 3: The "velocity of air change" is the peak back-and-forth velocity as the sound wave passes. According to Table 9-1, $R_0 = V/I$ of an electrical system becomes $R_0 = F/v$ in a mechanical system. Then $v = F/R_0 = 2.155 \times 10^{-4}$ m/s peak.

Row 4: The corresponding displacement of air is given by $v/2\pi f = 2.957 \times 10^{-8}$ m peak.

Row 5: The value $R_0 = 415$ kg/s/m^2 of air is for a column whose cross-sectional area is 1 m^2. It is assumed that the human ear canal has a diameter of 6 mm. Its characteristic resistance, therefore, is 0.01173 Ω.

Row 6: The motion of air molecules of course exerts a force on the ear drum. In this case $F = vR_0 = 2.529 \times 10^{-6}$ newton (N) peak. (The newton is named after Sir Isaac Newton [1642–1727]. One newton is approximately equivalent to 0.1 kg or 0.22 lb.)

Row 7: Because of the ossicular lever, the force at the oval window is 3.288×10^{-6} N peak.

Row 8: Because of the ossicular lever, the velocity of the oval window is velocity of air change/1.3 $= 1.658 \times 10^{-4}$ m/s peak.

Row 9: From $R_0 = F/v$, the characteristic resistance of the cochlear line is 0.01983 Ω. This is 1.3^2 times R_0 of the ear canal.

Row 10: Pressure = force/area. For the latter it is assumed that the area at the base of the cochlea is 3 mm^2. Since only half of this area is exposed to the oval window, we have 2.192 Pa peak $= 2.192 \times 10^{-5}$ bar peak. The increased pressure in the cochlea, 21.92 μbar compared to 0.8944 μbar outside, a ratio of 24.5, is mainly due to the small cross-sectional area of the cochlea compared to the area of the ear drum.

Row 11: Because of a weak resonance effect, the velocity at the 2-cm location, for 1160 Hz, is around 2.5 times the input velocity, $\cong 4 \times 10^{-4}$ m/s peak.

Row 12: Dividing v by $2\pi f$, we get $\cong 6 \times 10^{-8}$ m peak.

Figures 9-1 to 9-3 and Table 9-2 describe mammalian ears in general, although smaller animals are tuned to higher frequencies than humans because their ears are assembled out of smaller structures. The anatomy of the mammalian ear is complicated; it has evolved over many millions of years. Many things can go wrong, and do, as evolution continues to fine-tune succeeding generations.

9-2 Single-Tone Response of the Basilar Membrane

In this section we concentrate on the BM and its hair cells, but with a single-frequency (single-tone) input in order to simplify the discussion insofar as possible. For every BM location, there is a *single-tone* input, for which the vibration amplitude is maximum. This input is called the *characteristic frequency*, f_c, for that location.

Ever since the microscope revealed two sets of hair cells, neuroanatomists have wondered why the ear requires split sensory arrays. At the same time, especially since 1980, some amazing phenomena have been observed. It seems that the IHCs are the primary sensory receptors [A. J. Hudspeth, 1983], but many if not all of the OHCs are tiny muscle fibers [W. E. Brownell and B. Kachar, 1986; A. Flock, 1983; H. P. Zenner, U. Zimmerman & U. Schmitt, 1985]. The OHC motor units respond to APs that are brought in by efferent fibers from the auditory centers of the brain. Five of the phenomena are briefly reviewed as follows:

Tip Enhancement: A typical tuning response curve of the BM is shown in Fig. 9-4. Here, in an experiment, a vibration-measuring device is installed at the f_c = 4000-Hz location of the cochlea, while a tiny earphone feeds a single-tone audio input to the stapes. The animal model is usually a cat, monkey, guinea pig, chinchilla, or gerbil. Starting with a low input frequency f_{in}, say: as the frequency of f_{in} is increased, the vibration amplitude at the 4000-Hz location increases at a 6-dB/oct rate. In the vicinity of 4000 Hz f_{in} there is a sharp increase in amplitude response, as if caused by a high-Q tuned circuit. This adds a "tip" to the response curve of Fig. 9-4, so it is customary to call it "tip enhancement." Above 4000 Hz f_{in} the amplitude drops off at an unbelievably high rate—minus 300 dB/oct [W. S. Rhode, 1971; P. M. Sellick, R. Patuzzi & B. M. Johnstone, 1983]. A transmission-line model of the cochlea, considered later on in this section, does not display tip enhancement, and the dropoff above f_c is only around -100 dB/oct (at 60 dB down). Tip enhancement is explained by *positive* feedback; that is, the OHCs are part of a feedback loop in which efferent signals, as previously mentioned, are fed back to the cochlea. A hypothetical model, which seeks to explain frequency discrimination in the auditory system, is considered in Section 9-5.

Compression: The feedback system is nonlinear. Tip enhancement can amount to 40 dB at low SPLs, but it is "compressed" to practically zero at an SPL of 140 dB. This is in addition to 20-dB compression introduced by the bones of the middle ear. Therefore, as the SPL is increased from 0 to 140 dB, a factor of 10^7, the BM vibration amplitude at f_c only increases by 80 dB, a factor of 10^4 [H. Duifhuis, 1989].

Spontaneous Emission: About half of the population have the somewhat amusing characteristic that they emit sounds from their ears under very quiet conditions! For these people, at a few frequencies in the mid-frequency range, the previously mentioned 40-dB enhancement of tips can become infinite, and the system oscillates. Without external stimulation, one or more

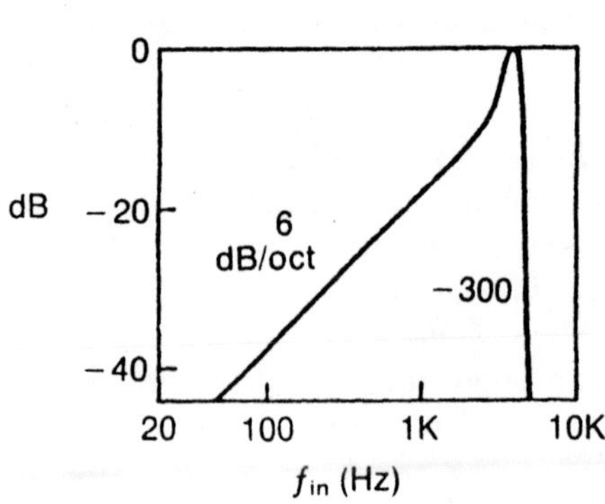

Fig. 9-4 Typical tuning response curve of the BM as measured at the characteristic frequency f_c = 4000 Hz location of the cochlea. The sharp increase in response in the vicinity of 4000 Hz is called "tip enhancement." (From Deutsch and M-Tzanakou, *Neuroelectric Systems*, New York Univ. Press, New York, 1987.)

OHC "muscles" pull so strongly on the BM, at an audible rate, that very weak sound is transmitted backwards to the ear drum. (Try to listen in on your neighbor's ear!) The phenomenon is called *spontaneous otoacoustic emission* [D. T. Kemp, 1978].

Combination Tones: Since the system is nonlinear, with single-tone stimulation we expect harmonics to be excited. With two-tone stimulation f_1 and f_2, we also expect sum and difference frequencies to be excited. We do not expect nonlinear effects to occur, however, under low-SPL conditions. With active feedback to the OHCs, nonlinearities can somehow show up also under low-SPL levels. An important class of resulting distortion products are the combination tones $2f_1 - f_2$, where $f_2 > f_1$. For example, if $f_1 = 750$ Hz and $f_2 = 1000$ Hz, the combination tone 500 Hz is generated. You can actually hear the tone, and it is physically present since it can be canceled by an external audio signal of suitable frequency, magnitude, and phase [J. L. Goldstein, 1967; G. F. Smoorenburg, 1972].

Two-Tone Suppression: Start with input f_1, and then add f_2. In a linear system, you will hear both of these tones. In the actual system, there are a great many data that record conditions under which the stronger input suppresses the weaker. Perhaps some form of lateral inhibition acts to suppress the weaker components. If $f_2 > f_1$, suppression is somewhat different from what it is for $f_2 < f_1$ [J. L. Hall, 1974; R. R. Pfeiffer, 1970; W. S. Rhode, C. D. Geisler & D. T. Kennedy, 1978].

The preceding five effects are complex, and are the basis for a great deal of research and many thesis topics. Julius L. Goldstein (1989), for example, has proposed a multiple-bandpass-nonlinearity (MBPNL) model that accounts for much of the data (but without pinpointing the physical processes that are involved). In this chapter we espouse the philosophy that one must understand the linear system before undertaking the nonlinear model. Accordingly, since this book is an introductory text, *only the linear model*, without OHC feedback (except for Sect. 9-5) is considered in what follows; the nonlinear effects are largely ignored.

Figure 9-5 shows a "motion-picture" sequence (frames frozen at equal intervals of time) of BM vibration during one audio cycle at a frequency of 1160 Hz. The cycle is broken up into 12 frames, each 30° apart. The full cycle takes place in 0.86 ms, so the frames are separated from each other in time by 0.072 ms. The thin outline in each frame is the vibration envelope, the outer limit reached as the BM vibrates; as the thin outlines show, maximum vibration amplitude occurs at the 2-cm location.

To get a "feel" for what is happening, let us follow a peak of the BM wave as it travels to the right. The locations of the positive peaks in each frame are indicated by black dots. Starting in the 0° frame at BM distance = 0, the relative BM displacement is at 0.37; that is, as a result of oval window movement at this instant, the BM at the base has reached its maximum positive movement with respect to its resting position. At first the peak moves rapidly to the right, more slowly later on. In the 30° frame it is

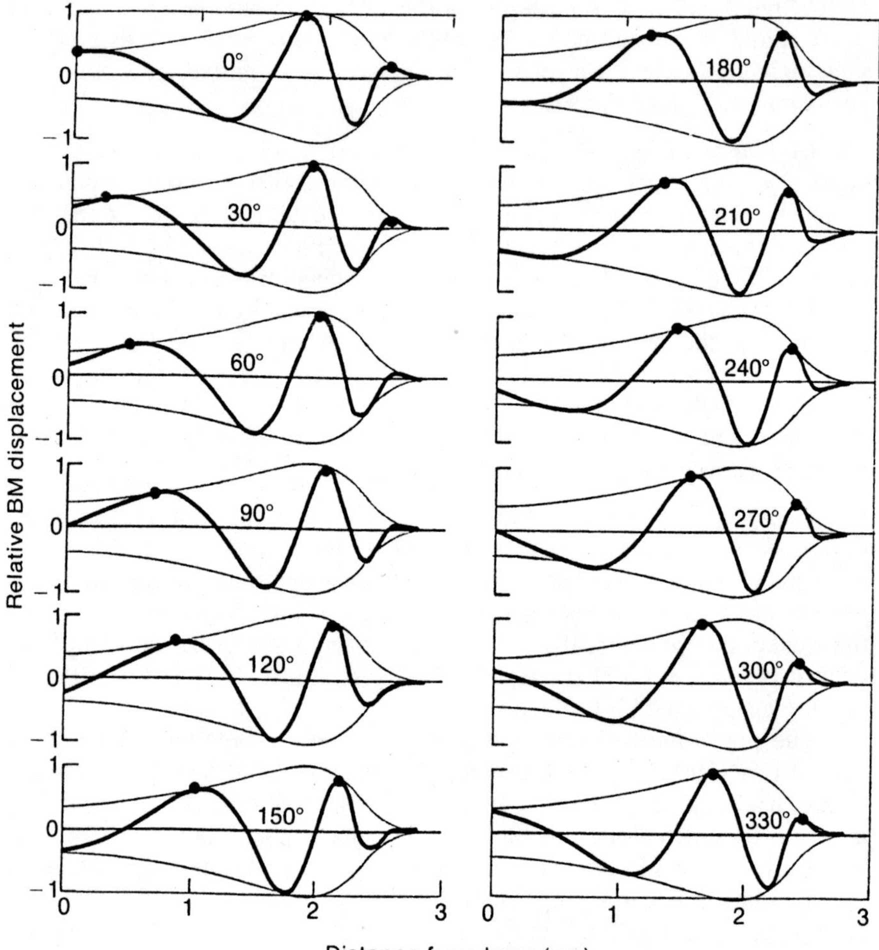

Fig. 9-5 "Motion-picture" sequence (that is, frames frozen at equal intervals of time) showing BM vibration during one cycle of an 1160-Hz audio input, tremendously magnified in the vertical direction. The cycle is broken up into 12 frames, each 30° apart. The thin outline in each frame is the vibration envelope, the outer limit reached as the BM vibrates. The locations of the positive peaks in each frame are indicated by black dots, and corresponding values 90° apart are entered in Table 9-3. The base is defined in Fig. 9-3(*a*). (From Deutsch and M-Tzanakou, *Neuroelectric Systems*, New York Univ. Press, New York, 1987.)

at BM distance = 0.27 cm, displacement = 0.42; in the 60° frame it is at BM distance = 0.50 cm, displacement = 0.48; and so forth. Movement of the peak is a "traveling wave." The BM only displaces up and down, of course, but the illusion of a wave traveling to the right is created. Exactly the same process occurs when a lake ripples in response to a stone thrown into it. In Fig. 9-5, in the 180° frame, the peak that we have been following is at BM distance = 1.19 cm, displacement = 0.73. If you follow this peak as it travels to the right, it continues again in the 0° frame, at BM distance = 1.85 cm, displacement ≅ 1. (The time is at t = 0.86 ms.) The peak reaches maximum displacement at BM distance = 2 cm in the 60° frame; after this it starts to fall as it moves to the right. After traveling for a bit over two cycles, the peak that we have been following dies out at BM distance = 2.6 cm. The traveling-wave peak locations and displacements at 0°, 90°, 180°, . . . are listed in Table 9-3.

Notice that, although the cochlea is 3.5 cm long, practically nothing remains of an 1160-Hz audio tone past the 2.6-cm location. The curves show that the BM is a crude filter. A good filter would vibrate with a sharp maximum at 2 cm from the base for an 1160-Hz tone; it would vibrate with a sharp maximum at 1 cm from the base for a 4700-Hz tone, and so forth. Instead, Fig. 9-5 shows that 75% of the total length is set into vibration by an 1160-Hz tone. Practically the full length vibrates for relatively low input frequencies.

Incidentally, each of the six waveforms at the right side of Fig. 9-5 is the inverse of the waveform at the left.

The curves of Fig. 9-5 are magnified tremendously in the vertical direction. At the level of normal conversation, according to Table 9-2, the peak displacement of the BM is 0.06 μm relative to the cochlea length of 3.5 cm, a ratio of 580,000 to 1. In Fig. 9-5 the ratio is only 7 to 1. Despite the minuscule BM vibration magnitude, painstaking work with microscopes and other sophisticated devices has enabled research workers to actually measure the displacements. The trailblazer was Georg von Békésy (1899–1972) who was awarded the Nobel Prize in medicine and physiology in 1961 [G. von Békésy, 1960].

The traveling wave of Fig. 9-5 is of no importance; it is the *vibration*

TABLE 9-3 The Traveling Wave of Fig. 9-5

	Peak is at	
Degrees of Cycle	Distance from Base (cm)	Relative BM Displacement
0	0	0.37
90	0.70	0.55
180	1.19	0.73
270	1.56	0.90
0	1.85	1
90	2.07	0.95
180	2.25	0.75
270	2.40	0.45
0	2.53	0.18

envelope, the maximum displacement of the basilar membrane, that is significant because it is this that stimulates the hair cells as they bend and unbend against the tectorial membrane.

Figure 9-6 is a plot showing all of the characteristic frequencies, f_c, from 15,000 Hz to 20 Hz, versus BM distance from the base, for the human ear [J. J. Zwislocki, 1965]. Figure 9-7 shows four of the vibration envelopes as follows (each of these should resemble the thin outline of Fig. 9-5, but it is customary to show only the upper half of the envelope since the lower half is symmetrical and contributes nothing new):

15,000 Hz: This is f_c at the base of the cochlea.

4700 Hz: As Fig. 9-6 indicates, this is f_c at BM distance = 1 cm from the base.

1160 Hz: This curve is a repeat of the vibration envelope of Fig. 9-5.

170 Hz: According to Fig. 9-6, this is f_c at BM distance = 3 cm from the base. Because of the crowding of low frequencies that is revealed in Fig. 9-6, the 170-Hz envelope falls off very steeply to the right of the peak at 3 cm.

In general, each of the vibration envelopes falls off steeply to the right of its peak. The falloff to the left is much gentler.

The curves of Fig. 9-7 represent an experimental scenario that is practically impossible to achieve. Suppose, for example, that we apply a tone

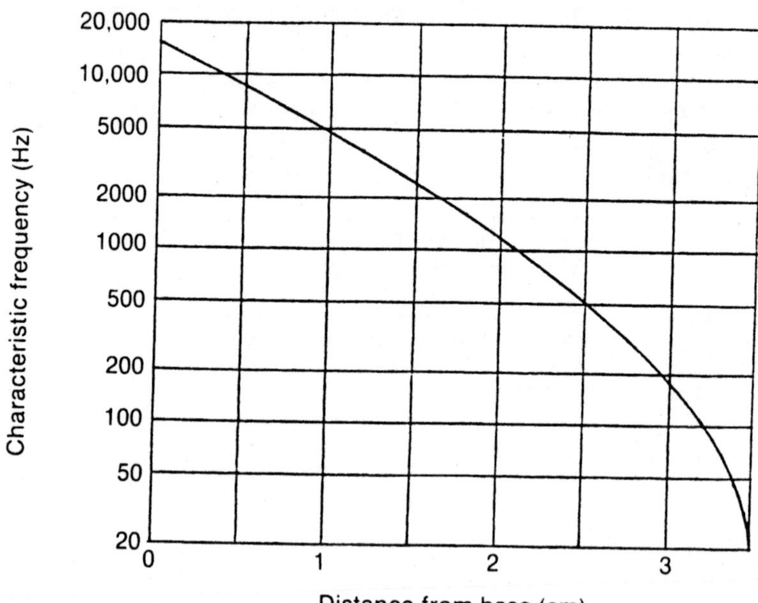

Fig. 9-6 Single-tone input at which BM vibration magnitude is maximum (characteristic frequency f_c) versus distance from the base of the cochlea.

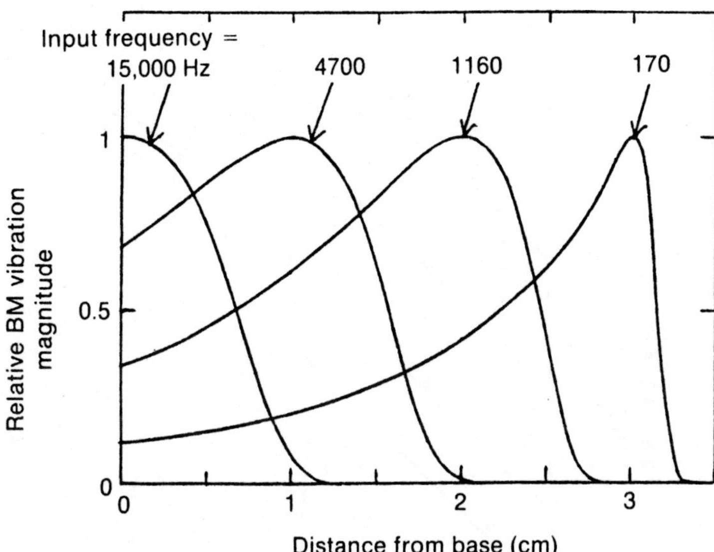

Fig. 9-7 Upper half of the envelope of BM vibration, versus distance from the base of the cochlea, for single-tone input frequencies at 170, 1160, 4700, and 15,000 Hz. These are the characteristic frequencies at distances 3, 2, 1, and 0 cm, respectively.

with a constant frequency (1160 Hz) to a small earphone in the ear of a suitable animal. To measure vibration envelope versus distance we would have to drill many tiny holes all along the cochlea. It is much easier to drill a *single* hole, at BM distance = 2 cm from the base, say, and measure BM displacement as the input frequency is increased from 20 Hz to 15,000 Hz. (Fortunately, the hearing apparatus functions normally while an animal is anesthetized.) This type of measurement yields the curve of Fig. 9-4. Similarly, Fig. 9-8 shows the results of making measurements at the 0-, 1-, 2-, and 3-cm locations. Notice that logarithmic scales are used in this figure.

It may seem that each curve in Fig. 9-8 should peak at f_c; that is, the "3-cm" curve should peak at 170 Hz, the "2-cm" curve at 1160 Hz, and so forth. This is only approximately correct. In Fig. 9-7 displacement is measured at a single frequency as *distance from the base* is varied, but in Fig. 9-8 it is measured at a single location as *input frequency* is varied, and these curves do not peak exactly at their f_c values. The "0-cm" curve, in fact, does not have a peak because, in effect, it is located directly at the input earphone that is vibrating the oval window. As the frequency fed to the earphone increases from 20 to 15,000 Hz, the BM vibration magnitude at the 0-cm location increases as shown.

It is obvious from Fig. 9-1, which represents cochlear morphology, that the cochlea is a transmission line. According to Table 9-1, inductance L is analogous to fluid mass, while capacitance C is analogous to the compliance

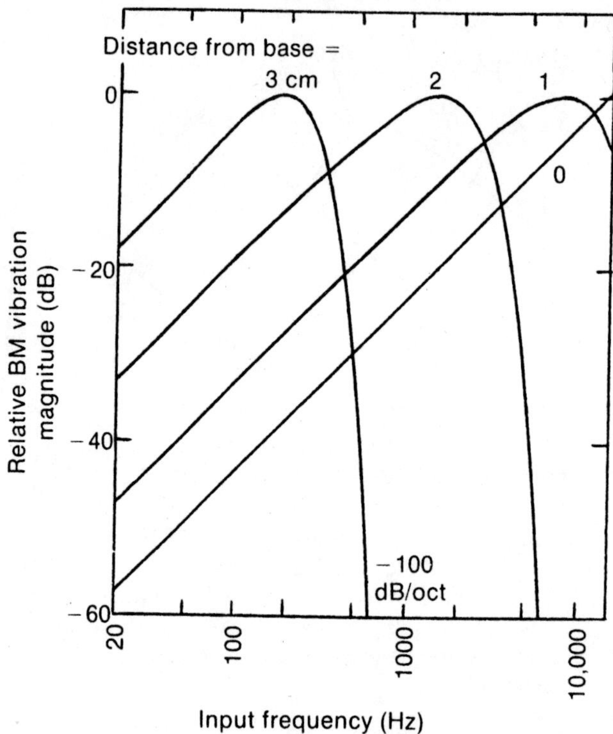

Fig. 9-8 Idealized envelope of BM vibration versus single-tone input frequency at distances 0, 1, 2, and 3 cm away from the base of the cochlea. The "0-cm" curve does not have a peak because, in effect, it is located directly at the input earphone feeding the oval window. The slope 60 dB down is around −100 dB/oct.

of the basilar membrane and resistance R is analogous to frictional losses of the hydrodynamic fluid and basilar membrane. The cochlea is not, however, an ordinary lossy transmission line. A conventional line that is terminated in its characteristic impedance is a "flat" line, devoid of resonance effects, whereas the cochlea is a frequency spectrum analyzer, because each section of the BM resonates—it tends to vibrate with maximum amplitude at its own characteristic frequency, f_c. Why is the cochlea line different? Because it is a *tapered* line; that is, the Ls, Cs, and Rs are functions of cochlear distance from the base, x. At the base, the high-frequency end, L and C are relatively small, but they become much larger as the apex, the low-frequency end, is approached [M. Furst and J. L. Goldstein, 1982].

The electrical equivalent of Fig. 9-3(a) appears in 9-3(b). Other configurations are possible, such as R in series with L, and G in parallel with C, but the network shown is simple and it yields a better fit to the experimental data. It is assumed that upper and lower halves of the mechanical transmission line are identical, so that each $L/2$ represents a differentially small section of the distributed line. Hydrodynamic fluid losses reside in each $R/2$. The compliance and losses of the basilar membrane are represented by C and G. The helicotrema is replaced by a terminating resistor R_0; whether or not this is accurate is mostly academic since, except for relatively low frequencies, a negligible portion of the incoming energy ever reaches the apical end of the cochlea. At any particular value of x, currents I (or fluid velocities v) are equal and opposite as shown [G. Zweig, R. Lipes & J. R. Pierce, 1976; J. J. Zwislocki, 1950].

The round window is depicted as a short circuit in Fig. 9-3(b). This is because sound vibrations travel from the fluid to air at this point; because of the huge mismatch between fluid and air, practically no force is transmitted. In accordance with Table 9-1, the electrical equivalent is a short circuit. The round window short circuit allows us to swing the lower half of the line into, and combine it with, the upper half as in Fig. 9-3(c). This major simplification does not change any of the currents or voltages. Inductance L now represents the fluid mass of the *total* cross section of the cochlea. Current $I(x)$ corresponds to longitudinal fluid velocity in either the upper or lower halves of the cochlea, and $V(x)$ corresponds to the force or pressure difference between the upper and lower surfaces of the BM.

Analysis starts out as it does for a conventional line. Using infinitesimally small branches in Fig. 9-3(a), so that "deltas" can be replaced by derivatives, we have from Ohm's law

$$\frac{dV}{dx} = -I(x)Z(x), \tag{9-2}$$

where $Z(x)$ is the impedance of $R(x)$ in parallel with $L(x)$:

$$Z(x) = \frac{j\omega L(x)R(x)}{R(x) + j\omega L(x)}. \tag{9-3}$$

Similarly, from Ohm's law,

$$\frac{dI}{dx} = -V(x)Y(x), \tag{9-4}$$

where $Y(x)$ is the admittance of $G(x)$ in series with $C(x)$:

$$Y(x) = \frac{j\omega C(x)G(x)}{G(x) + j\omega C(x)}. \tag{9-5}$$

Further analysis is difficult unless simplifying assumptions are made.

1. It is assumed that R_0 is everywhere constant:

$$R_0 = \sqrt{\frac{Z(x)}{Y(x)}} = 0.02\ \Omega \tag{9-6}$$

(see row 9 of Table 9-2). In other words, although $Z(x)$ and $Y(x)$ taper, their *ratio* remains constant.

2. It is assumed that the line is terminated in R_0, as shown in Fig. 9-3(*c*). As a result, everywhere on the line,

$$V = IR_0 \tag{9-7}$$

and

$$\frac{dV}{dx} = R_0 \frac{dI}{dx}. \tag{9-8}$$

3. To simplify the work of finding dV/dx and dI/dx, it is assumed that the $Z(x)$ and $Y(x)$ tapers are exponential.
4. The $Z(x)$ and $Y(x)$ values have to describe a transmission line that fits the curve of Fig. 9-6, which specifies the characteristic frequency f_c versus cochlear distance x.
5. Inductance L represents fluid mass. A reasonable assumption is that the cochlear cylinder fluid cross section is 3 mm² at around $x = 2.5$ cm. A cylinder of water 3 mm² in cross section has a mass of 0.003 kg/m, so L in Fig. 9-3(*c*) should be 0.003 H/m in the vicinity of $x = 2.5$ cm.

With the aid of the above five assumptions, curve fitting yields (MKS units are used in all of the following equations) [S. Deutsch and E. M-Tzanakou, 1987]

$$Z = 0.39 f_{\text{in}}^2\, \epsilon^{-100z} + j0.1325 f_{\text{in}} \epsilon^{-20z}\ \Omega/\text{m}, \tag{9-9}$$

where

$$z = \sqrt{0.035 - x} \tag{9-10}$$

TABLE 9-4 Various Values of Characteristic Frequency f_c versus Distance from the Base x[a]

x, cm	f_c, Hz	f_c, Hz	x, cm
0.25	13870	50	3.37
0.5	9942	100	3.21
0.75	7035	125	3.14
1	4906	167	3.05
1.25	3368	200	2.98
1.5	2272	250	2.87
1.75	1504	500	2.39
2	979.9	1000	1.99
2.25	634.6	1500	1.75
2.5	425.0	2000	1.58
2.75	306.9	5000	0.99
3	188.4	10000	0.50
3.25	85.7		

[a] As given by Eq. (9-25).

so that

$$\frac{dx}{dz} = -2z. \tag{9-11}$$

The 0.035 term in Eq. (9-10) is based on the 0.035-m length of the cochlea. Also, $Y(x)$ is uniquely determined from $\sqrt{Z(x)/Y(x)} = 0.02\ \Omega$.

The analysis is continued in Appendix A9-1. In addition to the tuning curves of Figs. 9-7 and 9-8, a characteristic frequency curve similar to that of Fig. 9-6 is derived. Various values of f_c versus x thus obtained are listed in Table 9-4.

9-3 Multitone Response of the Basilar Membrane

A single input tone is used in the previous section to minimize complexity. Only rarely, however, does the ear get a single tone. Music usually consists, from moment to moment, of a few fundamental frequencies and their harmonics. The harmonics are, of course, integer multiples of their fundamental frequency. For example, for a fundamental musical pitch of 440 Hz (musical A), the second harmonic is 880 Hz, the third harmonic is 1320 Hz, and so forth. (This should not be confused with octaves: one octave above 440 Hz the frequency is 880 Hz; two octaves above it is 1760 Hz; and so forth.) The harmonics are usually weaker than the fundamental, but not always. If a complex sound reveals strong 880- and 1320-Hz components, one would conclude that the fundamental is 440 Hz even if none of the latter is present. Several fundamental frequencies and their harmonics may be present simultaneously. Speech is of course a still more complex mixture, with many fundamentals and harmonics rapidly changing with time.

In this section we briefly consider two or more simultaneous input tones. The sound waveform can be become quite complex with as few as two frequency components, as shown in Fig. 9-9. In Fig. 9-9(a) and (b), we, respectively, have 1100-Hz and 1000-Hz inputs, each of peak magnitude 1, and their sum waveform in Fig. 9-9(c). To get the sum waveform we add the original values at every instant. The fourth peak from the left in Fig. 9-9(c) has a maximum height of +2 because the input components of Fig. 9-9(a) and (b) reinforce each other here; the ninth "peak" has zero height because the components cancel each other so that the "sum" wave vanishes. The sum wave of Fig. 9-9(c) is typical for two components that are relatively close in frequency. The slow rise and fall of the envelope is a beat frequency, in this case 100 times/s because 1100 − 1000 = 100 Hz. A low beat frequency, such as when 1000- and 1005-Hz tones combine to give a 5-Hz beat, adds an element of beauty to the perceived sound. The 100-Hz beat of Fig. 9-9(c) sounds somewhat discordant, such as when two adjacent piano keys are struck.

The 500-Hz wave of Fig. 9-9(d) is combined with Fig. 9-9(a) to give the sum wave of Fig. 9-9(e). In this case the beat frequency is 1100 − 500 = 600 Hz. Because the beat is approximately at the same frequency as one of

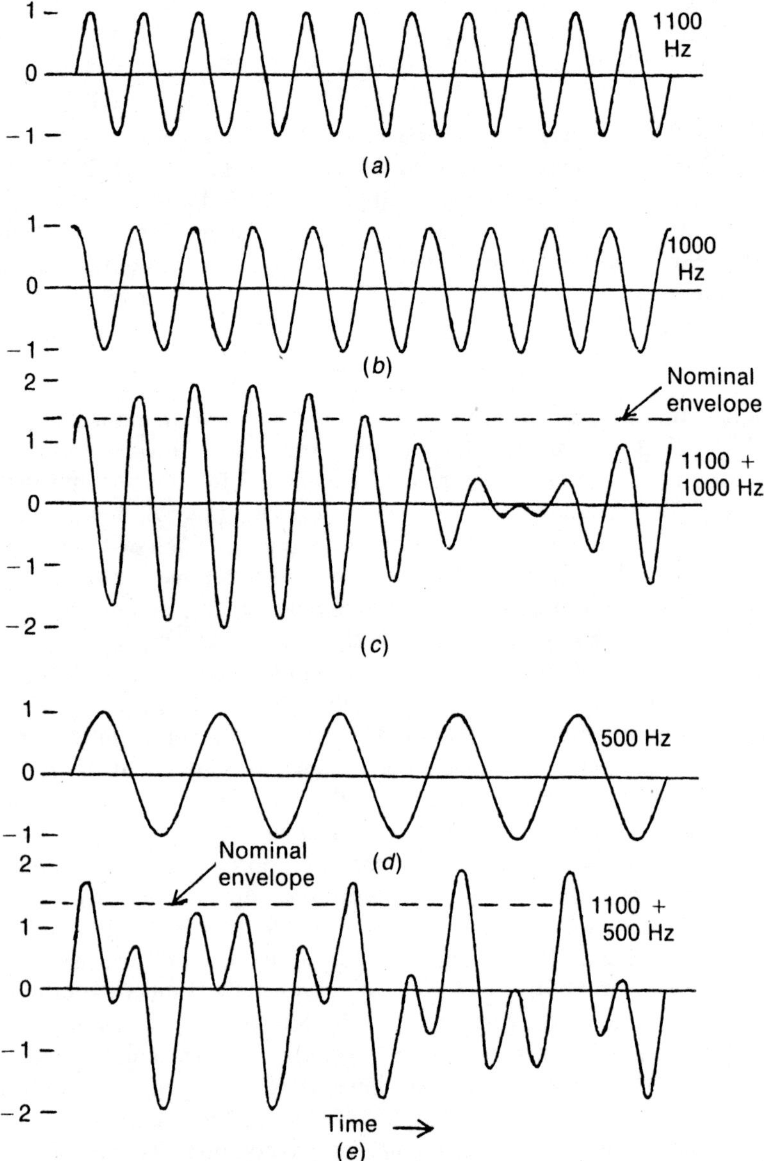

Fig. 9-9 The results of combining two audio tones of equal amplitude (peaks = 1) but of different frequencies. (*a*) 1100-Hz tone. (*b*) 1000-Hz tone. (*c*) Sum of parts (*a*) and (*b*). The waveform displays a 100-Hz beat frequency; that is, its envelope rises and falls 100 times per second. (*d*) 500-Hz tone. (*e*) Sum of parts (*a*) and (*d*). The waveform is that of a typical, somewhat chaotic, sound wave. The nominal envelopes of the "sum" waveforms are at level $(1^2 + 1^2)^{1/2} = 1.41$, as explained in the text.

the original components, the combined waveform of Fig. 9-9(*e*) is that of a typical, somewhat chaotic, sound wave.

Although the ear drum, ossicles, and windows vibrate in accordance with the "sum" waveforms of Fig. 9-9, the BM acts as a filter, tending to separate a complex input into its frequency components. To illustrate this, since in Fig. 9-7 we already have the BM response to four single-tone inputs, it is a simple matter to combine them into a complex four-tone input consisting of equal components at 170, 1160, 4700, and 15,000 Hz. (You cannot duplicate 15 kHz on musical instruments, which only go up to 5 kHz. Older persons cannot hear 15 kHz, and may be severely limited at 4700 Hz. Nevertheless, these components are valid for illustrating the BM filter.)

The envelope of the four-tone combination is plotted in Fig. 9-10. In this case the envelope is the *nominal* outer limit of motion as the BM vibrates up and down. The nominal envelope of the multitone signal of Fig. 9-10 is given by the square root of sum of squares of the individual component envelopes of Fig. 9-7. The "square root of sum of squares" rule is based

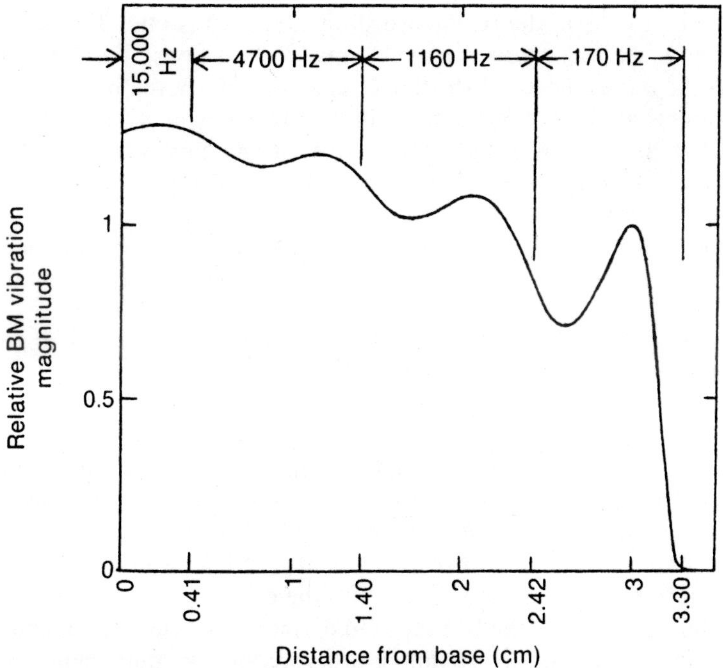

Fig. 9-10 Nominal envelope of BM vibration versus distance from the base of the cochlea for an input consisting of four equal amplitudes (peaks = 1) of 170, 1160, 4700, and 15,000 Hz. The vibration envelope of each tone by itself is shown in Fig. 9-7. With the combined input, BM vibration breaks up into bands, as indicated, in accordance with the dominant frequency in each band.

upon energy considerations; the "square" appears because the power contributed by a component is proportional to V^2/R.

One example is given to illustrate the calculation: at BM distance = 0.2 cm from the base, envelope values corresponding to 170, 1160, 4700, and 15,000 Hz in Fig. 9-7 are, respectively, 0.13, 0.38, 0.75, and 0.97. Then the nominal outer limit of motion is given by the curve in Fig. 9-10 as

$$\sqrt{0.13^2 + 0.38^2 + 0.75^2 + 0.97^2} = 1.29.$$

In Fig. 9-9 the value of each component envelope is 1, so the nominal envelopes of the sums are $(1^2 + 1^2)^{1/2} = 1.41$, as shown.

With an eye on the "sum" curve of Fig. 9-9(*e*), where only two components result in complex motion, the BM vibration in Fig. 9-10 must be psychedelic. Fortunately, the envelope is manageable as follows.

From BM distance 0 to 0.41 cm, according to Fig. 9-7, the 15-kHz component is stronger than the other three components. Referring to Fig. 9-10, at the 0.41-cm edge of this region the BM vibration is complex. Toward the base to the left of this edge, however, the 15-kHz rate dominates, and *all* of the hair cells in the region are stimulated at this rate. The other three components are present as relatively small "distortions" superimposed on the 15-kHz wave; these "distortions" are also felt by the hair cells. These minor ripples are taken into account in the preceding analysis. It shows that the 15-kHz motion, by itself, would have an envelope value of 0.97; with the addition of ripples due to the 4700-, 1160-, and 170-Hz components the envelope value increases to 1.29.

From BM distance 0.41 cm to 1.4 cm the 4700-Hz component dominates, so the BM up and down vibration is mostly at a 4700-Hz rate, as indicated in Fig. 9-10.

From BM distance 1.4 cm to 2.42 cm the 1160-Hz component dominates, so the BM up and down motion is at this rate, as indicated.

From BM distance 2.42 cm to 3.3 cm the 170-Hz component dominates, so the BM motion is at a 170-Hz rate. According to Fig. 9-7, in the right-hand portion extending from 2.8 cm to 3.3 cm, 170 Hz is the *only* component so that, in this region, the motion is that of a "pure" wave free of any wiggles due to the 1160-Hz component.

To summarize: The BM responds along its length to different components of air vibration in accordance with the characteristic frequency curve of Fig. 9-6. As a result, the BM tends to dissect a complex audio input signal into regions or bands corresponding to its frequency components, as illustrated in Fig. 9-10, in decreasing frequency from BM distance = 0 to distance = 3.5 cm from the base. At the edges of each band the BM vibration is complex, as in Fig. 9-9(*e*). If two components are relatively close in frequency, so that the BM does not succeed in separating them, the envelope of BM motion includes the difference (beat) frequency, as in Fig. 9-9(*c*). We have not considered cases in which one component is much weaker than another; in those events, as mentioned in Section 9-2 in connection with

OHC feedback, subjective listening experiments show that there are complex combinations in which one frequency component masks another.

The fact that each hair cell and its neuron are maximally sensitive to their characteristic frequency is the basis for hearing prostheses in which a cable with several tiny wires is inserted into the cochlea. A possible design has the following specifications: The cable consists of six platinum wires. (A 22-wire (!) implant is available [*Sci. Am.*, Nov. 1990].) The six wires end, respectively, at BM distance = 0.5, 1, 1.5, . . . , 3 cm. An audio signal picked up by a microphone is passed through electrical filters that break up the signal into six frequency components centered, respectively, on 8500, 4700, 2400, 1160, 500, and 170 Hz. Each of the wires is insulated electrically except for a tiny sphere at its tip; each tip stimulates the local hair cell neurons in accordance with the frequency components that fall into its band. The electrical stimulation has proved to be helpful for some people who have lost one or more of the mechanical portions of their hearing apparatus. The electrical signals can be fed through the intact skin behind the ear, via small transformer coils, to circuits embedded subcutaneously. It is an expensive surgical procedure, with daunting after-care, for a few adventuresome and/or desperate souls [I. J. Hochmair-Desoyer and E. S. Hochmair, 1980; G. E. Loeb, 1985; H. McDermott, 1989; R. L. White, 1982].

9-4 Signals Carried by the Acoustic Nerve

It would seem from the previous section that the role of the acoustic nerve is obvious—each axon simply carries APs that are synchronized with the local vibratory motion of the BM. This appraisal is incorrect for three reasons: first, in the absence of stimulation, each axon carries its own spontaneous AP discharge; second, the AP discharge rate of a sensory receptor increases, in general, if the stimulus amplitude increases. Suppose that a hair-cell neuron is generating a 100-Hz AP sequence in response to a single-tone stimulation pitch of 100 Hz. If the loudness of the source increases, how can the neuron generate 120 Hz, say, if the stimulus continues to emit 100 Hz? Third, complex sounds usually include frequency components that are much too high for an auditory neuron. How can a hair-cell neuron synchronize with the BM when the latter is moving up and down at a rate in excess of 1000 vibrations/s? In other words, how can the vibrations of the BM be translated into electrical neuronal signals that can be interpreted correctly by the brain's auditory cortex? [D. H. Johnson, 1980; M. C. Liberman, 1978; J. M. Miller et al., 1979; W. M. Siebert, 1970; E. Zwicker, 1974]

The answers to the preceding questions are given in the five parts—(*a*) to (*e*)—of Fig. 9-11. In each part a BM motion is given along with, underneath, the corresponding AP discharge of a hair-cell neuron that is experiencing the BM motion. For convenience in comparing the discharge sequences, the *average* discharge frequency of the neuron in each case is 100 Hz. Comments on each part follow.

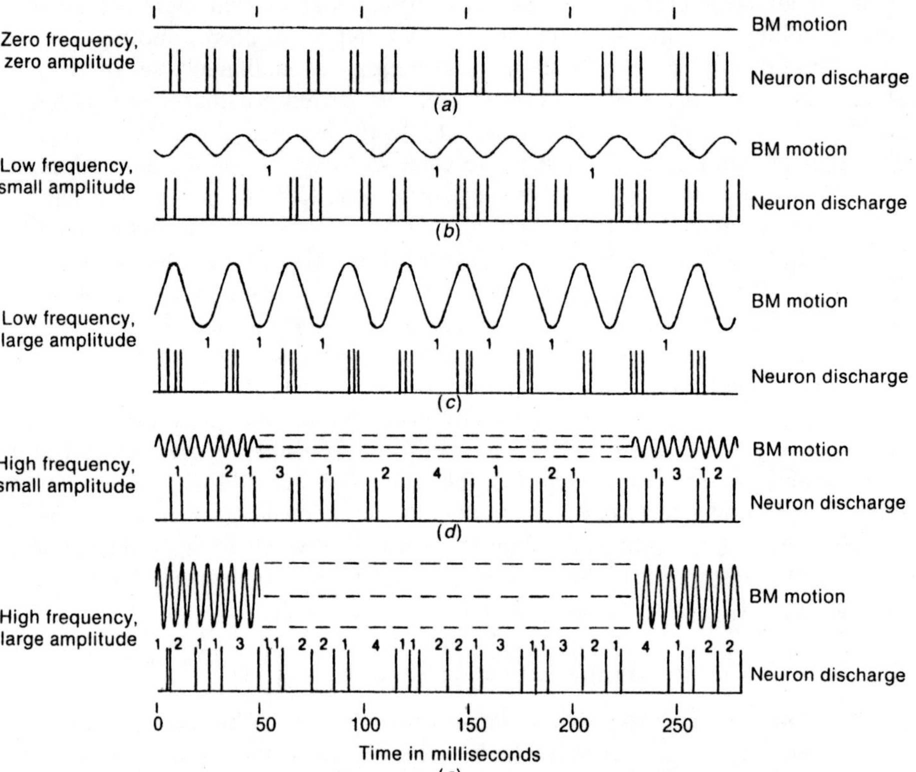

Fig. 9-11 Idealized BM motion and corresponding AP discharge of the local hair-cell neuron for five representative conditions. Average discharge frequency in each case is 100 spikes/s. (*a*) Zero motion. Spontaneous, random discharge. (*b*) 39-Hz, small-amplitude motion. The 1s indicate synchronized discharge intervals. (*c*) 36-Hz, large-amplitude motion. Almost fully synchronized discharge. (*d*) 170-Hz, small-amplitude motion. The 2s and 3s indicate discharge intervals 2 × BM period, 3 × BM period, and so forth. (*e*) 166-Hz, large-amplitude motion. Discharge almost fully synchronized with BM periods or integer multiples thereof.

Fig. 9-11(a): Zero stimulation. The BM is motionless, as indicated by a horizontal line. Underneath, the neuron is discharging spontaneously, similarly to that of Fig. 3-3(*a*). During the 280-ms sample shown, there are 28 AP spikes, so that the average frequency is 100 spikes/s. The discharge, however, is random; the interval between one spike and the next varies randomly. The sequence is drawn for a refractory period of 2.5 ms.

Fig. 9-11(b): The BM frequency is relatively low (39 Hz, period = 26 ms) and the amplitude is small. The neuron discharge *does* synchronize with the BM motion, but in only three intervals out of the total of 30 shown. The three correct intervals are indicated by 1s; the remaining intervals are more or less random, like those of Fig. 9-11(*a*). Somewhere in the auditory cortex, based on infrequent 26-ms intervals, it will be recognized that the ear is receiving a weak 39-Hz component.

Fig. 9-11(c): The BM frequency is relatively low (36 Hz, period = 28 ms) but the amplitude is large. When the BM moves up a correspondingly large distance, the hair-cell neuron discharges at a furious rate, with two, three, or four closely spaced spikes that retain some degree of randomness. When the BM moves down, the hairs are unbent and the neuron falls into silence. The net result is that, during the 280-ms sample, there are seven intervals at 28 ms as indicated by seven 1s, and a few relatively wide intervals of 28 ms more or less. The auditory cortex will interpret this as a strong 36-Hz component.

Fig. 9-11(d): The BM frequency is relatively high (170 Hz, period = 5.9 ms) but the amplitude is small. The neuron has a refractory period of 2.1 ms (maximum frequency 476 Hz), but it will discharge at this 476-Hz rate only for an extremely large BM motion; in fact, for a vibratory magnitude that is greater than the maximum tolerable. Why is the 170-Hz rate called "relatively high"? Because the neuron cannot synchronize consistently with its 5.9-ms period. Over the 280-ms sample shown, there are:

- Seven 5.9-ms intervals as indicated by 1s;
- Four intervals 2 × BM period = 11.8 ms wide as indicated by 2s;
- Two intervals 3 × BM period = 17.7 ms wide;
- One interval 4 × BM period = 23.6 ms wide.

The remaining 15 intervals are random, like those of Fig. 9-11(*a*). In other words, the output is a hodgepodge, a mixture of intervals that are integer multiples of 5.9 ms plus random intervals. Despite the hodgepodge, the auditory cortex recognizes that the ear is receiving a weak 170-Hz component.

Fig. 9-11(e): The BM frequency is relatively high (166 Hz, period = 6.0 ms) and the amplitude is large. Now, over the 280-ms sample, there are only two "random" discharges. There are:

- Thirteen intervals 6 ms wide;
- Eight intervals 12 ms wide;
- Three intervals 18 ms wide;
- Two that are 24 ms wide.

The neuron is fully synchronized; now, with practically all of the intervals an integer multiple of 6 ms, the auditory cortex will recognize that a strong 166-Hz auditory component is being heard.

From a different viewpoint, one can say that the sequence of Fig. 9-11(*e*) "looks like" that of Fig. 9-11(*a*), except that the intervals of Fig.

9-11(*e*) are integer multiples of 6 ms and are therefore ordered, not random, while those of Fig. 9-11(*a*) have no relation to 6 ms or to any other interval.

The answers to the questions posed at the beginning of this section are as follows.

A neuron stimulated at a frequency of 100 Hz (period 10 ms) generates a mixture of 10-ms intervals and random intervals. If the loudness of the source increases, the neuron generates an increasing proportion of 10-ms intervals and less of random intervals. If the neuron cannot follow (respond exactly to) the stimulus frequency, it generates integer multiples of the stimulus period [J. L. Goldstein and P. Srulovicz, 1977; P. Srulovicz and J. L. Goldstein, 1983].

In the extreme case of a 15,000-Hz stimulus (period 0.067 ms) the neuron may generate intervals $15 \times 0.067 = 1$ ms wide, $16 \times 0.067 = 1.07$ ms wide, $17 \times 0.067 = 1.13$ ms wide, and so forth (plus random intervals not related to 0.067 ms, depending on the loudness of the stimulus). (Such high-frequency synchronization has not, however, been verified experimentally.)

The response of a hair cell and its neuron depends on BM motion and is largely independent of its characteristic frequency. Suppose, for example, that the audio input is a single-tone 170-Hz signal. According to Fig. 9-7, practically the entire BM will vibrate at the 170-Hz rate, and a neuron at the base of the cochlea, having an f_c of 15,000 Hz, will generate intervals corresponding to 170 Hz. The characteristic frequency tells us only that the neuron at that location would generate a greater number of synchronized intervals for 15,000 Hz input compared to any other single-tone frequency of equal loudness.

The discussion in this section up to now concerns only individual neurons and their axonal AP discharges. It is important to get an overall picture of the entire 12,000-axon afferent portion of the acoustic nerve. To explore this holistic viewpoint, the amplitude of a single-tone 500-Hz sound (2-ms period) is gradually increased in Fig. 9-12. Here a *highly idealized* plot of "relative AP synchronization" versus BM distance is shown. "Relative AP synchronization," or "relative sync," refers to the AP discharge relative to BM motion. In Fig. 9-11(*c*) and (*e*) the relative AP sync is 1 because further increase in BM vibration magnitude will not materially alter the AP output sequence (the loudness or amplitude is already large); in Fig. 9-11(*b*) and (*d*) the relative AP sync is around 0.5 because approximately half of the maximum number of periods that can do so are synchronized; and so forth.

Figure 9-12 shows relative AP sync as the input 500-Hz tone is increased from 20 to 140 dB SPL. You will recall that human auditory perception starts at 0 dB SPL, normal conversation is at a level of 70 dB SPL, and damage to hearing may start to occur at 140 dB SPL. The following comments are appropriate to Fig. 9-12.

A level of 20 dB SPL is evidently a weak signal, with maximum value

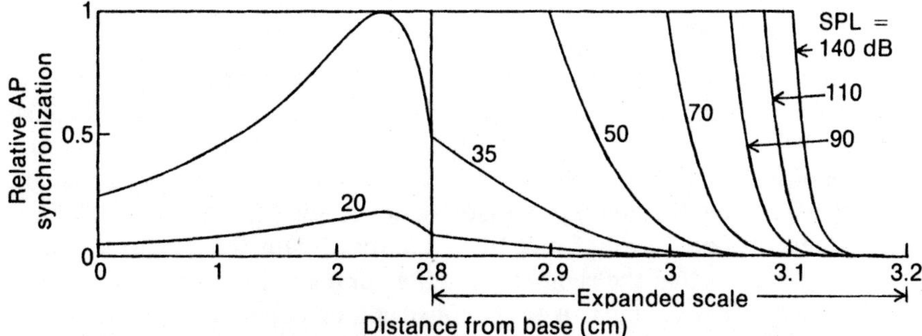

Fig. 9-12 Highly idealized relative AP synchronization (number of synchronized intervals/total number of intervals) versus distance from the base of the cochlea as amplitude of a single-tone 500-Hz input is gradually increased from 20 to 140 dB SPL. The middle ear introduces an attenuation of 1.8-to-1 at 110 dB SPL and 10-to-1 at 140 dB SPL. Notice that horizontal scale to the right of 2.8 cm is magnified by a factor of 10.

of 0.18 for relative AP sync. (The horizontal scale to the right of 2.8 cm is magnified by a factor of 10 to show the curves more clearly.)

At a level of 35 dB SPL the neurons at $x = 2.4$ cm are saturated, as implied by their relative AP sync level of 1. According to Fig. 9-6, 500 Hz is the characteristic frequency (approximately) of the 2.4-cm location.

From here on a very surprising condition is revealed: most of the neurons become saturated. At 50 dB SPL every neuron between 0 and 2.9 cm is saturated; at 70 dB SPL every neuron between 0 and 3 cm is saturated. In other words, a 500-Hz whistle at the level of normal conversation results in an AP discharge similar to that of Fig. 9-11(*e*) for every neuron between 0 and 3 cm. Each of these neurons will generate its own sequence of intervals 2, 4, 6, . . . ms wide in random order, but fully synchronized to the 500-Hz BM vibration.

If the input *is* normal conversation rather than a 500-Hz whistle, most of the neurons saturate momentarily, at vibration peaks, but different sections of the BM vibrate at different frequencies as the sound is dissected approximately into its frequency components.

As we continue to increase the 500-Hz sound level in Fig. 9-12, additional neurons are recruited into the "relative AP sync = 1" saturation category. It is assumed in plotting these curves that the middle-ear acoustic reflex (see Sect. 9-1) introduces an attenuation of 5 dB for every 10 dB of SPL input starting with 100 dB SPL. The "110-dB" curve therefore includes an attenuation of 5 dB (a ratio of 1.78 to 1). The "140-dB" curve, the maximum tolerable input, includes an attenuation of 20 dB (a ratio of 10 to 1). It is assumed that the compression introduced by OHC feedback (see Sect. 9-2) has no effect at the relatively small BM displacements of Fig. 9-12.

9-5 Frequency Discrimination in the Auditory Cortex

Tests show that a human with normal hearing can detect a change of 3 Hz at 1000 Hz. That is, if you are listening to a 1000-Hz tone, a sudden change to either 997 Hz or 1003 Hz is noticeable. The 3-Hz shift corresponds to a distance of less than 20 μm along the BM. There are two schools of thought with regard to how the audio processor achieves good frequency resolution. The "steep dropoff" school (not to be confused with the "high dropout" school) points out that the right sides of the curves in Figs. 9-7 and 9-8 drop off extremely rapidly. One can easily construct, on paper at least, a neuronal network that will use the steep dropoff of a single-tone input to perform fine frequency selectivity. The situation becomes more complicated for a multitone input, such as that of Fig. 9-10, but one can probably devise a suitable neuronal network in this case also.

The second conjecture is based on the fact that the neuronal output of auditory hair cells is partially synchronized to the incoming tones or to their subharmonics. In the remainder of this chapter we assume that the "partially synchronized action potential" hypothesis is correct only because it follows logically from the curves of Fig. 9-11. However, the "steep dropoff" conjecture may eventually turn out to be correct; we must emphasize over and over again that the remainder of this chapter is hypothetical [S. Deutsch, 1990; B. C. J. Moore and R. D. Patterson, 1986; A. C. Sanderson, 1975].

The curves demonstrate that the BM is a crude filter. The complex four-tone signal of Fig. 9-10 shows, however, that the BM performs a vital function. By breaking up this complex signal into its frequency components, we have one group of hair cells receiving a relatively clean 15,000-Hz wave, another 4700 Hz, another 1160 Hz, and a fourth group 170 Hz. Without BM filtering, each hair cell would receive the sum of all four components: a chaotic, complex waveform full of "noise" in the sense of unwanted signal components. The 15,000-Hz hair cells would be buffeted about by undesired 4700-, 1160-, and 170-Hz motions, and so forth.

We seek to explain the "tip enhancement" of Fig. 9-4, since the bandwidth here is sufficiently narrow so that it fits the sharp frequency resolution of the ear. It is not sufficient to simply say, however, that the narrow tip results from feedback to the OHCs. The feedback has to be associated with some kind of tuned circuit. We have to imagine 3000 tuned circuits, each resonating at the characteristic frequency (f_c) of its driving IHC. Since the auditory cortex does not contain inductances, a reverberatory type of circuit, which does not require Ls or Cs, is assumed to be valid, as defined in Fig. 9-13.

Figure 9-13 depicts the reverberatory circuit for the IHC whose f_c is 500 Hz. According to Fig. 9-6, it is located at $x \cong 2.4$ cm. The audio input vibrates the BM, which, in turn, excites the IHC with an equivalent input voltage component, V_{in}. The waveforms underneath the circuit are drawn for a single narrow audio input pulse, such as the sharp wavefront caused by an exploding firecracker (nowadays gunfire may be more appropriate).

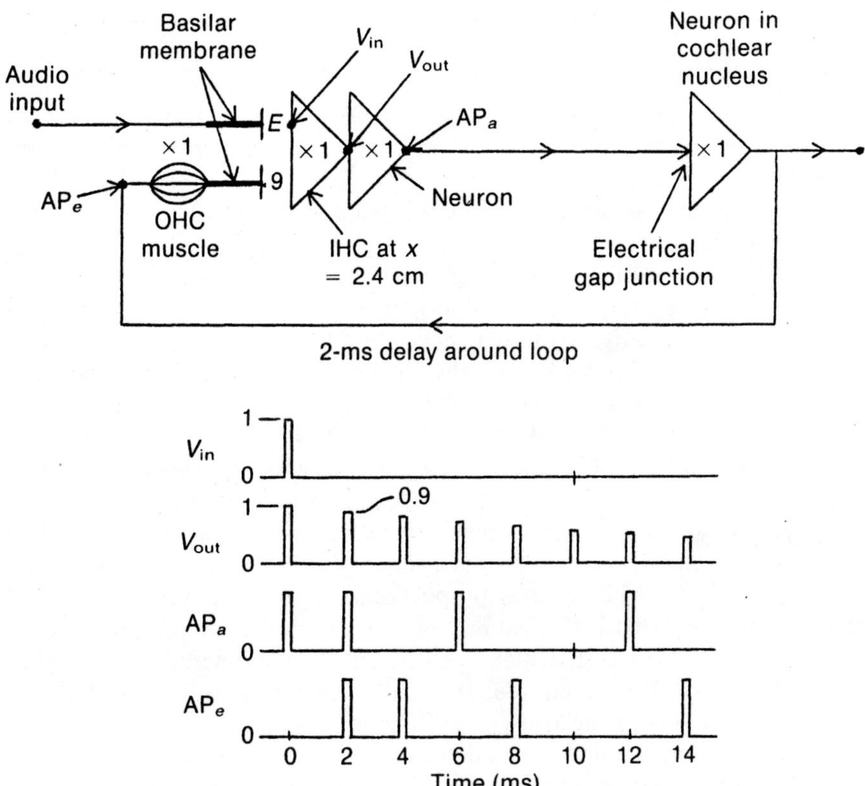

Fig. 9-13 Hypothetical reverberatory circuit that processes the output of inner hair cell located 2.4 cm from the base. The response to a single V_{in} audio input pulse is given below the circuit. The V_{in} signal enters a feedback loop that has a loop gain of 0.9; because of time delay of 2 ms around the loop, the characteristic frequency of the IHC is 500 Hz. The loop includes the outer hair cell as a muscle pulling upon the basilar membrane. The V_{out} reverberations decay in accordance with heights of 1, 0.9, 0.9^2, 0.9^3, ... AP_a and AP_e represent afferent and efferent 0.1-V high action potentials, respectively. They agree statistically with V_{out}: the probability that an AP_a spike will occur is proportional to the amplitude of the V_{out} pulse. AP_e is the AP_a waveform delayed 2 ms. The cochlear nucleus is in the medulla, the upper end of the spinal cord where it joins the brain.

In any event, the signal enters a feedback loop that yields a loop gain of 0.9 and time delay of 2 ms around the loop. The original single pulse reverberates around the loop as shown by the V_{out} waveform, its height decreasing in accordance with 1, 0.9, 0.9^2, 0.9^3, ... It is convenient to express this in columnar form (time is in ms units).

Time = 0	2	4	6	8	10	12	14	...	∞
V_{in} = 1									
V_{out} = 1	0.9	0.81	0.729	0.656	0.590	0.531	0.478	...	0

(The response to a steady 500-Hz input, which demonstrates resonance, is considered later on.)

Auditory nerve fibers have a high conduction speed. Assuming v = 100 m/s, the 2-ms delay requires a total loop length of 20 cm. This is consistent with the neuroanatomy of the ear, according to which the auditory nerve terminates in junctions located in the "cochlear nucleus" in the medulla, the upper end of the spinal cord where it joins the brain [I. Glass, 1983]. For this long journey the auditory signals must of course be in the form of APs. The IHC of Fig. 9-13 is therefore associated with a neuron that converts V_{out} into afferent action potentials (AP_a). These potentials are statistically in agreement with V_{out}, as explained in connection with Fig. 9-11: since V_{out} is a graded potential, whereas AP_a is a series of 0.1-V spikes, the probability that an AP_a spike will occur is proportional to the amplitude of the V_{out} pulse. If the input 500-Hz signal is "on" for a second, for example, we get 500 vibrations of the BM, which is sufficient for good agreement between AP_a spikes/s and V_{out} amplitude. But if the 500-Hz input is "on" for only 0.1 s, 50 vibrations of the BM is insufficient. In that event we can "hear" the 0.1-s tone, but it is difficult to detect a change from 500 to 501.5 or 498.5 Hz. (In general, regardless of frequency, the input tone has to be "on" for at least 0.1 s to be "heard" as a tone.)

The AP_a frequency information is fragile. It is lost if the pathway leads to an ordinary synaptic junction, since firing of the second neuron will not be synchronized with that of the first neuron. In Fig. 9-13, therefore, the AP_a information is fed to the "neuron in cochlear nucleus" via an electrical gap junction that *does* preserve the AP_a sequence. From here a branch carrying the efferent action potential (AP_e) returns to stimulate the OHC "muscle." The AP_a and AP_e waveforms are identical, except that the latter is delayed 2 ms by time delay around the loop.

The OHC vibration of the BM reaches the IHC via a hydrodynamic junction; that is, a mixture of forces transmitted by the cochlear fluid and/or BM motion. The junction is equivalent to a weighting factor of 0.9, as shown. The AP_e waveform does not exactly correspond to V_{out} because a statistical interpretation is again illustrated: over a reasonably long period of time, such as one second, the V_{out} amplitude *will* agree with the AP_e spike frequency.

Why do the AP_e signals feed back to OHCs? Because this is a more accurate way to preserve frequency information. It does not involve synaptic junctions, which lose frequency information. Also, V_{out} is the graded potential output of the IHC, and the BM vibration due to the OHC "muscle" is also a graded motion that is added to that of the audio input to yield V_{out}.

The net result is that, in the vicinity of $f_{in} = f_c$, the BM motion is enhanced, and the IHC response curve becomes relatively narrow, as shown in Fig. 9-4.

It is not at all obvious that the reverberatory circuit of Fig. 9-13 is sharply tuned to 500 Hz. To reveal resonance behavior, one must apply a *continuous* stream of V_{in} pulses that are 2 ms apart, as in Fig. 9-14(*a*). This is the idealized response of the IHC if 500-Hz pulses enter the system starting at $t = 0$.

Each new V_{in} pulse adds its contribution so that, instead of decaying, the V_{out} pulses increase in height. If the frequency of V_{in} pulses is only a few percentage points above or below 500 Hz, however, the feedback contributions do not add, so that the V_{out} pulses remain dwarfed in height. This points to the useful characteristic of a reverberatory circuit: it is sharply tuned to its input f_c, in this case 500 Hz.

Eventually, if a stream of 500-Hz V_{in} pulses continues without end, the V_{out} pulses will reach a height of 10, as shown in the "infinite" portion of Fig. 9-14(*a*). This corresponds to a BM tip enhancement of 20 dB. It is not necessary to wait too long for this to happen. After the thirtieth V_{in} pulse has gone by, V_{out} has reached a height of 9.576 (obtained easily on a cal-

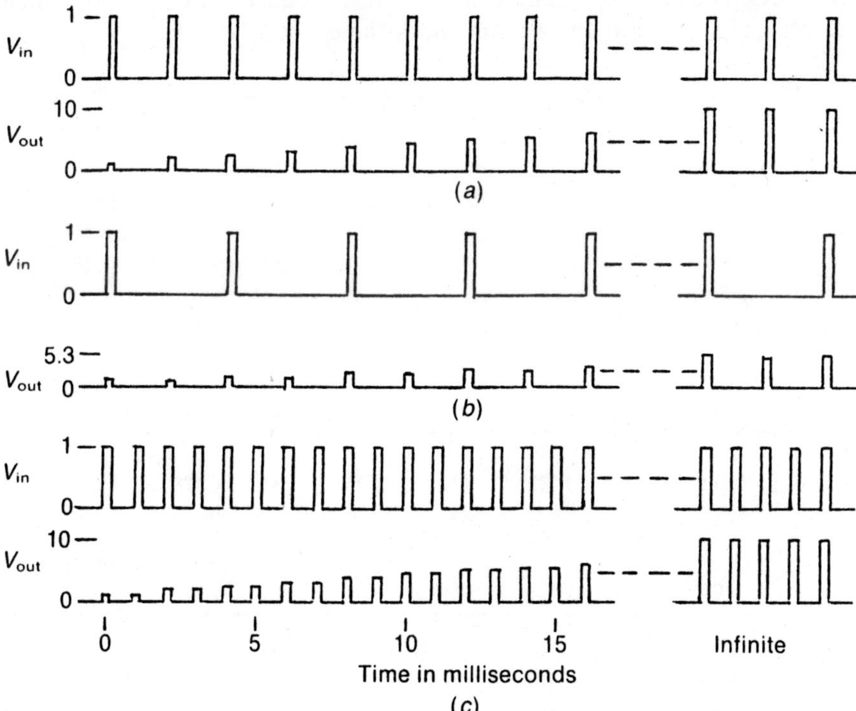

Fig. 9-14 Idealized V_{out} responses of the reverberatory circuit of Fig. 9-13, which is tuned to 500 Hz, to three representative V_{in} conditions. (*a*) V_{in} frequency = 500 Hz. (*b*) V_{in} frequency = 250 Hz. (*c*) V_{in} frequency = 1000 Hz.

culator). The fiftieth pulse results in $V_{out} = 9.948$; at 500 spikes/s, this happens in only 0.1 s. Presumably, this is the signal that goes to the IHC neuron.

In columnar form we have for Fig. 9-14(*a*):

Time = 0	2	4	6	8	10	12	14	· · ·	∞
V_{in} = 1	1	1	1	1	1	1	1	· · ·	1
V_{out} = 1	0.9	0.81	0.729	0.656	0.590	0.531	0.478	· · ·	
+	1	0.9	0.81	0.729	0.656	0.590	0.531	· · ·	
+		1	0.9	0.81	0.729	0.656	0.590	· · ·	
+			1	0.9	0.81	0.729	0.656	· · ·	
+				1	0.9	0.81	0.729	· · ·	
+					1	0.9	0.81	· · ·	
+						1	0.9	· · ·	
+							1	· · ·	
= 1	1.9	2.71	3.439	4.095	4.686	5.217	5.695	· · ·	10

How can we prove that the V_{out} pulses will eventually reach a height of 10? The "∞" column would show that $V_{out} = 1 + 0.9 + 0.9^2 + 0.9^3 + \cdots$, but one can easily demonstrate that this is equal to $1/(1 - 0.9)$, which is, of course, equal to 10. We can check the equation

$$1 + 0.9 + 0.9^2 + 0.9^3 + \cdots = \frac{1}{1 - 0.9}$$

as follows: Multiply both sides by $(1 - 0.9)$ and expand.

Although the contributions do not add if V_{in} is slightly different from 500 Hz, they *do* add if the input frequency is 500/2 = 250 Hz, or 500/3 = 167 Hz, and so forth. The 500/2 = 250-Hz case is illustrated in Fig. 9-14(*b*). Because only one out of every two V_{out} pulses is reinforced, the output only reaches a height of 5.3 rather than 10. In columnar form we have:

Time = 0	2	4	6	8	10	12	14	· · ·	∞
V_{in} = 1		1		1		1		· · ·	
V_{out} = 1	0.9	0.81	0.729	0.656	0.590	0.531	0.478	· · ·	
+		1	0.9	0.81	0.729	0.656	0.590	· · ·	
+				1	0.9	0.81	0.729	· · ·	
+						1	0.9	· · ·	
= 1	0.9	1.81	1.629	2.466	2.219	2.998	2.698	· · ·	5.263

These columns show that maximum height is given by $V_{out} = 1 + 0.81 + 0.81^2 + \cdots$, which is equal to $1/(1 - 0.81) = 5.263$.

In general, for the reverberatory circuit, we can construct Table 9-5. The table shows that the circuit discriminates against all V_{in} pulse frequencies that are below the optimum (f_c) for that particular location of the BM. The lower the V_{in} frequency, the lower the V_{out} maximum.

TABLE 9-5 Output of Reverberatory Circuit of Fig. 9-13 versus Input Audio Frequency[a]

V_{in} Frequency	V_{out} Maximum
f_c	$1/(1 - 0.9) = 10$
$f_c/2$	$1/(1 - 0.9^2) = 5.263$
$f_c/3$	$1/(1 - 0.9^3) = 3.690$
$f_c/4$	$1/(1 - 0.9^4) = 2.908$
$f_c/5$	$1/(1 - 0.9^5) = 2.442$
$f_c/6$	$1/(1 - 0.9^6) = 2.134$

[a] The optimum input frequency is the characteristic frequency, f_c.

As to V_{in} pulse frequencies *above* the frequency to which the circuit is tuned, it fails to discriminate if V_{in} is an integer multiple of the optimum frequency, such as $2f_c$ or $3f_c$. In Fig. 9-14(*c*) the case V_{in} pulse frequency $= 2f_c = 1000$ Hz is illustrated. A maximum V_{out} height of 10 is reached again. In columnar form we have:

Time = 0	1	2	3	4	5	6	7	. . .	∞
V_{in} = 1	1	1	1	1	1	1	1	. . .	1
V_{out} = 1		0.9		0.81		0.729		. . .	
+	1		0.9		0.81		0.729	. . .	
+		1		0.9		0.81		. . .	
+			1		0.9		0.81	. . .	
+				1		0.9		. . .	
+					1		0.9	. . .	
+						1		. . .	
+							1	. . .	
= 1	1	1.9	1.9	2.71	2.71	3.439	3.439	. . .	10

How can the 500-Hz reverberatory circuit avoid being confused if it responds equally well to 1000-,1500-, 2000-, . . . Hz inputs? The answer is that this is precisely the job that the BM takes over if it is not overwhelmed. Look again at the BM vibration envelope curves of Fig. 9-7: the response drops off sharply to the right of each peak. For example, the IHC at $x = 2$ cm leads to a (hypothetical) reverberatory circuit tuned to 1160 Hz. If $2 \times 1160 = 2320$ Hz enters the system, the hair cell at 2 cm is only weakly stimulated. The 4700-Hz curve shows that $4 \times 1160 = 4640$ Hz provides negligible stimulation at 2 cm. Is it a coincidence that the BM curves drop off sharply for BM distances above that of f_c, where the hypothetical reverberatory circuit has a "flat" response to harmonics? Probably not, since the BM did not *have* to evolve with this particular response characteristic.

Our hypothetical proposal, then, is that 3000 of the fibers in each acoustic nerve are subjected to signal processing by 3000 reverberatory circuits, each of which is tuned to the characteristic frequency of its input IHC. The full human audio range from 20 to 15,000 Hz, as plotted in Fig. 9-6, is thus covered. This calls for a smooth transition, from low to high frequencies, in the way auditory nerve fibers are terminated. In the anatomy of the cochlear nucleus (referred to in Fig. 9-13), high frequencies are represented along the middle, while low frequencies are represented along the lateral edge of the nucleus. Similar spatial orientation occurs throughout the auditory pathway as it leads to the cortex; this spatial orientation, for "tones," is called *tonotopic*. The orderly progression is reasonable since much of the processing that may be taking place, such as lateral inhibition, entails comparisons between axons that were originally contiguous, or nearly so [J. J. Guinan, B. E. Norris & S. S. Guinan, 1972].

Unfortunately, the circuit of Fig. 9-13 is not without major difficulties: first, since we have a reverberatory circuit that is tuned to its characteristic frequency, its V_{in}, V_{out}, and AP pulses should have a maximum width, say, of 10% of the f_c period. This is no problem in Fig. 9-13, where $0.1/f_c = 0.2$ ms. At 15,000 Hz, however, the rule calls for a width of 6.7 *microseconds*. Since the AP of a typical myelinated fiber has a width of 500 μs, how can one get an AP width of 6.7 μs? The circuit will continue to function with a wider AP, but with diminished frequency resolution. Diminished accuracy is normal, however, at the high-frequency end of the auditory spectrum.

Second, the frequency to which the reverberatory circuit is tuned is determined by the time delay around its feedback loop. The 2-ms delay of Fig. 9-13 is no problem, but an IHC tuned to 15,000 Hz requires a time delay of 67 μs. Does the auditory system employ accurate delays as short as 67 μs and, if yes, how is it accomplished?

Third, at the other extreme of frequency, a circuit tuned to 20 Hz requires a time delay of 50 ms. Does the auditory system employ accurate delays as long as 50 ms and, if yes, how is it done? Diminished frequency resolution is normal, however, at the low-frequency end of the audio spectrum. Short-term memory (see Chap. 12) is not related to auditory processing, but it also requires relatively long time delays; how this is accomplished is one of the mysterious attributes of the brain.

It is instructive to consider, in Fig. 9-15, the response of the array of 3000 reverberatory circuits to a single-tone 500-Hz input. Only the circuits tuned to multiples and submultiples of 500 Hz are of interest: 500, 1000, 1500, . . . and 250, 167,125 Hz. These are the frequencies labeled in Fig. 9-15. Frequencies below 125 Hz need not be included because the BM so effectively discriminates against them that, even with 140-dB SPL input, there is negligible response at the 500/5 = 100-Hz location.

The five plots show surprisingly little change as the input goes from a weak 20 dB SPL to the maximum tolerable 140 dB SPL. For 35 dB SPL and above, the 500-, 1000-, 1500-, . . . Hz reverberatory circuits (fed by IHCs that are located at $x = 2.4, 2.0, 1.8, \ldots$ cm, respectively) operate with a stream of APs that is fully synchronized to 500 Hz because the amplitude is large; their outputs are therefore given by Table 9-5. For the reverberatory circuits tuned to 250, 167, and 125 Hz (fed by IHCs that are

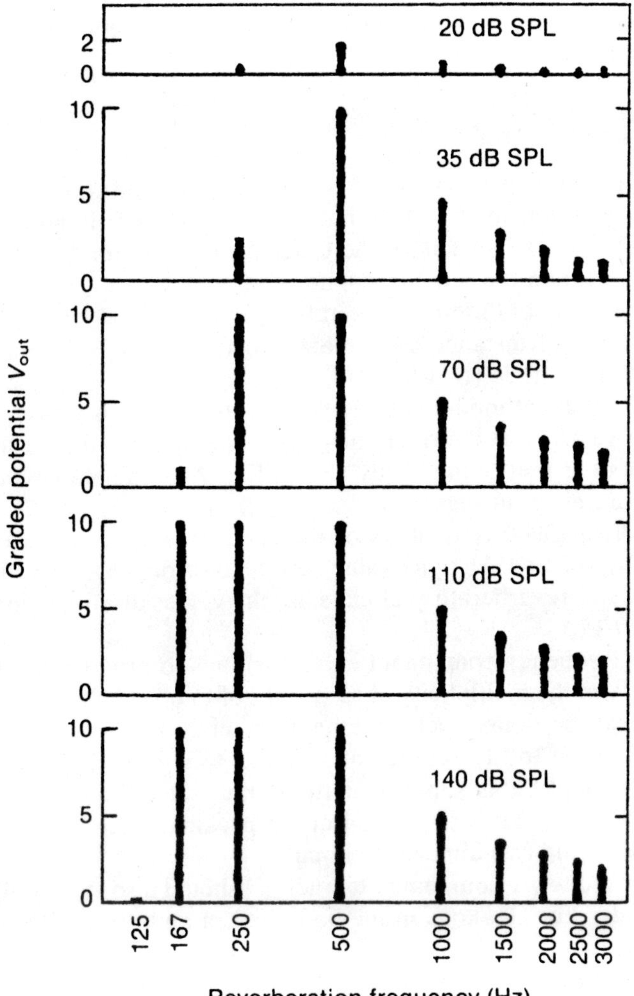

Fig. 9-15 Hypothetical response of array of reverberatory circuits to a 500-Hz single-tone input of amplitude 20, 35, 70, 110, and 140 dB SPL. For 35 dB SPL and above, values for IHCs at 500, 1000, 1500, . . . Hz are given in Table 9-5. Values for IHCs at 250, 167, and 125 Hz are taken from Fig. 9-12.

located at x = 2.87, 3.05, and 3.14 cm, respectively) one has to refer to the levels of Fig. 9-12 to get V_{out} values. If the relative AP sync is 1 in Fig. 9-12, the component appears with V_{out} = 10 in Fig. 9-15.

Although it is gratifying to find that the hypothetical array of reverberatory circuits responds without great change over a 140-dB range of input level, one should remember that the situation is much more complicated for a voiced sound that contains many frequency components [E. D. Young and M. B. Sachs, 1979].

9-6 Two-Dimensional Auditory Map

Sound is two-dimensional. We have concentrated on the measurement of frequency components because that seems to be more important and challenging, but we cannot ignore the other dimension, amplitude. Two audio inputs can contain the same frequency components, but sound subjectively different because of different spectral amplitudes. What we expect is a map in which vertical distance corresponds to frequency, say, and horizontal distance to amplitude [B. Blum, 1972].

A hypothetical model is given in Fig. 9-16. The two-dimensional map is the square array at the right-hand side. The figure is drawn for the same 500-Hz input as that of the 70 dB SPL of Fig. 9-15; the numerical values for Fig. 9-15 are given in Fig. 9-16.

Out of the 3000 IHC fibers of the acoustic nerve, nine are labeled as shown. The relative AP sync values are taken from Fig. 9-12. The outputs of the array of reverberatory circuits are the values discussed in connection with Fig. 9-15.

The "frequency component extractor" has to produce a single 500-Hz decision given the multiplicity of responses of Fig. 9-15. It is easy to imagine how this can be done: each reverberatory circuit of frequency f_c applies lateral inhibition to the neurons at $f_c/2$ and $f_c/3$. With 140 dB SPL in Fig. 9-15, for example, this suppresses the outputs at 250 and 167 Hz, leaving the output at 500 Hz as the "head of the pecking order." This is the component fed to the two-dimensional map.

The "frequency component extractor" should also supply an indication of amplitude. For all axons in the body except auditory, AP discharge frequency is a function of stimulus amplitude. The auditory system is different because of synchronization. In Fig. 9-11(*a*), (*b*), and (*c*) (or in (*a*), (*d*), and (*e*)), we progressively go from zero to small to large BM stimulation. For zero stimulation, in Fig. 9-11(*a*), we have a spontaneously discharging hair-cell neuron. As the BM vibration magnitude increases, the major change is the synchronization that takes place in Fig. 9-11(*c*) and (*e*). To the human eye, Fig. 9-11(*a*) and (*e*) appear to be similar, but they are actually very different. Here, a reliable measure of stimulus amplitude is not average discharge frequency but the degree of synchronization; that is, the "relative AP sync" of Fig. 9-12. To translate this into a graded potential, since the amplitude-induced changes occur at the low-frequency end of Fig. 9-15, we can simply add the V_{out} values at f_c, $f_c/2$, $f_c/3$, ... We then get, from Fig. 9-15, the listing shown in Table 9-6.

For our 500-Hz input, the Table 9-6 V_{out} total of 21.1 is fed to the two-dimensional map, where it corresponds to an input amplitude of 70 dB SPL. A neuron at the point representing 500 Hz and 70 dB SPL "lights up," as indicated in Fig. 9-16. Since the brain contains 10^{12} neurons, it is reasonable

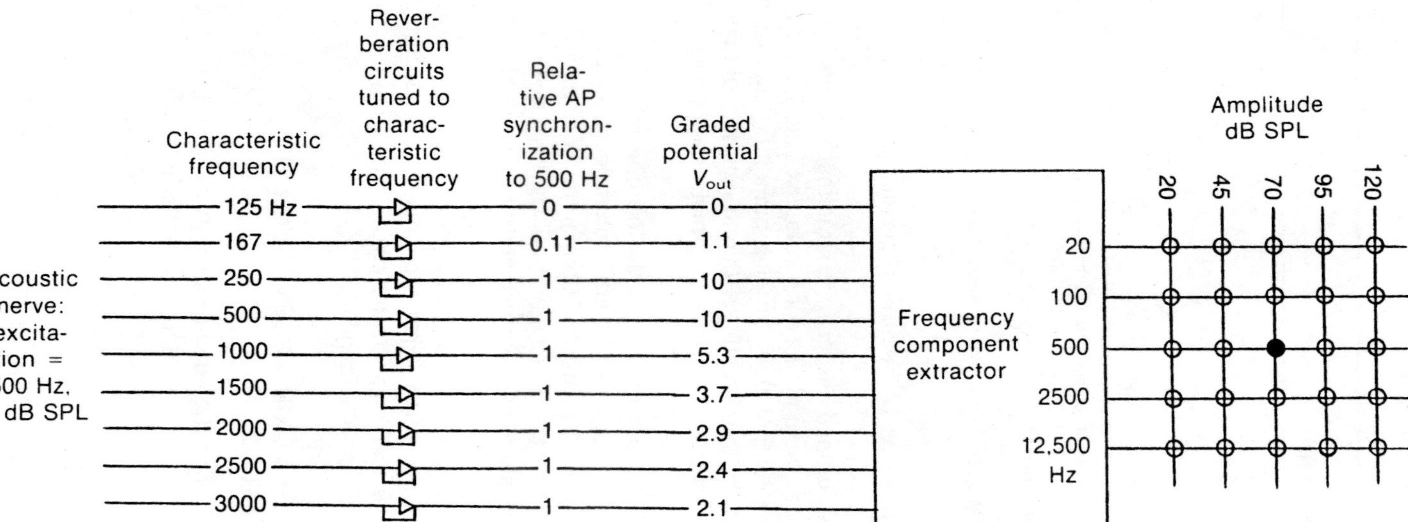

Fig. 9-16 Hypothetical generation in the auditory cortex of a two-dimensional map, of frequency versus amplitude information, for an auditory input signal. The figure is drawn for a 500-Hz single-tone input of amplitude 70 dB SPL. In the output map, the circles represent neurons that fire or "light up" so as to pinpoint each frequency component and its amplitude.

TABLE 9-6 Amplitude Information Taken from the V_{out} Values of Fig. 9-15[a]

SPL	f_c (500 Hz)		$f_c/2$ (250 Hz)		$f_c/3$ (167 Hz)		$f_c/4$ (125 Hz)		V_{out} Total
20 dB:	1.8	+	0.5					=	2.3
35 dB:	10	+	2.6					=	12.6
70 dB:	10	+	10	+	1.1			=	21.1
110 dB:	10	+	10	+	10			=	30.0
140 dB:	10	+	10	+	10	+	0.3	=	30.3

[a] Only the components supplied by characteristic frequencies equal to or less than the input frequency are used in the summation. (Values are obtained as follows: First, multiply curve supplied by Eq. (9-22) by a factor such that peak value at f_c is 10. This is the 35-dB SPL curve. Second, multiply by a factor corresponding to input dB SPL [including compression due to acoustic reflex] relative to 35 dB SPL. Third, clip maximum values at 10.)

to suppose that 1% of them, or 10^{10} neurons, are devoted to the auditory map.

The simple methods outlined earlier for extracting frequency and amplitude undoubtedly have to be modified for a complex voiced sound [S. A. Shamma, 1985]. It is conjectured that the map pinpoints each frequency component and its amplitude, thus creating a unique signature of that sound. Speculation on what the brain does with this information appears in Chapter 12.

The entire vitally important area of sound location has been omitted from the chapter to prevent excessive length. Sound location employs accurate *binaural* timing information circuits in the brain. With regard to man, Eric I. Knudsen wrote (1981): "Surprisingly, until the barn owl was tested, man was the species with the greatest known ability to locate the source of a sound; human beings are about as accurate as the owl in azimuth but are three times worse in elevation." Since man's life style is not particularly oriented toward nighttime predation, what accounts for his accurate timing ability? Probably the need to recognize the rapidly changing fundamental and formant frequencies of speech sounds.

A9-1 DERIVATION OF TUNING CURVES FOR THE AUDITORY SYSTEM

It is convenient to assume the boundary condition $I = 1 + j0$ A (fluid velocity $v = 1 + j0$ m/s) at $x = 0$. Because of terms introduced by this boundary condition, I appears to be much more complicated than it really is:

$$I = \exp[5.765 \times 10^{-10} f_{in}^2 + j3.725 \times 10^{-3} f_{in} - 0.0039 f_{in}^2 (1 + 100z)\epsilon^{-100z} - j0.033125 f_{in}(1 + 20z)\epsilon^{-20z}] \text{ A}, \quad (9\text{-}12)$$

from which

$$\frac{dI}{dx} = -I(19.5f_{\text{in}}^2\epsilon^{-100z} + j6.625f_{\text{in}}\epsilon^{-20z}) \text{ A/m}. \tag{9-13}$$

In deriving Eq. (9-13), use is made of $dI/dx = (dI/dz)(dz/dx)$, where dz/dx is taken from Eq. (9-11).

Do the preceding equations satisfy Eqs. (9-2) and (9-4)? From Eqs. (9-8) and (9-13),

$$\frac{dV}{dx} = -I(0.39f_{\text{in}}^2\epsilon^{-100z} + j0.1325f_{\text{in}}\epsilon^{-20z}), \tag{9-14}$$

which, with the aid of Eq. (9-9), agrees with Eq. (9-2). Similarly, one can show that Eq. (9-4) is satisfied.

How can we derive basilar membrane motion from the preceding equations? Assume that the membrane vibrates up and down, as depicted in Fig. 9-3(*a*). Now think of an infinitesimally small section; it is like a rectangular piston having a length dx and width W_m. The volume moved by this section is $v_m W_m\, dx$, where v_m is the velocity of the BM. This volume has to equal that of the fluid moved transversely above and below the piston, $A_c\, dv$, where A_c is *half* the cross-sectional area of the cochlear cylinder and v is the longitudinal velocity of the fluid. Then

$$v_m W_m\, dx = A_c\, dv. \tag{9-15}$$

Since dv is the mechanical equivalent of the electrical dI, however, we get

$$v_m = \frac{A_c}{W_m}\frac{dI}{dx}. \tag{9-16}$$

Because W_m is a function of x, the conversion factor A_c/W_m is not constant. It happens that W_m tapers linearly from approximately 0.1 mm at $x = 0$ to 0.5 mm at $x = 3.5$ cm. To simplify matters, however, we will use an in-between value, $W_m = 0.3$ mm. The half-cross-section value 1.5 mm^2 for A_c is reasonable. Then, using MKS units,

$$v_m = 0.005\frac{dI}{dx}. \tag{9-17}$$

We are more interested in displacement y_m than in velocity v_m. For sinusoidal motion,

$$y_m = \frac{v_m}{j2\pi f_{\text{in}}}. \tag{9-18}$$

Specifically, we wish to derive y_{rmd}, the *relative* membrane displacement, defined as

$$y_{\text{rmd}} = \frac{\text{basilar membrane displacement, } y_m}{\text{fluid displacement at the input, at } x = 0}. \tag{9-19}$$

From Eqs. (9-17) and (9-18), $y_m = (0.005/j2\pi f_{in})\,(dI/dx)$. The boundary values in Eq. (9-12) are such that the fluid velocity at the input is $1 + j0$. Then the fluid displacement at the input is $(1 + j0)/j2\pi f_{in}$. The $j2\pi f_{in}$ terms cancel and we are left with

$$y_{rmd} = 0.005\,\frac{dI}{dx}. \tag{9-20}$$

Substituting Eq. (9-13) for dI/dx,

$$y_{rmd} = -I(0.0975 f_{in}^2 \epsilon^{-100z} + j0.033125 f_{in} \epsilon^{-20z}). \tag{9-21}$$

Since Eq. (9-21) is complex, it includes amplitude and phase information. By inspection, the magnitude of membrane displacement is given by

$$|\,y_{rmd}\,| = |\,I\,|\sqrt{(0.0975 f_{in}^2 \epsilon^{-100z})^2 + (0.033125 f_{in} \epsilon^{-20z})^2} \tag{9-22}$$

where, from Eq. (9-12),

$$|\,I\,| = \exp\{f_{in}^2[5.765 \times 10^{-10} - 0.0039(1 + 100z)\epsilon^{-100z}]\}. \tag{9-23}$$

Equations (9-22) and (9-23) were employed to plot the curves of Figs. 9-7 and 9-8 with the modification that, for convenience, all curves are normalized so that the maximum value is 1. Before normalization, the peak values for the four curves of Fig. 9-7 are given as follows:

$$\begin{array}{lcllll} f_{in} & = & 170 & 1160 & 4700 & 15{,}000 \text{ Hz} \\ |\,y_{rmd}\,|_{max} & = & 1.298 & 2.457 & 5.492 & 12.04 \end{array} \tag{9-24}$$

Since $|\,y_{rmd}\,|$ is the BM displacement relative to that of the fluid at $x = 0$, the model indicates that there is some amplification of motion. This is due to a weak resonance effect.

How well do the peaks of Fig. 9-7 fit the f_c curve of Fig. 9-6? We can set the derivative of $|\,y_{rmd}\,|^2$ with respect to z equal to zero in order to locate the peak of the $|\,y_{rmd}\,|$ curve. This yields

$$f_c = \frac{1}{c_1}\sqrt{\frac{100c_1 - c_2 z + \sqrt{(100c_1 - c_2 z)^2 + 80c_1 c_2 z}}{2z}}, \tag{9-25}$$

where

$$c_1 = 39\epsilon^{-100z} \tag{9-26}$$

and

$$c_2 = (13.25\epsilon^{-20z})^2. \tag{9-27}$$

Table 9-4 lists various values of f_c versus x, as given by the preceding equations. Except for the extremes at $x = 0$ and 3.5 cm (which are therefore omitted), the errors are reasonably small relative to the experimentally derived curve, Fig. 9-6.

Finally, the traveling-wave representation of BM motion, in Fig. 9-5, shows instantaneous displacement if $f_{in} = 1160$ Hz. The curves are given by

$$y = |y_{rmd}| \sin(\theta_{yrmd} + \phi), \tag{9-28}$$

where $\phi = 0°, 30°, 60°, \ldots$, and the curve is modified so that the maximum value of the envelope is 1. According to Eq. (9-24), $|y_{rmd}|_{max}$ for $f_{in} =$ 1160 Hz is 2.457, so Eq. (9-28) is divided by 2.457 to get Fig. 9-5. Equation (9-21) gives

$$\theta_{yrmd} = \theta_I + \arctan\left(\frac{0.3397}{f_{in}}\epsilon^{80z}\right), \tag{9-29}$$

for the y_{rmd} phase shift, while Eq. (9-12), by inspection, indicates that the I phase shift is

$$\theta_I = f_{in}[0.003725 - 0.033125(1 + 20z)\epsilon^{-20z}]. \tag{9-30}$$

REFERENCES

B. Blum, Logic operations in the central nervous system—Implications for information transfer mechanism, *Kybernetik*, vol. 11, pp. 170–174, 1972.

E. Borg and S. A. Counter, The middle-ear muscles, *Sci. Am.*, vol. 261, pp. 74–80, Aug. 1989.

W. E. Brownell and B. Kachar, Outer hair cell motility: A possible electro-kinetic mechanism, in *Peripheral Auditory Mechanisms*, eds. J. B. Allen et al. New York: Springer-Verlag, 1986.

P. Dallos, *The Auditory Periphery*. New York: Academic Press, 1973.

S. Deutsch, On the determination of input sound frequencies by the auditory central processor, *IEEE Trans. Biomed. Eng.*, vol. 37, pp. 556–564, June 1990.

S. Deutsch and E. M-Tzanakou, *Neuroelectric Systems*. New York: New York Univ. Press, 1987.

H. Duifhuis, Current developments in peripheral auditory frequency analysis, in *Working Models of Human Perception*, eds. B. A. G. Elsendoorn and H. Bouma. London: Academic Press, 1989.

A. Flock, Hair cells, receptors with motor capacity? in *Hearing Physiological Basis and Psychophysics*, eds. R. Klinke and R. Hartmann. Berlin: Springer-Verlag, 1983.

A. Flock and S. Orman, Micromechanical properties of sensory hairs on receptor cells of the inner ear, *Hear. Res.*, vol. 11, pp. 249–260, 1983.

M. Furst and J. L. Goldstein, A cochlear nonlinear transmission-line model compatible with combination tone psychophysics, *J. Acoust. Soc. Am.*, vol. 72, pp. 717–726, 1982.

I. Glass, Tuning characteristics of cochlear nucleus units in response to electrical stimulation of the cochlea, *Hear. Res.*, vol. 12, pp. 223–237, 1983.

J. L. Goldstein, Auditory nonlinearity, *J. Acoust. Soc. Am.*, vol. 41, pp. 676–689, 1967.

J. L. Goldstein, Updating cochlear driven models of auditory perception: A new model for nonlinear auditory frequency analysing filters, in *Working Models of*

Human Perception, eds. B. A. G. Elsendoorn and H. Bouma. London: Academic Press, 1989.

J. L. Goldstein and P. Srulovicz, Auditory-nerve spike intervals as an adequate basis for aural frequency measurement, in *Psychophysics and Physiology of Hearing*, eds. E. F. Evans and J. P. Wilson. New York: Academic Press, 1977.

J. J. Guinan, B. E. Norris, and S. S. Guinan, Single auditory units in the superior olivary complex: Locations of unit categories and tonotopic organization, *Int. J. Neurosci.*, vol. 4, pp. 147–166, 1972.

J. L. Hall, Two-tone distortion products in a nonlinear model of the basilar membrane, *J. Acoust. Soc. Am.*, vol. 56, pp. 1818–1828, 1974.

I. J. Hochmair-Desoyer and E. S. Hochmair, An eight channel scala tympani electrode for auditory prostheses, *IEEE Trans. Biomed. Eng.*, vol. BME-27, pp. 44–50, Jan. 1980.

A. J. Hudspeth, The hair cells of the inner ear, *Sci. Am.*, vol. 248, pp. 42–52, Jan. 1983.

D. H. Johnson, The relationship between spike rate and synchrony in responses of auditory-nerve fibers to single tones, *J. Acoust. Soc. Am.*, vol. 68, pp. 1115–1122, 1980.

D. T. Kemp, Stimulated acoustic emissions from within the human auditory system, *J. Acoust. Soc. Am.*, vol. 64, pp. 1386–1391, 1978.

E. I. Knudsen, The hearing of the barn owl, *Sci. Am.*, vol. 245, pp. 82–91, Dec. 1981.

M. C. Liberman, Auditory-nerve responses from cats raised in a low-noise chamber, *J. Acoust. Soc. Am.*, vol. 63, pp. 442–455, 1978.

G. E. Loeb, The functional replacement of the ear, *Sci. Am.*, vol. 252, pp. 104–111, 1985.

H. McDermott, An advanced multiple channel cochlear implant, *IEEE Trans. Biomed. Eng.*, vol. 36, pp. 789–797, July 1989.

J. M. Miller and A. L. Towe, Audition: Structural and acoustical properties, in *Physiology and Biophysics: The Brain and Neural Function*, eds. T. Ruch and H. D. Patton. Philadelphia: Saunders, 1979.

J. M. Miller, A. L. Towe, B. E. Pfingst, B. M. Clopton, and J. M. Snyder, The auditory system: Transduction and central processes, in *Physiology and Biophysics: The Brain and Neural Function*, eds. T. Ruch and H. D. Patton. Philadelphia: Saunders, 1979.

B. C. J. Moore and R. D. Patterson, *Auditory Frequency Selectivity*. New York: Plenum, 1986.

R. R. Pfeiffer, A model for two-tone inhibition of single cochlear-nerve fibers, *J. Acoust. Soc. Am.*, vol. 48, pp. 1373–1378, Dec. 1970.

W. S. Rhode, Observations of the vibration of the basilar membrane in squirrel monkeys using the Mossbauer technique, *J. Acoust. Soc. Am.*, vol. 49, pp. 1218–1231, 1971.

W. S. Rhode, C. D. Geisler, and D. T. Kennedy, Auditory nerve fiber responses to wideband noise and tone combinations, *J. Neurophysiol.*, vol. 41, pp. 692–704, 1978.

A. C. Sanderson, Discrimination of neural coding parameters in the auditory system, *IEEE Trans. Syst. Man, Cybern.*, vol. SMC-5, pp. 533–542, Sept. 1975.

M. R. Schroeder, Models of hearing, *Proc. IEEE*, vol. 63, pp. 1332–1350, Sept. 1975.

P. M. Sellick, R. Patuzzi, and B. M. Johnstone, Comparison between the tuning properties of inner hair cells and basilar membrane motion, *Hear. Res.*, vol. 10, pp. 93–100, 1983.

S. A. Shamma, Speech processing in the auditory system, *J. Acoust. Soc. Am.*, vol. 78, pp. 1612–1632, 1985.

W. M. Siebert, Frequency discrimination in the auditory system: Place or periodicity mechanisms? *Proc. IEEE*, vol. 58, pp. 723–730, 1970.

G. F. Smoorenburg, Audibility region of combination tones, *J. Acoust. Soc. Am.*, vol. 52, pp. 603–614, 1972.

P. Srulovicz and J. L. Goldstein, A central spectrum model: A synthesis of auditory-nerve timing and place cues in monaural communication of frequency spectrum, *J. Acoust. Soc. Am.*, vol. 73, pp. 1266–1276, 1983.

G. von Békésy, *Experiments in Hearing*. New York: McGraw-Hill, 1960.

T. F. Weiss, A model of the peripheral auditory system, *Kybernetik*, vol. 4, pp. 153–175, 1966.

W. Welkowitz, S. Deutsch, and M. Akay, *Biomedical Instruments: Theory and Design*, 2nd ed. New York: Academic Press, 1992.

R. L. White, Review of current status of cochlear prostheses, *IEEE Trans. Biomed. Eng.*, vol. BME-29, pp. 233–238, Apr. 1982.

E. D. Young and M. B. Sachs, Representation of steady-state vowels in the temporal aspects of the discharge patterns of populations of auditory-nerve fibers, *J. Acoust. Soc. Am.*, vol. 66, pp. 1381–1403, 1979.

H. P. Zenner, U. Zimmerman, and U. Schmitt, Reversible contraction of isolated mammalian cochlear hair cells, *Hear. Res.*, vol. 18, pp. 127–133, 1985.

G. Zweig, R. Lipes, and J. R. Pierce, The cochlear compromise, *J. Acoust. Soc. Am.*, vol. 59, pp. 975–982, 1976.

E. Zwicker, On a psychoacoustical equivalent of tuning curves, in *Facts and Models in Hearing*, eds. E. Zwicker and E. Terhardt. Heidelberg: Springer-Verlag, 1974.

J. J. Zwislocki, Theory of the acoustical action of the cochlea, *J. Acoust. Soc. Am.*, vol. 22, pp. 778–784, 1950.

J. J. Zwislocki, Analysis of some auditory characteristics, in *Handbook of Mathematical Psychology*, vol. 3, eds. R. D. Luce et al. New York: Wiley, 1965.

Problems

1. Calculate Table 9-2 for the peak values of excitation in the ear canal and cochlea when the input is at 100 dB SPL at f_{in} = 500 Hz. Given for row 11: the velocity at the peak location, x = 2.4 cm, is 1.57 times the input velocity. [Ans.: (Row 1) 28.28 μbar; (3) 0.6815 cm/s; (4) 2.169 μm; (6) 79.95 μN; (7) 103.9 μN; (8) 0.5243 cm/s; (10) 692.9 μbar; (11) 0.8231 cm/s; (12) 2.62 μm.]
2. Given that the curve of Fig. 9-6, the characteristic frequency f_c versus distance from the base x, fits the equation $f_c = \alpha \exp[\beta(0.035 - x)^\gamma]$, and the curve goes through the following three points: $x = 0$, f_c = 15,000 Hz; x = 0.02 m, f_c = 1160 Hz; x = 0.035 m, f_c = 20 Hz: (a) Find $f_c(x)$; (b) find $x(f_c)$; (c) plot the curve. {Ans.: (a) $f_c = 20 \exp[45.79(0.035 - x)^{0.5769}]$; (b) $x = 0.035 - (0.02184 \log f_c - 0.06542)^{1.733}$.}
3. Given the cochlear line impedance Z of Eq. (9-9) and the equivalent $R \| L$ of Eq. (9-3): (a) Find R and L as functions of z and f_{in}. Find R and L at x = 1 cm if

(b) f_{in} = 170 Hz; (c) f_{in} = 15,000 Hz. At x = 2.5 cm and f_{in} = 170 Hz, (d) find R, L, and the Q; (e) find for the BM the equivalent G in series with C, and the Q. [Ans.: (a) $R = 0.04502\epsilon^{60z} + 0.39f_{in}^2\epsilon^{-100z}$ Ω/m, $L = 0.02109\epsilon^{-20z} + 0.1827f_{in}^2\epsilon^{-180z}$ H/m; (b) R = 593.6 Ω/m, L = 0.8927 mH/m; (c) R = 605.5 Ω/m, L = 0.9107 mH/m; (d) R = 18.67 Ω/m, L = 2.935 mH/m, Q = 5.957; (e) G = 46690 S/m, C = 7.337 F/m, Q = 5.957.]

4. Show that the boundary condition $I = 1 + j0$ at $x = 0$ is satisfied in Eq. (9-12).
5. Carry out the step-by-step derivation of Eq. (9-13).
6. Show that Eq. (9-4) is satisfied by Eqs. (9-12) and (9-13).
7. For distance from the base x = 2.5 cm, use Eqs. (9-22) and (9-23) to plot relative BM vibration magnitude, in decibels, versus input frequency using a log scale. At a point 60 dB down, construct an appropriate small triangle and find the slope, in dB/oct, from $\Delta y/\Delta x$.
8. Carry out the step-by-step derivation of Eq. (9-25).
9. For f_{in} = 1160 Hz, plot $|I|$ (fluid velocity) and $|y_{rmd}|$ (relative basilar membrane displacement) versus x (distance from the base) for $0 < x < 2.6$ cm. Underneath the magnitude curves, also plot their phase shifts, θ_I and θ_{yrmd}.
10. For f_{in} = 500 Hz: (a) From Eqs. (9-29) and (9-30), plot BM phase shift θ_{yrmd} versus distance from the base x. (b) With the aid of Eqs. (9-22) and (9-23), plot the "motion-picture" sequence, similar to that of Fig. 9-5, for ϕ = 0°, 90°, 180°, ... (c) Summarize the traveling wave by listing ϕ versus x and relative BM displacement y, similar to Table 9-3. [Ans.: (c) ϕ = 0°, x = 0, y = 0.25; ϕ = 90°, x = 1.32 cm, y = 0.55; ϕ = 180°, x = 2.00 cm, y = 0.86; ϕ = 270°, x = 2.39 cm, y = 1; ϕ = 360°, x = 2.64 cm, y = 0.86; ϕ = 450°, x = 2.84 cm, y = 0.36; ϕ = 540°, x = 3.02 cm, y = 0.01.]
11. (a) For f_{in} = 500 Hz, use Eqs. (9-22) and (9-23) to plot relative BM vibration magnitude $|y_{rmd}|$ versus distance from the base x. (b) Repeat for f_{in} = 1500 Hz. (c) Normalize the curves of parts (a) and (b) so that their peak value is 1; calculate and plot the nominal envelope if both tones are simultaneously received. Indicate the BM vibration bands in which each input tone is dominant. (d) For f_{in} = 1500 Hz, plot the family of curves showing relative AP sync versus distance from the base, similar to Fig. 9-12, using the same assumptions: 100% sync at the peak of the 35-dB SPL curve, and acoustic reflex compression starting at 100 dB SPL.
12. In Fig. 9-14(*b*), the V_{in} frequency to a reverberatory circuit tuned to 500 Hz is 250 Hz. If the V_{in} frequency is 166.7 Hz, (a) show the columnar addition; (b) draw the corresponding V_{out} waveform; (c) find V_{out} at t ≅ 0.1 s. [Ans.: (c) n = 16, V_{out} = 3.673.]
13. We wish to construct an amplitude information table, similar to that of Table 9-6, for f_{in} = 1500 Hz. (a) Use Eq. (9-25) to find x for characteristic frequencies 1500, 750, 500, 375, and 300 Hz. (b) Calculate the table entries using the procedure given in Table 9-6: "First, multiply curve supplied by Eq. (9-22) by a factor . . . clip maximum values at 10." [Ans.: (a) x = 1.75, 2.15, 2.39, 2.59, and 2.77 cm, respectively; (b) 20 dB SPL, V_{out} = 3.4; 35 dB SPL, V_{out} = 18.9; 70 dB SPL, V_{out} = 34.5; 110 dB SPL, V_{out} = 40; 140 dB SPL, V_{out} = 40.2.]

10

The Eye as a Transducer

ABSTRACT. Visible light is any electromagnetic wave that falls into the wavelength range extending from 380 nanometers (frequency = 7.9×10^{14} Hz, adjacent to ultraviolet), to 825 nm ($f = 3.6 \times 10^{14}$ Hz, adjacent to infrared). When we look at an object, the lens of the eye intercepts a cone of light from the object and focuses it on the retina. As the object being viewed moves closer to the eye, accommodation muscles change the shape of the lens (within limits) so as to maintain sharp focus.

At the center of the eye, the foveola, cone cell sensory receptors are about 3 μm in diameter. This corresponds to a visual angle of 0.7′. Including a factor of safety, it is customary to assume that the human eye cannot resolve objects that subtend an angle of less than 1′.

Cone photoreceptors handle daylight from the maximum tolerable illuminance level to "black," an amplitude range of approximately 4500 to 1, with 7 minutes required for dark adaptation. The visible range is extended an additional factor of around 22 by rod photoreceptors, with 30 minutes required for dark adaptation.

It is conjectured that the photoreceptors supply graded potentials proportional to the *logarithm* of visual stimulus rather than proportional to the stimulus itself.

The retina contains five layers: the *photoreceptors* feed a receptor potential to *bipolar* and *horizontal* cells. The latter gather in the receptor potentials over a local area and, in effect, measure the average local value. A fraction of the average receptor potential is subtracted from the bipolar cell input to yield an output that is fed to *ganglion* cells. All of the potentials leading up to the ganglion cells are graded; the ganglion cells translate the graded potentials into action potentials (APs). A layer of *amacrine* cells is involved with the detection of illuminance change.

In the process of subtracting local receptor potential from bipolar cells, the horizontal cells enhance contrast and improve the visibility of small objects. Several simple examples are given, including that of Mach bands.

Many of the ganglion cells are of a *transient* variety: they and amacrine cells respond to change (flicker) in the bipolar cell output. The amacrine cells gather in the flicker over a local area and, in effect, measure the average local flicker. The average flicker is subtracted from the transient ganglion cell input. This suppresses transient ganglion cell output due to massive flicker and improves the detection of flicker (movement) of small objects.

About 65% of the cones respond maximally to yellow wavelengths and are called *red*; 33% respond maximally to green and are called *green*; 2% respond maximally to indigo and are called *blue*. The subjective sensation of color (hue and saturation) depends on the red:green:blue ($C_R:C_G:C_B$) cone output ratio (before logarithmic compression). For *white*, the values are 90:45:1. Four different methods of generating the "white" ratio are presented.

The standard chromaticity diagram issued by the International Commission on Illumination is discussed. It turns out that if we plot $0.9 \log C_R + 0.1 \log C_B - \log C_G$ versus $\log C_G - \log C_B$, over the visible wavelengths, we get a chromaticity diagram that is topologically similar to the standard diagram. It is conjectured that the visual cortex forms thousands of two-dimensional color maps, with *illuminance* forming a third dimension. The input to every group of contiguous red, green, and blue cones corresponds to a point in the map derived from that group.

The derived chromaticity diagram is modified to represent three extreme types of color blindness: no red; no green; or no blue cone photoreceptors.

In black-and-white contrast enhancement, a bright center is surrounded by a region that is darker than the normal background. In color contrast enhancement, a colored center is surrounded by a region that is its complement, or "opponent." If a bright green center is seen against a dark gray background, the surround will appear to be slightly reddish. A blue center can have a yellowish opponent surround; and so forth. A hypothetical retinal model for the generation of red and green opponent-surround receptive fields is considered.

10-1 Some Characteristics of the Eye

All of us are acquainted with magnets, those small rectangular parallelepipeds that hold promissory notes against the sides of the refrigerator. We are also acquainted with electric fields via lightning discharges and pieces of paper that are attracted to a good insulator on a dry day. The magnet and insulator demonstrate magnetic and electric fields that change slowly, if at all. But let a magnetic or electric field change rapidly, and we get a different creature, an electromagnetic wave. In a vacuum, all electromagnetic waves propagate at a speed of 3×10^8 meters per second (m/s). Visible light is an electromagnetic wave in a particular and relatively narrow range of frequencies.

Visible light is generated by electron motions that span atomic distances. In effect, the atoms and molecules act like tiny antennas. The lowest visible frequency, adjacent to infrared, has a wavelength of 8250 Å (8.25×10^{-7} m), so the lowest visible frequency is 3.6×10^{14} Hz. The highest visible frequency, adjacent to ultraviolet, has a wavelength of 3800 Å (3.8×10^{-7} m), and its frequency is 7.9×10^{14} Hz.

People who work with visible light prefer to use wavelength rather than frequency, and they prefer nanometers (nm) rather than angstroms, perhaps because the wavelength in nm is one digit shorter. Accordingly, in this chapter we will use nanometers: visible light extends from a wavelength of 380 nm to 825 nm.

Why has the eye evolved so as to receive the range 380 to 825 nm? Our natural energy source—the sun—bathes the earth with electromagnetic radiation in this range of wavelengths. The sun generates a very wide spectrum, from radio waves to X-rays, but the earth's atmosphere does not allow much of the infrared and ultraviolet to reach the living planet. Some radio waves do get through, but plant and animal cells are much too small to act as antennas for radio waves. They are just about the right size, however, for visible light. Size alone is insufficient; plant cells had to evolve chemical cycles, such as that which uses chlorophyll, in which the synthesis of organic compounds is energized by visible light. Similarly, animals had to evolve specialized photoreceptor cells that contain organic compounds, such as rhodopsin, that can transduce visible light into graded receptor potentials via a suitable electrochemical cycle.

A highly simplified cross section through the human eye is depicted in Fig. 10-1. The orientation is indicated by "temporal side" (toward the side of the head) and "nasal side" (toward the nose). A visual object is shown in the form of an arrow pointing in the temporal direction. A small cone of light from the arrowhead is intercepted by the lens, as shown, and focused on the retina some 15 mm (typically) behind the center of the lens. The rays of light bend as they leave the rarer medium, air, and enter the denser lens material. (If you swim underwater with your eyes open, vision is blurred not because water pressure flattens the lens but because the rays suffer insufficient bending in going from water to the eye's lens material.) As the object being viewed moves closer to the eye, the lens has to be squeezed by the *accommodation* muscles of Fig. 10-1 so that it bulges more at the center in order for the object to continue to focus 15 mm behind the lens center. For a person with "normal" vision at any age, the relaxed lens, looking at a distant object, correctly focuses 15 mm behind the lens center. A 10-year-old can squeeze the lens so as to change its focal length by 2.3 mm, sufficient to properly focus on an object only 8 cm away. But the lens of a 70-year-old is so rigid that he or she can only change the focal length by 0.22 mm, sufficient for an object 1 m away. The focal-length change values

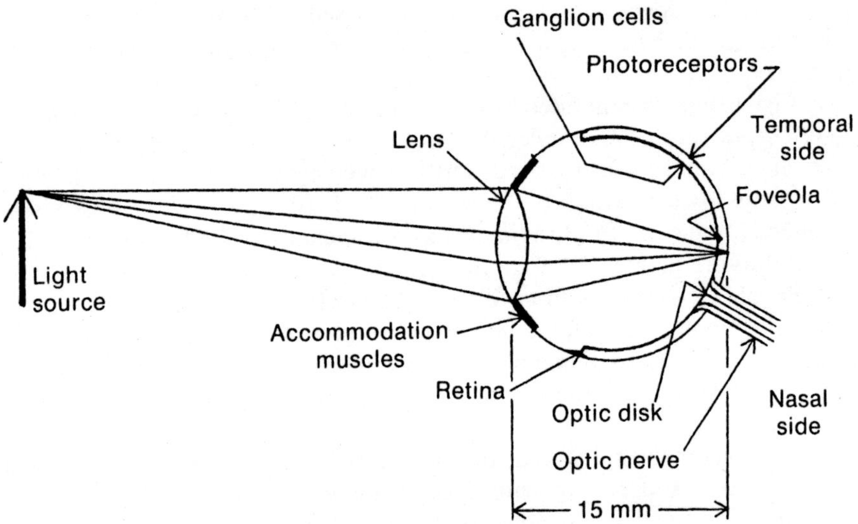

Fig. 10-1 Highly simplified cross section through the human eye. The cornea and iris are omitted. The orientation is given by "temporal side" (toward the side of the head) and "nasal side" (toward the nose). The cone of light from an arrowhead is traced out as it passes through the lens and through the retina to focus on the photoreceptors some 15 mm behind the lens center.

of 2.3 mm for a 10-year-old and 0.22 mm for a 70-year-old are relatively invariant limits. Thus, a 70-year-old whose relaxed lens focuses at 1 m can squeeze the lens sufficiently to focus at ½ m; for objects beyond 1 m or less than ½ m the person needs bifocal-lens eyeglasses [G. S. Brindley, 1970; T. N. Cornsweet, 1971; C. Graham, 1966; D. Marr, 1982; C. R. Michael, 1969; R. W. Rodieck, 1973; T. C. Ruch, 1979; J. L. Schnapf and D. A. Baylor, 1987; L. Stryer, 1987].

The retina contains five cell layers, but the two most important are the photoreceptors, where light energy is converted into a graded receptor potential, and the ganglion cells, where graded potentials are converted into APs for transmission, via the optic nerve, to the brain. Here it seems as if there is an error on the drawing because light has to shine through the ganglion cells to reach the photoreceptors! Alas, it is an example in which evolution started off on the wrong foot and was not able to turn back. (One of our life-scientist reviewers insists that photoreceptors are toward the rear because they "require a high oxygen tension" which, presumably, is maintained because the photoreceptors are closer to the blood supply.) The fact remains that entering light has to cross ganglion, amacrine, bipolar, and horizontal layers to reach the photoreceptors, so all of these cells have to be essentially transparent. Furthermore, the million or so axons that leave the ganglion cells come together at a single location, the "optic disk," and leave by way of the optic nerve through a hole in the retina. The hole forms a blind spot about 15° toward the nasal side of the foveola. Because light rays cross over as shown, the optic disk on the nasal side of the retina is responsible for a blind spot on the temporal side of the visual scene. (To "see" the blind spot, close your right eye, say, and look straight ahead with the left eye. Hold a pencil horizontally, with eraser pointing temporally, and move the pencil slowly to the left. About 15° to the left of center the eraser will disappear. What does the blind spot look like? Is it a black hole? Not at all—it seems to blend in with its surroundings. Viewed against a red sheet of paper, the disappearing pencil eraser seems to be replaced by the red sheet of paper. One cannot be certain, however, because visual acuity (sharpness of vision) 15° off to the side is far less than at the foveola.)

The foveola is the central area used when we "look at" an object. Here some 34,000 photoreceptors are crowded together in a diameter of 0.6 mm. On an average basis, each receptor has a diameter of around 3 μm; the receptors in the center are even a bit smaller than this (3 μm = 3000 nm, a diameter that corresponds to only a few wavelengths of red light). Visual acuity is limited by the diameter of foveal receptors. At a distance of 15 mm behind the lens, a receptor diameter of 3 μm corresponds to the following visual angle:

$$\frac{3\ \mu\text{m}}{15{,}000\ \mu\text{m}} \times \frac{360^\circ}{2\pi} = 0.011^\circ = 0.7'.$$

Tests show that it is better to be conservative and use 1′ rather than 0.7′ for the limiting resolving power of the eye. You can try the following experiment: look at an area covered by parallel black-and-white strips. If each strip subtends an angle of 1′ or more at your eye, you should be able to see the individual strips. If the angle is less than 1′, you will probably see a uniform gray field.

A ubiquitous example is a TV screen. The U.S. picture is synthesized out of 495 horizontal strips. This is not quite the same as in the previous paragraph, where we have an equally wide sequence of white, black, white, black, . . . strips. For the TV picture, how far away do we have to sit so that each strip will subtend an angle of 1′? At this distance the edges between strips should disappear. The calculation follows:

$$\text{Viewing distance} = \left(\frac{\text{picture height}}{495}\right)\left(\frac{60 \times 360'/\text{circle}}{2\pi}\right) = 7 \times \text{height}.$$

For a picture height of 0.4 m (16 in.), the edges between horizontal strips should disappear if you sit 2.8 m (9 ft) away. Yet, on many sharply focused TV pictures, one can clearly see horizontal scanning strips from a viewing distance of 7 × picture height. For technical reasons, adjacent scan strips tend to form pairs so that the spacing between strips 1 and 2, 3 and 4, . . . is slightly less than the spacing between strips 2 and 3, 4 and 5, ... In other words, we actually have 495 strips, but they may come across as 247 strips of double width. For a picture height of 0.4 m, therefore, one may have to sit 5.6 m away before the lines between strips vanish.

There are two different types of photoreceptors: rod-shaped or *rod* receptors that are very sensitive dim-light transducers, and cone-shaped or *cone* receptors that dominate and inhibit the rods under "daylight" conditions. The cones are responsible for color vision, while rod vision is free of color. The rods are maximally sensitive to green at a wavelength of 512 nm [M. C. Cornwall, E. F. MacNichol, Jr. & A. Fein, 1984], but the subjective impression is that of dark-gray. Although some young people would have it otherwise, the human animal is *not* nocturnal; in fact, there are no rods at all in the foveola, so one must actually look slightly off center to see those dim nighttime forms. The density of rods increases away from the foveola, but rod acuity at best is poor—only about 7% as good as that of foveal cone vision. The rods are specialized for high sensitivity to light and thereby sacrifice resolution; the subjective response is blurred and dim, as well as free of color.

Color vision is considered later on in this chapter; for now, it will be assumed that cone vision is black-and-white. The relative time responses of cones and rods are given, in Fig. 10-2, versus light intensity L_z on a logarithmic scale.

The International System (SI) unit for illuminance is the *lux*. This is a power unit that is inconvenient for deriving gradient voltages. Therefore, in Fig. 10-2 as well as in the remainder of this chapter, we employ $\sqrt{\text{lux}}$, which is a measure of amplitude rather than power. The word "brightness" is not encouraged in scientific work because it is a subjective response that one cannot measure precisely. The lux, on the other hand, is precisely defined: it is the illuminance corresponding to one lumen/m^2. The lumen is defined in terms of a precisely specified "standard candle" source of light, called a *candela* [S. Sherr, 1970].

Figure 10-2 shows cone and rod sensitivity phases as the light intensity is reduced from maximum tolerable, 320, to the minimum at which light can be detected, 0.0032. The maximum/minimum ratio is 100,000—not quite as good as the 10^7 ratio for the ear, but an impressive evolutionary accomplishment nevertheless. (The ear uses 20 dB of acoustic reflex compression plus 40 dB of nonlinear basilar membrane compression to reduce the actual vibration ratio to 10,000.)

If the light intensity is suddenly decreased from 320 to 0.5, a ratio of around 600, the subjective response is that of a sudden decrease from maximum tolerable white to black. This is represented by the thick vertical line, from 320 to 0.5, at $t = 0$. The pupil, which is maximally closed to protect the eye at the 320 level, now starts to open. The pupil (omitted in Fig. 10-1) is an opaque membrane, with an expandable opening, that almost covers the lens. With the sudden decrease of intensity at $t = 0$, the cones' pupillary reflex allows the opening to slowly increase to maximum so that, after 7 minutes, the sensitivity increases to the point at which level 0.07 is judged

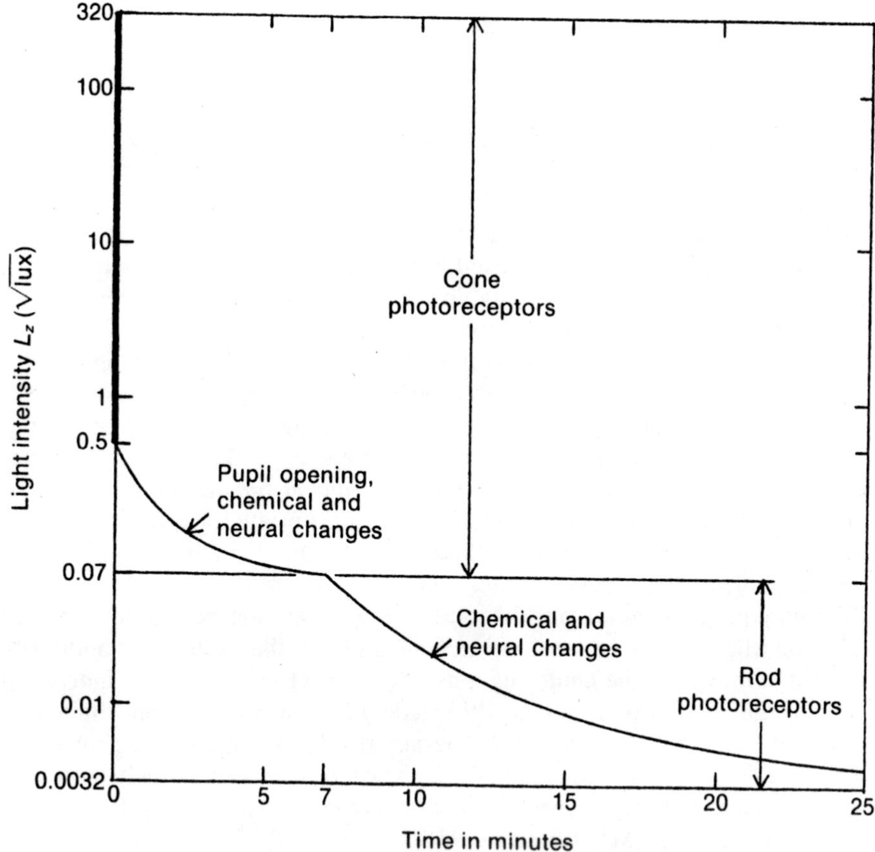

Fig. 10-2 Cone and rod sensitivity phases as relative light intensity (amplitude, not power) is reduced from maximum tolerable white to minimum at which light can be detected. The thick vertical line at zero time represents a sudden decrease from maximum tolerable white to black. The black level is given by the lower boundary of the diagram.

to be the complete absence of light. As indicated in Fig. 10-2, some chemical and neural changes are also implicated in the increase of sensitivity [S. Hecht and S. Shlaer, 1938; L. W. Stark, 1984]. Some of these changes are discussed in the next section.

At light intensity level 0.07 the cones cease to inhibit the rods, which have their own curve of increasing sensitivity. Here the changes are entirely due to slow chemical and neural factors; 30 minutes after the initial maximum level of 320, the sensitivity level 0.0032 is approximately reached. The eye thus becomes fully dark adapted in around 30 minutes.

The remainder of the chapter is devoted to cone excitation, with light intensity input L_z between levels 0.5 and 320.

10-2 Retinal Response to Steady Light

The retina is committed to more than the transduction of visible light energy into APs. This alone can be accomplished with only two layers: photoreceptors that yield graded potentials, and ganglion cells that convert the latter into APs. Perhaps the eyes of some primitive animals exhibit two layers, but the mammalian retina also enhances contrast and signals the presence of motion in the visual scene; this requires three additional layers.

The detection of motion is discussed in Section 10-5. The present section is devoted to a steady-state visual input. The neural model is given in Fig. 10-3, with the "wiring" of the motion detectors omitted.

Signal processing starts, of course, with the layer of photoreceptors (P). A cone receptor has to cope with a 600-to-1 dynamic input range; it does this by continuously "measuring" the average light intensity. The output versus input characteristic is then modified by "chemical and neural changes," as noted in Fig. 10-2, so that the *transfer function* (ratio of output graded potential V_P to input stimulus L_z) is appropriate for the average light intensity. Exactly how the cone "knows" the average intensity is a controversial subject; there may be lateral communication between cones and/or they may get feedback from the next retinal layer, the horizontal cells. In addition, the eyes are in constant *saccadic* motion; that is, they involuntarily move randomly, as the centering muscles contract in response to individual APs, so that each cone wanders within a field equivalent to 50 cone diameters in about 4 s. (All of us are thus characterized, to some extent, by shifty eyes.) In this way, the light usually seen by each cone continuously increases and decreases with respect to the local average intensity [D. A. Baylor, M. G. F. Fuortes & P. M. O'Bryan, 1971; R. Siminoff, 1983].

After the cone receptor measures the average light intensity, it compresses its output because a 600-to-1 ratio is excessive. Without compression we would expect an AP output frequency of 12,000 Hz, say, for a very bright stimulus and 20 Hz for a very dim input. A frequency of 12,000 Hz is, of course, out of the question. With compression, the very bright input may result in an AP frequency of 500 Hz, versus 20 Hz for a very dim stimulus.

What is a reasonable shape for the output versus input compression curve? The ganglion AP output in Fig. 10-3 carries color information, and we know that subjective color is only weakly dependent upon brightness. This implies that the nervous system extracts *ratios* such as green input divided by red input, and blue input divided by green input. Although evolution, given sufficient time, could stumble on a way to actually calculate ratios, it is much more natural to suppose that the retina generates *logarithms*, via a simple nonlinearity, followed by subtraction. As a numerical example, suppose that red and green stimuli are, respectively, 270 and 135 (in relative units). Logarithmic subtraction yields $\log 270 - \log 135 = 5.60 - 4.91 = 0.69$. If there is a 10-to-1 reduction in illuminance intensity, we get $\log 27 - \log 13.5 = 3.29 - 2.60 = 0.69$ again, thus resulting subjectively in the same hue. (Although we use natural logarithms in this chapter, the

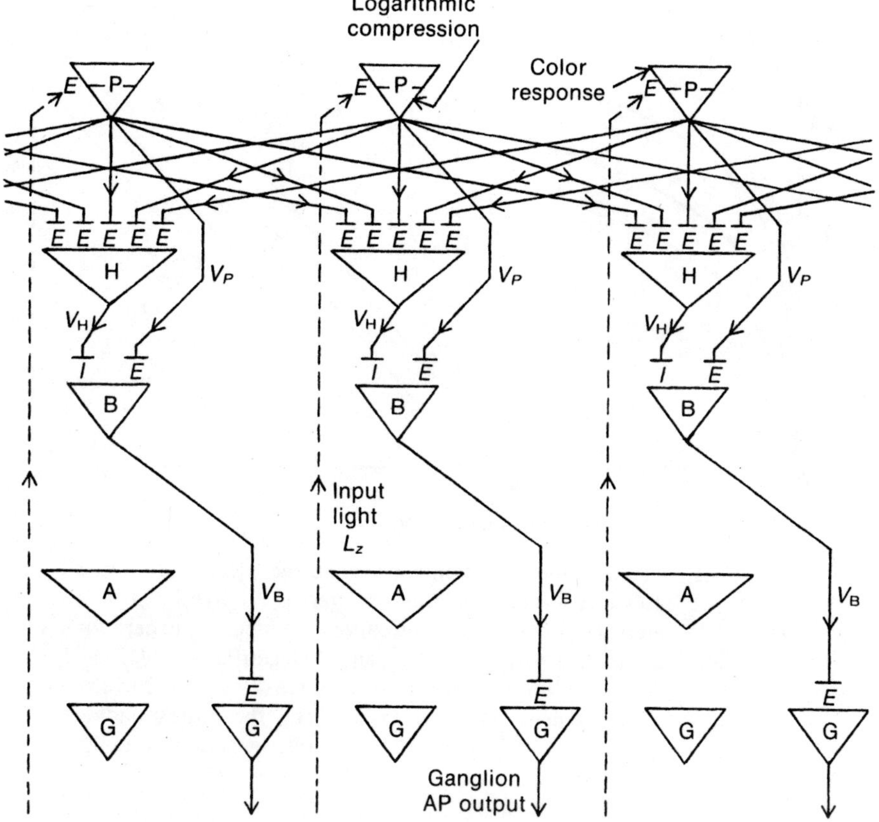

Fig. 10-3 Idealized neural model of the retinal elements involved in signal processing of a steady-state visual input. The A and G cells ("wiring" not shown) are involved with the detection of motion. Input light enters at the bottom of the page and passes through transparent layers of G, A, B, and H cells until it reaches the P photoreceptors. Key: P = photoreceptor, H = horizontal, B = bipolar, and G = ganglion cells; V_P, V_H, and V_B are graded potentials; ganglion cell outputs are action potentials.

invariance does not depend on the logarithmic base.) [G. Buchsbaum, 1980; R. A Normann and I. Perlman, 1979; R. A. Normann et al., 1983]

Logarithmic compression is illustrated by the family of curves in Fig. 10-4, which show how the cones' transfer characteristics change as average intensity changes. Each curve obeys the following equation (voltages in this chapter are given with respect to the resting potential):

$$V_P = 16 + 5 \log\left(\frac{L_z}{L_{z0}}\right), \qquad 0.0408 < \frac{L_z}{L_{z0}} < 24.5 \quad \text{(mV units)} \qquad (10\text{-}1)$$

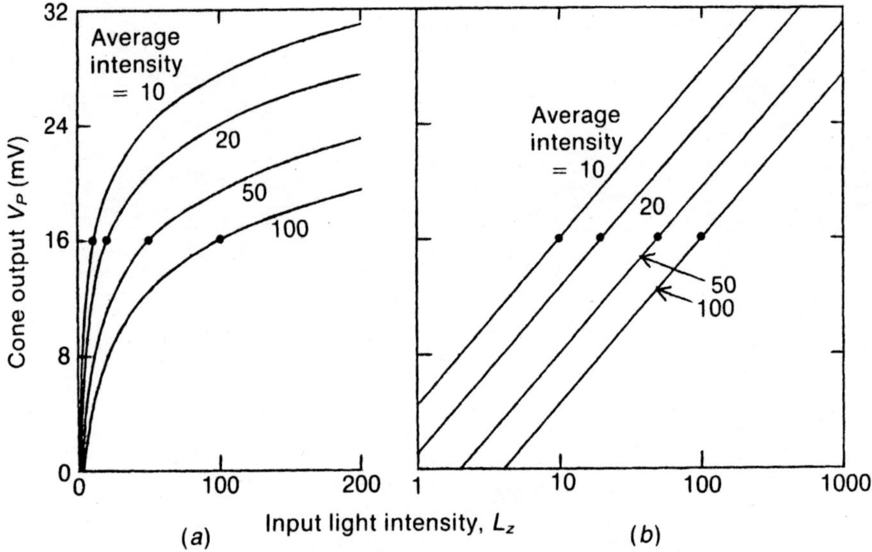

Fig. 10-4 Idealized output potential of cone receptors, before and after logarithmic compression, versus relative input light intensity. Both families of curves are identical except for (*a*) uniform horizontal scale, and (*b*) logarithmic scale. The black dots indicate where input intensity is equal to average intensity over the visual field; the corresponding cone output is 16 mV. The cone output saturates at the 32-mV level.

where V_P = cones' output potential,
L_z = input light intensity,
L_{z0} = average of local light intensities.

The units of light intensity are not pertinent because we always work with ratios, such as L_z/L_{z0}. This equation is valid over a restricted range: L_z/L_{z0} cannot be greater than 24.5 or less than 0.0408, as indicated. This agrees with the 600-to-1 dynamic input range because $24.5/0.0408 = 600$. Also, 24.5 and 0.0408 are reciprocals, and $24.5 = \sqrt{600}$.

In Eq. (10-1), if the input light intensity is the maximum value, we get $V_P = 16 + 5 \log 24.5 = 32$ mV. This is shown as the upper limit in Fig. 10-4(*b*). For L_z/L_{z0} greater than 24.5, the cone output saturates at the 32-mV value. If the input light intensity is the minimum value, we get $V_P = 16 + 5 \log 0.0408 = 0$. For L_z/L_{z0} less than 0.0408, the cone output remains at the zero (black) level. If the input light intensity is the same as the average value, we get $V_P = 16$ mV; this is shown by the series of black dots.

The family of curves in Fig. 10-4(*a*) and (*b*) are identical except that Fig. 10-4(*b*) is plotted using a log horizontal scale that emphasizes the symmetry inherent in Eq. (10-1). The curves of Fig. 10-4(*a*) are better at showing that the cone output severely compresses high input light intensities.

The "average intensity = 50" curve is used in illustrative examples in the remainder of this chapter.

Returning to Fig. 10-3, the P cells are shown as if they first perform the preceding logarithmic compression; that is, they first respond to light intensity followed by a response to wavelength (color). The horizontal (H) cells are "wired up" to measure the average photoreceptor output over a relatively wide area. They are shown as receiving excitatory inputs *from* all of the local P cells. Actually, in a cross section seen through a microscope, H cells appear with dendrites reaching out *to* all of the local P cells and, if the orientation is the same as that of Fig. 10-3, the dendrites look like horizontal arms (hence the name). It is more convenient for our purposes, however, to show the excitatory junctions as being lined up against the body of the H cell.

The next layer consists of bipolar (B) cells. The B cell is excited by the P photoreceptor but inhibited by the H cell above it. The purpose of the B cell is to engage in this interplay between E and I inputs so as to yield contrast enhancement, as illustrated later on in this chapter. Specifically, it will be assumed that

$$V_B = V_P - 0.5V_H; \tag{10-2}$$

here an inhibitory weighting factor of 0.5 yields reasonable results. In a cross section seen through a microscope, the bipolar cell appears as an approximately spherical body with dendrites reaching upward and an axon reaching downward; that is, it has a bipolar structure [A. L. Gilchrist, 1979; H. K. Hartline, F. Ratliff & W. H. Miller, 1961; B. W. Knight, J. I. Toyoda & F. A. Dodge, 1970; M. V. Srinivasan, S. B. Laughlin & A. Dubs, 1982].

The fourth layer in the retina consists of amacrine (A) cells. These are concerned with motion, which is discussed in Section 10-5.

Finally, the last layer consists of ganglion (G) cells. As indicated, these convert the graded potential output of bipolar cells into a corresponding AP output. Half of the G cells in Fig. 10-3 are "transient" ganglia that are turned on by motion and are therefore omitted from the "steady-state" schematic diagram [B. G. Cleland, M. W. Dubin & W. R. Levick, 1972; J. C. Curlander and V. Z. Marmarelis, 1983; C. Enroth-Cugell and J. G. Robson, 1966; Y. Fukada and H. Saito, 1971; H. Ikeda and M. J. Wright, 1972; S. W. Kuffler, 1953; J. J. Kulikowski and D. Tolhurst, 1973; H. Noda, 1975; R. M. Shapley and J. D. Victor, 1979].

10-3 Contrast Enhancement

In this section various "visual scenes" are presented to the eye, and the resulting bipolar cell output V_B is derived. In each case the *average* input light intensity L_{z0} is 50, so the cone output V_P is given by the "50" curves of Fig. 10-4. We assume a one-to-one correspondence between the bipolar graded potential output and the steady-state ganglion AP output frequency.

Figure 10-5: Our first visual scene is almost blank. It is a gray back-

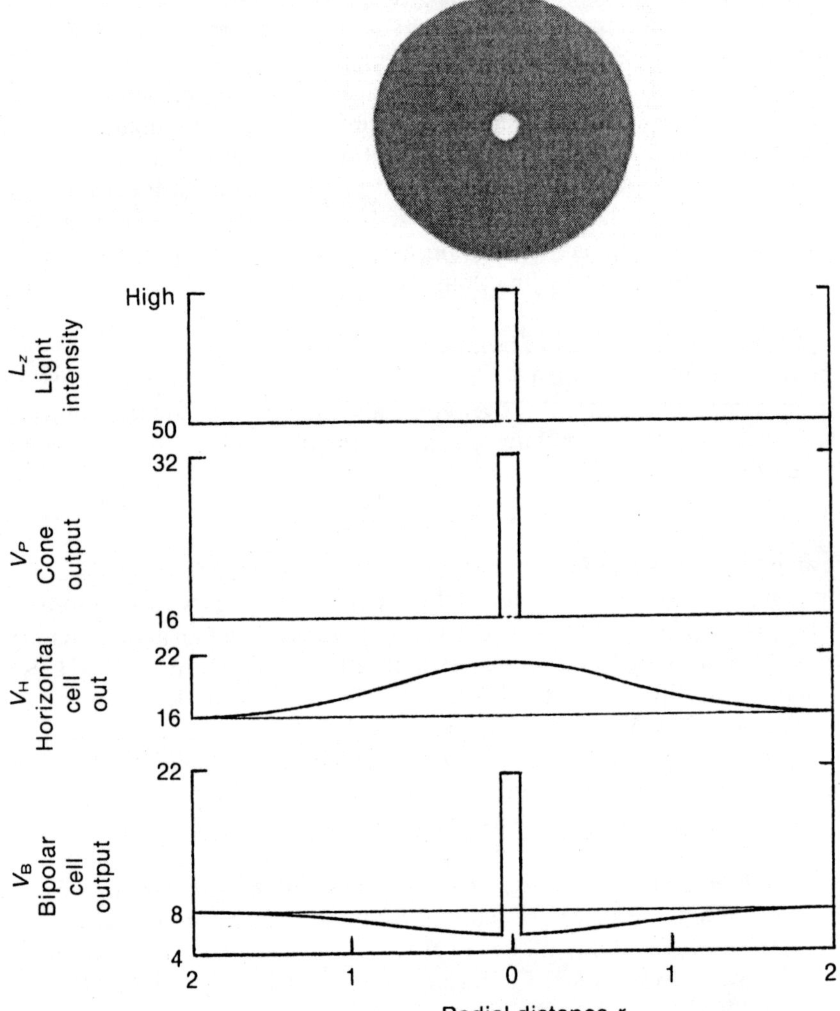

Fig. 10-5 Visual scene consisting of a gray field, of light intensity 50, with a bright light in its center. Assuming that the eye is stationary, idealized cone (V_P), horizontal (V_H), and bipolar (V_B) cell outputs are shown, in millivolts, versus radial distance from the center of the image of the bright light. The cells at the center respond to this stimulation with $V_P = 32$, $V_H = 21$, and $V_B = 21.5$ mV; the cells far off to the side with $V_P = 16$, $V_H = 16$, and $V_B = 8$ mV. The bipolar output should demonstrate contrast enhancement.

ground except for a bright light in the center, in accordance with the display at the top of Fig. 10-5. Because the bright light is too small to appreciably affect the average intensity, the gray background is shown as having a value $L_z = 50$. The corresponding V_P, according to Eq. (10-1) and Fig. 10-4, is 16 mV. At the center, in response to the bright light, the cones are saturated with an output of 32 mV.

The H cells far from the center do not "see" the bright light, so their output is the same as that of the peripheral cones, 16 mV. At the center, however, the high cone output, $V_P = 32$ mV, is responsible for $V_H \cong 21$ mV in the H cells directly underneath. As we move away from the center, the H-cell output gradually decreases from 21 mV to 16 mV. The reason is that, although the H cell dendrites funnel local P potentials into each H cell, the more remote cone information suffers more attenuation because it travels a longer dendritic distance before reaching the H cell. As a result, H cell output is biased in favor of its nearest cone neighbors.

The curves of Fig. 10-5 are based on a mathematical model that assumes that

$$\Delta V_H = \frac{\Delta V_P}{\pi} \epsilon^{-r^2}, \tag{10-3}$$

where ΔV_H = horizontal cell output relative to the background level,
ΔV_P = change in photoreceptor (cone) output relative to the background level,
r = radial distance from the light source center.

This "Gaussian distribution" model holds only if the light source has a relatively small radius, such as the $r = 0.1$ value illustrated in Fig. 10-5. The reason for the π in the denominator is that the volume under the ΔV_H curve is then numerically equal to ΔV_P because

$$\int_0^\infty \left(\frac{1}{\pi} \epsilon^{-r^2}\right) (2\pi r\, dr) = 1. \tag{10-4}$$

This leads to some simplification later on.

In Fig. 10-5, where $\Delta V_P = 16$ mV, we have $\Delta V_H = 16 \exp(-r^2)/\pi$ mV. Since the background level is 16 mV, we get

$$V_H = 16 + 5.093\epsilon^{-r^2} \quad \text{(mV units)}. \tag{10-5}$$

The final payoff is that contrast is enhanced because the bipolar cell output displays a voltage depression immediately surrounding the bright-light output. At the center, in accordance with Eq. (10-2), we have $V_B = 32 - (0.5)(21.093) = 21.45$ mV. The voltage depression is at level $V_B = 16 - (0.5)(21.093) = 5.45$ mV. Far away from center we have $V_B = 16 - (0.5)(16) = 8$ mV. The subjective impression should be that of a bright light surrounded by a dark region that radially and gradually approaches a gray background level, rather than a bright light surrounded by a uniform gray background.

You might remark that *your* subjective impression does not agree with this optimistic prognostication. Alas, it is difficult to demonstrate some of these visual illusions with small-area large-contrast examples.

Figure 10-6: Our next visual scene is a bright vertical line surrounded

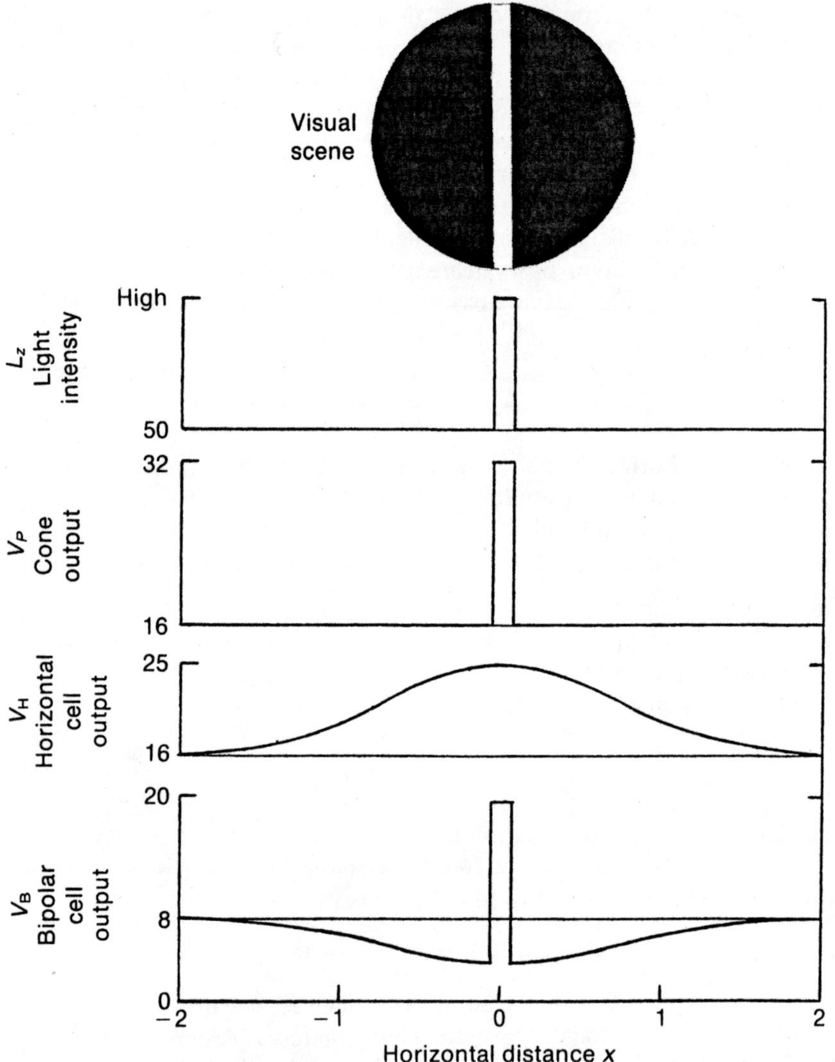

Fig. 10-6 Visual scene consisting of a gray field, of light intensity 50, with a bright line crossing its center. Assuming that the eye is stationary, idealized cone, horizontal, and bipolar cell outputs are shown, in millivolts, versus horizontal distance from the center of the image of the bright line. The bipolar output should demonstrate contrast enhancement.

by a featureless, gray background, as depicted at the top of Fig. 10-6. The spatial graded potential cross section is similar to that of Fig. 10-5 except that the radial distance of Fig. 10-5 is replaced by horizontal distance in Fig. 10-6. The average light intensity L_{z0} is 50, with negligible contribution from the relatively narrow vertical line of light. The V_P cone output is 16 mV except for a central line of 32 mV. The corresponding V_H output can be obtained from Eq. (10-3) as follows.

Think of the vertical line as an infinite array of the bright spots of Fig. 10-5. At a particular point x units away from the center of the vertical line, the contribution from this infinite array of bright spots is obtained by integrating Eq. (10-3) in the y direction, with $r^2 = x^2 + y^2$:

$$\Delta V_H = \frac{\Delta V_P}{\pi} \int_{-\infty}^{\infty} \epsilon^{-(x^2+y^2)}\, dy = \frac{\Delta V_P}{\sqrt{\pi}} \epsilon^{-x^2}. \tag{10-6}$$

This answer is easily obtained because it is based upon a standard definite integral. Next, the area under the ΔV_H curve is numerically equal to ΔV_P because

$$\int_{-\infty}^{\infty} \frac{1}{\sqrt{\pi}} \epsilon^{-x^2}\, dx = 1. \tag{10-7}$$

In Fig. 10-6, where $\Delta V_P = 16$ mV, $\Delta V_H = 16 \exp(-x^2)/\sqrt{\pi}$ mV. Since the background level is 16 mV, we get

$$V_H = 16 + 9.027\epsilon^{-x^2} \quad \text{(mV units)}. \tag{10-8}$$

The central line is a more effective stimulant, of course, than the dot of Fig. 10-5, so the H-cell output ranges from 16 mV to 25 mV in Fig. 10-6. Contrast is enhanced as the bipolar-cell output is 19.49 mV at the center, 3.49 mV at the valleys to either side of center, and 8 mV far from the central line. The subjective impression should be that of a bright line flanked by dark valleys that gradually approach a gray background level.

Figure 10-7: Next is the step transition of Fig. 10-7. The left half of the visual scene is black, the right half white. To maintain an average light intensity of $L_{z0} = 50$, the left half is at 2.04 mV while the right half is at 97.96 mV. According to Eq. (10-1), the V_P cone outputs are 0 and 19.36 mV, respectively. The H-cell outputs are also 0 and 19.36 mV, respectively, far from the central transition line. Between the two extremes the H-cell outputs smoothly increase from 0 to 19.36 mV as shown.

The curves of Fig. 10-7 can be obtained by integration from Fig. 10-6. In integrating, one starts at $x = -\infty$ and proceeds to $x = +\infty$, plotting the area under the curve that has already been swept out as one proceeds from left to right. For the ΔV_P curve of Fig. 10-6, the area is 0 for $-\infty < x < 0$; at $x = 0$ we pick up an area of A, and this value does not change for $0 < x < +\infty$. The integration thus yields an A step (19.36 mV high in Fig. 10-7). Since we are dealing with a linear system (except for the L_z curves), if integration of ΔV_P of Fig. 10-6 yields ΔV_P in Fig. 10-7 (except for a scale

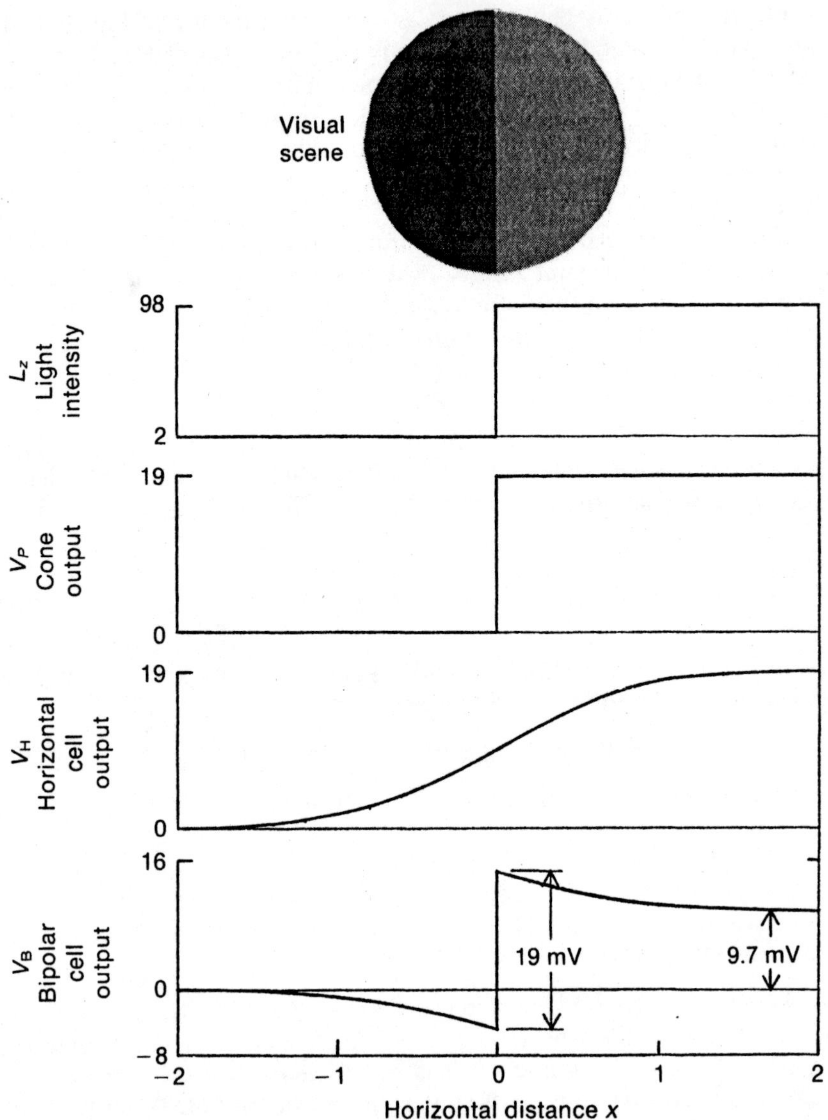

Fig. 10-7 Visual scene consisting of a dark-gray field on the left, of light intensity 2, and light-gray field on the right, of light intensity 98, so that the average intensity is 50. Assuming that the eye is stationary, idealized cone, horizontal, and bipolar cell outputs are shown, in millivolts, versus horizontal distance from the image of the step of light. The bipolar output should demonstrate contrast enhancement.

change), then integration of ΔV_H and ΔV_B of Fig. 10-6 yields ΔV_H and ΔV_B in Fig. 10-7 (except for the scale change).

The ΔV_H curve of Fig. 10-7 is accordingly given by integrating the right-hand side of Eq. (10-6):

$$\Delta V_{\mathrm{H}} = \frac{\Delta V_P}{\sqrt{\pi}} \int_{-\infty}^{x} \epsilon^{-u^2}\, du. \tag{10-9}$$

Because the ΔV_{H} curve has odd symmetry around $x = 0$, we can rewrite Eq. (10-9) for positive values of x as

$$\Delta V_{\mathrm{H}} = \Delta V_P\left(\frac{1}{2} + \frac{1}{\sqrt{\pi}} \int_{0}^{x} \epsilon^{-u^2}\, du\right). \tag{10-10}$$

This is evidently related to the *error function* (also called the *error integral* or *probability integral*), which is defined by

$$\operatorname{erf} x = \frac{2}{\sqrt{\pi}} \int_{0}^{x} \epsilon^{-u^2}\, du, \tag{10-11}$$

so that

$$\Delta V_{\mathrm{H}} = \frac{\Delta V_P}{2}(1 + \operatorname{erf} x). \tag{10-12}$$

Table 10-1 lists some values of erf x for positive x, but it is obvious from the table that $\operatorname{erf}(-x) = -\operatorname{erf} x$. (Also, $\operatorname{erf} \infty = 1$.) It therefore turns out that Eq. (10-12) is correct for negative as well as positive values of x.

It is easy to calculate the bipolar cell output, V_{B}, using Eq. (10-2). Contrast of the step transition is enhanced as V_{B} starts off at 0 and decreases to -4.84 mV to the left of the step. On the right, going from right to left, the potential starts off at 9.68 mV and increases to 14.52 mV. (The 2.04-to-97.96 range of L_z is mathematically convenient because of the zero V_P, V_{H}, and V_{B} baselines, but it is excessive. In a real-life situation one would go from 20 to 80, say, to avoid negative values in the bipolar cell output curve.

TABLE 10-1 Some Values of the Error Function (Also Called the Error Integral or Probability Integral)

x	erf x	x	erf x	x	erf x
0	0	0.5	0.5205	1	0.8427
0.05	0.0564	0.55	0.5633	1.1	0.8802
0.1	0.1125	0.6	0.6039	1.2	0.9103
0.15	0.1680	0.65	0.6420	1.3	0.9340
0.2	0.2227	0.7	0.6778	1.4	0.9523
0.25	0.2763	0.75	0.7112	1.6	0.9763
0.3	0.3286	0.8	0.7421	1.8	0.9891
0.35	0.3794	0.85	0.7707	2	0.9953
0.4	0.4284	0.9	0.7969	2.5	0.9996
0.45	0.4755	0.95	0.8209	3	1.0000

Note that $\operatorname{erf}(-x) = -\operatorname{erf} x$, and $\operatorname{erf} \infty = 1$.

Video signal excursions that are more negative than the black level are called *blacker-than-black* levels.)

Another way to look at the contrast enhancement is to observe that edges are preserved while background changes are reduced by a factor of 2 [because of the inhibitory weight of 0.5 in Eq. (10-2)]. In Fig. 10-7, the edge in the V_B curve has a height of 19.36 mV, the same as the ΔV_P output. At a distance of two units away from the edge, however, the V_B background change is only from 0 on the left to 9.68 mV on the right. The V_B output first dips and then overshoots so as to enhance contrast.

Figure 10-8: The next visual scene, that of Fig. 10-8, allows the reader to participate in an experiment that demonstrates the power of edge enhancement. We have a photograph of five gray vertical strips and, at the right, four spatial curves of cross sections taken through the photographs. The first cross section, Fig. 10-8(*a*), is "subjective" in that it illustrates our perception: five vertical gray strips, each with slightly lighter shading on the left and slightly darker shading on the right (except, of course, at the left and right edges of the photograph). The cross section in Fig. 10-8(*c*) is also subjective: we see a uniform gray area crossed by four narrow vertical black strips. The cross section in Fig. 10-8(*b*) shows the true conditions when the photograph was made: the *original* consisted of five vertical strips of *unequal* shades of gray, with the darkest strips on the left and lightest on the right. (Here there is some unintentional reversal of shading because of our amateurish photographic ability and equipment, but it does not destroy the edge

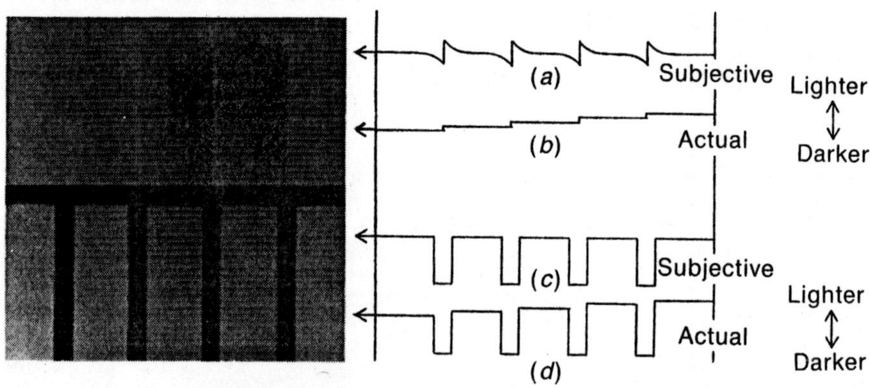

Fig. 10-8 Photograph of five gray vertical strips on the left and the same area plotted, on the right, with idealized ordinates representing brightness. (*a*), (*b*), (*c*), and (*d*) are cross sections taken through the photograph as shown. What we see looking across, with the benefit of contrast enhancement, is given by the "subjective" curves in (*a*) and (*c*). The true brightness variation is given by the "actual" curves in (*b*) and (*d*). (Some unintentional variation of shading was introduced by the photography process.)

enhancement.) The cross section in Fig. 10-8(*d*) shows the actual intensity levels after the edges were covered with four narrow vertical black strips.

Here we are the beneficiary of an optical illusion. The edge enhancement in Fig. 10-8(*a*) is subjective. Actually, the shading of each strip does not vary in a horizontal direction. The subjective response agrees with that of Fig. 10-7.

Why do we not perceive edge enhancement in the lower half of the photograph? It is probably there, but "swamped out" by the large changes in light intensity between gray and black strips, so that it is not noticeable. Contrast enhancement is of obvious survival value in an environment in which animals are camouflaged to blend in with their surroundings. (The human animal also uses camouflage, of course, to foster military deception.)

Figure 10-9: Our next example, that of Fig. 10-9, does not result in edge or contrast enhancement but, instead, is responsible for a well-known incorrect assessment of small-area brightness. To the left of $x = -2$, the visual field is dark gray with $L_z = 10$. A medium-gray vertical strip, of $L_z = 30$, is located at $x = -6$. To the right of $x = 2$ the visual field is light gray with $L_z = 90$. A medium-gray vertical strip, again of $L_z = 30$, is located at $x = 6$. The average light intensity, L_{z0}, is 50. Equation (10-1) yields $V_P = 7.95$ mV on the left and 18.94 mV on the right, with the vertical strips at 13.45 mV. The strips are narrow, so much so that they are not "seen" by the H cells, whose output is therefore the same as that of V_P: 7.95 mV on the left and 18.94 mV on the right. The net result of applying Eq. (10-2) is that V_B is 9.47 mV at the left strip, but only 3.98 mV at the right strip. Thus, although the strips are of equal L_z, the subjective impression is that the strip seen against a dark background seems to be lighter than the strip seen against a light background. (The values in Fig. 10-9 are for a special case in which the left strip has the same subjective brightness as the right background, and the right strip looks as dark as the left background.)

Figure 10-10: In the next visual scene, Fig. 10-10, the entire visual field is covered by vertical strips of light, called *sinusoidal gratings*, which increase and decrease from an L_z minimum of 10 to maximum of 90, with average L_{z0} of 50. Looking at Fig. 10-10(*a*), two spatial cycles extend from $x = -5$ to $+5$. (The horizontal distance units are the same as for Figs. 10-5, 10-6, 10-7, and 10-9.) Since we have two cycles in 10 units of distance, the spatial frequency is 0.2 cycles/unit distance. In Fig. 10-10(*b*), on the other hand, two spatial cycles extend from $x = -1$ to $+1$, so that the spatial frequency is 1 cycle/unit distance. We have a relatively low spatial frequency in Fig. 10-10(*a*) and high frequency in (*b*).

For both spatial frequencies the cone output V_P varies from a minimum of 7.95 mV to a maximum of 18.94 mV. The light intensity in Fig. 10-10(*a*) is given by

$$L_z = 50 - 40 \cos(72x^\circ). \tag{10-13}$$

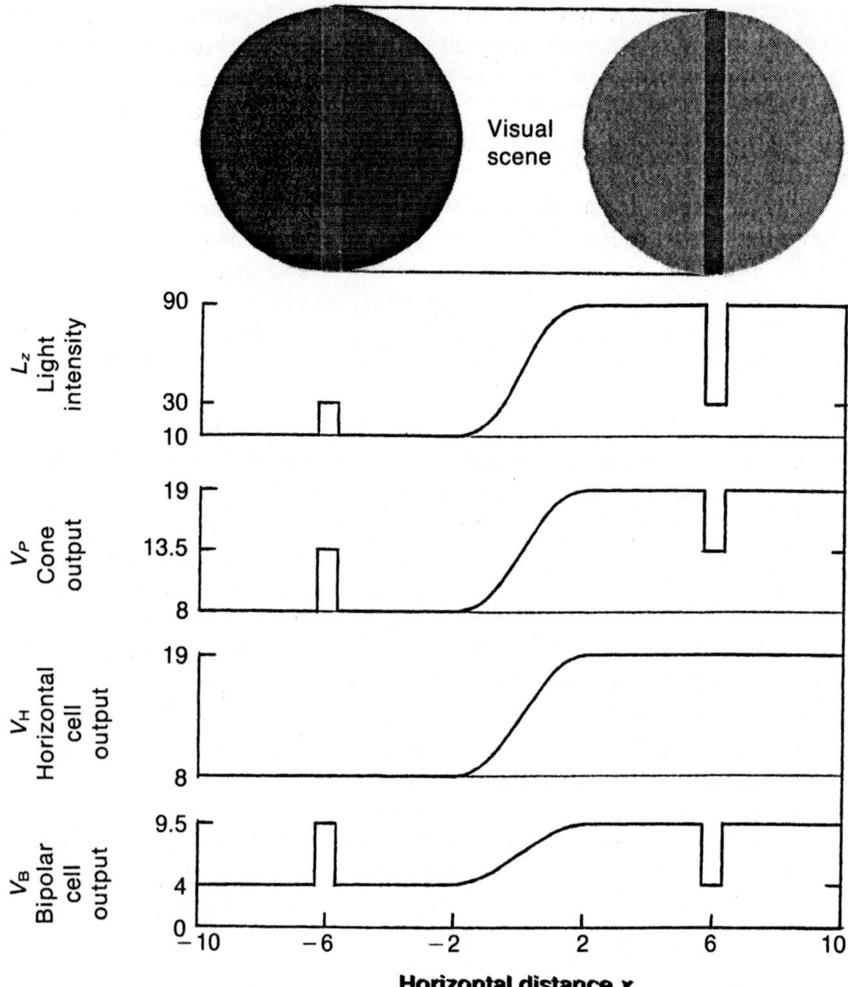

Fig. 10-9 Visual scene consisting of a gray vertical line against a dark background on the left, and the *same* vertical line against a light background on the right. Assuming that the eye is stationary, idealized cone, horizontal, and bipolar cell outputs are shown, in millivolts, versus horizontal distance from the center of the visual image. In the bipolar output, the line seen against a dark background seems to be lighter than the same line seen against a light background.

From Eq. (10-1), therefore,

$$V_P = 16 + 5 \log[1 - 0.8 \cos(1.257x)], \qquad (10\text{-}14)$$

where $1.257x = 72x°$. The V_P curve is flattened because of cone compression of high light intensities, of course, and this distortion introduces spatial har-

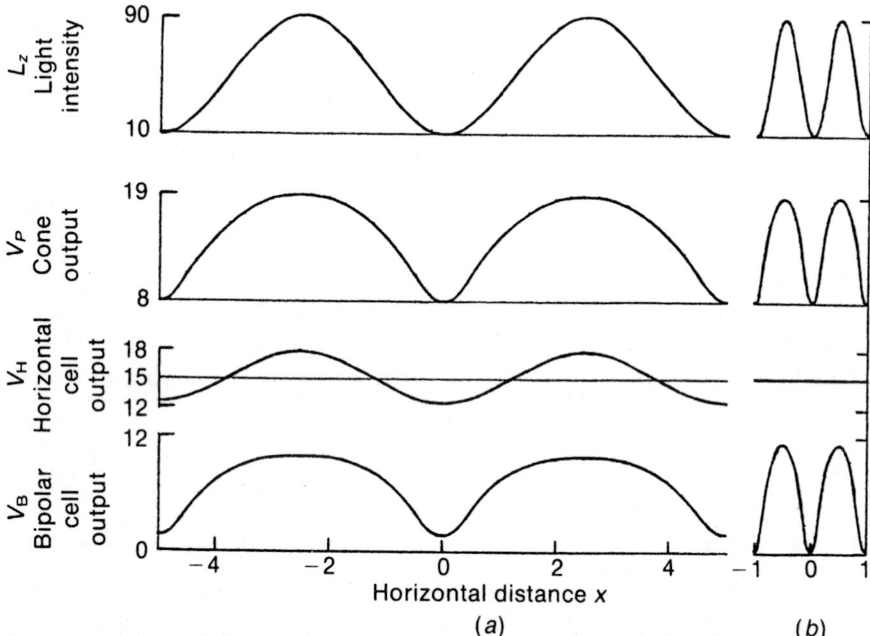

Fig. 10-10 Visual scene (not shown because it is technically difficult to reproduce) consisting of sinusoidal spatial variations of illuminance. On the left (*a*), we have 0.2 cycle per unit distance; on the right (*b*), 1 cycle per unit distance. Assuming that the eye is stationary, idealized cone, horizontal, and bipolar cell outputs are shown, in millivolts, versus horizontal distance from the center of the visual image. The peak-to-peak bipolar output is 8.3 mV in (*a*) and 11 mV in (*b*), showing that there is a greater response to small objects compared to large objects.

monics. In reaching out to the photoreceptor cells over a wide area, however, the H-cell output tends to smooth out these V_P variations. In other words, the second and higher spatial harmonics of V_{H} will be relatively small and can be ignored. In Fig. 10-10, therefore, it is assumed that V_{H} consists only of a dc and fundamental component.

The dc component is the same as that of the V_P curve:

$$V_{P0} = V_{\mathrm{H}0} = \frac{1}{\pi}\int_0^{\pi} [16 + 5\log(1 - 0.8\cos\theta)]\, d\theta \text{ mV}. \quad (10\text{-}15)$$

This result is evaluated with the aid of a standard definite integral,

$$\int_0^{\pi} \log(a \pm b\cos\theta)\, d\theta = \pi \log\left(\frac{a + \sqrt{a^2 - b^2}}{2}\right), \quad (10\text{-}16)$$

which yields $V_{P0} = V_{\mathrm{H}0} = 14.88$ mV.

A simple way to get an approximate value for the fundamental component of cone output, V_{P1}, is to use $\log(1 + u) \cong u$, which is valid for small u:

$$5 \log(1 - 0.8 \cos \omega x) \cong 5(-0.8 \cos \omega x) \text{ mV} \tag{10-17}$$

so that

$$V_{P1} = -4 \cos \omega x \text{ mV}. \tag{10-18}$$

For the fundamental component of V_{H}, since the pattern of Fig. 10-10 is vertically oriented, we can seek guidance from the curves of Fig. 10-6, where a ΔV_P line of area 1 at $x = 0$ would result in a ΔV_{H} curve that is described by $\exp(-x^2)/\sqrt{\pi}$. If we regard V_P of Fig. 10-10 as a curve that is traced out by an infinite number of vertical lines of excitation, the corresponding $V_{\mathrm{H}1}$ is found by integrating Eq. (10-6):

$$V_{\mathrm{H}1} = \frac{V_{P1}}{\sqrt{\pi}} \int_{-\infty}^{\infty} \epsilon^{-(u-x)^2} \cos \omega u \, du, \tag{10-19}$$

where $V_{P1} \cos \omega x$ is the fundamental component of cone output. This leads to a standard definite integral,

$$\int_0^{\infty} \epsilon^{-a^2u^2} \cos \omega u \, du = \frac{\sqrt{\pi}}{2a} \epsilon^{-\omega^2/(4a^2)}, \tag{10-20}$$

resulting in

$$V_{\mathrm{H}1} = V_{P1}\epsilon^{-\omega^2/4}. \tag{10-21}$$

For Fig. 10-10(a), where $\omega = 1.257$, Eqs. (10-18) and (10-21) indicate that

$$V_{\mathrm{H}1} = -2.695 \cos 1.257x \text{ mV} \tag{10-22}$$

but for Fig. 10-10(b), where $\omega = 6.283$, we get

$$V_{\mathrm{H}1} = -0.0002 \cos 6.283x \text{ mV}. \tag{10-23}$$

In Fig. 10-10(b), in other words, the horizontal cells do not "see" the spatial sinusoidal grating, so the drawing only shows V_{H} as a dc component, 14.88 mV.

Finally, as shown in the drawing, the application of Eq. (10-2) yields a bipolar-cell output V_{B} of 8.3 mV peak to peak for the low spatial frequency and 11 mV peak to peak for the high spatial frequency. The net effect is that our eyes are more sensitive to high spatial frequencies (small objects) than to low frequencies (large objects). This is true only up to a point, however; when the small objects approach one minute of arc, which is the limiting spatial acuity determined by foveal cone diameters, we can no longer resolve the objects.

You can judge the preceding effects for yourself from page 156 of *Mach Bands*, by Floyd Ratliff (1965). It shows a sinusoidal grating test pattern (but it is a good idea to place a black sheet of paper behind the pattern to keep the printed material on page 155 from showing through on page 156).

Figure 10-11: The last "visual scene" in this section is the ramp transition of Fig. 10-11. Here the light intensity L_z in the left half of the visual field is a black level of 10. Between $x = -1$ and $+1$ the light intensity linearly increases to a white level of $L_z = 90$. As usual, because the stationary retinal image sees the entire visual scene (in this case from $x = -3$ to $+3$), the average, L_{z0}, is 50. From Eq. (10-1), the cone output V_P is around 8 mV in the left half of the field and around 19 mV in the right half.

The H-cell output can be found through a tabular procedure rather than through integration. (It is fully described in S. Deutsch and E. M-Tzanakou, 1987.) First, the horizontal distance scale is broken up into strips and steps. For Fig. 10-11, from $x = -\infty$ to $x = -3$, the cone output V_P is an 8-mV-high step. Between $x = -3$ and $x = +3$, the horizontal scale is broken up

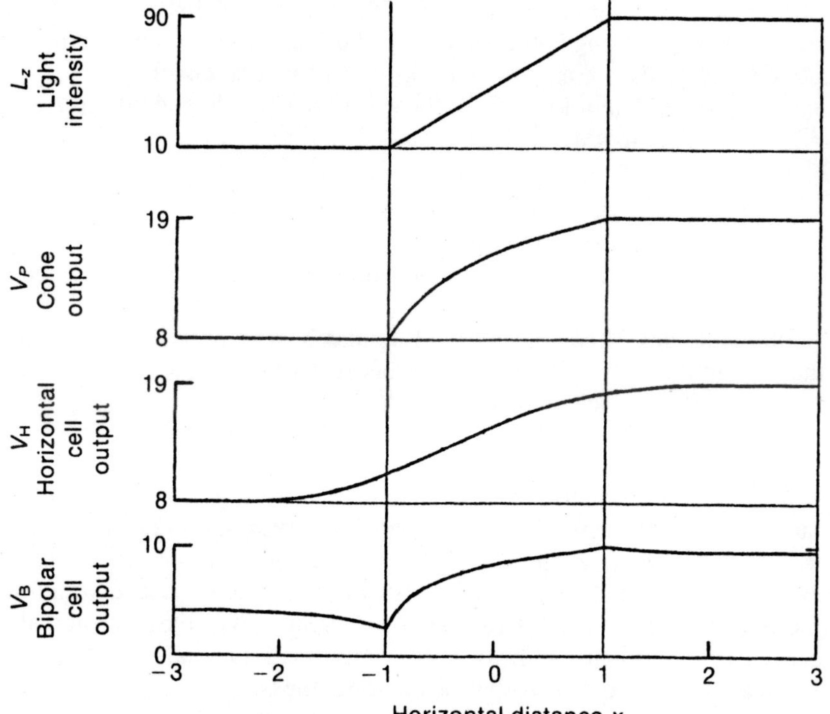

Fig. 10-11 Visual scene (not shown because it is technically difficult to reproduce) consisting of a ramp-step spatial variation of illuminance. Assuming that the eye is stationary, idealized cone, horizontal, and bipolar cell outputs are shown, in millivolts, versus horizontal distance from the center of the visual image. The bipolar output demonstrates contrast enhancement.

into strips 0.2 unit wide (strip from -3 to -2.8, strip from -2.8 to -2.6, . . . , strip from $+2.8$ to $+3$), so that the 6-unit-wide section yields 30 vertical strips. From $x = +3$ to $x = +\infty$, V_P is a 19-mV-high step. The central idea is that each horizontal retinal cell receives an input from *each* of these steps and strips, and the summation of inputs determines the horizontal cell output, $V_H(x)$.

From Eq. (10-12), the H-cell response to a step of excitation V_P high is given by

$$V_H = \frac{V_P}{2}(1 - \operatorname{erf} \Delta x) \tag{10-24}$$

where Δx is the horizontal distance measured from the *edge* of the step to an H-cell located at x. (A minus sign is needed because, in this definition, the horizontal distance is always a positive quantity.)

The H-cell response to a V_P strip is approximated by the response to a vertical line of excitation. Figure 10-6 shows the curve if V_P is a narrow line. The V_H curve is approximately flat in the range $-0.1 < x < +0.1$; this implies that there is negligible change in the V_H curve if V_P is a strip 0.2 unit wide, rather than a narrow line, provided the area under the V_P curve remains the same. From Eq. (10-6), therefore, the H-cell response to a strip V_P high and 0.2 unit wide is given by

$$V_H = \frac{0.2V_P}{\sqrt{\pi}} \epsilon^{-(\Delta x)^2}, \tag{10-25}$$

where Δx is the distance from the *center* of the strip to an H cell located at x.

Although each H cell receives 32 inputs (2 steps plus 30 strips), most of the contributions are practically zero because $\operatorname{erf} \Delta x$ and $\exp[-(\Delta x)^2]$ fall off rapidly with distance.

The smooth V_H curve of Fig. 10-11 is drawn "by eye" through the calculated points located at $x = -2.9, -2.7, \ldots, +2.9$. Equation (10-2) is then applied to get the bipolar cell output, V_B. The latter displays contrast enhancement: at the left end of the light-intensity ramp there is a cusp of decreased brightness, while at the right end there is a cusp of increased brightness. The subjective impression should be that the slope discontinuities at the left and right edges of the ramp are greater than they are actually. Because high-input light intensities are compressed, the high-brightness enhancement is much less than that of low-brightness.

10-4 Mach Bands

When Ernst Mach (1838–1916) looked at a stationary visual scene corresponding to that of Fig. 10-11, he noticed contrast enhancement. Since he was the first person to report on the phenomenon, the "Mach bands" came to be named after him [A. Fiorentini and T. Radici, 1958].

Typical experimentally generated Mach bands are illustrated in Fig. 10-12(*a*). For the reader who is ambitious, the example of cubist art given in Fig. 10-12(*b*) represents a disk that, if rapidly rotated, will generate the stimulus of Fig. 10-12(*a*). If the disk is rotated at 60 revolutions per second (rps) or faster, persistence of vision (i.e., the time taken for photoreceptors to respond to an input change) will yield smooth ramp transitions between gray and white levels, without flicker. (It is highly recommended that a photocopy of Fig. 10-12(*b*) should be used rather than tearing the page out of the book.)

According to the stimulus curve of Fig. 10-12(*a*), the middle of the rotating disk has a step that is black for 135°, a ramp that gradually changes from 135° to 45° black, and a step that is black for 45°. (The reader can easily check this for him- or herself by projecting down from the stimulus curve to the disk.) The corresponding visual response of Fig. 10-12(*a*) displays a bump at each edge of the ramp; these are the Mach bands. The visual effect is as follows: As we look at the rapidly rotating disk, along the ramp from left to right, it gradually becomes lighter. At the 45° level, where the ramp flattens off, we see a narrow white band between the ramp and the light-gray region to the right. Similarly, as we look at the rotating disk, along the ramp from right to left, it gradually becomes darker. At the 135° level, where the ramp flattens off, we see a narrow black band between the ramp and the dark-gray region to the left. The subjective white and black bands highlight, of course, the transition from ramp to constant spatial light intensity; these are the Mach bands.

The ramps to the left and right of the central stimulus are calibration ramps: we match the Mach bands of the central ramp against the gray transitions of the left and right calibration ramps. From the location of the match, we can estimate the size of each Mach-band bump on a drawing of the response curve such as that of Fig. 10-12(*a*). The effect is entirely subjective; there is no way to use a light meter or any other light-measuring device to outline the Mach bands.

As we have previously noted, it is difficult to reproduce some of these visual illusions. Ratliff (1965) shows two pages of "photographs" of ramp transitions, on pages 44 and 53, but had to have them "touched up" to exaggerate the white and black Mach-band effects.

In the previous "visual scenes" of this chapter we have a high-contrast input stimulus, such as light intensity ranging from 10 to 90 in Figs. 10-9, 10-10, and 10-11, a maximum/minimum ratio of 9. In Fig. 10-12, however, the contrast is low. It is in such low-contrast situations that the Mach bands are needed to improve contrast, just as the edge enhancements of Fig. 10-8 become apparent in a low-contrast environment.

Although the Mach bands are similar to the cusps on the bipolar cell output curve for the ramp-step visual scene of Fig. 10-11, judging from Ratliff's book the following substantial differences remain.

(a) The calculated bipolar-cell output curve of Fig. 10-11 retains sharp

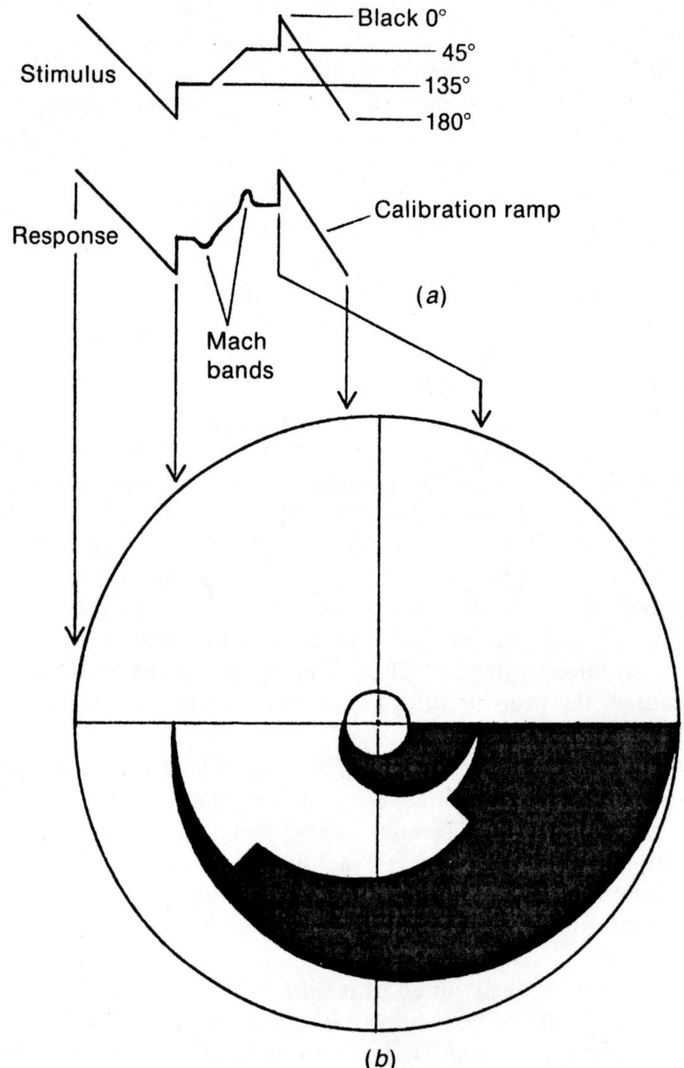

Fig. 10-12 A Mach-band generator. The disk of (*b*), if rapidly rotated, generates the visual stimulus of (*a*). To check this for yourself, project down from the stimulus curve to the disk. According to the stimulus curve, the middle of the rotating disk has a step that is black for 135°, a ramp that gradually changes from 135° to 45° black, and a step that is black for 45°. The corresponding visual response of (*a*) displays Mach bands as shown. The two outside ramps, which gradually change from 0° to 180° black, can be used to calibrate the Mach bands: the viewer compares the brightness of each Mach band with that of the nearest calibration ramp, locating the ramp point whose brightness equals that of the Mach band.

transition edges, whereas the edges are replaced by rounded "bands" in Fig. 10-12(*a*).

(b) The effect of changing the slope of the ramp while maintaining constant light-intensity maximum and minimum values (by changing to a different rotating disk, for example) is most unusual. With a small slope the Mach bands are of course not observed. As the slope is increased, the amplitude of the Mach bands increases. When the ramp becomes almost vertical, the amplitude decreases! With a step transition only the weak and very broad overshoots and undershoots of the bipolar-cell output of Fig. 10-7 remain.

(c) Mach bands are only seen with the central, foveal region of each eye. If you look a few degrees to the side, the bands disappear. Compare this with the contrast enhancement of Fig. 10-8, which is fairly constant over the entire retina as you look off to the side.

[S.D. writes: "Tom Smith of the Electrical Engineering Department, University of South Florida, glued a photocopy of the disk of Fig. 10-12(*b*) to a plate attached to a small motor. The disk rotated at close to 3600 rpm. The Mach bands proved to be relatively weak, and would not be recognized if we did not know what to expect. I disagree with the claim that they are only observed with the central region of the eye; it seemed to me that their "disappearance" when looking off to the side was caused by decreased visual acuity, and not by a change in the way visual images are processed."]

The Mach band processor modulates the visual image before it is "projected" upon the surface that constitutes our subjective experience. In an ordinary viewing situation, we ascribe Mach bands to the interaction between reflectance of the object viewed and ambient lighting conditions, unaware of the sophisticated image processor.

10-5 Response to Flicker and Motion

Consider now the detection of moving "bugs" in which a small part of the visual field starts to change; motion detection is obviously an important survival strategy.

John E. Dowling and Frank S. Werblin (1969) used the ingenious experimental technique of a rotating many-bladed "windmill," such as that of Fig. 10-13. The center of the windmill was deliberately left unobstructed, and one could change the illuminance of this central "dot." As the windmill rotated, it obscured one area of the visual field as it exposed another, so that a constant total input over a second, say, was received by the horizontal cells in the surround region. In other words, the horizontal-cell output was independent of the speed of windmill rotation [V. Torre, W. G. Owen & G. Sandini, 1983; F. S. Werblin, 1973].

As an experimental subject, Werblin and Dowling chose the mudpuppy salamander, which has relatively large retinal cells. They were able to get microelectrodes into the transient ganglion cells, the G cells that are not

Fig. 10-13 "Windmill" used as a visual scene in the study of amacrine and transient ganglion cells (see text). The *total* stimulus received by horizontal cells is independent of windmill rotation speed. Illuminance of the central "dot" can be varied as the windmill is rotated. Changes in the central dot are perceived when the windmill is either stationary or moving so fast that it appears to be a uniform gray source. For an in-between rotation speed, when the windmill appears to be a flickering source, the perception of changes in the central dot is inhibited. (From Deutsch and M-Tzanakou, *Neuroelectric Systems*, New York Univ. Press, New York, 1987.)

"wired up" in Fig. 10-3. The experimental scenarios and results can be summarized as follows:

Windmill Motion	Central-Dot Input	Central-Dot Transient Ganglion
(a) Stationary	Varied	Active
(b) Slow rotation (flicker)	Varied	Inhibited
(c) Rapid rotation (no flicker)	Varied	Active

In (a), the windmill was stationary, but the transient ganglion cells that were associated with the central dot generated APs in response to central-dot illuminance variation. The same was true in (c), where the windmill was rotated so rapidly that it was equivalent to a stationary gray blur. But in (b), where the retina surround saw flicker because of slow rotation, the central-dot transient ganglion APs were inhibited. The conclusion is that large-area change (represented by the flickering windmill) inhibits the transient ganglion cells, but small-area change due to a moving "bug" (such as change in the light intensity of the central dot) is fully seen if the surrounding visual field is quiet. The subjective effect is that we do not notice the moving "bug" when its surroundings are changing.

The transient ganglion circuits are shown, fully wired up, in Fig. 10-14. This diagram may look like an awful mess at first, but a bit of patience will show that it can be deciphered. The symbol F stands for a response to "flicker," or change. The bipolar (B) cell outputs feed amacrine (A) cells and transient ganglion cells as well as the steady-state ganglion cells of Fig. 10-3. The amacrine cells are portrayed as being topologically similar to horizontal (H) cells. The amacrine cells measure the bipolar output over a relatively wide area. They are shown as receiving flicker inputs *from* all of the

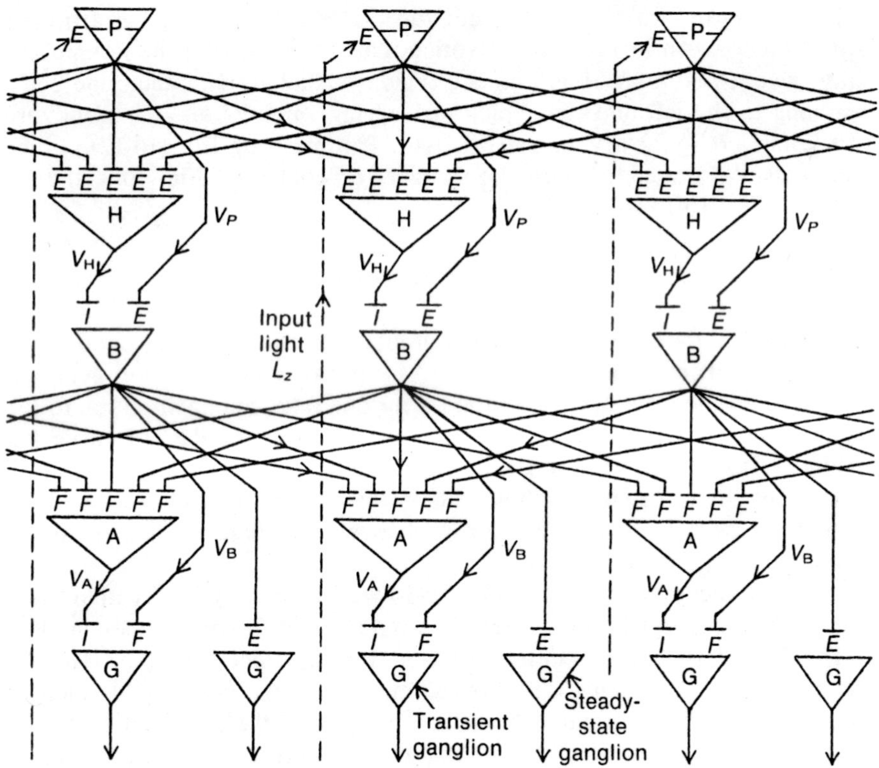

Fig. 10-14 Neural model of Fig. 10-3 including elements involved in signal processing of a transient visual input. Key: A = amacrine cells, F indicates that the cell responds to flicker, V_A is a graded potential that inhibits the transient ganglion cell.

local B cells. Actually, in a cross section seen through a microscope, A cells appear with dendrites reaching out *to* all of the local B cells. It is more convenient, however, to show the flicker junctions as being lined up against the body of the A cell.

In the last layer, the transient ganglion cell is excited by flicker from the bipolar cells, but is inhibited by the amacrine cell above it. Here we can assume that

$$f_{\mathrm{TG}} = V_{\mathrm{BF}} - V_{\mathrm{A}} \text{ (Hz, mV units)} \tag{10-26}$$

where f_{TG} = transient ganglion cell output frequency,
V_{BF} = bipolar cell output flicker component,
V_{A} = amacrine cell graded potential output.

Two examples are given that illustrate how this system operates. First, in Fig. 10-15, the left half of the visual scene is gray, while the right half is flickering; that is, it is alternating between black and white, say, ten times per second. This is the kind of psychedelic stimulus that can drive a susceptible person into an epileptic convulsion, but these dramatic attributes are of no concern in Fig. 10-15. Notice that V_{BF} is zero in the nonmoving gray region, and is at 20 mV in the region of flicker. The amacrine cells, reaching to the left and right, pick up varying amounts of V_{BF}, from zero at the far left to 20 mV at the far right. Paraphrasing Eq. (10-12), which describes V_{H} associated with the spatial step change of Fig. 10-7, the V_{A} curve of Fig. 10-15 is given by

$$V_{\mathrm{A}} = \frac{\Delta V_{\mathrm{BF}}}{2}(1 + \operatorname{erf} x), \tag{10-27}$$

where ΔV_{BF} is the bipolar cell flicker output relative to the background level.

The result of applying Eq. (10-26) is that transient ganglion cells immediately to the right of $x = 0$ fire at a furious rate, calling attention to the transition between quiet and flickering regions. For the dashed curve immediately to the left of $x = 0$, the net potential is negative, indicating inhibition of the transient ganglion cells in this region. The subjective effect is that we become aware of intense change immediately adjacent to the nonmoving gray area.

As the second example, in Fig. 10-16 we have a gray field with a flickering or moving dot in the center. The bipolar cell flicker component V_{BF} is 15 mV underneath the central dot, and 0 in the quiet surround. Because the central dot represents only a small element in the visual field, such as the local disturbance created by a moving "bug," the contribution to the amacrine cell layer is a negligibly small bump as shown. Here we can paraphrase Eq. (10-3), which describes V_{H} associated with the single bright light of Fig. 10-5. The V_{A} curve of Fig. 10-16 is given by

$$V_{\mathrm{A}} = \frac{\Delta V_{\mathrm{BF}}}{\pi} \epsilon^{-r^2}. \tag{10-28}$$

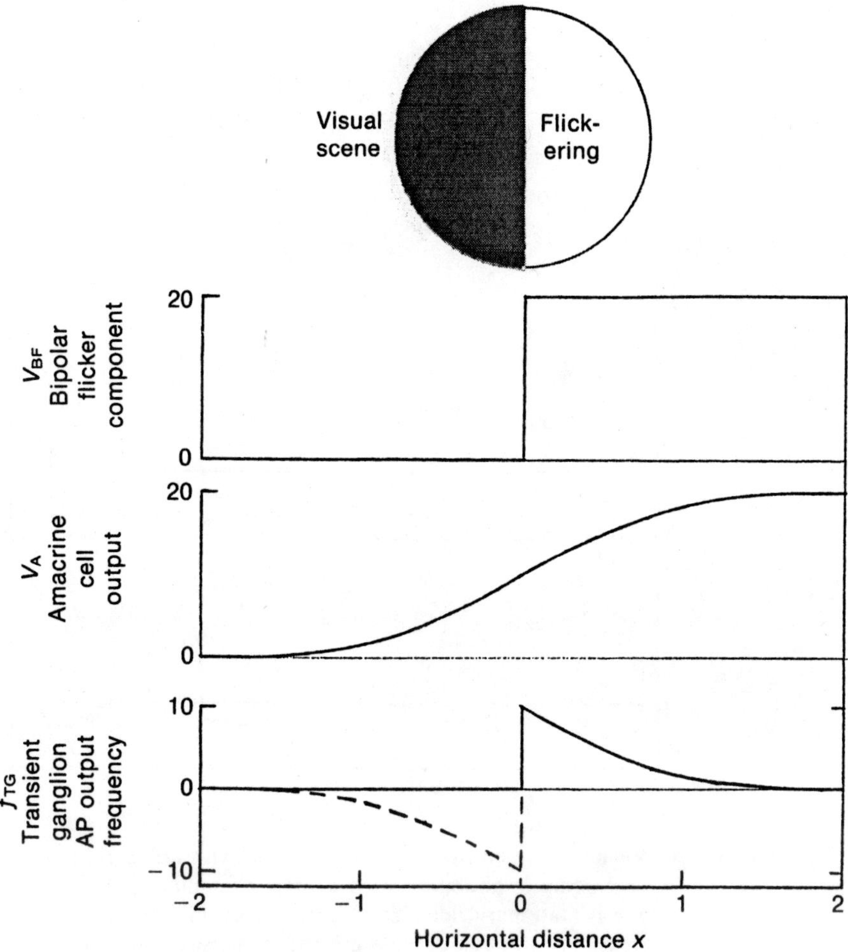

Fig. 10-15 Visual scene consisting of a gray field on the left and flickering field, between white and black, on the right. Assuming that the eye is stationary, idealized bipolar, amacrine, and transient ganglion cell outputs are shown versus horizontal distance from the dividing line between steady and flickering images. The transient ganglions demonstrate a response to flicker.

As shown, application of Eq. (10-26) leaves the transient ganglion cells underneath the flickering or moving dot with practically full excitation. We again conclude that "small-area change due to a moving 'bug' is fully seen if the surrounding visual field is quiet."

The eyes are in constant saccadic motion, but this is a relatively small-angle ongoing activity that is insufficient to stimulate the amacrine cells. There are many instances, however, in which massive flicker occurs nat-

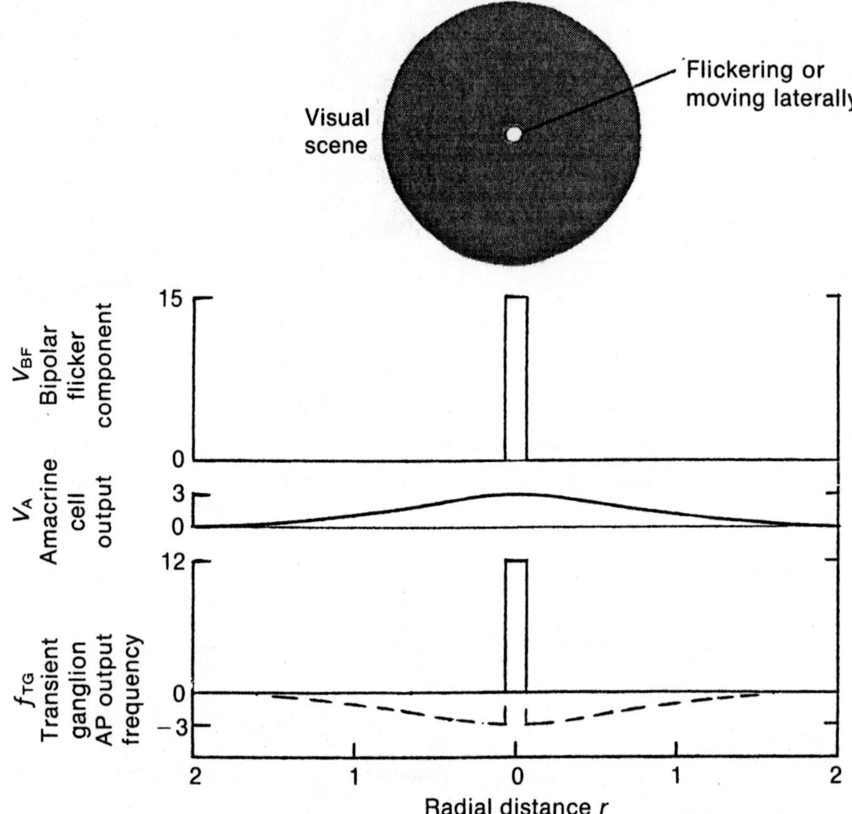

Fig. 10-16 Visual scene consisting of a gray field with a flickering or moving bright spot in the center. Assuming that the eye is stationary, idealized bipolar, amacrine, and transient ganglion cell outputs are shown versus radial distance from the center of the image of the bright spot. The transient ganglions demonstrate a response to motion as well as flicker.

urally: when the eye blinks, or the eye rapidly scans, or clouds move in and out of the sun's rays, or leaves vibrate to and fro in the wind. In each of these cases amacrine cells are stimulated by the large-area flicker or movement. They inhibit the transient ganglion cells to prevent them from reporting a moving object in the visual field when there actually is none [H. B. Barlow, R. M. Hill & W. R. Levick, 1964; Y. J. Lettvin et al., 1959; A. J. Van Doorn and J. J. Koenderink, 1983].

10-6 Subjective Color Characteristics

The discussion in the chapter up to this point has ignored the fact that cones come in three varieties: red, green, and blue. It has been assumed that the stimulus is white light. Actually, about 65% of the cones are "red," 33%

are "green," and only 2% are "blue." In each contiguous group, individual cone outputs interact eventually to create the subjective sensation of color.

A great deal of experimental work has been done in an effort to determine the relative sensitivities of the three cone receptors. We will use the results of Pieter L. Walraven and his colleagues as depicted in Fig. 10-17. These curves show the response of the three "primaries" near threshold level; that is, the illuminance is the lowest level at which full color sensations are experienced. The level is so low that nonlinearity due to the logarithmic compression of Fig. 10-4 is negligible. Each curve is determined by a pigment that gives the cone its nominal "color," so P_R stands for the red pigment response, P_G the green pigment response, and P_B the blue response. One

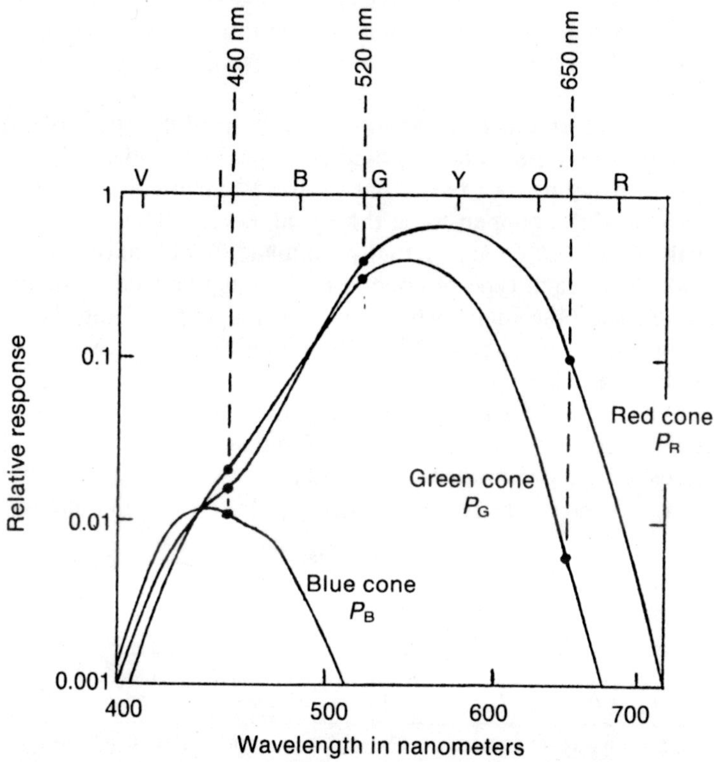

Fig. 10-17 Relative response versus wavelength for red, green, and blue cones of the human eye, according to data of J. J. Vos and P. L. Walraven (1971). Illuminance is at lowest level at which full color sensations are experienced, so that nonlinearity due to logarithmic compression is negligible. Representative ROYGBIV wavelengths are indicated by ticks. Also shown by dashed lines are three monochromatic wavelengths (650, 520, and 450 nm) used to synthesize practically any hue and saturation, including white, as explained in the text.

cannot measure directly the graded potential output of cone cells because they are only 3 μm in diameter in humans and primates. Therefore, the curves of Fig. 10-17 are psychophysically obtained; that is, experimental subjects (including people who lack red, or green, or blue responses) are exposed to various light stimuli versus matching combinations, and their reactions are recorded [P. Brou et al., 1986; F. M. De Monasterio and P. Gouras, 1975; E. H. Land, 1977; J. J. Vos and P. L. Walraven, 1971; J. J. Vos, 1978; G. Wald, 1964; G. Wyszecki and W. S. Stiles, 1982].

In Section 10-1 it is pointed out that electromagnetic waves in the visible range lie between wavelengths of 380 and 825 nm. Figure 10-17 shows relative responses between 400 and 720 nm. Seven nominal "ROYGBIV" colors are indicated by ticks at the top of the figure and are listed in Table 10-2. These ROYGBIV wavelengths have been chosen so that the ticks on the log wavelength scale are equally spaced; the ratio between adjacent wavelengths is 1.089. These are monochromatic (single-frequency, single-hue) inputs.

Figure 10-17 demonstrates at least three "unsightly" features: first, the red cone peaks at yellow and the blue cone peaks at indigo. Only a purist would insist on saying, however, that we have "yellow, green, and indigo" cones because these happen to be their peak colors. The color designation is nominal, since each cone is actually stimulated by *all* of the visible wavelengths, although each type of cone is most sensitive to a particular wavelength. Even the blue cone, whose relative response is not shown below 10^{-3} in Fig. 10-17, is listed by Vos–Walraven as having relative response 1.133×10^{-12} at 825 nm!

Since the days of Sir Isaac Newton we have been combining red, green, and blue lights to generate the subjective impression of "white." Color television tubes and color photography, similarly, use red, green, and blue primary colors. When it was relatively recently discovered that color vision is

TABLE 10-2 The Seven Nominal ROYGBIV Colors and White

			Fig. 10-17			Fig. 10-19 (*a*)		Fig. 10-19 (*b*)	
Hue	Name	λ (nm)	Red P_R	Green P_G	Blue P_B	Horizontal comp	Vertical comp	Relative x	Relative y
R	Red	682	0.0144	0.000574	1.79×10^{-8}	1.86	10.38	0.325	0.707
O	Orange	626	0.280	0.0299	1.06×10^{-6}	0.99	10.25	0.282	0.707
Y	Yellow	575	0.628	0.287	1.44×10^{-5}	−0.29	9.90	−0.029	0.706
G	Green	528	0.474	0.361	0.000346	−0.45	6.95	−0.477	0.674
B	Blue	485	0.0794	0.0862	0.00429	−0.37	3.00	−0.098	0.197
I	Indigo	445	0.0140	0.0169	0.0115	−0.20	0.38	0.101	0.132
V	Violet	409	0.00228	0.00145	0.00312	0.49	−0.77	0.115	0.134
Total			1.492	0.7830	0.01927				
W	White		90	45	1	0.24	3.81	0	0.471

Note: λ is the wavelength. For Fig. 10-19(*a*), Horiz comp = $0.9 \log C_R + 0.1 \log C_B - \log C_G$ and Vert comp = $\log C_G - \log C_B$.

mediated by three different cone types, it was natural to call them "red, green, and blue."

Second, the red and green cone responses are practically identical for wavelengths below 530 nm. It is obviously inefficient to have two different sensory receptors responding in the same way to the same stimuli.

Third, the blue cone is much less sensitive than the green or red cone. Here it is obviously inefficient to have a "weak" sensory receptor.

One can only assume that the curves of Fig. 10-17 represent the best that evolution was able to accomplish given the organic pigments and other light-absorbing, light-sensitive substances that are available to organisms for photoreception. From an engineering perspective, if we had three evenly spaced curves of equal peak sensitivity it would improve our "signal-to-noise" ratio for color reception; that is, it would allow color vision at much lower levels of light intensity [G. Buchsbaum and A. Gottschalk, 1983]. One can understand also why only 2% of cone area is devoted to the blue species. Why waste precious space on an insensitive receptor? At low light levels the B, I, and V colors may be inaccurate, but survival is better served by replacing the blue cones with sensitive red cones. (At the end of this chapter we will see that the color vision of a tritanope, a person whose blue cones are inoperative, is almost normal.)

In order to calculate the response to various monochromatic inputs, it is necessary to modify Eq. (10-1). Also, each photoreceptor of Fig. 10-3 is separated into color response and logarithmic compression "compartments," as indicated. First,

$$\begin{aligned} C_{\mathrm{R}} &= L_{z1}P_{\mathrm{R1}} + L_{z2}P_{\mathrm{R2}} + \cdots; \\ C_{\mathrm{G}} &= L_{z1}P_{\mathrm{G1}} + L_{z2}P_{\mathrm{G2}} + \cdots; \\ C_{\mathrm{B}} &= L_{z1}P_{\mathrm{B1}} + L_{z2}P_{\mathrm{B2}} + \cdots, \end{aligned} \qquad (10\text{-}29)$$

where L_{z1} = light intensity at wavelength λ_1,
L_{z2} = light intensity at wavelength λ_2, and so forth,
P_{R1} = red pigment response (Fig. 10-17) to λ_1,
P_{R2} = red pigment response to λ_2, and so forth,
C_{R} = total *internal* response of the red cone,
C_{G} = total internal response of the green cone, and so forth.

(Equation (10-29) can be more compactly expressed using matrix multiplication.) Second, logarithmic compression yields

$$V_{P\mathrm{R}} = \log C_{\mathrm{R}}; \qquad V_{P\mathrm{G}} = \log C_{\mathrm{G}}; \qquad V_{P\mathrm{B}} = \log C_{\mathrm{B}}, \qquad (10\text{-}30)$$

where the V_Ps are the photoreceptor graded potential outputs.

It is instructive to consider how the sensation of "white" is produced. Four methods are presented as follows.

(a) Isaac Newton passed white light through a prism and thereby broke the light up into a continuum of wavelengths extending from 380 to

825 nm. (Raindrops accomplish the same feat when they give rise to a rainbow.) The continuum consists of all wavelengths and, indeed, electron motions in the white-hot filament of a tungsten lamp (light bulb) are sufficiently chaotic so that practically all visible wavelengths are generated. Imagine, then, that the three curves of Fig. 10-17 are bathed by this continuum of wavelengths. The response of each cone will simply be proportional to the area underneath its curve. Relative to the blue cone, these areas are [J. J. Vos, 1978]

$$C_R = 90, \qquad C_G = 45, \qquad C_B = 1. \tag{10-31}$$

(Because of the log scale, it does not look as if the red cone curve has twice as much area as the green cone curve, and 90 times as much area as the blue cone curve, but we would see that this is so if a linear scale is used.)

The 90:45:1 ratio of Eq. (10-31) is the "magic formula" for achieving the subjective sensation of white. It does not matter how it is accomplished: if for the red cone $C_R = 90C_B$ while at the same time, for the green cone, $C_G = 45C_B$, we will see white (or gray for a region of relatively low light level). Any appreciable departure from this set of ratios is judged to be a color other than white or gray.

(b) The continuum can be approximated by using seven monochromatic light beams at the ROYGBIV wavelengths (682 nm, 626 nm, . . . , 409 nm, as listed in Table 10-2). If we look at these seven beams combined, with approximately equal intensities, values taken from the Fig. 10-17 curves (see the "Total" row in Table 10-2) show that the red:green:blue cone internal responses will have the ratios 77:41:1. This is close enough to the ideal 90:45:1 ratios so as to give the subjective impression of white.

(c) The 90:45:1 ratios can be achieved with only two monochromatic beams of light. For example, if we look at 650 nm (red) combined with 494 nm (green) of the correct intensity, we will see white; similarly, 569 nm (green) combined with 450 nm (indigo). Given a monochromatic beam between 650 and 569 nm, it can be combined with the proper monochromatic beam somewhere between 494 and 450 nm to create the sensation of white. This statement is based on a chromaticity diagram that is discussed in the next section. However, although two monochromatic beams of light can thus yield white, they can produce little else. For example, if we start with a beam at 587 nm (orange), gradually reduce the intensity of the orange as we add a beam at 485 nm (blue), we will successively see the monochromatic orange (monochromatic hues are called *saturated*), then unsaturated orange (a mixture of orange and white), then white, then unsaturated blue (a mixture of blue and white), and finally saturated blue. With orange and blue it is not possible to synthesize red, yellow, green, indigo, or violet.

(d) With *three* properly chosen monochromatic beams, however, it is possible to generate practically any color sensation. This is how it is done in a color television tube and in color photography. We illustrate with monochromatic beams of light at 650 nm (red) combined with 520 nm (green) and 450 nm (indigo), as indicated by dashed lines in Fig. 10-17. To synthesize white with these three beams instead of the tungsten lamp continuum, we have determined mathematically that the beams should be of somewhat unequal intensities (normally accomplished by blocking out part of a beam with a shutter); namely, the red:green:indigo beams should be of L_z intensity ratios 372:130:86. If the three arrows at the top of Fig. 10-17 have these intensities, how will the cones respond? The beam at 650 nm, of intensity $L_{z1} = 372$, strikes a red cone that has a relative response $P_{R1} = 0.101$ in Fig. 10-17. According to Eq. (10-29), then, $C_{R1} = 372 \times 0.101 = 38$. Similarly, the beam at 520 nm of intensity $L_{z2} = 130$ interacts with the red cone in accordance with $C_{R2} = 130 \times 0.393 = 51$; the beam at 450 nm of intensity $L_{z3} = 86$ interacts with the red cone in accordance with $C_{R3} = 86 \times 0.016 = 1$. The total red cone response is given by $C_R = 38 + 51 + 1 = 90$. In tabular form, the calculations for the three cones are listed in Table 10-3. Since the desired cone response ratios red:green:blue = 90:45:1 are thus achieved, the subjective sensation is white.

Calculations of the type just made are valid if the system is linear. In the present instance this is so because the input light intensities are relatively low, as previously noted.

TABLE 10-3 How the Sensation of White (C_R:C_G:C_B = 90:45:1) Is Obtained[a]

	Input Monochromatic Beam		Red Cone		Green Cone		Blue Cone	
	λ	L_z	P_R	C_R	P_G	C_G	P_B	C_B
(1)	650 nm	372	0.101	38	0.006	2	0	0
(2)	520 nm	130	0.393	51	0.317	41	0.001	0.1
(3)	450 nm	86	0.016	1	0.020	2	0.011	0.9
Total				90		45		1

[a] Here we use three monochromatic beams of light at 650 nm (red), 520 nm (green), and 450 nm (indigo); L_z is the relative intensity of each beam; the *P*s are the cone pigment responses of Fig. 10-17; the *C*s are given by Eq. (10-29).

10-7 The Standard Chromaticity Diagram

The bible for people who design colorimetry and photometry equipment is the chromaticity diagram of Fig. 10-18. The reason for discussing this diagram here is that it very much mirrors what is going on in the nervous system when one sees color; in the next section, in fact, we will derive something akin to the chromaticity diagram. Like the cone response curves of Fig. 10-17, the chromaticity diagram is derived experimentally. It was drawn in 1931 by the International Commission on Illumination, which examined large amounts of data. The diagram summarizes in a useful way the color characteristics of the human eye as explained next.

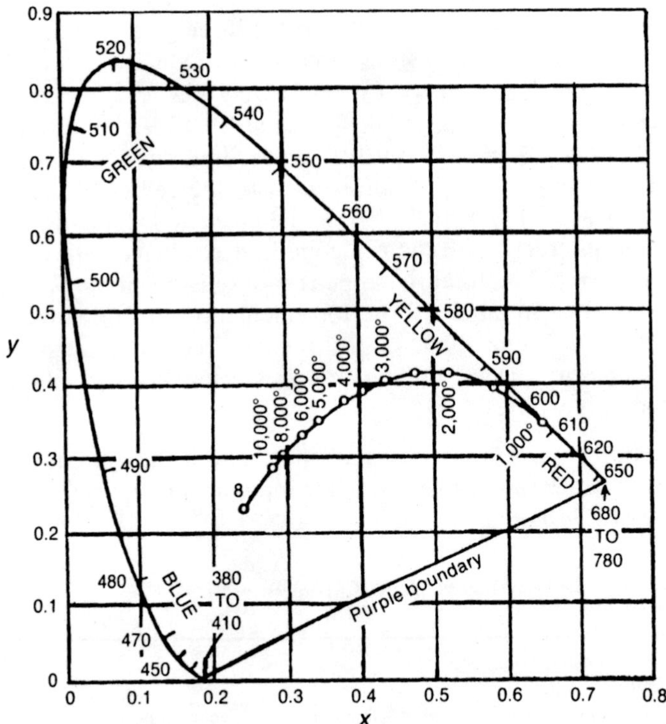

Fig. 10-18 Standard chromaticity diagram issued by the International Commission on Illumination. Every possible hue and saturation to which the human eye responds can be represented by a point on or inside the triangle, which extends from short visible (380 nm) to medium (520 nm) to long wavelengths (780 nm). Derivation of the *x* and *y* scales is beyond the scope of this book. The curved line inside the triangle is the locus generated as an ideal filament is heated to an infinite temperature. "White" is represented by the surface temperature of the sun, 6000 K.

Monochromatic hues are traced out by the left and right edges of the triangle. Starting in the red corner and moving counterclockwise, the boundary is traced out as monochromatic input wavelengths are decreased from 780 nm to green at 520 nm to blue at 380 nm. (The hues that we have been calling blue, indigo, and violet are all loosely labeled "blue" on the diagram.) The base of the triangle represents various combinations of monochromatic 680 nm (red) and 410 nm ("blue") light beams, which are subjectively perceived as various shades of purple. The inside of the triangle represents various degrees of saturation; that is, various mixtures of monochromatic hues plus white.

The curved line inside the triangle is the locus generated when a tungsten filament, say, is heated. The locus starts at 1000 K, which is equal to 727°C or 1341°F. At this temperature, our subjective judgment is that the filament is a monochromatic source of 607-nm (red) light; actually, it is generating a broad continuum of wavelengths. As the temperature rises, our judgment is that the source becomes unsaturated because the locus swings away from the boundary. At 2000 K = 1727°C = 3141°F, the source looks like a mixture of white and orange light because it is located opposite orange at 587 nm. If we continue to increase the temperature, the filament will burn out at 3643 K when the melting temperature of tungsten has been reached. The surface of the sun has a temperature of 6000 K, so this is the standard point taken to represent daylight or white light. The locus continues into the unsaturated blue region, which is the color of stars much hotter than the sun. The locus ends at an infinite temperature point that has been calculated but is, of course, unattainable.

It is instructive to take an imaginary tour in and around the chromaticity diagram. We start in the lower-right corner with a 650-nm (red) source shining on a white sheet of paper in an all-white room. There is no other source of illumination. We gradually reduce the intensity of the red source and at the same time increase the intensity of a 520-nm (green) source. Subjective color is given by a line directly joining the 650-nm and 520-nm points on the diagram. The hue of the sheet of paper will change from red to orange to yellow to green, indistinguishable from corresponding monochromatic sources of light.

To continue the tour, gradually reduce the intensity of the green source and at the same time increase the intensity of a 450-nm (indigo) source. Subjective color is given by a line joining the 520-nm and 450-nm points on the diagram. The hue of the sheet of paper will change from green to blue to indigo, indistinguishable from corresponding monochromatic sources (except for some lack of saturation where the locus is an appreciable distance away from the left boundary of the triangle).

Colors along the base of the triangle are synthesized if we gradually reduce the intensity of the indigo source and at the same time increase the intensity of the red source.

As an example of travel across the inside of the triangle, let us start with a 587-nm (orange) source and gradually reduce its intensity while at the same time increasing the intensity of a 485-nm (blue) source. A line joining the two sources crosses the white (6000 K) point. Therefore, the color of the sheet of paper will change from orange to unsaturated orange to white to unsaturated blue to blue.

As pointed out in the previous section, one can also generate white by mixing 650 nm (red) and 494 nm (green), or by combining 569 nm (green) and 450 nm (indigo). In each case the locus passes through the white (6000 K) point.

The creation of subjective white via three beams at 650 nm (red), 520 nm (green), and 450 nm (indigo) is more complicated. Suppose that we join these three points on Fig. 10-18 with straight lines so as to define a triangle within the chromaticity diagram. It is obvious, from the previous discussion, that one can synthesize any hue and saturation within this triangle by properly adjusting the relative intensities of the three sources. This is the task of the electrical signals applied to the picture tube of a color television receiver. If we look at the picture tube through a magnifying glass when it is displaying a white image, we will see that the picture is actually composed of small red, green, and indigo (blue) dots. The face of the tube does not contain a single white dot!

10-8 Two-Dimensional Color Map

The response curves of Fig. 10-17 describe the behavior of each tiny group of contiguous color cones under low-level stimulation. The red, green, and blue cone in each group, or signals derived therefrom, interact so as to generate subjective color. What can we say about the form this signal processing may take?

First, it should be simple because many thousands of "color groups" are involved. As stated at the beginning of Section 10-6, some 65% of cones are C_R, 33% are C_G, and only 2% are C_B. Each blue cone must therefore be surrounded by and shared by several contiguous red and green cones. From the diameter of the cones and foveola (3 and 600 μm, respectively), it is easy to calculate that each foveola alone contains 5000 "color groups," and it is not reasonable to suppose that each of the "color groups" feeds a complex processor consisting of a large number of neurons. The relative C_R, C_G, and C_B distribution must be approximately uniform because color seems to be invariant across the visual field: if you look to the side, peripheral vision sees the same color as foveal vision. (As we approach the periphery, however, cone visual acuity decreases as rods proliferate. In the foveola, on the other hand, as pointed out in Sect. 10-1, there are no rods at all, so one must actually look slightly off center to see dim nighttime forms.)

Second, color is two-dimensional in the sense that two vectors are required to specify a color in Fig. 10-18. This implies that each retinal color group feeds a two-dimensional map, similar to the hypothetical matrix in the right-hand side of Fig. 9-16. The latter map is a single structure containing

10^{10} neurons, but it is conjectured that the visual cortex derives color information via many thousands of individual maps, each containing a relatively small number, such as 1000, neurons.

Third, the color signal processor has to calculate *ratios* between cone sensory receptor stimuli. This is the only way to explain how color remains invariant over a 600-to-1 change of input light intensity. Some of the discussion pertinent to this is repeated from Section 10-2: "Although evolution could stumble on a way to actually calculate ratios, it is much more natural to suppose that the retina generates logarithms, via a simple nonlinearity, followed by subtraction." In other words, the conjecture is that the visual system calculates $\log C_R - \log C_G$ and, for the second dimension, $\log C_G - \log C_B$.

The two-dimensional chromaticity diagram thus synthesized is shown in Fig. 10-19(*a*) with a minor change: For the horizontal dimension, instead of $\log C_R - \log C_G$, 10% blue has been introduced so that the horizontal dimension is $0.9 \log C_R + 0.1 \log C_B - \log C_G$. Adding a small amount of blue in this way improves the shape of the diagram, but one must compensate by taking away an equal amount of red in order to preserve invariance with respect to light intensity. In other words, the algebraic sum of log coefficients has to be zero: in the preceding case we have $0.9 + 0.1 - 1$. Other possibilities are $0.8 \log C_R + 0.2 \log C_B - \log C_G$, or $\log C_R + \log C_B - 2 \log C_G$, and so forth.

The calculations for Fig. 10-19(*a*) are summarized as follows:

$$\begin{aligned} \text{Horizontal component} &= 0.9 \log C_R + 0.1 \log C_B - \log C_G \\ \text{Vertical component} &= \log C_G - \log C_B. \end{aligned} \tag{10-32}$$

As an example, in the upper-right "corner" of Fig. 10-19(*a*), the location of the *R* dot is calculated as follows: The wavelength is 682 nm, so from the Vos–Walraven data (curves of Fig. 10-17) we get $C_R = 0.0144$, $C_G = 0.000574$, and $C_B = 1.79 \times 10^{-8}$. In the horizontal direction, then,

$$0.9 \log C_R + 0.1 \log C_B - \log C_G = 1.86.$$

In the vertical direction,

$$\log C_G - \log C_B = 10.38.$$

There is an inconvenient feature associated with these numbers. Because all of the curve values in Fig. 10-17 are less than 1, all of the log values are negative. Since the curves of Fig. 10-17 show *relative* response, however, we can first multiply all values by 10^8, say, to get numbers that are greater than 1, and then apply logarithmic compression. Let us assume, in other words, that the *actual* photoreceptor output V_P in Eq. (10-30) is given by

$$\text{Cone output, mV} = \log(10^8 \times \text{Vos–Walraven data}). \tag{10-33}$$

Consider how dramatically the values for *R* (682 nm) given previously change

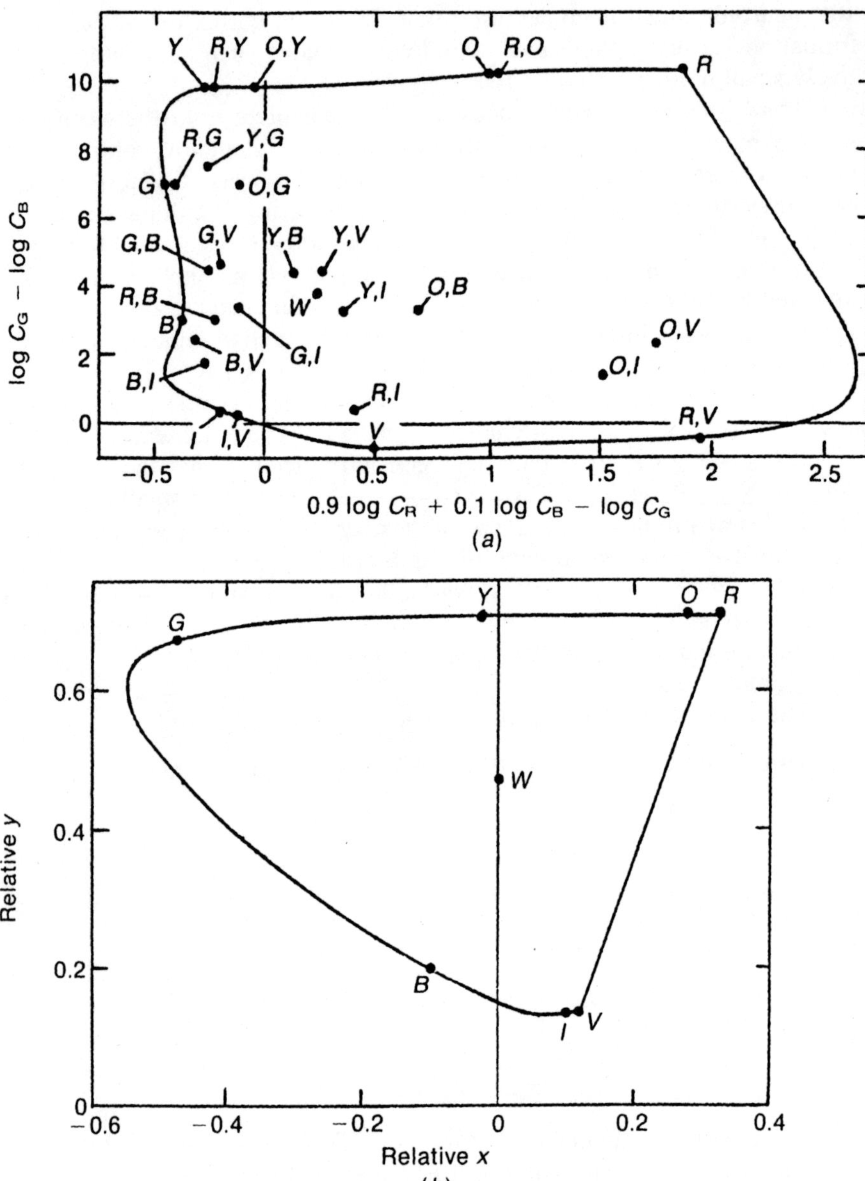

Fig. 10-19 Hypothetical chromaticity diagrams. (*a*) Two-dimensional color map derived from the curves of Fig. 10-17 for wavelengths from 682 nm (*R*) to 409 nm (*V*). Logarithms of green, red, and blue cone outputs are plotted in a way that yields a diagram independent of light intensity; that is, points on the diagram do not change as input brightness to cone receptors changes. Starting with *R* in the upper-right "corner" and moving counterclockwise, the boundary is traced as monochromatic (single-

if this transformation is applied:

Vos–Walraven Data	Photoreceptor Output
$C_R = 0.0144$	$V_{PR} = 14.18$ mV
$C_G = 0.000574$	$V_{PG} = 10.96$ mV
$C_B = 1.79 \times 10^{-8}$	$V_{PB} = 0.58$ mV

Notice that all of the V_P values are positive. Using these values, the location of the R dot in Fig. 10-19(a) is given by

$$0.9V_{PR} + 0.1V_{PB} - V_{PG} = 1.86$$

and

$$V_{PG} - V_{PB} = 10.38.$$

Such is the power of logarithmic compression that, despite multiplication by 100,000,000, the two output values have not changed one iota!

Continuing in this fashion, starting with R and moving counterclockwise, the boundary of Fig. 10-19(a) is traced out as $0.9 \log C_R + 0.1 \log C_B - \log C_G$ and $\log C_G - \log C_B$ are calculated from data given in the Vos–Walraven papers. The boundary calculation ends at V (409 nm). Along the way, dots indicate the locations of the ROYGBIV colors. The calculated values are listed in the "Fig. 10-19(a)" columns of Table 10-2.

To close the right-hand side of Fig. 10-19(a): Starting at R and moving clockwise, the boundary is traced out as the amount of R (682 nm) is gradually decreased to 0 while V (409 nm) is increased. For each mixture, values are calculated and plotted in accordance with Eq. (10-32). The dot labeled R,V locates the 50% red, 50% violet point, which should appear to be purple.

To show that the chromaticity diagram of Fig. 10-19(a) is similar to that of Fig. 10-18, the latter has been rotated 45° counterclockwise. This converts the R to O to Y locus into a horizontal line, in approximate agreement with Fig. 10-19(a). The rotated diagram is depicted in Fig. 10-19(b), with dots

hue) input wavelengths are decreased from 682 nm to O, Y, G, B, I, and finally V (409 nm). Starting with R and moving clockwise, the boundary is traced as the amount of 682 nm (R) is gradually decreased to zero while 409 nm (V) is increased. Also located are equal amounts of all possible pair combinations formed by the ROYGBIV hues. Also located is white (W). (b) The International Commission on Illumination diagram of Fig. 10-18 is rotated counterclockwise 45° to better show that it resembles (a). The implication is that the visual cortex actually forms two-dimensional logarithmic maps similar to (a) for the determination of color.

indicating the locations of the ROYGBIV colors. The calculated values are listed in the "Fig. 10-19(b)" columns of Table 10-2.

In Section 10-6, we discussed the cone responses to white (W) light; namely, $C_R:C_G:C_B = 90:45:1$. The corresponding location for W is shown in Fig. 10-19(a) and listed in Table 10-2. In Fig. 10-18, W (6000°) is located at $x = ⅓$, $y = ⅓$. Since this forms a 45° angle with the origin at $x = 0$, $y = 0$, the rotation 45° counterclockwise lands the W point directly above the origin in Fig. 10-19(b).

In Fig. 10-19(a), all possible pair combinations formed by the ROYGBIV hues are located, mixed in equal amounts. For example, the Y,B point (which should appear to be a shade of white) is determined as follows: From Table 10-2, adding the Y and B values (since equal amounts of each are involved) we get $C_R = 0.707$, $C_G = 0.374$, and $C_B = 0.00431$. Then

$$0.9 \log C_R + 0.1 \log C_B - \log C_G = 0.13$$

and $\log C_G - \log C_B = 4.46$, as shown.

The preceding discussion regarding a hypothetical chromaticity diagram born out of logarithmic bits and pieces is incomplete without an equally hypothetical model for generating the two-dimensional map of Fig. 10-19(a). As explained below, the model also shows how every point in Fig. 10-19(a) can translate into a specific *zero-potential depression corresponding to every hue and saturation* that the normal visual system can comprehend.

The hypothetical model is illustrated in Fig. 10-20. On the left we have red, green, and blue logarithmic inputs from the corresponding photoreceptors of Fig. 10-3. The latter schematic diagram shows the cone outputs going to horizontal and bipolar cells; bipolar cell outputs in turn feed to ganglion cells. To simplify matters, photoreceptor outputs go directly to retinal ganglion cells in Fig. 10-20, but one can certainly use a more exact model if this is important.

The numerical values are for a white input. Cone responses before logarithmic compression are given by $C_R:C_G:C_B = 270:135:3$, which of course satisfies the 90:45:1 requirement for white. The respective ganglion cell inputs are $\log 270 = 5.60$, $\log 135 = 4.91$, and $\log 3 = 1.10$ mV as shown. The single ganglion cells of Fig. 10-3 are augmented by three interacting cells as follows: N2 supplies an achromatic, brightness, or *luminance* component, involved with the perception of shape and movement, not color. Here it is simplest to assume that the N2 output is the average of the three cone inputs. This would be fine if we were dealing with C_R, C_G, and C_B, but it distorts the output in dealing with their logarithms because C_B is weak and the log of zero is minus infinity! In Fig. 10-20, problems with $\log C_B$ are avoided by omitting this input; as shown, N2 is the average of the $\log C_R$ and $\log C_G$ inputs only, so that $0.5(5.60 + 4.91) = 5.25$ mV. N2 should also include an input (not shown) from the transient ganglion cells of Fig. 10-14. The N2 output goes to the many "pattern recognition" regions in the brain.

N1 is connected so as to supply $0.9 \log C_R + 0.1 \log C_B - \log C_G =$

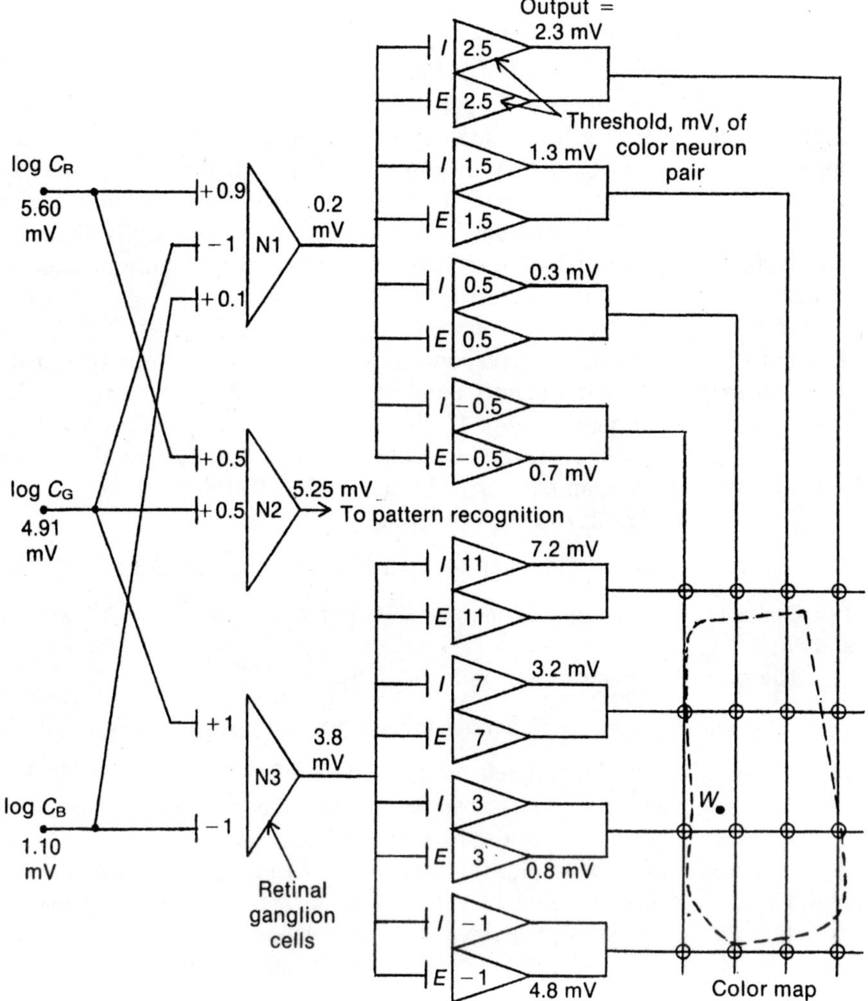

Fig. 10-20 Hypothetical model for generating the two-dimensional color map (chromaticity diagram) of Fig. 10-19(*a*). The numerical values represent a "white" input such that the red cone (C_R) response is 270, green cone (C_G) response is 135, and blue cone (C_B) response is 3. Actually, logarithms are generated so that the cone outputs are 5.60, 4.91, and 1.10, respectively. N2 is involved with the perception of shape and movement, not color, N1 feeds the horizontal scale of Fig. 10-19(*a*), while N3 feeds the vertical scale. In this figure, the vertical column of eight pairs of neurons converts its graded potential inputs into spatial distribution outputs in which the potentials form a V. The valleys of the V are located (ideally) at zero output level. For the "white" input, the valleys intersect at the point labeled *W*, producing a "circular" depression around *W*.

0.24 mV (simplified to 0.2 mV in Fig. 10-20), while N3 supplies log C_G − log C_B = 3.81 mV (simplified to 3.8 mV) for white.

The vertical column of eight pairs of neurons does the job of converting a graded potential input into a spatial distribution output. In each pair, the upper neuron has an inhibitory (I) input junction, while the lower has an excitatory (E) junction. Each pair feeds a single axon that ends in a "color map." The dashed-line phantom in the color map is none other than the chromaticity diagram of Fig. 10-19(a), compressed horizontally to make it fit into the lower-right corner of Fig. 10-20.

The upper set of neuron pairs operates with threshold levels 2.5, 1.5, 0.5, and −0.5 mV. These correspond to the four vertical axons that cross the color map; they span the horizontal scale of Fig. 10-19(a). Similarly, the lower set of neuron pairs in Fig. 10-20 operate with threshold levels 11, 7, 3, and −1 mV. These correspond to the four horizontal axons that cross the color map; they span the vertical scale of Fig. 10-19(a).

The output of each I neuron is given by

$$\text{Output} = \text{threshold level} - \text{input if positive; otherwise 0.} \quad (10\text{-}34)$$

For example, for the topmost neuron, Output = 2.5 − 0.2 = 2.3 mV as shown.

The output of each E neuron is given by

$$\text{Output} = \text{input} - \text{threshold level if positive; otherwise 0.} \quad (10\text{-}35)$$

For example, for the second neuron, Output = 0.2 − 2.5 = −2.3 mV, but since this is a negative value its actual output is 0. The signal fed to the color map is the nonzero output of the pair, in this case 2.3 mV.

As an additional example, for the lowest (E) neuron, we have Output = 3.8 − (−1) = 4.8 mV. For the next-to-lowest (I) neuron, we have Output = −1 − 3.8 = −4.8 mV, but since this is a negative value its actual output is 0. The signal fed to the color map is the nonzero output of the pair, in this case 4.8 mV.

In the color map, the circles represents "crossover" neurons, neurons that calculate the product of, or multiply, the graded potentials carried by the two axons at the crossover point. In Fig. 10-20, if you multiply the upper four voltages (2.3, 1.3, 0.3, and 0.7) by the lower four (7.2, 3.2, 0.8, and 4.8), you will get the following values for the crossover neurons:

5.04	2.16	9.36	16.56
2.24	0.96	4.16	7.36
0.56	0.24	1.04	1.84
3.36	1.44	6.24	11.04

The smallest value in this matrix, 0.24, belongs to the crossover neuron nearest the W location in Fig. 10-20. Notice how the model produces a "circular" depression whose location, in this case, corresponds to W. The accuracy of the scheme is improved, of course, as the number of neuron pairs and their axons and crossover neurons is increased. For a crossover neuron directly *at* the *W* point, the horizontal and vertical axons would each carry a graded potential of 0, so the earlier multiplication would display 0 at the *W* point.

According to the model, the crossover neuron or neurons near the minimum point will, based upon the location of the point, define a certain hue and saturation. If the *minimum* point moves to a different location, it will signal a different hue and saturation.

Hue and saturation are coupled to information relative to luminance (from neuron N2 in Fig. 10-20) so as to yield the net subjective impression of hue, saturation, and intensity of the source. With the addition of brightness, one can regard color as a three-dimensional phenomenon.

An attractive feature of the "color map" model is that the threshold values and density of color neuron pairs can be altered to get a much better fit to the International Commission on Illumination diagram of Fig. 10-19(*b*). In the horizontal direction, *R* to *O* and *V* to *I* are compressed, while *Y* to *G* and *B* to *G* are greatly expanded. In the vertical direction, the upper portion of the diagram is compressed relative to the lower portion. We have not worked out the details for these compressions and expansions, but it is obvious that they can be approximated by changing the threshold levels, number of color neuron pairs, and spacing between crossover neurons in Fig. 10-20.

What shall we say about the evidence that colors are perceived relative to one another so that, at sunset, a camera photograph shows a reddish tint whereas our visual sense sees a normal color balance? This phenomenon is easily explained as an example of adaptation. With a reddish input, the crossover neurons in the vicinity of *R* in Fig. 10-19(*a*) are at "zero voltage" level for an abnormally large proportion of time. Relative to the *R* crossover neurons, all of the *other* neurons are measuring high levels for a correspondingly large proportion of time. They simply adapt or habituate, reducing their graded potential or action potential frequency output to more normal levels.

10-9 Color Blindness

Our final exercise involving the color map of Fig. 10-19(*a*) is illustrated in Fig. 10-21. The diagram depicts the three principal categories of color blindness—people who lack red, or green, or blue cone information, usually caused by some deficiency in the photochemical pigment needed to absorb a particular range of visible wavelengths [A. C. Guyton, 1986; W. A. H. Rushton, 1975].

Consider first the lack of red cone information, known as *protanopia*. Without C_R, the $0.9 \log C_R + 0.1 \log C_B - \log C_G$ scale becomes meaningless because the log of 0 is minus infinity; all of the N1 neurons of Fig. 10-20, which are associated with the horizontal scale, will presumably atrophy. The

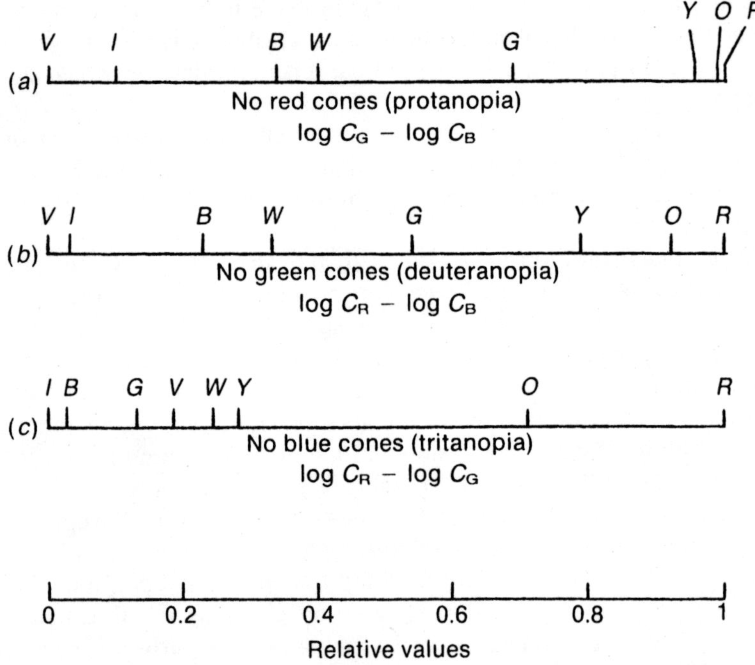

Fig. 10-21 Hypothetical collapse of the normal two-dimensional chromaticity diagram of Fig. 10-19(*a*) into a one-dimensional "map" if one of the three pigments responsible for color vision is absent. (*a*) Without C_R, the horizontal dimension of Fig. 10-19(*a*) vanishes, leaving the log C_G − log C_B scale. (*b*) Without C_G, the horizontal dimension vanishes again. In the other direction, C_G is replaced by C_R to yield log C_R − log C_B. (*c*) Without C_B, the vertical dimension vanishes. In the other direction, 0.9 log C_R increases to log C_R, yielding log C_R − log C_G.

log C_G − log C_B vertical scale then yields the one-dimensional protanopia line of Fig. 10-21.

If a person lacks green cone information, *both* of the scales in Fig. 10-19(*a*) becomes useless. Here it is reasonable to suppose that the red cones take the place of their silent green companions in the vertical scale, creating log C_R − log C_B The latter calculation yields the one-dimensional *deuteranopia* line of Fig. 10-21.

We see that the protanopia and deuteranopia lines are similar. As is well known, a person suffering from these forms of color blindness has trouble distinguishing between red, orange, and yellow. The explanation offered by the model is that *R*, *O*, and *Y* values practically coincide on the one-dimensional color map. If the upper portion of Fig. 10-19(*a*) is compressed relative to the lower portion, as proposed in the previous section,

G moves much closer to the Y, O, and R group. Green is then easily confused with yellow–orange–red because of a lack of spatial separation.

Color-blind people sometimes blame their disability when they receive a traffic ticket for passing a red light. Their color blindness is a rather weak excuse because there *is* a perceptual difference between red and green; furthermore, in a traffic-light assembly, the red light is uppermost. Incidentally, color blindness is a common genetic defect that is usually sex-linked. It is estimated that 2% of males are protanopes, 6% are deuteranopes, and only 0.4% of females are either red or green color blind.

If a person lacks blue cone information, the vertical scale of Fig. 10-19(*a*) becomes meaningless. The $0.9 \log C_R + 0.1 \log C_B - \log C_G$ scale becomes $\log C_R - \log C_G$ because, as explained in the previous section, the algebraic sum of log coefficients must equal zero. The $\log C_R - \log C_G$ calculation is shown as the one-dimensional *tritanopia* line in Fig. 10-21. As expected from the relatively modest contribution of blue cones to normal vision, the tritanope confuses violet with green but is normal otherwise. Is this conclusion borne out by actual data? Nobody knows for sure. Perhaps our ancestors had to know the color violet accurately for survival, because tritanopia turns out to be an extremely rare condition. Nowadays, most people who have tritanopia probably go through life unaware of their genetic defect. Only a person who suspects that he or she has a color deficiency is normally subjected to tests, unless there is an occupational requirement, and the visual tests are mostly aimed at detecting protanopia or deuteranopia.

In color blindness, the pigments responsible for color vision may be present, but in below-normal amounts, so there can be an infinite number of gradations of color deficiency. In general, however, the model of Fig. 10-21, which is derived from Fig. 10-19, offers a logical explanation for the subjective phenomena.

10-10 Color Contrast Enhancement

The discussion in Section 10-3 regarding contrast enhancement deals only with black-and-white images. The manner in which the system works is summarized in Fig. 10-5. Here the center of the visual field sees a bright light of relatively small diameter. The horizontal cell responses are depicted by a smooth curve that has a peak at the center and gradually drops off at increasing distance from the center. Since horizontal cells are inhibitory on bipolar cells, the latter demonstrate a dip corresponding to the horizontal cell rise. The dip represents contrast enhancement: in response to a bright center, we have a surround that is darker than the actual background.

One would expect the system to function in a similar manner if the visual scene is colored. Suppose that we have a bright-green light surrounded by a dull-green background: We expect color contrast enhancement to yield a green surround that is darker than the normal green background. This expectation is only partly correct, because evolution has improved on the scenario. Of greater contrast than bright green versus dark green is bright green versus dark *red*. If the center of the visual field sees a bright-green light of relatively small diameter, against a dark-gray background, contrast is enhanced because the surround will appear to be slightly reddish.

In general, the center and surround are complementary to each other. Roughly speaking, red and green are complementary, as are blue and yellow. To find the exact complementary color, locate the given hue on the standard chromaticity diagram of Fig. 10-18 and draw a line from the given hue through white (6000 K); the line will point to the complementary wavelength. According to this system, green and purple are complementary. Color contrast enhancement is a weak effect, however, so the subjective appearance is described as a green center versus a reddish surround, or a red center versus a greenish surround, and so forth.

Neurophysiologists have succeeded in recording objective data that substantiate the subjective reports: The nerve fibers leaving the retina proceed to the lateral geniculate nuclei (LGN) in the brain. Using monkeys, whose visual system is similar to that of man, the neurophysiologist probes the LGN with a microelectrode, and is able to record the AP activity of a single cell (or tiny group of cells). From the AP response to moving spots of light in the visual field of the animal, the experimenter is able to locate the visual center corresponding to the LGN neuron. The experimenter tests the LGN neuron for maximum sensitivity to red, green, or blue light; or it may be achromatic, responding to luminance, like N2 of Fig. 10-20. If it is a "green" neuron, shining red light into what corresponds to the adjacent surround will inhibit the AP discharge. If it is a "red" neuron, shining green light into the surround will inhibit the AP discharge [R. Shapley, E. Kaplan & R. Soodak, 1981].

In recording subjective human data, the incoming light source is nonchanging or constant while the viewer "scans" with his or her "neurons," reporting the color seen and its intensity. In recording objective data, the monkey's LGN neuron is fixed or stationary, while the experimenter scans light sources, of various hues and intensities, across the neuron. Thus, the subjective and objective situations are not identical; nevertheless, it is considered that they support each other.

Neurophysiologists prefer the word "opponent" to "complementary." The preceding contrast enhancement is called an *opponent surround* receptive field, or a field with opponent color coding.

The microelectrode measurements reveal a great deal more than can be gleaned from subjective data. For example, it is found that many LGN neurons can be turned *off* by light; instead of the usual "on center–off surround" situation of Fig. 10-5, we have an "off center–on surround" receptive field. The AP frequency of these neurons increases if their surround is illuminated. Broad illumination of both the center and surround of "on center–off surround" or "off center–on surround" neurons results in a feeble or negligible response.

Consider finally a simple hypothetical retinal arrangement, that of Fig.

10-22, for the generation of red and green opponent-surround color-receptive fields. To simplify matters, the blue cone receptor C_B has been omitted, and only the center of the receptive field has been "wired up." Light entering the center is absorbed by C_R and C_G (remember that each of these has a broad response, as given by Fig. 10-17, so that *both* are excited by an incoming beam of *any* color). Because of logarithmic compression, the graded potentials generated by C_R and C_G correspond to log C_R and log C_G, respectively, as indicated. The opponent-surround characteristic comes about

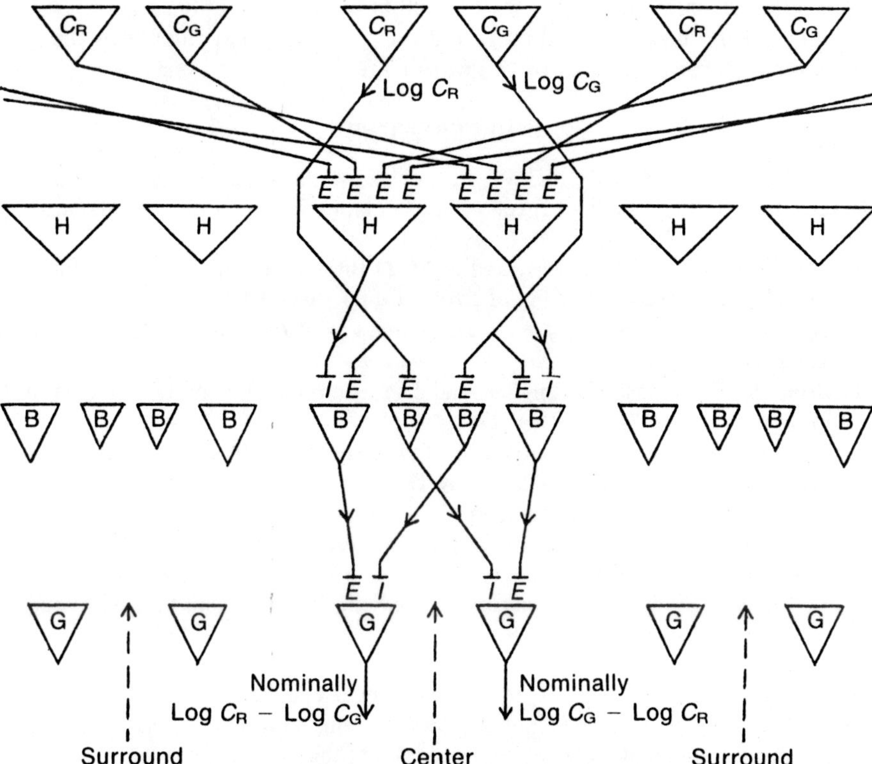

Fig. 10-22 A hypothetical model of the retina for the generation of red and green opponent-surround color receptive fields. The opponent-surround characteristic comes about because the H cell underneath C_R picks up only surround C_G potentials; similarly, the H cell underneath C_G picks up only surround C_R potentials. The ganglion cells yield logarithmic differences to get a measure for color that is independent of brightness. The ganglion cell underneath C_R is a "red center on–green surround off" type, while the cell underneath C_G is a "green center on–red surround off" type. Key: C_R = "red" cone receptor, C_G = "green" receptor; H = horizontal, B = bipolar, and G = ganglion cells.

because the H cell underneath the "red" C_R only reaches out to surrounding "green" C_G; therefore, the H cell underneath C_R inhibits one of its B cells in accordance with C_G in the surround. Similarly, the H cell underneath C_G only reaches out to the surrounding C_R, so that it inhibits one of its B cells in accordance with C_R in the surround.

If the H-cell outputs are omitted, you can trace the connections to verify that the G cell underneath C_R yields $\log C_R - \log C_G$. This is a "red center on–green surround off" cell that also calculates the logarithmic difference (instead of the ratio C_R/C_G) to get a measure for color that is independent of brightness. Similarly, the other G cell is wired to yield $\log C_G - \log C_R$; it is a "green center on–red surround off" cell that also calculates the logarithmic difference instead of the ratio C_G/C_R.

REFERENCES

H. B. Barlow, R. M. Hill, and W. R. Levick, Retinal ganglion cells responding selectively to direction and speed of image motion in the rabbit, *J. Physiol.*, vol. 173, pp. 377–407, 1964.

D. A. Baylor, M. G. F. Fuortes, and P. M. O'Bryan, Receptive fields of cones in the retina of the turtle, *J. Physiol.*, vol. 214, pp. 265–294, 1971.

G. S. Brindley, *Physiology of the Retina and Visual Pathway*, 2d ed. Baltimore: Williams & Wilkins, 1970.

P. Brou, T. R. Sciascia, L. Linden, and J. Y. Lettvin, The colors of things, *Sci. Am.*, vol. 255, pp. 84–91, Sept. 1986.

G. Buchsbaum, An analytical derivation of visual nonlinearity, *IEEE Trans. Biomed. Eng.*, vol. BME-27, pp. 237–242, May 1980.

G. Buchsbaum and A. Gottschalk, Trichromacy, opponent colours coding and optimum colour information transmission in the retina, *Proc. R. Soc. London*, vol. B220, pp. 89–113, 1983.

B. G. Cleland, M. W. Dubin, and W. R. Levick, Sustained and transient neurons in the cat's retina and lateral geniculate nucleus, *J. Physiol.*, vol. 217, pp. 473–496, 1972.

T. N. Cornsweet, *Visual Perception*. New York: Academic Press, 1971.

M. C. Cornwall, E. F. MacNichol, Jr., and A. Fein, Absorptance and spectral sensitivity measurements of rod photoreceptors of the tiger salamander, *ambystoma tigrinum, Vision Res.*, vol. 24, pp. 1651–1659, 1984.

J. C. Curlander and V. Z. Marmarelis, Processing of visual information in the distal neurons of the vertebrate retina, *IEEE Trans. Syst., Man, Cybern.*, vol. SMC-13, pp. 934–943, Sept./Oct. 1983.

F. M. De Monasterio and P. Gouras, Functional properties of ganglion cells of the rhesus monkey retina, *J. Physiol.*, vol. 251, pp. 167–195, 1975.

S. Deutsch and E. M-Tzanakou, *Neuroelectric Systems*. New York: New York Univ. Press, 1987.

J. E. Dowling and F. S. Werblin, Organization of retina of the mudpuppy, *J. Neurophysiol.*, vol. 32, pp. 315–355, May 1969.

C. Enroth-Cugell and J. G. Robson, The contrast sensitivity of retinal ganglion cells of the cat, *J. Physiol.*, vol. 187, pp. 517–552, 1966.

A. Fiorentini and T. Radici, Brightness, width and position of Mach bands as a function of the rate of variation of the luminance gradient, *Atti Fond Giorgio Ronchi*, vol. 13, pp. 145–155, 1958.

Y. Fukada and H. Saito, The relationship between response characteristics to flicker stimulation and receptive field organization in the cat's optic nerve fibers, *Vision Res.*, vol. 11, pp. 227–240, 1971.

A. L. Gilchrist, The perception of surface blacks and whites, *Sci. Am.*, vol. 240, pp. 112–124, March 1979.

C. Graham, ed., *Vision and Visual Perception*. New York: Wiley, 1966.

A. C. Guyton, *Textbook of Medical Physiology*. Philadelphia: Saunders, 1986.

H. K. Hartline, F. Ratliff, and W. H. Miller, Inhibitory interaction in the retina and its significance in vision, in *Nervous Inhibition*, ed. E. Flory. New York: Pergamon, 1961.

S. Hecht and S. Shlaer, An adaptometer for measuring human dark adaptation, *J. Opt. Soc. Am.*, vol. 28, pp. 269–275, 1938.

H. Ikeda and M. J. Wright, Receptive field organization of "sustained" and "transient" retinal ganglion cells which subserve different functional roles, *J. Physiol. London*, vol. 227, pp. 769–800, 1972.

B. W. Knight, J. I. Toyoda, and F. A. Dodge, A quantitative description of the dynamics of excitation and inhibition in the eye of Limulus, *J. Gen. Physiol.*, vol. 56, pp. 421–437, Oct. 1970.

S. W. Kuffler, Discharge patterns and functional organization of mammalian retina, *J. Neurophysiol.*, vol. 10, pp. 37–68, 1953.

J. J. Kulikowski and D. Tolhurst, Psychophysical evidence for sustained and transient neurones in human vision, *J. Physiol.*, vol. 232, pp. 149–162, 1973.

E. H. Land, The retinex theory of color vision, *Sci. Am.*, vol. 237, pp. 108–128, Dec. 1977.

Y. J. Lettvin, H. R. Maturana, W. S. McCulloch, and W. H. Pitts, What the frog's eye tells the frog's brain, *Proc. IRE*, vol. 47, pp. 1940–1951, 1959.

D. Marr, *Vision*. San Francisco: Freeman, 1982.

C. R. Michael, Retinal processing of visual images, *Sci. Am.*, vol. 220, pp. 104–114, May 1969.

H. Noda, Sustained and transient discharges of retinal ganglion cells during spontaneous eye movements of cat, *Brain Res.*, vol. 84, pp. 515–529, 1975.

R. A. Normann and I. Perlman, The effects of background illumination on the photoresponses of red and green cones, *J. Physiol.*, vol. 286, pp. 491–507, 1979.

R. A. Normann, B. S. Baxter, H. Ravindra, and P. J. Anderton, Photoreceptor contributions to contrast sensitivity: Applications in radiological diagnosis, *IEEE Trans. Syst., Man, Cybern.*, vol. SMC-13, pp. 944–953, 1983.

F. Ratliff, *Mach Bands*. San Francisco: Holden-Day, 1965.

R. W. Rodieck, *The Vertebrate Retina: Principles of Structure and Function*. San Francisco: Freeman, 1973.

T. C. Ruch, The eye as an optical instrument, vision and the retina, visual fields and central visual pathways, in *Physiology and Biophysics*, eds. T. Ruch and H. D. Patton. Philadelphia: Saunders, pp. 435–562, 1979.

W. A. H. Rushton, Visual pigments and color blindness, *Sci. Am.*, vol. 232, pp. 64–74, March 1975.

J. L. Schnapf and D. A. Baylor, How photoreceptor cells respond to light, *Sci. Am.*, vol. 256, pp. 40–47, Apr. 1987.

R. M. Shapley and J. D. Victor, Nonlinear spatial summation and the contrast gain control of cat retinal ganglion cells, *J. Physiol.*, vol. 290, pp. 141–161, 1979.

R. Shapley, E. Kaplan, and R. Soodak, Spatial summation and contrast sensitivity of X and Y cells in the lateral geniculate nucleus of the macaque, *Nature*, vol. 292, pp. 543–545, Aug. 1981.

S. Sherr, *Fundamentals of Display System Design*. New York: Wiley–Interscience, 1970.

R. Siminoff, Systems analysis of an analog model of the vertebrate cone retina, *IEEE Trans. Syst., Man, Cybern.*, vol. SMC-13, pp. 1021–1028, Sept./Oct. 1983.

M. V. Srinivasan, S. B. Laughlin, and A. Dubs, Predictive coding: A fresh view of inhibition in the retina, *Proc. R. Soc. London*, vol. B216, pp. 427–459, 1982.

L. W. Stark, The pupil as a paradigm for neurological control systems, *IEEE Trans. Biomed. Eng.*, vol. BME-31, pp. 919–924, Dec. 1984.

L. Stryer, The molecules of visual excitation, *Sci. Am.*, vol. 257, pp. 42–50, July 1987.

V. Torre, W. G. Owen, and G. Sandini, The dynamics of electrically interacting cells, *IEEE Trans. Syst., Man, Cybern.*, vol. SMC-13, pp. 757–765, Sept./Oct. 1983.

A. J. Van Doorn and J. J. Koenderink, The structure of the human motion detection system, *IEEE Trans. Syst., Man, Cybern.*, vol. SMC-13, pp. 916–922, Sept./Oct. 1983.

J. J. Vos, Colorimetric and photometric properties of a 2° fundamental observer, *Color Res. Appl.*, vol. 3, pp. 125–128, 1978.

J. J. Vos and P. L. Walraven, On the derivation of the foveal receptor primaries, *Vision Res.*, vol. 11, pp. 799–818, 1971.

G. Wald, The receptors for human color vision, *Science*, vol. 145, pp. 1007–1017, 1964.

F. S. Werblin, The control of sensitivity in the retina, *Sci. Am.*, vol. 228, pp. 70–79, Jan. 1973.

G. Wyszecki and W. S. Stiles, *Color Science*, 2d ed. New York: Wiley, 1982.

Problems

1. Carry out the step-by-step derivation of Eq. (10-6).
2. Carry out the step-by-step derivation of Eq. (10-12) and plot the horizontal (V_H) and bipolar (V_B) cell curves of Fig. 10-7.
3. Given that the spatial output of a photoreceptor is a unit V_P step, $V_P = 0$ for $x < 0$, and $V_P = 1$ for $x > 0$ (similar to V_P of Fig. 10-7). If we integrate the unit step, we get a unit ramp: $V_P = 0$ for $x < 0$, and $V_P = x$ for $x > 0$. Since the system is linear, we can integrate the unit step's V_H curve (which is similar to V_H of Fig. 10-7) to get the V_H curve corresponding to the V_P ramp. For $-1 < x < +1$, plot the spatial V_P, V_H, and V_B curves of the V_P ramp. Reference: $\int \operatorname{erf} x \, dx = (1/\sqrt{\pi}) \exp(-x^2) + x \operatorname{erf} x$. {Ans.: $V_H = 0.5[x + (1/\sqrt{\pi}) \exp(-x^2) + x \operatorname{erf} x]$.}
4. In Fig. 10-9, light intensity L_z is 20 in the left half of the visual field and 80 on the right. The gray vertical line has an L_z of 50. Calculate and plot the L_z, V_P, V_H, and V_B curves.
5. For the sinusoidal grating input of Fig. 10-10, carry out the step-by-step deri-

vation of $V_{P0} = V_{H0} = 14.88$ mV following Eq. (10-16). Also carry out the step-by-step derivation of Eqs. (10-21) to (10-23).

6. Carry out the step-by-step derivation of the curves of Fig. 10-11 at $x = -2.9, -2.7, \ldots, +2.9$.
7. Using the ROYGBIV cone pigment response values given in Table 10-2, replot the curves of Fig. 10-17 using a linear relative response scale rather than the logarithmic scale. Does it "look" as if the areas under the curves agree with the "white" ratios $C_R:C_G:C_B = 90:45:1$?
8. In Table 10-3, instead of using $\lambda = 650$, 520, and 450 nm, a color synthesizer uses $\lambda = 626$, 528, and 445 nm, respectively. The cone pigment response values are given in Table 10-2. Using the L_z values of Table 10-3, find the resulting $C_R:C_G:C_B$ ratios. [Ans.: $C_R = 167$, $C_G = 59.5$, $C_B = 1.03$, $C_R:C_G:C_B = 162:58:1$.]
9. The sensation of infinite temperature (∞) in Fig. 10-18 is given by $C_R:C_G:C_B = 15:11:1$. This ratio can be obtained by using three monochromatic beams of light at 650 nm (red), 520 nm (green), and 450 nm (indigo). Find L_z, the relative intensity of each beam, similar to the calculations for Table 10-3. [Ans.: $\lambda = 650$ nm, $L_z = 22.87$; $\lambda = 520$ nm, $L_z = 28.70$; $\lambda = 450$ nm, $L_z = 88.30$.]
10. Derive all of the horizontal and vertical component values given in Table 10-2 for Fig. 10-19(*a*) from the data given in the "Fig. 10-17" columns.
11. Given the following x and y coordinates for the ROYGBIV colors of the standard chromaticity diagram of Fig. 10-18—$R = 0.730, 0.270$; $O = 0.699, 0.301$; $Y = 0.479, 0.520$; $G = 0.140, 0.814$; $B = 0.070, 0.209$; $I = 0.164, 0.022$; $V = 0.176, 0.013$—derive all of the relative x and y values given in Table 10-2 for Fig. 10-19(*b*).
12. The sensation of infinite temperature (∞) in Fig. 10-18 is given by $C_R:C_G:C_B = 15:11:1$. This is the input in Fig. 10-20 with values $C_R = 45$, $C_G = 33$, and $C_B = 3$. (a) Find N1, N2, and N3. (b) Find the outputs of the vertical columns of eight pairs of neurons. (c) Find the values for the crossover neurons in the color map. [Ans.: (a) N1 = 0.039, N2 = 3.66, N3 = 2.4 mV. (b) 2.461, 1.461, 0.461, 0.539, 8.6, 4.6, 0.6, and 3.4 mV.

(c) 4.64	3.96	12.56	21.16
2.48	2.12	6.72	11.32
0.32	0.28	0.88	1.48
1.83	1.57	4.97	8.37.]

11

Visual Pattern Recognition, Neural Networks, and "Household Chores"

ABSTRACT. The visual field of each eye is split in half. Approximately half of the axons from each eye cross over so that the left hemisphere of the brain processes the right half of the visual fields while the right hemisphere processes the left half.

The optic nerves lead to the lateral geniculate nuclei and then proceed to the primary visual (striate) cortex. Here, according to the research of Hubel and Wiesel, groups of neurons are organized into short lines that fire when an edge, corresponding to that particular orientation, enters the visual field. Half of the neurons process signals that originated in the left eye (L layers) and the other half process signals from the right eye (R layers); the L and R layers are interleaved. The visual image is thus dissected into many short edges, the first step in the hierarchical buildup of more complex geometrical shapes. All possible edge orientations are detected by a layer of simple (S) cells; these feed a layer of complex (C) cells which bestow invariance with respect to lateral displacement.

A one-dimensional model is presented in which a 10-element visual signal is analyzed by an S_1C_1 pair of layers feeding an S_2C_2 pair. The model looks at (1) a white strip gradually lowered over a black background; (2) a white bar moving downward; and (3) a white bar that expands until it covers the entire black background.

Some rules for designing one-dimensional pattern-recognition models are derived.

Self-organization of the cortex is considered. A scheme is presented by which one cell in a group of cells captures the input for itself and suppresses all of the other neurons. A somewhat more complicated neural network is examined as it goes through learning, organizing, and testing phases.

Because the same techniques are common to real and artificial neural networks (ANNs), a section is devoted to ANNs. Associative memory, the task-assignment problem, and back propagation are considered.

Various "household chores" associated with vision are also discussed as follows: the constant jerky movements of the eye (saccadic motion) and compensation for it; control of the lens opening (pupillary reflex); distance accommodation; and binocular convergence.

11-1 Some Characteristics of the Visual Cortex

Reduced to bare essentials, the brain of an organism is a structure that has a built-in memory (instinct) plus a learned memory, and it subsequently uses its instincts and/or what it has learned to recognize patterns. Needless to say, the subjects of memory and pattern recognition are well represented in the literature [J. R. Anderson and G. H. Bower, 1973; C. Aoki and P. Siekevitz, 1988; M. A. Arbib, 1972; D. O. Hebb, 1949; G. Hinton and J. A. Anderson, 1981; T. Kohonen, 1978; M. Minsky and S. Papert, 1969; E. M-Tzanakou, 1988; N. J. Nilsson, 1965; F. Rosenblatt, 1962; F. O. Schmitt et al., 1981; D. E. Wooldridge, 1979].

In the present chapter, we first concentrate on what the brain does with the signals generated by the retina of the eye, although only a few of the details have been elucidated. For an animal, it is obvious that the most important goal of visual signal processing is the recognition of patterns: Does the image represent food, friend, or foe? For a human the image can also represent written message symbols, or it can be a beautiful painting without any pattern in particular that need be recognized. A few references to the visual cortex are given: [G. S. Brindley and W. S. Lewin, 1968; S. E. Brodie, B. W. Knight & F. Ratliff, 1978; L. N. Cooper, 1981; R. O. Duda and P. E. Hart, 1973; W. J. Freeman, 1991; J. P. Frisby, 1980; M. Kabrinsky, 1966; J. Szentagothai, 1973; A. Treisman, 1986; J. M. Wolfe, 1983; U. Yinon and E. Auerbach, 1975].

After looking at artificial neural networks, the chapter ends with a consideration of some "household chores." These are some of the less glamorous operations performed by the brain as a result of the retinal signals.

The visual pathways in the human brain are schematically shown in the cross section of Fig. 11-1. We are mainly concerned with information processing of signals that originate in the high-acuity region of each eye, the fovea. The fovea has about twice the diameter of the foveola and includes the latter region; each fovea is about 1.5 mm = 5° visual angle in diameter, with area around 1.8 mm^2, and contains around 75,000 cone sensory receptors ("red," "green," and "blue" cones whose responses nominally peak, respectively, in the red, green, and blue regions of the spectrum) plus some rods.

If we follow the lines in Fig. 11-1 carefully we see a remarkable bifurcation: Approximately half of the axons from each eye cross over, so that one of the lateral geniculate nuclei (LGN) ends up with fibers from the left

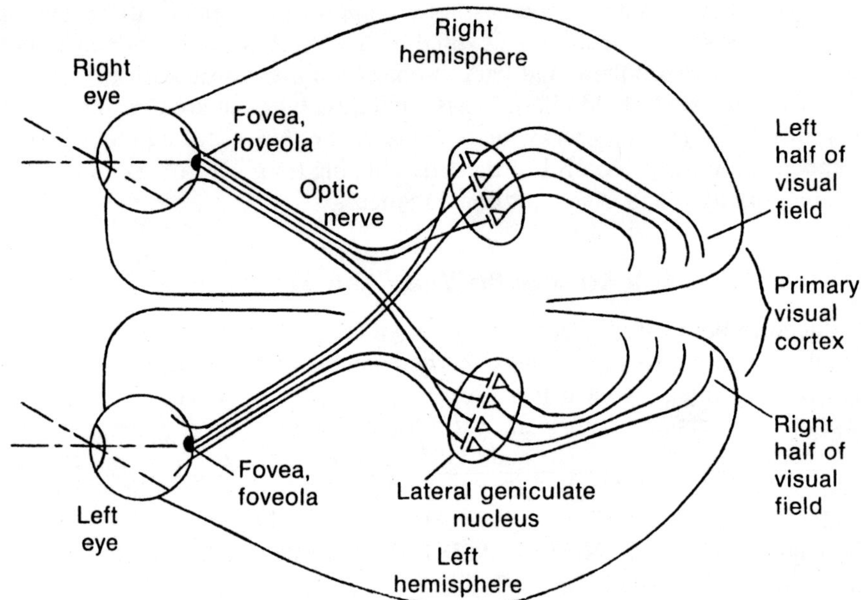

Fig. 11-1 Highly simplified cross section showing visual pathways of the human brain. Each fovea is about 1.5 mm = 5° visual angle in diameter. Rays from the right (left) half of the visual field cross over and are focused on the left (right) half of each retina. Approximately half of the axons from each eye cross over, as shown, so that the primary visual cortex in the left (right) hemisphere processes the right (left) half of the visual fields. (From Deutsch and M-Tzanakou, *Neuroelectric Systems*, New York Univ. Press, New York, 1987.)

half of the visual field and the other receives fibers from the right half. From the LGNs, the signal proceeds to the primary visual cortex. In other words, the visual field is split in half and the two halves are separately processed without our being aware of the bifurcation. An object that moves from right to left in the visual field instantaneously "hops" across the gap between the left and right hemispheres. The neural networks are obviously continuous although they are physically separated. The system is an excellent illustration of the futility of introspection in trying to determine how the brain functions.

It is "natural" for the left hemisphere to handle the right half of the visual field, and vice versa, because the right half of the visual field appears in the left half of each eye. This comes about because rays of light cross over in passing through the lens of the eye. Surprisingly enough, much of the rest of the nervous system crosses over as if to match the crossover of the eye lenses. Afferent fibers from the right half of the body cross over, before they enter the brain, and end up in the left hemisphere; efferent fibers from the left hemisphere cross over to the right side of the body after they leave the brain.

Our model for the initial stages of pattern recognition is based on the research of David H. Hubel and Torsten N. Wiesel, and others. After some 25 years of painstaking work, Hubel and Wiesel were awarded the Nobel Prize in physiology and medicine (shared with Roger W. Sperry) in 1981 [M. Constantine-Paton and M. I. Law, 1982; D. H. Hubel and T. N. Wiesel, 1959, 1979; B. B. Lee et al., 1977; C. R. Michael, 1981; V. B. Mountcastle, 1957; B. R. Payne, N. Berman & E. H. Murphy, 1981; T. P. S. Powell and V. B. Mountcastle, 1959; I. D. Thompson and D. J. Tolhurst, 1981].

Briefly, Hubel and Wiesel probed the visual system of an experimental animal, mostly the cat and macaque monkey, with microelectrodes. The animal was anesthetized in a way that left the visual responses relatively intact. At the same time, a visual input in the form of flashing or moving spots of light, and of flashing or moving bars of various angular orientations, was presented to the animal.

Figure 11-2 depicts a highly idealized cross section of the Hubel–Wiesel primary visual cortex. Proceeding away from the surface, the cortex consists of four layers of neurons, labeled C (complex), S (simple), G (geniculate input), and again C. Each layer is approximately 0.5 mm thick, so that the total thickness is 2 mm, as shown. Half of the neurons process signals that originated in the left eye (L layers) and the other half process signals from the right eye (R layers). The separation into L and R layers is known as "ocular dominance." Here we have another surprise that illustrates the futility of introspection: the L and R layers are interleaved as shown! The cortex here is organized into repeating units: in the left-to-right direction, a pair of L,R layers repeats every millimeter. Similarly, in the front-to-back direction, although it is not shown, the fine structure repeats every millimeter. There are approximately 50,000 neurons/mm^3, so that in the volume shown in Fig. 11-2, which is $2 \times 2 \times 3 = 12$ mm^3, there are 600,000 neurons as indicated [J. J. Kulikowski, S. Marcelja & P. O. Bishop, 1982].

To avoid additional complexity, not shown in Fig. 11-2 are cylindrical or pillarlike structures, 0.2 mm in diameter and centers spaced 0.5 mm apart, that are embedded in the center of each L and R layer. The neurons in these pillars seem to be engaged in the processing of color information.

The surface area of the primary visual cortex is about 15 cm^2, so that the cortex contains some 100,000 neurons/mm^2 $\times$ 1500 mm^2 $= 150 \times 10^6$ neurons. Most of them are devoted to the processing of foveal signals. Since there is a total of 150,000 foveal cones, we get a very approximate ratio of 1000 cortical neurons/cone.

The processing in the primary visual cortex consists mainly of an *orientation* response; that is, a response determined by the directions of lines and edges in the visual image. The short lines in Fig. 11-2 symbolize groups of neurons that fire when an edge, such as the top of a T, which corresponds to that particular orientation, enters that area of the visual field. When this horizontal edge is seen, those neurons sensitive to the horizontal orientation

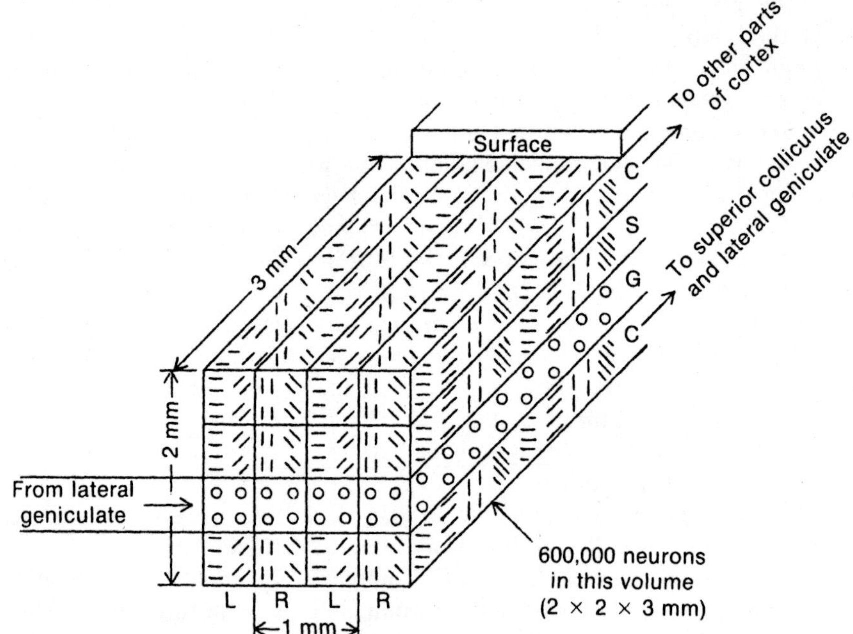

Fig. 11-2 A highly idealized three-dimensional cross section of a small volume of the primary visual cortex. The symbols stand for C (complex), S (simple), G (geniculate input), L (left eye), and R (right eye). The short lines represent groups of neurons that fire when an edge, corresponding to that particular orientation, enters the visual field. It is conjectured that each edge subtends only a few contiguous red, green, and blue cones in the fovea. The G layer contains interneurons that do not have an orientation bias. The C layer output is largely independent of lateral displacement of objects in the visual field because each C cell receives inputs from a relatively large area of the previous S or G layer. Not shown are pillarlike structures, 0.2 mm in diameter and centers spaced 0.5 mm apart, embedded in the center of each L and R layer, that seem to be concerned with color information.

will respond, as explained below. For ease in drawing, the short lines in Fig. 11-2 are shown as horizontal, or vertical, or at an angle of +45°, or at −45°, representing groups of neurons that respond to that specific orientation of an object in the visual field. In an actual cortex, however, *all possible orientations* are represented; furthermore, the transition from one orientation to the next is gradual. (Neurons in the previously mentioned pillars that are embedded in the center of each L and R layer are not orientation selective.)

How many neurons does each of the short lines of Fig. 11-2 represent? If we assume that neurons are identical closely packed cubes, each of them has a volume of 1 mm^3/50,000 = 20,000 μm^3, which corresponds to an edge length of $(20{,}000)^{1/3}$ = 27 μm. Each millimeter in Fig. 11-2 subtends 1000 μm/27 μm = 37 neurons, and there are around four short lines/mm, so each of the short lines is only nine neurons long. The calculation may be crude, but the conclusion is that the primary cortex dissects the visual image into many small ensembles, each of which corresponds to only a few contiguous red, green, and blue retinal cones (as well as rods outside the foveolas, which only contain cones). The short edges, such as the parts of the alphanumeric symbols you are reading at the present time, are the first step in a hierarchical buildup of more complex geometrical shapes. For example, the tip of the nose in a face is a short edge that will eventually be recognized as being part of a series of edges making up a nose.

Each part of the visual image is thus dissected and then reconstructed in the visual cortex: If the visual image is a face, the image is broken down into small ensembles that are first "seen" as tiny pieces of eyes, ears, nose, and so forth; later on, neuronal ensembles see eyes, ears, and a nose; finally, the complete face of your grandmother, say, is recognized. All in a fraction of a second!

A somewhat hypothetical view of what each of the short S lines may be doing is presented in Fig. 11-3. For simplicity, the short lines are shown as having a length of only five rather than nine neurons. In Fig. 11-3(*a*), three adjacent neuronal lines, shown as black dots, feed a single short horizontal edge detector (SHED) neuron. The short horizontal edge is shown, moving upwards, beneath the group of S neurons (which were previously shown in Fig. 11-2). When the horizontal edge reaches the bottom row of pertinent S neurons (that is, when the eye sees a horizontal edge moving upwards in this region), the SHED neuron gets an inhibitory input of -5 (five weighting factors, each of value -1 as shown, are excited). When the edge reaches those cones feeding into the middle row of S neurons, the SHED neuron gets a net input of $+5$ (five additional weights, each of value $+2$ as shown, are recruited). Accordingly, this indicates to the neurons in the next stage of analysis that a short horizontal edge has been detected. When the edge reaches the upper row of those cones feeding into the pertinent S neurons, the SHED neuron gets a net input of zero, because the input junction weights (ten, each of value -1, plus five, each of value $+2$) form a zero-sum array.

In Fig. 11-3(*b*), we get more for less. With only two rows of pertinent S neurons, we have a short horizontal edge moving up detector (SHEMUD) neuron. The SHEMUD receives $+5$ when the edge reaches the first row of pertinent S neurons; if the edge is moving down, the SHEMUD receives -5 and remains dormant. Therefore, whereas the SHED neuron detects only horizontal edges, the SHEMUD neuron also detects edge movement, but only in an upward direction. A companion detector, a short horizontal edge moving down detector (SHEMDD) neuron, is presumably also coupled to the same group of pertinent S neurons.

Returning to Fig. 11-2, the G layer contains neurons that do not have

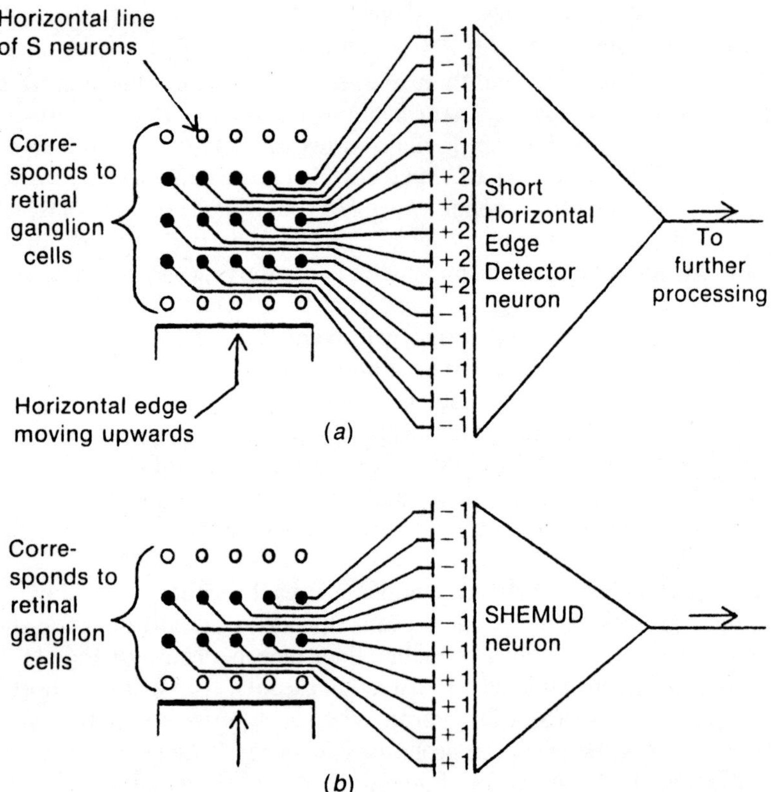

Fig. 11-3 Hypothetical neuronal circuits associated with short S lines of Fig. 11-2, here shown as horizontal lines five neurons long. A short horizontal edge is moving upwards relative to the S lines, as indicated. (*a*) The pertinent S neurons feed a SHED neuron. (*b*) The pertinent S neurons feed a SHEMUD neuron.

an orientation bias. They are therefore shown as small individual circles. Their function is to act as interneurons, feeding the input to adjacent S and C layers.

In the top layer of C cells, each cell (or group of synergistic, cooperating cells) interfaces with many S cell groups of a particular orientation. It is here that invariance or independence with respect to lateral displacement is acquired. For example, a 45° edge that moves across the visual field excites in succession one group of 45° S cells after another, but only the single 45° C cell (or small group of 45° C cells) that is monitoring 45° S cells in that area of the cortex will fire. It is a good example of how the nervous system discards less important information, since it can only handle a small portion of the huge amount of information it is receiving. Initially, the specific locations of the 45° edge are seen, but the upper C layer takes an average, in effect, that ignores information about precise positions. The retained message concerns a 45° edge, independent of its specific location in a particular region of the visual field.

From the top layer of C cells, axons project to one or more other parts of the cortex. It is assumed that a hierarchical buildup of geometrical shapes thus begins. One possible model argues that the primary visual cortex "looks at" edges that are four cone groups long, the next patch processes geometrical patterns that fit into a diameter of eight cone groups, the third patch processes a diameter of 16 cone groups, and so forth. Seven or eight of these brain patches are needed to cover the entire foveas, each of which has a diameter of only 300 cones. Neuroanatomical studies show, as a matter of fact, that a number of distinct cortical stations are connected in a hierarchical sequence that begins at the striate cortex and ends in the inferior temporal cortex. More is said about this in Chapter 12.

The bottom layer of C cells in Fig. 11-2 is similar to the top layer in that each cell receives inputs from a relatively large area of the previous layer, in this case, the G cells. The bottom layer of C cells projects to the superior colliculus part of the brain and feeds back to the lateral geniculate. It is considered that these pathways are not part of the visual pattern-recognition hierarchy. The superior colliculus seems to be involved with head and eye movements, an attention response toward an unexpected sound or touch or visual intrusion. It may also be involved with compensation for saccadic eye movements, as discussed in Section 11-6. As to feedback to the lateral geniculate, we do not know how this fits into the scheme of brain activity.

11-2 One-Dimensional Model

The cortex of Fig. 11-2 is lifeless unless we can construct a suitable model, apply some visual inputs, and show that reasonably effective pattern recognition takes place.

A computer simulation based on the Hubel–Wiesel primary visual cortex was demonstrated in 1980 at the NHK Broadcasting Science Research Laboratories in Tokyo. Called a *neocognitron*, it was the brain child of Kunihiko Fukushima (1980) and his colleagues. The neocognitron was trained to recognize 24 simple patterns regardless of where they appeared on a 16 × 16 element "visual field." If a pattern was shifted laterally, or made slightly larger or smaller, or tilted by a small amount, the neocognitron continued to correctly identify the pattern.

In this section, a one-dimensional model is demonstrated; it is a model that can recognize the difference between line patterns such as – ——, – –, —— –, and so forth. In effect, we are examining the initial processing of information when a simple pattern with a single orientation is seen. Even a relatively modest two-dimensional endeavor, such as that of Fukushima, requires computer simulation. The advantage of a one-dimensional "pencil and paper" model is that one can easily decipher its inner workings [S. Deutsch, 1981].

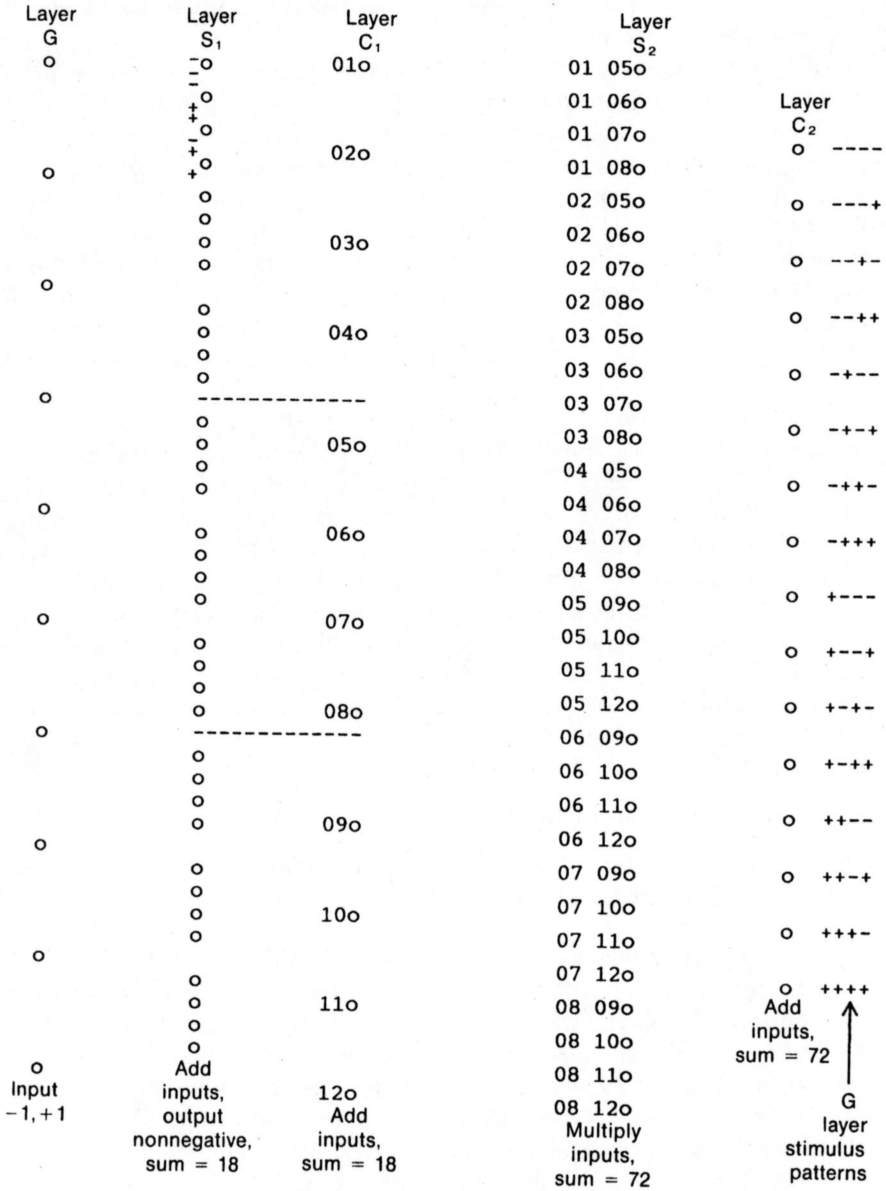

Fig. 11-4 Hypothetical ten-input-element one-dimensional model of the first two input layer pairs of the visual cortex. Model is "unwired" (axons are shown in Fig. 11-5). The G layer inputs correspond to -1 (black visual area) or $+1$ (white visual area). There are four types of S_1 detectors corresponding to multiplicative weighting factors, reading downward, as follows: $-1-1$ (short black line); $-1+1$ (black to white edge); $+1-1$ (white to black edge); and $+1+1$ (short white line). The S_1 outputs are positive; zero

In the neocognitron, Fukushima assumed that the C cells from one cortical patch lead to a pair of S,C (simple, complex) cell layers in the next patch. His 1980 computer simulation contained three pairs of S,C cell layers. Because all areas of the visual cortex seem to be similarly organized, his S_2C_2 and S_3C_3 pairs resembled the S_1C_1 pair. This viewpoint is repeated in the model presented here; it is somewhat conjectural, however, since only the S_1C_1 pair has been neurophysiologically explained by Hubel–Wiesel and other investigators.

Our naked, unwired one-dimensional model is shown in Fig. 11-4. It is a 10-input-element structure of minimum complexity that is still able to illustrate the main features of visual pattern recognition. The input layer at the left corresponds to the G layer of Fig. 11-2. The S_1,C_1 layers correspond to the upper S,C pair of Fig. 11-2. A further stage of signal processing is modeled by the S_2,C_2 layers.

The "visual input" stimulus consists of 10 black or white dashes of light entering the G layer. This stimulus, you may recall, comes from the cones via the retinal ganglion cells and lateral geniculate nuclei. Since we wish to simplify matters as much as possible, only black (-1) or white ($+1$) inputs are considered. No light-gray ($+0.5$) or gray (0) or dark-gray (-0.5) inputs; you can, of course, try them on your own. It turns out that if the inputs are *exactly* -1 or $+1$, a surprising simplification occurs: the summation of outputs for each layer is constant, regardless of the distribution of input -1s and $+1$s. This offers a convenient "parity check" on numerical calculations. As given in Fig. 11-4, the summation of outputs is 18 (mV units, say) for the S_1 and C_1 layers, and 72 mV for the S_2 and C_2 layers.

Since the model is one-dimensional, the orientation detectors in the S_1C_1 layers can respond only to vertical patterns; that is, only to black and white dashes in a vertical line. There are four types of vertically oriented S_1 detector neurons, as indicated by the weighting factors "$--$, $-+$, $+-$, $++$." Reading downward, the first S_1 neuron is a detector of short black lines $-1-1$. It "looks at" the first two G inputs (a short line) and multiplies this pair of inputs by $-1-1$, respectively. The output is equal to the sum (the instruction at the bottom of layer S_1 should say "add inputs after multiplying by the junction weighting factor," but this has been shortened to

←

otherwise. The C_1 neurons divide the visual field into upper, middle, and lower regions. Neurons in the S_2 layer detect all possible combinations taken from adjacent visual input regions; the numbers indicate C_1 input pairs. Each C_2 neuron tends to respond to its particular four-input-element pattern regardless of where it occurs in the visual field. The "G layer stimulus patterns" list the 16 different ways of forming four-element patterns from -1s and $+1$s. Regardless of input pattern, the sum of S_1 and C_1 output values is 18; the sum of S_2 and C_2 output values is 72.

"add inputs"). If the G signals happen to be $-1-1$ (black dashes), the first S_1 neuron output is 2 mV; if $G = -1+1$, (black-to-white pair) or $+1-1$ (white-to-black pair), the first S_1 neuron output is 0; if $G = +1+1$ (white dashes), the first S_1 neuron output is again 0 because negative outputs are forbidden as indicated by "nonnegative." Similarly, the second S_1 neuron detects black-to-white edges $-1+1$; the third detects white-to-black edges $+1-1$; the fourth detects short white lines $+1+1$. The 10 G signals are thus analyzed by nine groups of S_1 neurons. (In the neocognitron, the weights of S cells are modified in a self-organizing mode during which the "uneducated" cortex is "learning" input patterns, while the weights of C cells are fixed. Hypothetical models for self-organization are considered later in this chapter.)

The fully wired model, with all axons in place, is shown in Fig. 11-5. This illustration is admittedly dangerous because the casual reader will conclude that the nervous system is hopelessly complicated. Electrical engineering students, accustomed to seeing a maze of wires both schematically and physically, should realize that the complexity in this case is superficial. The complete schematic diagram is useful for reference in that it shows which neurons are connected to one another.

[S.D. writes: "I sometimes ask my students to trace out the active neurons and their respective outputs corresponding to a given input pattern, and they experience little difficulty in solving these homework and examination problems."]

A neurophysiologist looking at a "simple" creature such as the snail *Aplysia* is also confronted by a galaxy of neurons; as a first step in deciphering the maze, the neurophysiologist stimulates a single neuron, say, and follows its associated responses.

Similarly, the workings of our model are illustrated in Fig. 11-6 for an input pattern consisting of two $+1$s followed by eight -1s (a white line two units long followed by a black line eight units long). Only axons that contribute to final C_2 output are shown. In the first group of S_1 neurons, the $+1+1$ detector has an output of 2 mV. The next group "sees" the edge $+1-1$. Each of the following seven groups of S_1 neurons dissects the input into short black lines $-1-1$.

The C_1 layer, as previously described, lends invariance with respect to lateral motion. There are three groups, each consisting of four C_1 neurons, that divide the visual field into upper, middle, and lower regions. In each of these groups (of four C_1 neurons), reading downward, the individual C_1 neurons respond to S_1 neurons $-1-1$, $-1+1$, $+1-1$, and $+1+1$, respectively. Each of the C_1 neurons adds its S_1 inputs, as indicated. In Fig. 11-6, five of the C_1 neurons are stimulated by the S_1 cells as shown.

The C_1 neurons are saying, in effect, "We are only capable of seeing short lines. In the upper region of the visual field we see black–black, white–black, and white–white lines; in the middle region we see only black–black lines; and in the lower region, also, we see only black–black lines."

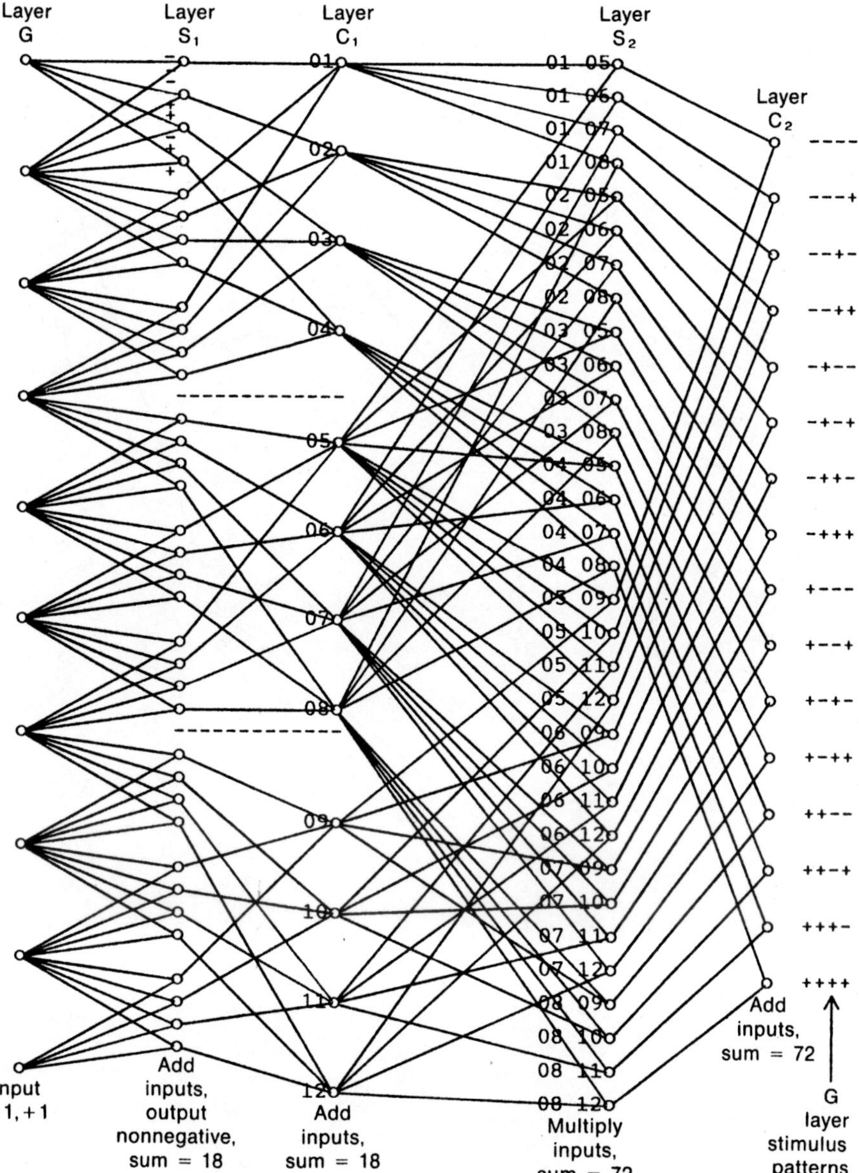

Fig. 11-5 The $S_1C_1S_2C_2$ model fully "wired up." This schematic diagram is useful for reference in that it shows which neurons are connected to one another.

The next step in the hierarchical chain is the S_2 layer. Its neurons detect all possible combinations taken from *adjacent* visual input regions (upper and middle, or middle and lower regions). For example, the second S_2 neuron in Fig. 11-5 is stimulated by $-1-1$ from the upper C_1 region (C_1 cell 01)

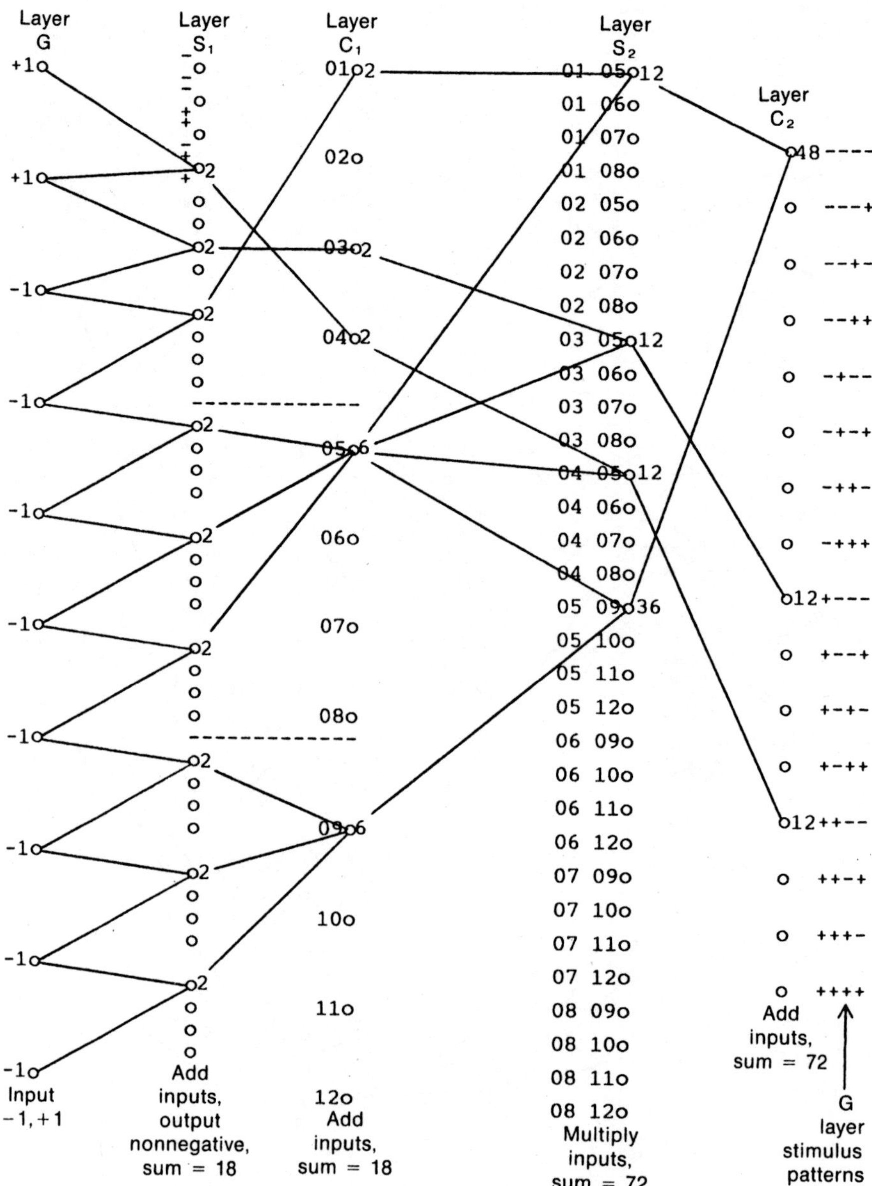

Fig. 11-6 The $S_1C_1S_2C_2$ model when the input pattern consists of two +1s followed by eight −1s. Only the axons that contribute to final C_2 output are shown. Its output potential (in millivolts, say) is given next to each neuron.

and by $-1+1$ from the middle C_1 region (C_1 cell 06), so that it responds to the input sequence $-1-1-1+1$. The eighteenth S_2 neuron is stimulated by $-1-1$ from the middle C_1 region (C_1 cell 05) and by $-1+1$ from the lower C_1 region (C_1 cell 10), so that it also responds to the input sequence

$-1-1-1+1$. The second C_2 neuron is stimulated by either the second or eighteenth S_2 neuron (or both), so that it tends to respond to the input sequence $-1-1-1+1$ *anywhere* in the visual field. It sees the vertical black–black–black–white pattern no matter where it occurs. In other words, C_2 responses are largely independent of a lateral shift in the visual pattern.

Each of the S_2 neurons *multiplies* its C_1 inputs, as indicated. Multiplication eliminates a great deal of insignificant information. If either one of the C_1 outputs is zero, the S_2 output is zero. For example, none of the S_2 neurons will respond to a two-element visual input such as $-1-1$ or $-1+1$, but one of them *will* respond to the four-element input $-1-1-1+1$. In a living cell, multiplication can be approximated if the input synaptic junctions occur on the cell body rather than on a remote dendrite, and if the output is a graded potential.

In Fig. 11-6, the multiplication of C_1 outputs stimulates four of the S_2 neurons as shown; the outputs reading down are 12, 12, 12, and 36 mV, which add up to 72 mV in agreement with the check sum.

The output of each of the C_2 neurons is the sum of S_2 inputs. Since there are 16 different ways of forming four-element patterns from the two-element pairs $-1-1$, $-1+1$, $+1-1$, and $+1+1$, there are 16 C_2 neurons. Its four-element visual input pattern is given next to each of the C_2 neurons in the model (the -1 and $+1$ sequence is abbreviated by using $-$ and $+$, respectively, without the 1s).

The final result for the example depicted in Fig. 11-6 is that the $-1-1-1-1$ input string is recognized with an output level of 48 mV; the $+1-1-1-1$ sequence is weakly recognized with output level 12 mV; and the $+1+1-1-1$ edge is weakly recognized with output level 12 mV. None of the other four-element regions of the input signal elicits an output response. The descriptive message is "I see mostly black lines four units long, but I also see the patterns white–black–black–black and white–white–black–black."

The characteristics of the model are best illustrated by gradually replacing an input string of -1s with $+1$s. One can think of this maneuver as the gradual lowering of a white strip against a black background. The results are summarized in Table 11-1, where the input sequences are shown as columns at the top of the table versus the C_2 cell layer outputs, which are listed as rows at the left side of the table. The preceding example is entered in the "Percent white = 20," "Ckt in Fig. 11-6" column. The "Visual input G-layer pattern" shows a white bar two units long above a black bar eight units long. Underneath we have 48 in the C_2 cell layer output $----$ row, 12 in the $+---$ row, and 12 in the $++--$ row. The sum, of course, is 72.

With regard to the remainder of Table 11-1, at first, with a solid -1 background, only the $-1-1-1-1$ output cell is stimulated, with an output level of 72 mV. The axons that contribute to final C_2 output are depicted in Fig. 11-7. Table 11-1 shows that the output of the $-1-1-1-1$ cell linearly decreases as the white strip is lowered. For a single $+1$ input, the model is

TABLE 11-1 For the $S_1C_1S_2C_2$ Model

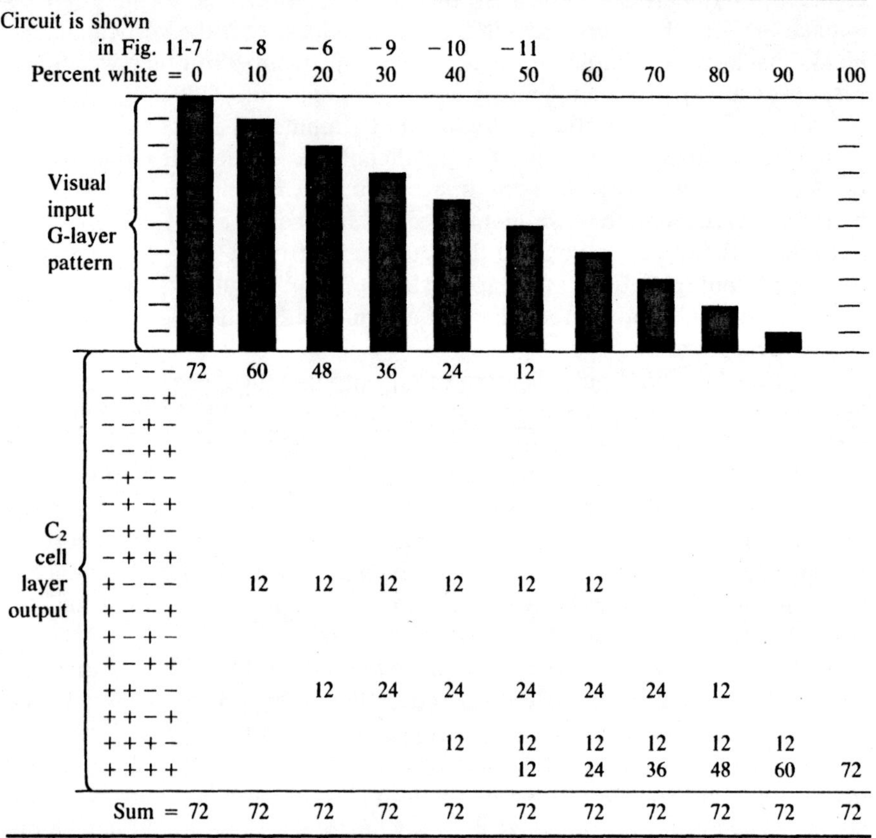

Circuit is shown in Fig.	11-7	-8	-6	-9	-10	-11					
Percent white =	0	10	20	30	40	50	60	70	80	90	100
Visual input G-layer pattern											
C_2 cell layer output											
− − − −	72	60	48	36	24	12					
− − − +											
− − + −											
− − + +											
− + − −											
− + − +											
− + + −											
− + + +											
+ − − −		12	12	12	12	12	12				
+ − − +											
+ − + −											
+ − + +											
+ + − −			12	24	24	24	24	24	12		
+ + − +											
+ + + −					12	12	12	12	12	12	
+ + + +						12	24	36	48	60	72
Sum =	72	72	72	72	72	72	72	72	72	72	72

Note: C_2 cell layer outputs (− represents black, + represents white) versus visual input patterns if a white strip is gradually lowered over a black background (percent white = 0, 10, 20, . . . , 100). Where applicable, the corresponding figure number is given.

shown in Fig. 11-8; for +1+1+1, input the model is shown in Fig. 11-9; for +1+1+1+1 input, the model is shown in Fig. 11-10; for a half-white input, the model is shown in Fig. 11-11.

In the 60% white column (G layer input = + + + + + + − − − −), the output of C_2 cell − − − − becomes zero. This column therefore demonstrates what one may call "poor peripheral vision," since the model does not respond to the − − − − sequence at the tail end of the G layer input.

The C_2 cell + − − − shows a weak, constant 12-mV output as the strip is lowered from 10% white to 60% white. The important edge +1+1−1−1 elicits a C_2 cell + + − − output which rises to 24 mV and remains constant at this level, independent of lateral (downward) shift, from 30% to 70% white coverage. The C_2 cell + + + − row is symmetrical with the + − − − row; there is a weak, constant output of 12 mV as the white input goes from 40% to 90%. Finally, the C_2 cell + + + + row is symmetrical with the − − − −

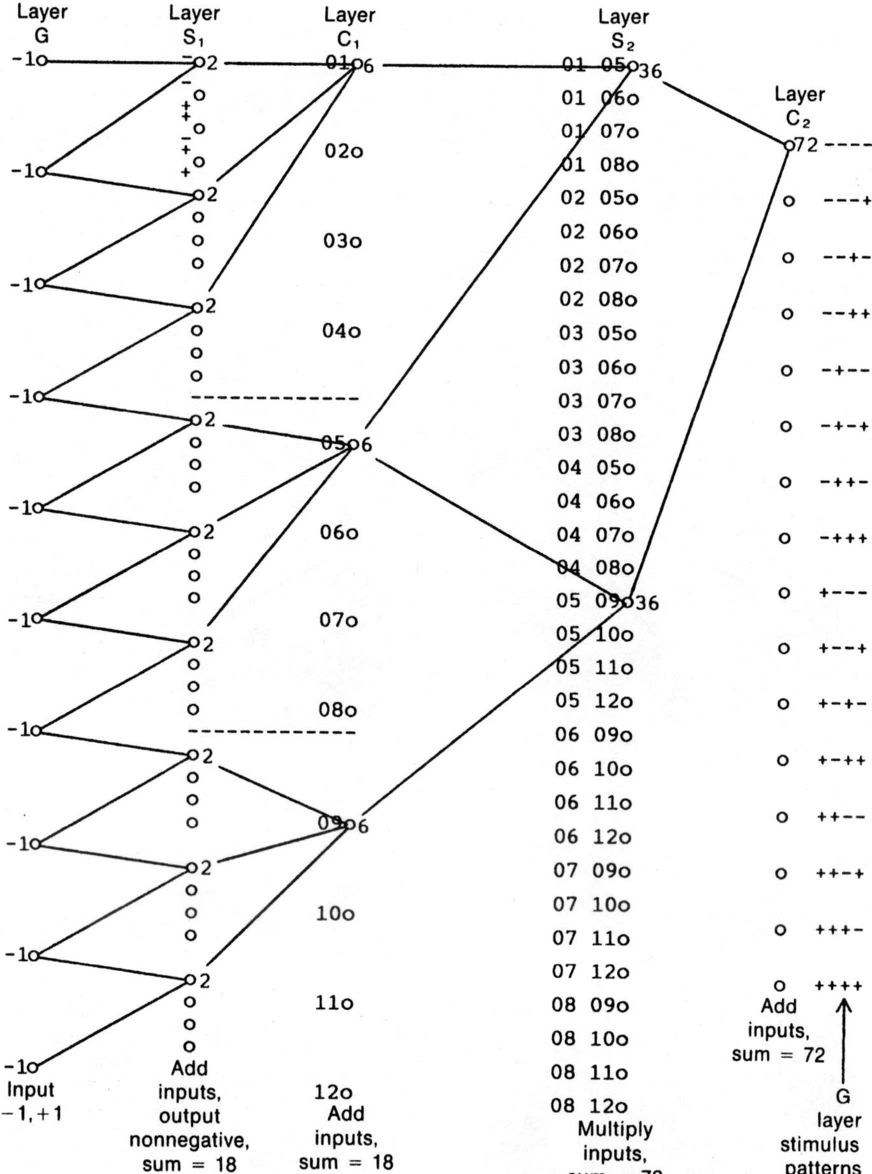

Fig. 11-7 The $S_1C_1S_2C_2$ model when the input pattern consists of a solid −1 background. Only the axons that contribute to final C_2 output are shown.

row; its output linearly rises from 0 to 72 mV as the input strip is lowered from the 40% to 100% white level.

The 11 inactive rows of Table 11-1 correspond to visual input patterns that never occur in this table, but they may of course occur with other input

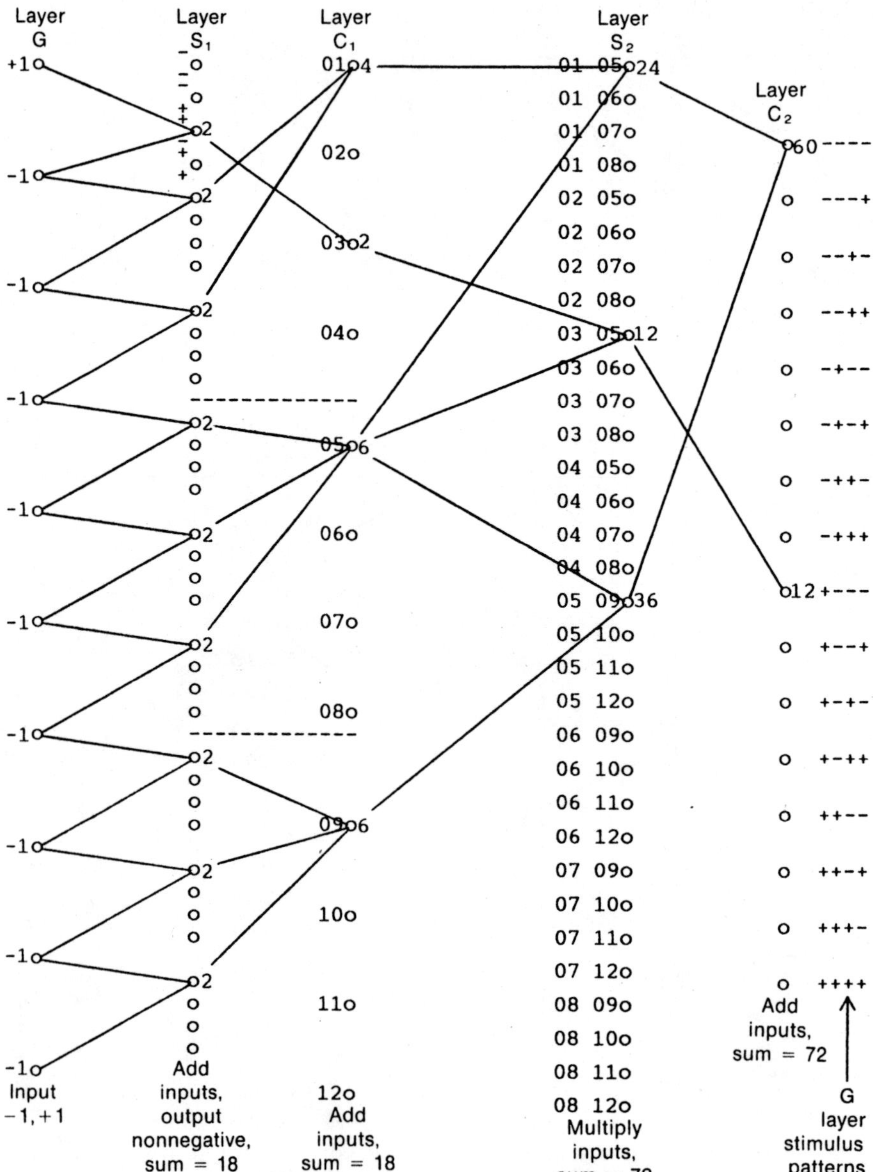

Fig. 11-8 The $S_1C_1S_2C_2$ model when the input pattern consists of +1 followed by nine −1s. Only the axons that contribute to final C_2 output are shown.

patterns. The lack of responses to false input combinations is just as important as are the valid responses in Table 11-1.

To illustrate the response of the model to *lateral shift*, a string of six +1s is applied in Table 11-2(a). As this "white bar" moves downward, the C_2 cell + + + + output remains constant at 24 mV, thus demonstrating in-

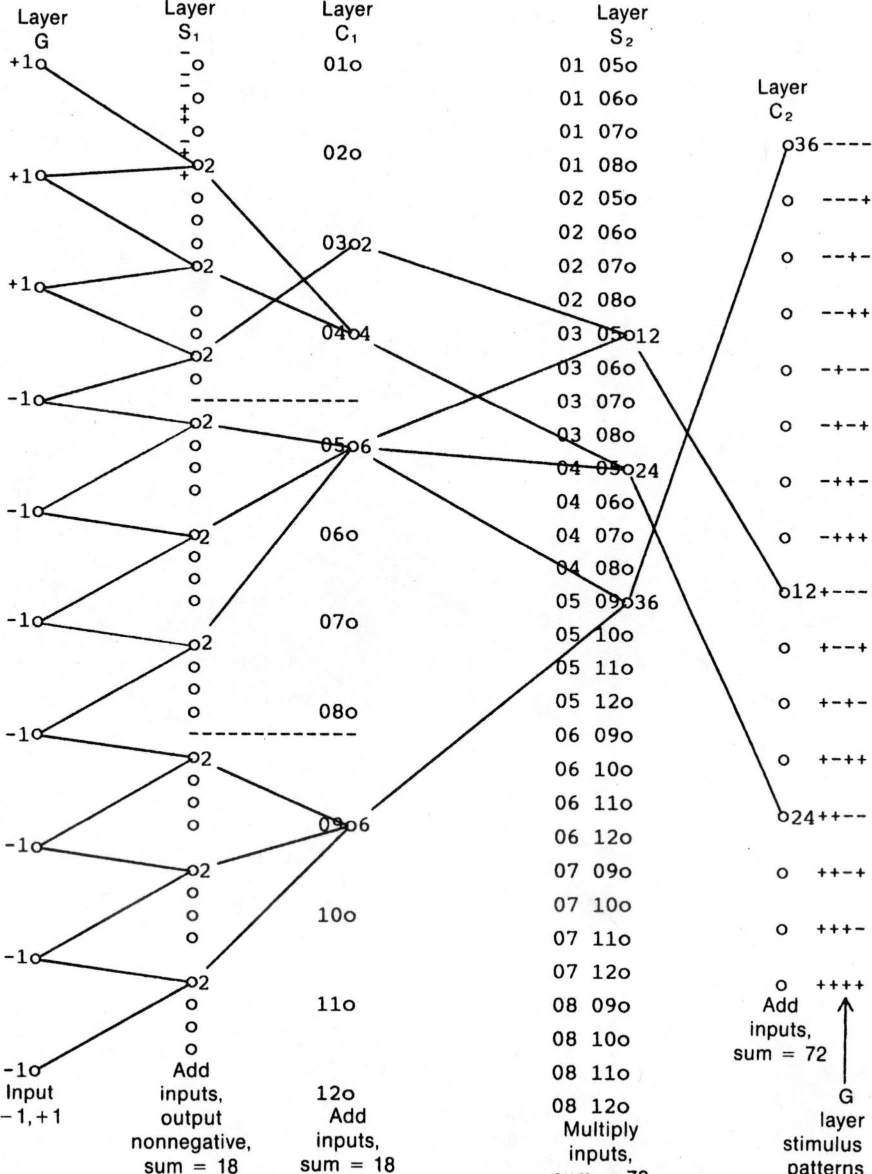

Fig. 11-9 The $S_1C_1S_2C_2$ model when the input pattern consists of three +1s followed by seven −1s. Only the axons that contribute to final C_2 output are shown.

sensitivity with respect to lateral displacement. At first the edge $+1+1-1-1$ is also seen with an output of 24 mV; when the white bar is at the end of its movement, the edge $-1-1+1+1$ is seen symmetrically, and so forth.

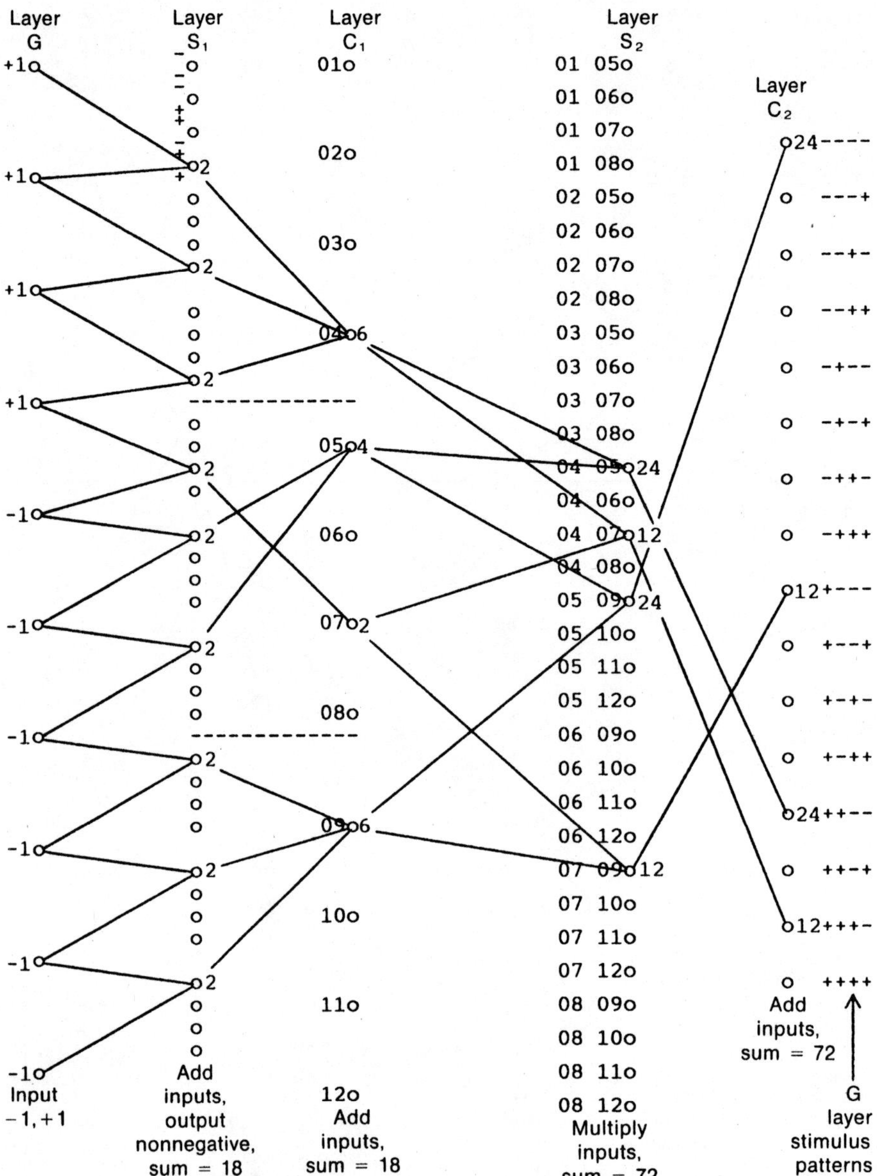

Fig. 11-10 The $S_1C_1S_2C_2$ model when the input pattern consists of four +1s followed by six −1s. Only the axons that contribute to final C_2 output are shown.

To illustrate the response of the model to a pattern *change of size*, a white bar in Table 11-2(*b*) is allowed to expand until it covers the entire black background. Here there are several surprises. First, the model turns out to be human: it makes mistakes! In the second column, where the G layer input is − − − − + + − − − −, the model sees − − + − (12-mV out-

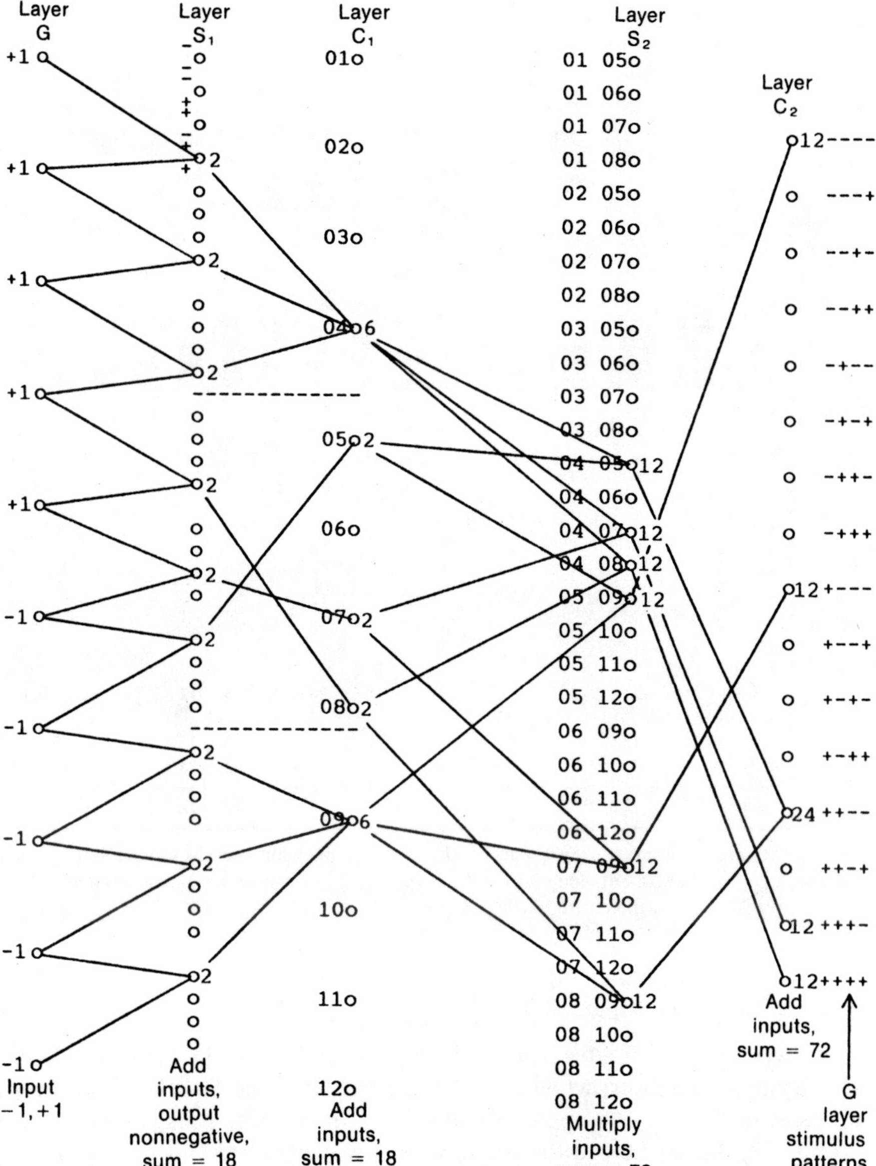

Fig. 11-11 The $S_1C_1S_2C_2$ model when the input pattern consists of five +1s followed by five −1s. Only the axons that contribute to final C_2 output are shown.

put) and, symmetrically, − + − −, although these four-element patterns are obviously not present in the input. The third column, where the white bar is three elements long (G layer input − − − + + + − − − −), is even more bizarre: it makes a mistake in that it sees − − + − (8-mV output), but it also spews forth output values such as 4, 8, and 16 mV. These are unusual

TABLE 11-2 For the $S_1C_1S_2C_2$ Model

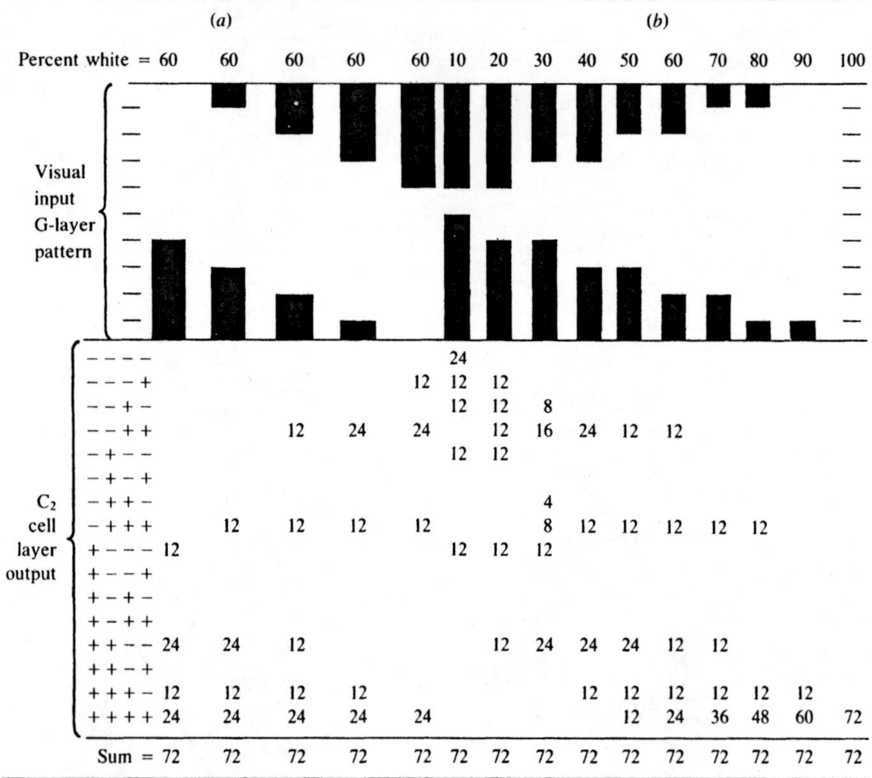

	(a)					(b)									
Percent white =	60	60	60	60	60	10	20	30	40	50	60	70	80	90	100
Visual input G-layer pattern															
C_2 cell layer output − − − −						24									
− − − +					12	12	12								
− − + −						12	12	8							
− − + +			12	24	24		12	16	24	12	12				
− + − −						12	12								
− + − +															
− + + −								4							
− + + +		12	12	12	12			8	12	12	12	12	12		
+ − − −	12					12	12	12							
+ − − +															
+ − + −															
+ − + +															
+ + − −	24	24	12				12	24	24	24	12	12			
+ + − +															
+ + + −	12	12	12	12					12	12	12	12	12	12	
+ + + +	24	24	24	24	24					12	24	36	48	60	72
Sum =	72	72	72	72	72	72	72	72	72	72	72	72	72	72	72

Note: C_2 cell layer outputs (− represents black, + represents white) versus visual input patterns if (*a*) a white bar six elements long moves downward (percent white = 60), and (*b*) a white bar expands until it covers the entire black background (percent white = 10, 20, 30, . . . , 100).

because all of the other inputs in Tables 11-1 and 11-2 have output values that are integer multiples of 12.

[S.D. writes: "I will admit that I calculated the values in Tables 11-1 and 11-2 using a computer; when the third column in Table 11-2(*b*) came along, I was sure that my computer program was incorrect. Alas, checking the problem by hand showed that the program was correct. This experience so unnerved me that I recalculated *all* of the outputs by hand!"]

The most important result of Table 11-2(*b*) is that, as the bar expands, the C_2 cell + + + + output linearly increases from 0 when the G-layer input is 40% white to 72 mV when it reaches a length of 100% white.

To summarize: In Table 11-1, what is the outstanding pattern that *you* see looking at the G-layer patterns from left to right? At first the pattern is all black, − − − − − − − − − − − −; then you see increasing white, + − − − − − − − − − − −, + + − − − − − − − − − −, and so forth. The corresponding C_2 cell − − − − output is 72, 60, 48, . . . , 0 mV. In the right-hand half of the table, you see the G-layer pattern increasing from 40% white (+ + + + − − − − − −) to 100% white (+ + + + + + + + + +). The corresponding C_2 cell + + + + output is 0, 12, 24, . . . , 72 mV.

In Table 11-2(*a*), the outstanding pattern that you see is a white bar 60% long that moves downward: + + + + + + − − − −, − + + + + + + − − −, to − − − − + + + + + +. The corresponding C_2 cell + + + + output remains constant at 24 mV.

In Table 11-2(*b*), what is the outstanding pattern that you see looking from left to right? At first the pattern contains 10% white: − − − − + − − − − −; then you see increasing white: − − − − + + − − − −, − − − + + + − − − −, and so forth to 100% white. The corresponding C_2 cell + + + + output is 0, 0, 0, 0, 12, 24, 36, 48, 60, and 72 mV.

In other words, for the simple one-dimensional model, which approximately agrees with the Hubel–Wiesel concept for the primary visual cortex, the numerical value of the output is an indication of the importance of a particular input pattern, and the numerical value is insensitive to where in the visual field a particular input pattern occurs.

11-3 Topological Representation

To examine the characteristics of more complicated structures, one can try to extend the one-dimensional model of Fig. 11-5 by adding an S_3,C_3 pair of layers. Unfortunately, the model becomes too unwieldy. The S_3,C_3 layers can recognize G input sequences that are eight elements long; a complete C_3 layer should therefore contain $2^8 = 256$ cells. If still another pair of layers is added, S_4,C_4, it should be able to recognize G input sequences that are 16 elements long, so that a complete C_4 layer should contain $2^{16} = 65{,}536$ cells.

The situation is much worse for the model of a real brain, which is two-dimensional. The C_2 layer should then be able to recognize any pattern in a $4 \times 4 = 16$-element array. Since each element can be either -1 or $+1$, the C_2 layer should contain $2^{16} = 65{,}536$ cells. A C_3 layer should be able to recognize any pattern in an $8 \times 8 = 64$-element array, so the C_3 layer should contain $2^{64} = 1.8 \times 10^{19}$ cells! Yes, in an 8×8 array,

```
X X X X X X X X
X X X X X X X X
X X X X X X X X
X X X X X X X X
X X X X X X X X
X X X X X X X X
X X X X X X X X
X X X X X X X X
```

if each element can be -1 or $+1$, then 1.8×10^{19} different patterns are possible. Since this is much more than the total number of neurons in the nervous system, there is something radically wrong with the model.

Perhaps the pejorative "radically wrong" is too strong. Out of the total of 1.8×10^{19} different patterns, only a few would actually occur during the lifetime of the model. (In an 80-year lifetime there are 2.5×10^9 seconds.) In an 8×8 foveal array, one would expect to see straight or curved edges or corners at various angles—a few thousand patterns at the most. It is unreasonable to expect the brain to accommodate in advance intricate patterns that will probably never occur. The number of neurons in S and C layers is far less than what one would expect from a representation of all possible patterns.

The mammalian cortex does not display the explosive increase in the number of cells discussed earlier as one progresses through the pattern-recognition hierarchy. In other words, there must be a limit to the number of different "grandmothers" that one can recognize [D. H. Hubel and T. N. Wiesel, 1979].

If one does not wish to actually construct more complicated models, there is a shorthand way of representing the topology of models such as that of Fig. 11-5. This is shown in Fig. 11-12(*a*) for the model of Fig. 11-5. Between the G and S_1 layers, and between any C layer and the next S layer, the topology is that of a zigzag, with the number of terminals on the right one less than the number on the left. (Except for the G layer, the dots in Fig. 11-12 denote *terminals* rather than neurons.) Between any S layer and the next C layer the topology is that of a convergent group of branches. In Fig. 11-12(*a*) there is a 3-to-1 convergence in going from S_1 to C_1, so the number of terminals on the right (three) is one-third the number on the left (nine). For 2-to-1 convergence, such as in going from S_2 to C_2, the number of terminals on the right is half the number on the left.

By comparing Fig. 11-5 (neurons) with its topological representation in Fig. 11-12(*a*) (terminals) one comes to the following conclusions:

1. The number of G neurons is the same as the number of G terminals.
2. For a particular pair of S,C layers, the S/C neuron ratio is the same as the S/C terminal ratio.
3. For C layers, the number of terminals is equal to the number of regions into which the visual field is divided. For example, the C layers of Fig. 11-12(*a*) show that the visual field is divided into three and one region, respectively.
4. For C layers, the number of terminals multiplied by the number of different patterns recognized by that layer is equal to the number of neurons in the model. For example, the C_1 layer of Fig. 11-12(*a*) can recognize $--$, $-+$, $+-$, and $++$, and is represented by three terminals. Accordingly, there are $3 \times 4 = 12$ C_1 neurons in Fig. 11-5.

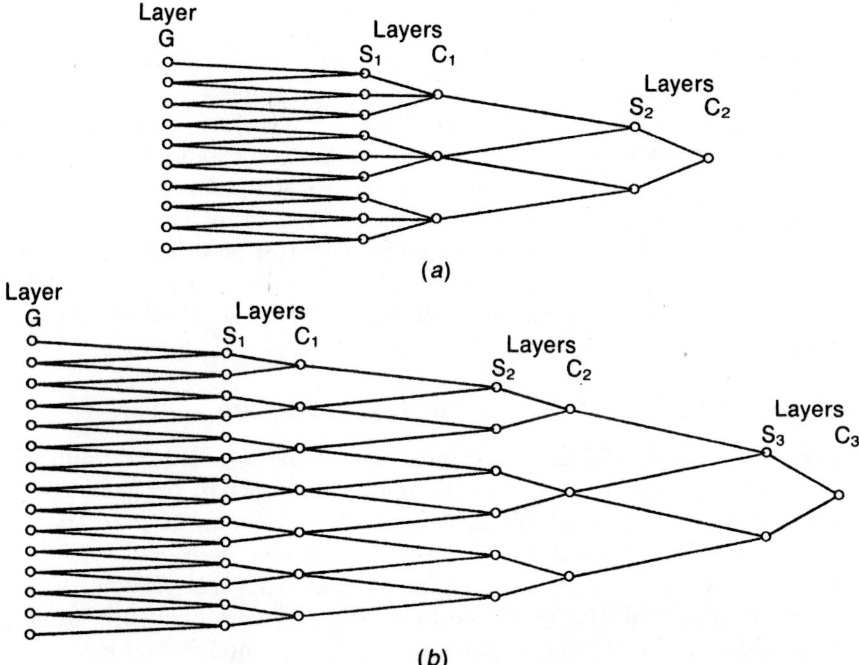

Fig. 11-12 Topological representation of (*a*) $S_1C_1S_2C_2$ model; (*b*) $S_1C_1S_2C_2S_3C_3$ model. Between the G and S_1 layers, and between any C layer and the next S layer, the topology is that of a zigzag. For C layers, the number of terminals is equal to the number of regions into which the visual field is divided. (From Deutsch and M-Tzanakou, *Neuroelectric Systems*, New York Univ. Press, New York, 1987.)

Given these rules and the topological representation, one can design more complicated models without actually drawing a complete schematic such as that of Fig. 11-5. For example, the drawing of Fig. 11-12(*b*) includes an S_3,C_3 pair of layers. We are given that layer C_1 recognizes four different patterns and layer C_2 accommodates $2^4 = 16$ different patterns. A complete C_3 layer should contain $2^8 = 256$ cells, as previously mentioned, but let us suppose that an examination of the patterns that could actually occur during the lifetime of the model reveals that 64 patterns are sufficient. Design of the model can then be summarized in the form of a table:

	G	S_1	C_1	S_2	C_2	S_3	C_3
Number of patterns			4		16		64
Number of terminals	15	14	7	6	3	2	1
Number of neurons	15	56	28	96	48	128	64

One has to know the 64 viable output patterns to draw the schematic diagram. The 128 neurons of the S_3 layer are a rather large number for a pencil-and-paper exercise, but the model can of course be completed given sufficient patience.

The real challenge and practical application comes about in designing a two-dimensional model, such as Fukushima's computer-aided neocognitron. One can continue to add layers, but the model rapidly becomes complicated. Nevertheless, given a sufficient number of picture elements, orientation detectors, and cortical layers, the Hubel–Wiesel model seems to be capable of duplicating many of the feats of human visual pattern recognition.

11-4 Neural Networks That Can Learn (and Forget)

The neural network of Fig. 11-5 is incapable of learning. It is an inflexible, fully wired structure, whose connections are determined by the genetic blueprint. The circuit is the model for an instinctive, built-in, unlearned response.

In this section several neural networks that can learn (and forget) are considered. They are simple representatives of what has become the impressive new field of "neural networks." *Artificial* neural networks are discussed in Section 11-5. In the present section, only models that have some relation to actual animal neurophysiology, although the relation may be admittedly vague, are considered. Figure 11-5 falls into this category.

Referring to the C_2 layer of Fig. 11-4, how can a partially random array of cortical neurons (partially random because of the influence of "noise" in growth based on the genetic blueprint) organize themselves so that one cell responds to the equivalent of $- - - +$, another to $- - + -$, and so forth? Although many theories have been proposed, none has been neurophysiologically verified, so that the discussion that follows is conjectural.

As Fukushima (1981) wrote, "Among the post-synaptic cells situated in a small area, only the one which is yielding the maximum output (i.e., maximum firing frequency) has the chance to have its input synapses reinforced. In such a situation, not all input synapses of the cell are reinforced, but only those through which impulses are actually arriving." In other words, we have a situation that is analogous to Darwinian evolution in that the "fittest" neuron is reinforced.

Fukushima proposes two quite different mechanisms for reinforcement: glial cells or lateral inhibition. Brain neurons are surrounded by glial cells that, although not themselves excitable tissue, are thought to "take care" of the neurons much as a governess watches over a child. Fukushima (1981) proposes the following:

> Let us suppose, for instance, that for the reinforcement of a synapse a certain chemical substance (nutrient or DNA) must be supplied to the postsynaptic cells, and that a certain kind of glial cell provides this chemical substance. . . . The glial cell supplies the nutrient exclusively to the single cell which has yielded a maximum output within the competition area.

The "lateral inhibition" mechanism is illustrated in Fig. 11-13. Here we have a group of five almost identical neurons. The neurons receive hundreds or thousands of synaptic junction inputs, and they extend inhibitory output branches (I) to each other. What sets this group apart from all others is that one particular input, C, feeds an excitatory signal (E) that is common to all five neurons. In general, the output of each neuron is described by

$$\text{Output} = \frac{1 + \text{summation of excitatory inputs}}{1 + \text{summation of inhibitory inputs}} - 1,$$

$$\text{Output positive, otherwise 0.} \quad (11\text{-}1)$$

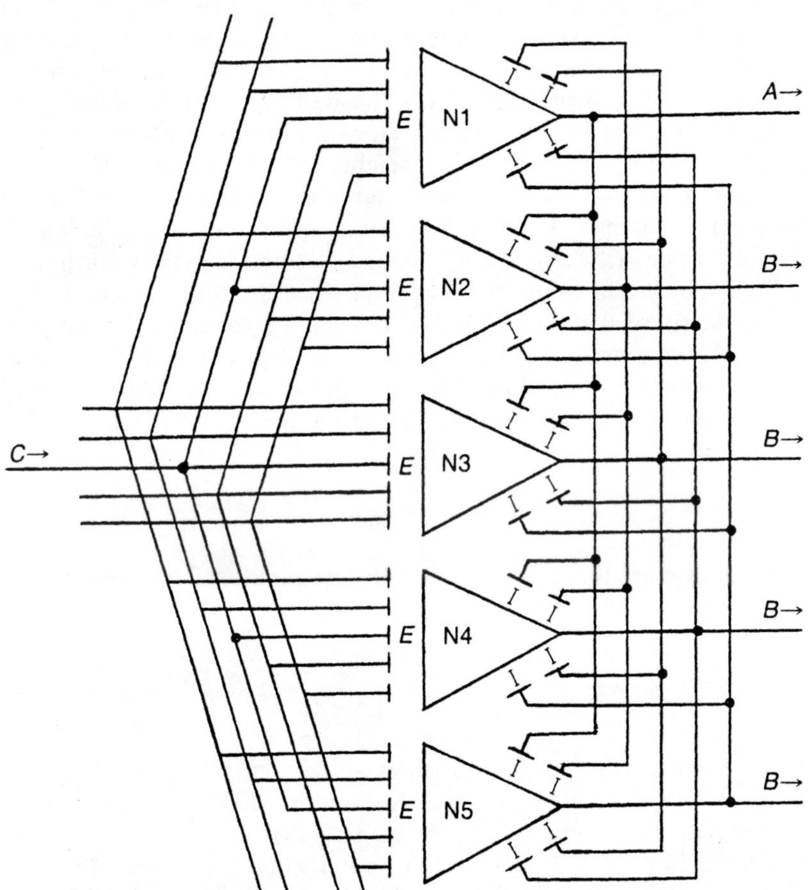

Fig. 11-13 A hypothetical model that illustrates self-organization of the cortex. The five neurons have a common input, C, and are almost identical, but output A is slightly greater than, or its output slightly precedes in time, that of output B. If inhibitory weighting factor I is greater than 1, A demonstrates "ruthless greediness" (winner-take-all) in capturing the C signal for itself and suppressing all of the B neurons so that their outputs are zero.

Notice that this equation is structured so that the output is zero if there are no inputs, or if the summation of excitatory inputs is equal to the summation of inhibitory inputs. But if the sum of excitatory inputs is greater than the sum of inhibitory inputs, the latter exerts a strong influence on the output because it is in the denominator. This is realistic since *I* junctions, as discussed in Section 5-3, are located on or near the soma (body) of the neuron, whereas *E* junctions tend to be farther away on the dendritic tree.

Mathematical analysis of the circuit of Fig. 11-13 shows that it is the epitome of ruthless greediness. If *I* is numerically greater than 1, one of the neurons captures the *C* signal for itself and suppresses all of the other neurons. Because of the considerable degree of randomness that is associated with growth of the nervous system, one of the neurons inevitably has greater output than, or its output slightly precedes in time, that of the other neurons. Let us suppose that neuron N1 is this "head of the pecking order," and that $I = 1.5$ (in this chapter, all weight values are positive, but we distinguish between excitatory and inhibitory weights because *E*s are entered in the numerator of Eq. (11-1) while *I*s are entered in the denominator). As output *A* increases it inhibits all of the *B* neurons, which decreases the inhibitory effect of *B* upon *A*, which further increases output *A*, and so forth.

Some numerical values will illustrate what is meant by "ruthless greediness": Suppose that all of the *B* neurons are suppressed so that $B = 0$, and that the common input $C = 2$ mV, while the excitation weight $E = 3$. Neuron *A* has four *I* junctions, but the input to each of them is zero. Substituting into Eq. (11-1), we accordingly have for neuron N1

$$A = \frac{1 + (3)(2)}{1 + (4)(1.5)(0)} - 1 = 6 \text{ mV}. \qquad (11\text{-}2)$$

Each *B* neuron has three *I* junctions with zero input, and one *I* junction with 6-mV input. We therefore get for neurons N2 to N5

$$B = \frac{1 + (3)(2)}{1 + (3)(1.5)(0) + (1)(1.5)(6)} - 1 = -0.3 \text{ mV}, \qquad (11\text{-}3)$$

but, since negative values are not allowed, the actual *B* output is zero. We thus verify that all of the *B* neurons are suppressed. To complete this model of self-organization, one must assume that all of the *C* junctions on the *B* neurons atrophy since the *B*s never respond when *C* is active. Eventually, in this way, the *A* neuron captures the *C* input all for itself. This type of circuit is known as a "Winner-Take-All" network [D. Wang and M. A. Arbib, 1990].

Each of the *B* neurons would presumably be "head of the pecking order" for its own unique set of input signal axons, laterally suppressing competing neurons. It is conjectured that this may be how learning takes place in the mammalian cortex. For a human, at the end of the hierarchical stages for visual pattern recognition, it is somewhat facetiously claimed that a single cell or small group of cells would thus learn to fire when the individual's "grandmother" is seen.

As a somewhat more complicated example, consider the neural network of Fig. 11-14. Here the inhibitory (*I*) feedback junctions are similar to those of Fig. 11-13. The purpose of the network is to replace, via self-organization, the *upper 16 neurons* of the S_2 layer of Fig. 11-5 [G. M. Edelman, 1987].

It is pointed out in the previous section that the number of neurons in an S or C layer is equal to the number of different patterns that are likely to be processed by that layer during the "lifetime" of the model. For the case at hand, Table 11-1 shows that, if a white strip is gradually lowered over a black background, only five C_2 neurons are activated (− − − −, + − − −, + + − −, + + + −, and + + + +). These are the five neurons of Fig. 11-14 that eventually become identified with these patterns, as indicated.

Each of the five neurons receives the same input stream: From the C_1 layer of Fig. 11-5, signals come from the upper third of the 10-element one-dimensional visual field via 01 (− −), 02 (− +), 03 (+ −), and 04 (+ +). Signals also come from the middle third of the visual field via 05 (− −), 06 (− +), 07 (+ −), and 08 (+ +). With the aid of a reasonable set of algorithms, the weighting factors of synaptic junctions 01 to 40 adjust so that one neuron eventually recognizes the four-element spatial sequence − − − −, another recognizes + + − −, and so forth.

Self-organization can be conveniently broken down into three phases: learning, organizing, and testing. During the *learning* phase, each of the synaptic junctions starts out with a neutral value, weight = 1. Gradually, based upon input "events" and a set of algorithms, the weight values change as the network learns. The *organizing* phase takes place in the absence of input stimuli, perhaps when the brain is asleep (perhaps the organizing phase is one of the important reasons for sleep). The neural network "looks at" the history of input patterns versus dominant neurons that occurred during the learning phase. Synaptic junction weights of 2 are assigned for the patterns that were recognized. The remaining junctions are assigned weights of 0.2 rather than 0; with 0.2, they are held in readiness so that, if an emergency occurs such as damage to some vital neurons because of a stroke or accident, other neurons will be capable of forgetting one pattern as they learn a different pattern during rehabilitation therapy. Finally, during the *testing* phase, we examine the response of the network to various input stimulus patterns. Specifically, for Fig. 11-14:

Learning Phase: The work sheet is displayed in Table 11-3. On the left are listed the eight input axons ((− − upper) to (+ + middle)) and the 40 synaptic junction (SJ) numbers. The initial weight (Wt) of each junction is 1 at birth. At the top of the table are listed the first 14 events that shape the lives of our five neurons. The entries (V_{in}) are the C_1 layer potentials corresponding to the gradual lowering of a white strip over a black background. For example, in Fig. 11-5, for 0% white, the outputs of (− − upper) and (− − middle) are 6 mV each, while all − +, + −, and + + outputs are zero.

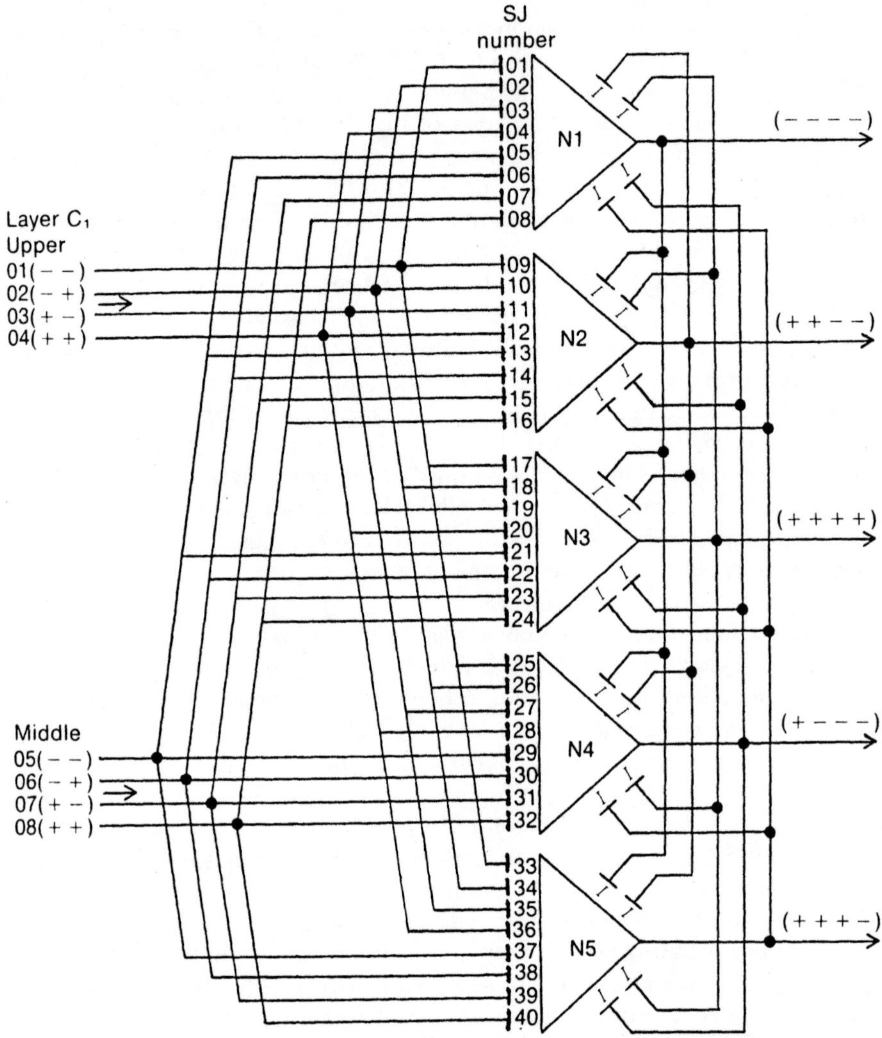

Fig. 11-14 Hypothetical neural network that, through self-organization, can learn (and forget). The five neurons are fed by the upper and middle portions of the C_1 layer of Fig. 11-5. During a "learning" phase, the 40 synaptic junction weights are adjusted in accordance with various algorithms. The work sheet is shown in Table 11-3. During a subsequent "organizing" phase, based on what each neuron has learned, a pattern is assigned as indicated by − − − −, + + − −, and so forth. The weights are readjusted to agree with these patterns. The work sheet for a "testing" phase is shown in Table 11-4.

The corresponding event no. 1 sequence in Table 11-3 is 6 0 0 0 6 0 0 0.

The percent white as well as figure numbers are given under corresponding event numbers. The table does not go beyond percent white = 70 (event no. 8) because the upper and middle portions of the visual field are all white (0 0 0 6 0 0 0 6) at this point. Instead, the visual cycle starts to repeat with event no. 9 = 0% white (6 0 0 0 6 0 0 0).

Event no. 1 is taken from Fig. 11-7. The C_l layer potentials result in V_{in} = 6 0 0 0 6 0 0 0. The first algorithm follows.

A1. The V_{out} contribution of each synaptic junction is given by

$$V_{out} = V_{in} \times \text{Wt}. \tag{11-4}$$

(It is convenient to assume that the units of potential are mV, although this may be unrealistic.) Since each Wt associated with event no. 1 is 1, the V_{out} contributions of SJ1, SJ5, SJ9, . . . , SJ37 are 6 mV each as shown. Inactive junction contributions are blank.

A2. The *total* output of each neuron is given by the *product* of the individual *active* synaptic junction contributions. As shown, the total output of each neuron is 36 mV.

A3. The *dominant* neuron, given in the bottom row of Table 11-3, is the one with the greatest V_{out} total. Since V_{out} total = 36 mV for *each* neuron following event no. 1, we require the following step.

A4. If two or more V_{out} totals are the same, the one nearest the top of the page shall take precedence. In accordance with this rule, neuron N1 is dominant.

A5. The *strongest two* SJ contributions of the *dominant* neuron shall be underlined. For the case at hand, there are only two SJ contributions, so both are underlined as shown (SJ1 = $\underline{6}$ and SJ5 = $\underline{6}$).

A6. The basis for learning and forgetting follows: Underlined SJ contributions shall have their Wts *increased* by 0.1 unit. This "reward" or reinforcement, because these junctions responded to the stimulus, is a form of learning. At the same time, all of the other Wts of the dominant neuron shall be *decreased* by 0.1 unit. This "punishment" for failing to respond is a form of forgetting. (Anthropomorphic terms such as "reward" and "punishment" are commonly used in connection with neural networks.) The preceding steps taken to focus upon a particular input pattern are called the *selectivity* property. The net result is that the Wts of the second mV column become 1.1, 0.9, 0.9, 1.1, 0.9, 0.9, as shown.

A7. In all of the subservient neurons, Wts of SJs corresponding to the underlined contributions shall be *decreased* by 0.05 unit. At the same time, all of the other Wts of the subservient neurons shall be *increased* by 0.05

TABLE 11-3 Work Sheet for "Learning" Phase of the Neural Network of Fig. 11-14

			Event Number 1		2		3		4		5		6		7		8		9		10		11		12		13		14
			% White 0		10		20		30		40		50		60		70		0		10		20		30		40		50
			Fig. 11-7		11-8		11-6		11-9		11-10		11-11		—		—		11-7		11-8		11-6		11-9		11-10		11-11
(− − upper)	V_{in}		6		4		2		0		0		0		0		0		6		4		2		0		0		0
(− + upper)	V_{in}		0		0		0		0		0		0		0		0		0		0		0		0		0		0
(+ − upper)	V_{in}		0		2		2		2		0		0		0		0		0		2		2		2		0		0
(+ + upper)	V_{in}		0		0		2		4		6		6		6		6		0		0		2		4		6		6
(− − mid)	V_{in}		6		6		6		6		4		2		0		0		6		6		6		6		4		2
(− + mid)	V_{in}		0		0		0		0		0		0		0		0		0		0		0		0		0		0
(+ − mid)	V_{in}		0		0		0		0		2		2		2		0		0		0		0		0		2		2
(+ + mid)	V_{in}		0		0		0		0		0		2		4		6		0		0		0		0		0		2
	SJ	Wt	mV	Wt	mV	Wt	mV	Wt	mV	Wt	mV	Wt	mV	Wt	mV	Wt	mV	Wt	mV	Wt	mV	Wt	mV	Wt	mV	Wt	mV	Wt	mV
N1 V_{out}	1	1	6	1.1	4.4	1.2	2.4	1.25		1.3		1.35		1.4		1.45		1.5	9	1.6	6.4	1.7	3.4	1.75		1.8		1.85	
N1 V_{out}	2	1		0.9		0.8		0.85		0.9		0.95		1		1.05		1.1		1		0.9		0.95		1		1.05	
N1 V_{out}	3	1		0.9	1.8	0.8	1.6	0.75	1.5	0.8		0.85		0.9		0.95		1		0.9	1.8	0.8	1.6	0.75	1.5	0.8		0.85	
N1 V_{out}	4	1		0.9		0.8	1.6	0.85	3.4	0.8	4.8	0.75	4.5	0.7	4.2	0.65	3.9	0.6		0.5		0.4	0.8	0.45	1.8	0.4	2.4	0.35	2.1
N1 V_{out}	5	1	6	1.1	6.6	1.2	7.2	1.15	6.9	1.1	4.4	1.05	2.1	1.1		1.15		1.2	7.2	1.3	7.8	1.4	8.4	1.35	8.1	1.3	5.2	1.25	2.5
N1 V_{out}	6	1		0.9		0.8		0.85		0.9		0.95		1		1.05		1.1		1		0.9		0.95		1		1.05	
N1 V_{out}	7	1		0.9		0.8		0.85		0.9	1.8	0.95	1.9	0.9	1.8	0.95		1		0.9		0.8		0.85		0.9	1.8	0.95	1.9
N1 V_{out}	8	1		0.9		0.8		0.85		0.9		0.95	1.9	1	4	0.95	5.7	0.9		0.8		0.7		0.75		0.8		0.85	1.7
Totals			36		52		44		35		38		34		30		22		65		90		37		22		22		17
	SJ	Wt	mV	Wt	mV	Wt	mV	Wt	mV	Wt	mV	Wt	mV	Wt	mV	Wt	mV	Wt	mV	Wt	mV	Wt	mV	Wt	mV	Wt	mV	Wt	mV
N2 V_{out}	9	1	6	0.95	3.8	0.9	1.8	0.8		0.7		0.6		0.65		0.7		0.75	4.5	0.7	2.8	0.65	1.3	0.7		0.6		0.5	
N2 V_{out}	10	1		1.05		1.1		1		0.9		0.8		0.85		0.9		0.95		1		1.05		1.1		1		0.9	
N2 V_{out}	11	1		1.05	2.1	1.1	2.2	1.2	2.4	1.1		1		1.05		1.1		1.15		1.2	2.4	1.25	2.5	1.2	2.4	1.1		1	
N2 V_{out}	12	1		1.05		1.1	2.2	1	4	1.1	6.6	1.2	7.2	1.15	6.9	1.1	6.6	1.05		1.1		1.15	2.3	1.2	4.8	1.3	7.8	1.4	8.4
N2 V_{out}	13	1	6	0.95	5.7	0.9	5.4	1	6	1.1	4.4	1.2	2.4	1.25		1.3		1.35	8.1	1.3	7.8	1.25	7.5	1.2	7.2	1.3	5.2	1.4	2.8
N2 V_{out}	14	1		1.05		1.1		1		0.9		0.8		0.85		0.9		0.95		1		1.05		1.1		1		0.9	
N2 V_{out}	15	1		1.05		1.1		1		0.9	1.8	0.8	1.6	0.75	1.5	0.8		0.85		0.9		0.95		1		0.9	1.8	0.8	1.6
N2 V_{out}	16	1		1.05		1.1		1		0.9		0.8	1.6	0.85	3.4	0.8	4.8	0.75		0.8		0.85		0.9		0.8		0.7	1.4
Totals			36		45		47		58		52		44		35		32		36		52		56		83		73		53

	SJ	Wt		mV	Wt	mV	Wt	mV	Wt	mV	Wt	mV	Wt	mV	Wt	mV	Wt	mV	Wt	mV	Wt	mV	Wt	mV	Wt	mV	Wt	mV	Wt	mV
	17	1		6	0.95	3.8	0.9	1.8	0.95		1		1.05		0.95		0.85		0.75	4.5	0.7	2.8	0.65	1.3	0.7		0.75		0.8	
	18	1			1.05		1.1		1.15		1.2		1.25		1.15		1.05		0.95		1		1.05		1.1		1.15		1.2	
	19	1			1.05	2.1	1.1	2.2	1.05	2.1	1.1		1.15		1.05		0.95		0.85		0.9	1.8	0.95	1.9	0.9	1.8	0.95		1	
	20	1			1.05		1.1	2.2	1.15	4.6	1.1	6.6	1.05	6.3	1.15	6.9	1.25	7.5	1.35		1.4		1.45	2.9	1.5	6	1.45	8.7	1.4	8.4
N3	21	1	V_{out}	6	0.95	5.7	0.9	5.4	0.85	5.1	0.8	3.2	0.75	1.5	0.65		0.55		0.45	2.7	0.4	2.4	0.35	2.1	0.3	1.8	0.25	1	0.2	0.4
	22	1			1.05		1.1		1.15		1.2		1.25		1.15		1.05		0.95		1		1.05		1.1		1.15		1.2	
	23	1			1.05		1.1		1.15		1.2	2.4	1.25	2.5	1.35	2.7	1.25		1.15		1.2		1.25		1.3		1.35	2.7	1.4	2.8
	24	1			1.05		1.1		1.15		1.2		1.25	2.5	1.15	4.6	1.25	7.5	1.35		1.4		1.45		1.5		1.55		1.6	3.2
Totals				36		45		47		49		51		59		86		56		12		12		15		19		23		30
	SJ	Wt		mV	Wt	mV	Wt	mV	Wt	mV	Wt	mV	Wt	mV	Wt	mV	Wt	mV	Wt	mV	Wt	mV	Wt	mV	Wt	mV	Wt	mV	Wt	mV
	25	1		6	0.95	3.8	0.9	1.8	0.95		1		1.05		1.1		1.15		1.2	7.2	1.15	4.6	1.1	2.2	1		1.05		1.1	
	26	1			1.05		1.1		1.15		1.2		1.25		1.3		1.35		1.4		1.45		1.5		1.4		1.45		1.5	
	27	1			1.05	2.1	1.1	2.2	1.05	2.1	1.1		1.15		1.2		1.25		1.3		1.35	2.7	1.4	2.8	1.5	3	1.55		1.6	
	28	1			1.05		1.1	2.2	1.15	4.6	1.1	6.6	1.05	6.3	1	6	0.95	5.7	0.9		0.95		1	2	0.9	3.6	0.85	5.1	0.8	4.8
N4	29	1	V_{out}	6	0.95	5.7	0.9	5.4	0.85	5.1	0.8	3.2	0.75	1.5	0.8		0.85		0.9	5.4	0.85	5.1	0.8	4.8	0.9	5.4	0.85	3.4	0.8	1.6
	30	1			1.05		1.1		1.15		1.2		1.25		1.3		1.35		1.4		1.45		1.5		1.4		1.45		1.5	
	31	1			1.05		1.1		1.15		1.2	2.4	1.25	2.5	1.2	2.4	1.25		1.3		1.35		1.4		1.3		1.35	2.7	1.4	2.8
	32	1			1.05		1.1		1.15		1.2		1.25	2.5	1.3	5.2	1.25	7.5	1.2		1.25		1.3		1.2		1.25		1.3	2.6
Totals				36		45		47		49		51		59		75		43		39		63		59		58		47		56
	SJ	Wt		mV	Wt	mV	Wt	mV	Wt	mV	Wt	mV	Wt	mV	Wt	mV	Wt	mV	Wt	mV	Wt	mV	Wt	mV	Wt	mV	Wt	mV	Wt	mV
	33	1		6	0.95	3.8	0.9	1.8	0.95		1		1.05		1.1		1.15		1.2	7.2	1.15	4.6	1.1	2.2	1.15		1.2		1.25	
	34	1			1.05		1.1		1.15		1.2		1.25		1.3		1.35		1.4		1.45		1.5		1.55		1.6		1.65	
	35	1			1.05	2.1	1.1	2.2	1.05	2.1	1.1		1.15		1.2		1.25		1.3		1.35	2.7	1.4	2.8	1.35	2.7	1.4		1.45	
	36	1			1.05		1.1	2.2	1.15	4.6	1.1	6.6	1.05	6.3	1	6	0.95	5.7	0.9		0.95		1	2	1.05	4.2	1	6	0.95	5.7
N5	37	1	V_{out}	6	0.95	5.7	0.9	5.4	0.85	5.1	0.8	3.2	0.75	1.5	0.8		0.85		0.9	5.4	0.85	5.1	0.8	4.8	0.75	4.5	0.7	2.8	0.65	1.3
	38	1			1.05		1.1		1.15		1.2		1.25		1.3		1.35		1.4		1.45		1.5		1.55		1.6		1.65	
	39	1			1.05		1.1		1.15		1.2	2.4	1.25	2.5	1.2	2.4	1.25		1.3		1.35		1.4		1.45		1.5	3	1.55	3.1
	40	1			1.05		1.1		1.15		1.2		1.25	2.5	1.3	5.2	1.25	7.5	1.2		1.25		1.3		1.35		1.4		1.45	2.9
Totals				36		45		47		49		51		59		75		43		39		63		59		51		50		67
Dominant				N1		N1		N2		N2		N2		N3		N3		N3		N1		N1		N4		N2		N2		N5

On the left are listed the eight input axons and 40 synaptic junction numbers. The initial Wt of each junction is 1 at birth. At the top are listed the 14 events of the learning phase. The V_{in} values are the C_I layer potentials of Fig. 11-5 corresponding to the gradual lowering of a white strip over a black background. The V_{out} values (in millivolts) are given by $V_{in} \times$ Wt, where the Wt values are adjusted in accordance with various algorithms given in the text. "Total" outputs are given by the *product* of individual *active* synaptic junction contributions.

unit. These maneuvers discourage the subservient neurons from recognizing 6 0 0 0 6 0 0 0, and encourage them to recognize some other pattern. This is called the *dispersion* property. The Wts of the second mV column in Table 11-3 accordingly become 0.95, 1.05, 1.05, 1.05, 0.95, 1.05, 1.05, 1.05, as shown for N2, N3, N4, and N5. (To achieve these changes, the feedback portion of the neural network of Fig. 11-14 should be much more extensive than depicted, with *E* as well as *I* junctions, but this would excessively complicate the drawing.)

For event no. 2, taken from Fig. 11-8, the C_1 layer potentials result in V_{in} = 4 0 2 0 6 0 0 0. Again N1 is dominant, and the strongest two of the three active SJ contributions are 4.4 and 6.6.

For event no. 3, taken from Fig. 11-6, the C_1 layer potentials result in V_{in} = 2 0 2 2 6 0 0 0. A change takes place as N2 becomes dominant. It has four active SJ contributions: 1.8, 2.2, 2.2, and 5.4 mV. We can paraphrase A4 with the following entry.

A8. If two or more SJ contributions are the same, the one nearest the top of the page shall take precedence. This underlining rule yields 1.8, 2.2, 2.2, and 5.4 mV, respectively.

Here N2 remains dominant during event nos. 3, 4, and 5, and N3 is dominant during event nos. 6, 7, and 8. The visual cycle starts to repeat at event no. 9. N1 becomes dominant during event nos. 9 and 10. But the response to event no. 11 does not repeat that of event no. 3: because of the self-organizing schedule of reinforcement (learning) and weakening (forgetting), N4 becomes dominant during event no. 11. Similarly, N5 becomes dominant during event no. 14. The *learning* phase ends with event no. 14.

Organizing Phase: We first require:

A9. The *organizing* phase proceeds in sequence from N1, N2, to N5;

A10. A neuron is inhibited from "learning" a pattern if the latter has already been learned by one of the other neurons; and

A11. Synaptic junctions of learned patterns are assigned Wt = 2; the other junctions are assigned Wt = 0.2.

Looking at the Table 11-3 entries, with emphasis placed upon the more recent events, it is clear that N1 recognizes

$$_{-}6_{-}\quad 0\quad 0\quad 0 \mid {}_{-}6_{-}\quad 0\quad 0\quad 0$$

and

$$_{-}4_{-}\quad 0\quad {}_{+}2_{-}\quad 0 \mid {}_{-}6_{-}\quad 0\quad 0\quad 0.$$

(The vertical bar separates the upper portion of the visual field at the left from the middle portion at the right.) The most prominent four-element visual combination of the upper and middle portions is, by far, $- - - -$. This, then, is the pattern learned by N1. It is entered as shown in Table 11-4. For N1, the pattern $- - - -$ corresponds to SJ1 and SJ5, so these are assigned Wt = 2. The other six SJs are assigned Wt = 0.2.

Next, N2 is dominant during

$$0 \quad 0 \quad {}_{+}2_{-} \quad {}_{+}4_{+} \mid {}_{-}6_{-} \quad 0 \quad 0 \quad 0$$

and

$$0 \quad 0 \quad 0 \quad {}_{+}6_{+} \mid {}_{-}4_{-} \quad 0 \quad {}_{+}2_{-} \quad 0.$$

Here the most prominent four-element combination is $+ + - -$. This pattern corresponds to SJ12 and SJ13. These get Wt = 2; the other SJs get Wt = 0.2.

The two most recent dominances for N3 are

$$0 \quad 0 \quad 0 \quad {}_{+}6_{+} \mid 0 \quad 0 \quad {}_{+}2_{-} \quad {}_{+}4_{+}$$

and

$$0 \quad 0 \quad 0 \quad {}_{+}6_{+} \mid 0 \quad 0 \quad 0 \quad {}_{+}6_{+}.$$

Here the most prominent four-element combination is $+ + + +$. This pattern corresponds to SJ20 and SJ24. These get Wt = 2; the other SJs get Wt = 0.2.

N4 recognizes

$${}_{-}2_{-} \quad 0 \quad {}_{+}2_{-} \quad {}_{+}2_{+} \mid {}_{-}6_{-} \quad 0 \quad 0 \quad 0.$$

Here there are three equally likely combinations: $- - - -$, $+ - - -$, and $+ + - -$. Since $- - - -$ has been preempted by N1 and $+ + - -$ by N2, according to A10 we are left with $+ - - -$ for N4. This pattern corresponds to SJ27 and SJ29. These get Wt = 2; the other SJs get Wt = 0.2.

Finally, N5 recognizes

$$0 \quad 0 \quad 0 \quad {}_{+}6_{+} \mid {}_{-}2_{-} \quad 0 \quad {}_{+}2_{-} \quad {}_{+}2_{+}.$$

Here again there are three equally likely combinations: $+ + - -$, $+ + + -$, and $+ + + +$. Since $+ + - -$ has been preempted by N2 and $+ + + +$ by N3, according to A10 we are left with $+ + + -$ for N5. This pattern corresponds to SJ36 and SJ39. These get Wt = 2; the other SJs get Wt = 0.2.

This reasoning with regard to which pattern should be claimed by each neuron should be expressed in the form of additional algorithms. This is left as an exercise for the reader.

Testing Phase: As shown, the work sheet of Table 11-4 is completed by using, as visual input stimuli, the same events (numbers 1 to 8) as are employed in Table 11-3. It turns out that this self-organized neural network does very well indeed, as a pattern recognizer, as follows. In each case, you should compare the visual input with the neural network response.

Event no. 1: N1 strongly sees – – – – (V_{out} = 144 mV).

TABLE 11-4 Work Sheet for "Testing" Phase of the Neural Network of Fig. 11-14

	SJ	Wt		Event Number 1	2	3	4	5	6	7	8
				Figure 11-7	11-8	11-6	11-9	11-10	11-11	—	—
Visual input pattern				–	+	+	+	+	+	+	+
				–	–	+	+	+	+	+	+
				–	–	–	+	+	+	+	+
				–	–	–	–	+	+	+	+
				–	–	–	–	–	+	+	+
				–	–	–	–	–	–	+	+
				–	–	–	–	–	–	–	+
(– – upper)			V_{in}	6	4	2	0	0	0	0	0
(– + upper)				0	0	0	0	0	0	0	0
(+ – upper)				0	2	2	2	0	0	0	0
(+ + upper)				0	0	2	4	6	6	6	6
(– – middle)				6	6	6	6	4	2	0	0
(– + middle)				0	0	0	0	0	0	0	0
(+ – middle)				0	0	0	0	2	2	2	0
(+ + middle)				0	0	0	0	0	2	4	6
	SJ	Wt		mV	mV	mV	mV	mV	mV	mV	mV
N1	1	2	V_{out}	12	8	4					
	2	0.2									
	3	0.2			0.4	0.4	0.4				
	4	0.2				0.4	0.8	1.2	1.2	1.2	1.2
– – – –	5	2		12	12	12	12	8	4		
	6	0.2									
	7	0.2						0.4	0.4	0.4	
	8	0.2							0.4	0.8	1.2
Totals				144	38.4	7.68	3.84	3.84	0.768	0.384	1.44
	SJ	Wt		mV	mV	mV	mV	mV	mV	mV	mV
N2	9	0.2	V_{out}	1.2	0.8	0.4					
	10	0.2									
	11	0.2			0.4	0.4	0.4				
	12	2				4	8	12	12	12	12
+ + – –	13	2		12	12	12	12	8	4		
	14	0.2									
	15	0.2						0.4	0.4	0.4	
	16	0.2							0.4	0.8	1.2
Totals				14.4	3.84	7.68	38.4	38.4	7.68	3.84	14.4

(continued)

TABLE 11-4 (*Continued*)

			Event Number	1	2	3	4	5	6	7	8
			Figure	11-7	11-8	11-6	11-9	11-10	11-11	—	—
	SJ	Wt		mV	mV	mV	mV	mV	mV	mV	mV
	17	0.2		1.2	0.8	0.4					
	18	0.2									
	19	0.2			0.4	0.4	0.4				
N3	20	2				4	8	12	12	12	12
+ + + +	21	0.2	V_{out}	1.2	1.2	1.2	1.2	0.8	0.4		
	22	0.2									
	23	0.2						0.4	0.4	0.4	
	24	2							4	8	12
Totals				1.44	0.384	0.768	3.84	3.84	7.68	38.4	144
	SJ	Wt		mV	mV	mV	mV	mV	mV	mV	mV
	25	0.2		1.2	0.8	0.4					
	26	0.2									
	27	2			4	4	4				
N4	28	0.2				0.4	0.8	1.2	1.2	1.2	1.2
+ − − −	29	2	V_{out}	12	12	12	12	8	4		
	30	0.2									
	31	0.2						0.4	0.4	0.4	
	32	0.2							0.4	0.8	1.2
Totals				14.4	38.4	7.68	38.4	3.84	0.768	0.384	1.44
	SJ	Wt		mV	mV	mV	mV	mV	mV	mV	mV
	33	0.2		1.2	0.8	0.4					
	34	0.2									
	35	0.2			0.4	0.4	0.4				
N5	36	2				4	8	12	12	12	12
+ + + −	37	0.2	V_{out}	1.2	1.2	1.2	1.2	0.8	0.4		
	38	0.2									
	39	2						4	4	4	
	40	0.2							0.4	0.8	1.2
Totals				1.44	0.384	0.768	3.84	38.4	7.68	38.4	14.4

Note: On the left are listed the five neurons and their patterns assigned during the organizing phase. Weights of 2 are assigned for these patterns; the remaining junctions are assigned weights of 0.2. At the top are listed the eight events of the testing phase. In each case, one should compare the visual input pattern at the top with the V_{out} totals.

Event no. 2: N1 sees − − − − and N4 sees + − − − (V_{out} = 38.4 mV).

Event no. 3: The neurons respond weakly. N1 sees − − − −, N2 sees + + − −, and N4 sees + − − − (V_{out} = 7.68 mV).

Event no. 4: N2 sees + + − − and N4 sees + − − − (V_{out} = 38.4 mV).

Event no. 5: N2 sees + + − − and N5 sees + + + − (V_{out} = 38.4 mV).

Event no. 6: The neurons respond weakly. N2 sees + + − −, N3 sees + + + +, and N5 sees + + + − (V_{out} = 7.68 mV).

Event no. 7: N3 sees + + + + and N5 sees + + + − (V_{out} = 38.4 mV).

Event no. 8: N3 strongly sees + + + + (V_{out} = 144 mV).

The simple neural network employed in the preceding example has been chosen because, following only 14 events and a short list of algorithms that do not require the aid of a computer, it illustrates self-organization, learning, and forgetting. Other algorithms are, of course, possible.

Table 11-3 depicts a spatial sequence such as a white-to-black edge, + + − −. With somewhat different input transducers, the *same* table could just as easily stand for a time sequence such as a white surface changing to black, + + − −. According to the model, the brain's visual memory code is based on a particular axon representing a particular spatial or time sequence. If the graded potential (or action potential frequency) of this axon rises, the memory stack gets the message "At this particular location on the retina, a white-to-black edge appears," or "a white surface is changing to black," and so forth. Actually, each retinal location has to correspond to several axons that represent color (cone) as well as nighttime (rod) information. Imagine the tremendous number of spatial and temporal changes that have to be processed as you read this material!

It is tempting to apply the same memory code to the auditory system. The preceding model implies the following: starting with the amplitude-versus-frequency map of Fig. 9-16, a particular axon leaving this map represents a particular amplitude and frequency location. If the graded potential of this axon rises, the memory stack gets the message "At this particular amplitude and frequency, the amplitude is increasing," or "the frequency is increasing," and so forth.

In general, for any sensory modality, a particular memory axon represents a particular sensory level and location. In other words, your grandmother is spread among axons for facial features, hairdo, typical dress, voice frequency components, and the many other ways by which we characterize a person [J. G. Daugman, 1989; S. Grossberg, E. Mingolla & D. Todorovic, 1989; M. Porat and Y. Y. Zeevi, 1989; A. Rosenfeld, 1989; A. B. Watson and A. J. Ahumada, 1989].

11-5 Artificial Neural Networks

The Institute of Electrical and Electronics Engineers began to publish the *IEEE Transactions on Neural Networks* in March 1990. Before that, in 1988, the International Neural Network Society began publishing *Neural Networks* magazine. At the time of this writing, the new field is dominated by four "gurus": Stephen Grossberg (1988); John J. Hopfield [J. J. Hopfield and D. W. Tank, 1986]; Teuvo Kohonen (1988); and David E. Rumelhart [D. E. Rumelhart, J. L. McClelland & PDP Research Group, 1986]. One should also mention books by R. Hecht-Nielsen (1990); E. Sánchez-Sinencio and C. G. Lau (1992); and two special issues of the *Proceedings of the IEEE* edited by Clifford G. Y. Lau and Bernard Widrow (1990).

For most of the networks discussed in these publications, there is no attempt to imitate actual human-brain or lower-animal circuits. One reason is that, except for a few simple creatures such as the snail *Aplysia*, the circuits are mostly unknown. If the network "looks like" that of Fig. 11-5—simple geometric shapes called *neurons* with a generous sprinkling of connections between the neurons—it is a "neural network." The purpose of the network may be to recognize some kind of pattern, such as a tank on the photograph of a battlefield. Usually, the networks are too complicated to be built; instead, the computational procedures (algorithms) describing the behavior of each layer of neurons are written as a computer program. The network is usually capable of learning so that, if presented with a specific pattern during a training phase, recognition can be accomplished during subsequent stimulus-versus-response presentations. Thousands of cycles of calculation may be necessary (and are acceptable if the recognition time is reasonably short) [D. L. Alkon, 1983; S. I. Amari, 1983].

Since we have known for a long time that the mammalian nervous system is a parallel processor, why did neural networks wait until the 1980s before giving birth to the new field? First, relatively inexpensive high-powered parallel-processing computers became available. Second, it became possible to construct *analog* very-large-scale integration parallel processors that can implement some of the networks. Third, some ideas came from the advances of neurobiologists. Fourth, new mathematical concepts made it possible to solve some useful problems.

Parallel processing is sometimes called *collective computation* or *parallel distributed processing* (see *Parallel Distributed Processing*, by David E. Rumelhart, J. L. McClelland & PDP Research Group, 1986). An "oldtimer" will recognize that many of the neurocomputing elements are interconnected multivibrators. In a simple bistable "multi," two identical saturable amplifiers *A* and *B* are connected to form a feedback loop. When *A* is ON its output, which is inverted, is clipped at level 0; this turns *B* OFF so that its output is clipped at level 1; the latter, fed back to *A*, keeps *A* ON. An inhibitory input that exceeds some threshold level will cause the states to switch by turning *A* OFF; its output then turns *B* ON (or an excitatory input to an OFF amplifier can turn it ON, causing the states to switch, and so on). A neurocomputing element usually has many synaptic junction inputs, just as a bistable multi can have many inputs. The chief difference is that the artificial neuron's output usually goes to many other neurons; if there is feedback, the assembly may have many stable states compared to the two of a bistable multi.

In general, each new class of problem that can be solved by an artificial neural network (ANN) requires a different circuit. In the present brief introduction to ANNs, we look at three of the most important paradigms, those of associative memory, the task-assignment problem, and back propagation.

Associative Memory: How can neurons store items of memory? The following scheme is taken from an example by David W. Tank and John J. Hopfield in *Scientific American*, 1987. It is called an *associative memory* because the neurons are interconnected so that they can "recall" stored patterns even if some of the new input data are missing (within reason, of course).

Six neurons form the associative memory assembly. Each neuron can have any of *three* outputs: +1 representing one aspect of memory, −1 for a contrasting aspect, and 0 if the neuron is unstimulated, resting. The states are, respectively, ON, OFF, and ZERO. The features assigned to the memory neurons are given by:

	N1 Name	N2 Height	N3 Age	N4 Weight	N5 Hair	N6 Eyes
Neuron ON, $V_{out} = +1$:	Jones	Short	Young	Fat	Blond	Brown
Neuron OFF, $V_{out} = -1$:	Smith	Tall	Old	Thin	Brown	Blue

(For reasons unknown to us, the most important distinguishing feature, sex, has been omitted from the list. We have to assume, therefore, that this associative memory is devoted exclusively to, say, males.) So much for V_{out}. What about V_{in}? This aspect is determined by features of the particular people the robot meets. In the present example, the robot can remember three people, A, B, and C, who leave their imprints on the six neurons as follows:

$$\begin{array}{rrrrrrr} & \text{N1} & \text{N2} & \text{N3} & \text{N4} & \text{N5} & \text{N6} \\ A= & +1 & +1 & +1 & -1 & -1 & -1 \\ B= & +1 & -1 & +1 & +1 & -1 & +1 \\ C= & +1 & +1 & -1 & +1 & -1 & -1 \end{array} \tag{11-5}$$

In other words, A is named Jones, he is short, young, and thin, with brown hair and blue eyes; B is also named Jones, but he is tall, young, and fat, with brown hair and brown eyes; and C is another Jones, who is short, old, and fat, with brown hair and blue eyes. Since we apparently are in Jonestown, where all of the people have brown hair, neurons N1 and N5 could more efficiently be devoted to some other distinguishing feature, such as sex, but let us continue with the example as presented.

The learned memory is stored in the form of synaptic junction weights in the network of Fig. 11-15. The weights are determined as follows: we

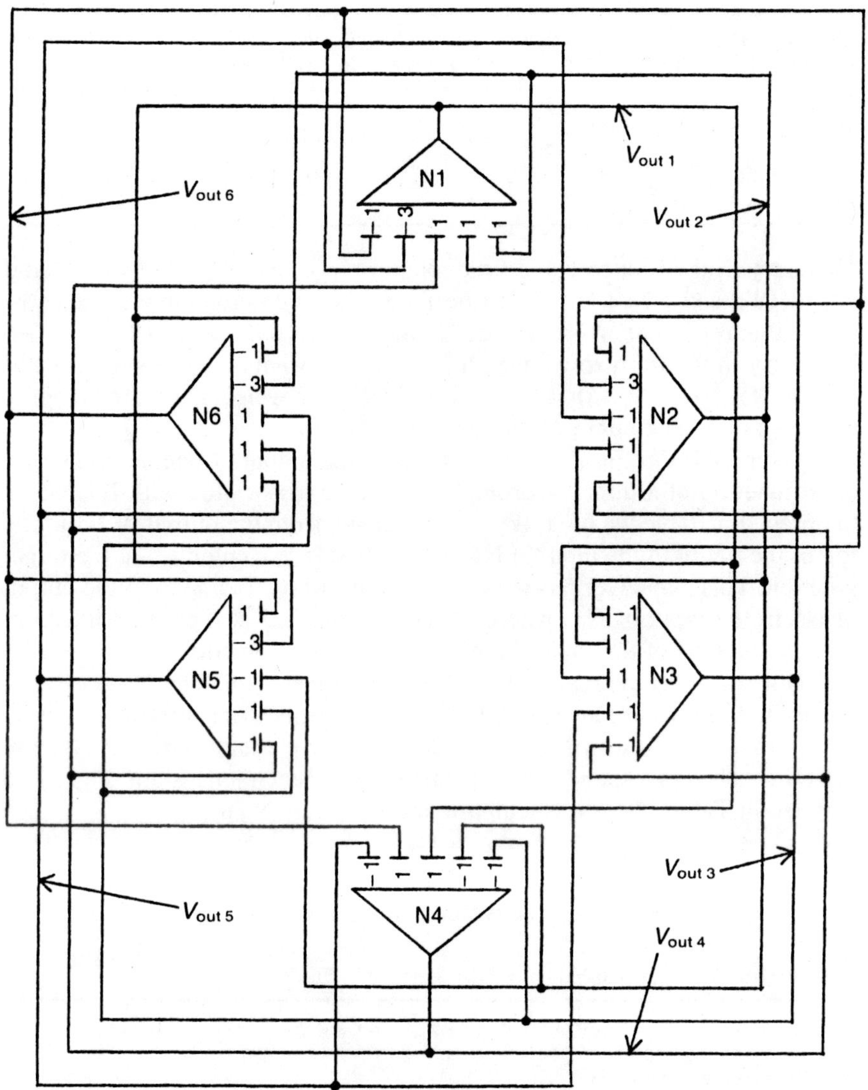

Fig. 11-15 An ANN for the Associative Memory example discussed in the text. Each neuron has six synaptic junction inputs with multiplicative weights as shown. The circuit is capable of recognizing three different people based on six of their features. It can recognize a pattern even if some of the information is missing (within reason).

multiply all possible pair combinations of Eq. (11-5) to form the following table:

	N1 N2	N1 N3	N1 N4	N1 N5	N1 N6	N2 N3	N2 N4	N2 N5	N2 N6	N3 N4	N3 N5	N3 N6	N4 N5	N4 N6	N5 N6
A	+	+	−	−	−	+	−	−	−	−	−	−	+	+	+
B	−	+	+	−	+	−	−	+	−	+	−	+	−	+	−
C	+	−	+	−	−	−	+	−	−	−	+	+	−	−	+
W	+1	+1	+1	−3	−1	−1	−1	−1	−3	−1	−1	+1	−1	+1	+1

The entry in the first row and first column represents $A = +1$ for N1 and $+1$ for N2, so $(+1)(+1) = +1$ (the 1s are omitted in the table). The entry below this represents $B = +1$ for N1 and -1 for N2, so $(+1)(-1) = -$. The entry in the third row beneath N3, N5 represents $C = -1$ for N3 and -1 for N5, so $(-1)(-1) = +$, and so forth. The weights, W, are given by the sums of each column, as shown.

To enter Ws in Fig. 11-15, simply trace the output of one neuron to the input junction of a target neuron. For example, associated with N1, N5 of the preceding table we have $W = -3$. Tracing from the output of N1 at the top of the figure to the input of N5 at the left side, we enter -3 at the target junction. This maneuver has to be complemented by tracing from the output of N5 to the input of N1, where we again enter -3 at the target junction. The 15 columns of the table therefore represent 30 junctions.

Our robot is now ready to stroll around Jonestown. The first person it meets is A. In accordance with Eq. (11-5), circuits from the robot's visual cortex (not shown) turn N1, N2, and N3 ON while they turn N4, N5, and N6 OFF. The corresponding V_{in} contribution of each junction in Fig. 11-15 is given in Table 11-5. In calculating these values, we use

$$V_{in\ i} = WV_{out\ j}. \tag{11-6}$$

Therefore, if $V_{out\ j}$ is negative (as is the case for N4, N5, and N6), and W is also negative, we get a positive contribution.

TABLE 11-5 The V_{in} Contribution of Each Junction of Fig. 11-15

	$V_{in\ 1}$	$V_{in\ 2}$	$V_{in\ 3}$	$V_{in\ 4}$	$V_{in\ 5}$	$V_{in\ 6}$
N1 ON, $V_{out\ 1} = 1$	·	1	1	1	−3	−1
N2 ON, $V_{out\ 2} = 1$	1	·	−1	−1	−1	−3
N3 ON, $V_{out\ 3} = 1$	1	−1	·	−1	−1	1
N4 OFF, $V_{out\ 4} = -1$	−1	1	1	·	1	−1
N5 OFF, $V_{out\ 5} = -1$	3	1	1	1	·	−1
N6 OFF, $V_{out\ 6} = -1$	1	3	−1	−1	−1	·
V_{in} total =	5	5	1	−1	−5	−5

Note: This table is true if the robot meets person A, and circuits from its visual cortex turn N1, N2, and N3 ON while they turn N4, N5, and N6 OFF. The sign of V_{out} should agree with that of V_{in} total.

The algebraic sum of all contributions is given by V_{in} total. The neurons of Fig. 11-15 saturate, or clip, at -1 and $+1$. Where V_{in} total $= +5$, for example, the neuron's V_{out} should be $+1$; that is, the *sign* of V_{out} should agree with the sign of V_{in} total, as is indeed the case in Table 11-5.

It is left as an exercise for the reader to determine that the signs of V_{out} agree with those of V_{in} total if the robot encounters *B* or *C*.

What is so special about this scheme for associative memory? It is that the network can recognize a pattern even if some of the information is missing. This is illustrated in the "partial data" of Table 11-6(*a*). Here our robot sees a person off in the distance. From a large placard the man is carrying aloft, the robot can make out that the name is Jones (N1), and it can see that Jones is short (N2) and thin (N4), but it cannot tell Jones's age (N3) or the color of his hair (N5) or eyes (N6). Which Jones is it—*A*, *B*, or *C*? From the known data and Eq. (11-5), the robot's visual cortex turns neurons N1 and N2 ON and N4 OFF. Neurons N3, N5, and N6 remain at ZERO, unstimulated. Only the columns belonging to the missing data, $V_{in\,3}$, $V_{in\,5}$, and $V_{in\,6}$ are shown in Table 11-6(*a*). As the column sums indicate, V_{in} total $=$ 1, -3, and -5, respectively.

In predicting the outcome, it is prudent to allow a single change to occur as a single event in accordance with the strongest V_{in} total value—in this case, $V_{in\,6} = -5$. (In case of a tie, operate on the unstimulated neuron nearest the top of the table.) Since the $V_{in\,6}$ sign is negative, this cuts off N6, and the calculations are continued as in Table 11-6(*b*). The *assumed* data indicate that V_{in} total $= 0$ and -4, respectively. In accordance with the strongest value, $V_{in\,5} = -4$, we allow N5 to turn OFF. The calculations now move to Table 11-6(*c*). Here we get $V_{in\,3}$ total $= 1$, which turns N3 ON. Based on the partial data, therefore, the robot concludes that the memory neuron signature of the man in the distance is $+++---$, and that it is *A*.

TABLE 11-6 The V_{in} Contribution of Each Junction of Fig. 11-15

		(*a*) Partial Data			(*b*) N6 Cuts Off			(*c*) N5 Cuts Off	
		$V_{in\,3}$	$V_{in\,5}$	$V_{in\,6}$		$V_{in\,3}$	$V_{in\,5}$		$V_{in\,3}$
N1	ON	1	-3	-1	ON	1	-3	ON	1
N2	ON	-1	-1	-3	ON	-1	-1	ON	-1
N3	ZERO	·	0	0	ZERO	·	0	ZERO	·
N4	OFF	1	1	-1	OFF	1	1	OFF	1
N5	ZERO	0	·	0	ZERO	0	·	OFF	1
N6	ZERO	0	0	·	OFF	-1	-1	OFF	-1
V_{in} total =		1	-3	-5		0	-4		1

Note: This table is true if the robot only gets partial data. (*a*) Circuits from its visual cortex turn N1 and N2 ON and N4 OFF. Strongest V_{in} total, $V_{in\,6}$, turns N6 OFF in (*b*). Strongest V_{in} total, $V_{in\,5}$, turns N5 OFF in (*c*). The latter turns N3 ON, thus completing full recognition.

The associative memory ANN shows how a dedicated neuron can store conflicting data (short and tall, young and old, and so forth) and yet have the ability to supply missing data. One should refer to *Parallel Models of Associative Memory*, by Geoffrey E. Hinton and James A. Anderson (1981), and *Self-Organization and Associative Memory*, by Teuvo Kohonen (1988).

Task-Assignment Problem: There is a certain class of optimization problem that, for want of a better name, can be called the *task-assignment* problem. The traveling salesman problem, in which the salesman has to minimize total travel distance between many cities, is an example. However, we illustrate below with a relatively simple "book shelving" problem.

The task is to reshelve the physics, chemistry, and history books of a library. We have three assistants, Jessica, George, and Tim. Jessica is easily distracted by physics books, and can only shelve four of them per minute; she is not interested in chemistry, so nine of them get put away per minute; she is only mildly interested in history, shelving seven of them per minute. Each of the assistants has different likes, dislikes, and abilities, of course, which are reflected in their shelving rates. The latter are summarized as follows:

	Physics	Chemistry	History
Jessica	4	9	7
George	8	3	6
Tim	6	4	1

Only a single assistant can be assigned to a single class of books. The problem is—which arrangement maximizes the number of books shelved per minute?

For the simple example just cited, there are six possible solutions. Omitting the column headings, they are

$$\begin{array}{r} \text{Jessica} = \\ \text{George} = \\ \text{Tim} = \\ \text{Total/min} = \end{array} \begin{array}{c} \begin{vmatrix} 4 & \cdot & \cdot \\ \cdot & 3 & \cdot \\ \cdot & \cdot & 1 \end{vmatrix} \\ 8 \end{array} \begin{array}{c} \begin{vmatrix} 4 & \cdot & \cdot \\ \cdot & \cdot & 6 \\ \cdot & 4 & \cdot \end{vmatrix} \\ 14 \end{array} \begin{array}{c} \begin{vmatrix} \cdot & 9 & \cdot \\ 8 & \cdot & \cdot \\ \cdot & \cdot & 1 \end{vmatrix} \\ 18 \end{array} \begin{array}{c} \begin{vmatrix} \cdot & \cdot & 7 \\ 8 & \cdot & \cdot \\ \cdot & 4 & \cdot \end{vmatrix} \\ 19 \end{array} \begin{array}{c} \begin{vmatrix} \cdot & 9 & \cdot \\ \cdot & \cdot & 6 \\ 6 & \cdot & \cdot \end{vmatrix} \\ 21 \end{array} \begin{array}{c} \begin{vmatrix} \cdot & \cdot & 7 \\ \cdot & 3 & \cdot \\ 6 & \cdot & \cdot \end{vmatrix} \\ 16 \end{array}$$

In the optimum solution, Jessica shelves nine chemistry books/min, George shelves six history books/min, and Tim shelves six physics books/min, for a total of 21 books/min.

The key to setting up an ANN to solve the problem is that each row and column has only a single entry. Therefore, each matrix element can be represented by a neuron that, by means of lateral inhibition, tries to "knock out" all of the other neurons in the *same* row and *same* column. The circuit, for the preceding example, is depicted in Fig. 11-16. Here it is rather far-fetched to claim that this is in any way correlated with an actual mammalian brain.

It is convenient to employ the row–column pairs of matrix notation in Fig. 11-16:

$$\begin{vmatrix} V_{11} & V_{12} & V_{13} \\ V_{21} & V_{22} & V_{23} \\ V_{31} & V_{32} & V_{33} \end{vmatrix}. \qquad (11\text{-}7)$$

In Fig. 11-16, if we trace $V_{\text{out } 11}$ from N11, we see that it has a horizontal branch that is inhibitory (I) on N12 and N13, and a vertical branch that is inhibitory on N21 and N31. Similarly, $V_{\text{out } 12}$ from N12 has a horizontal branch that is inhibitory on N11 and N13, and a vertical branch that is inhibitory on N22 and N32.

We envisage a circuit or parallel processor that, given the shelving rate matrix values of input voltages V_{in} just cited, can generate a V_{out} matrix corresponding to the optimum solution.

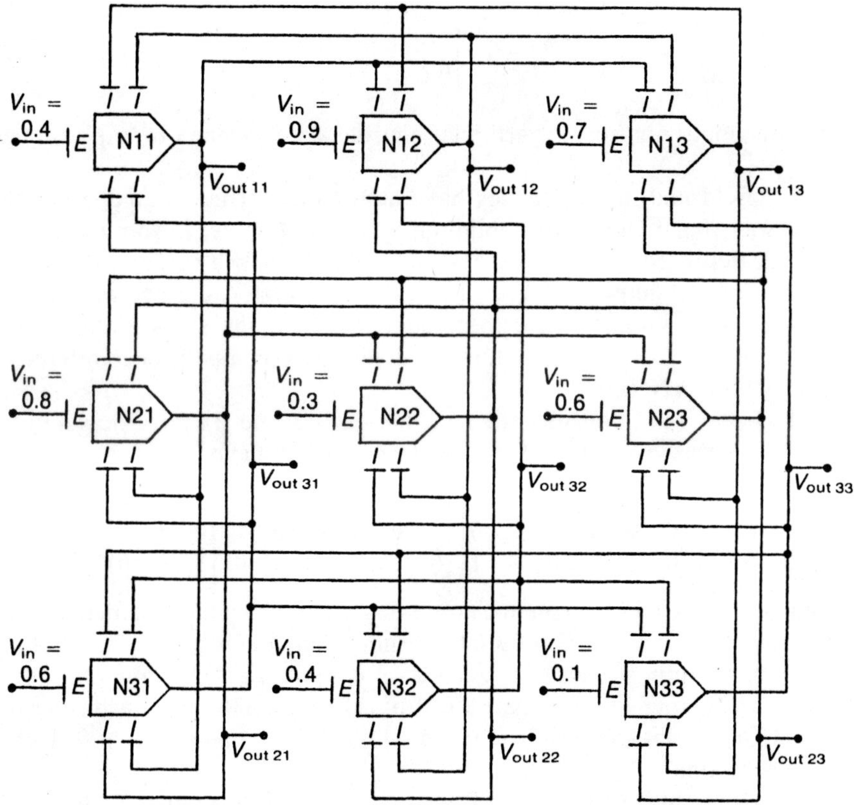

Fig. 11-16 An ANN for the Task-Assignment Problem discussed in the text. Each neuron is inhibitory to all of the other neurons in the *same* row and *same* column. The V_{in} values are taken from the numerical example. Based on V_{out} values, the circuit is capable of indicating the optimum solution to the problem.

The task can be simplified if all voltages lie between the limits 0 and 1. For the input matrix, this is easily accomplished by dividing the shelving rate matrix by 10. Accordingly,

$$V_{\text{in}} = \begin{vmatrix} 0.4 & 0.9 & 0.7 \\ 0.8 & 0.3 & 0.6 \\ 0.6 & 0.4 & 0.1 \end{vmatrix}, \tag{11-8}$$

and these are the values shown in Fig. 11-16.

Another simplification is obtained if the solution is presented as

$$V_{\text{out}} = \begin{vmatrix} 0 & 1 & 0 \\ 0 & 0 & 1 \\ 1 & 0 & 0 \end{vmatrix} \tag{11-9}$$

rather than

$$V_{\text{out}} = \begin{vmatrix} \cdot & 0.9 & \cdot \\ \cdot & \cdot & 0.6 \\ 0.6 & \cdot & \cdot \end{vmatrix} \tag{11-10}$$

since our multivibrators or artificial neurons are designed to clip at 0 and 1.

Hopfield and his colleagues have actually constructed analog parallel processors that indicate the optimum solution. For us, in the absence of sophisticated hardware, a programmable calculator will suffice to yield the solution for the simple example of Eq. (11-8). (Some 450 program steps are required.) It is important for the reader to follow the steps outlined next because they illustrate procedures and pitfalls commonly encountered in working with ANNs.

To begin with, consider the fact that each of the six possible solutions is stable. Recast with 0s and 1s, we, respectively, have

$$\begin{vmatrix} 1 & 0 & 0 \\ 0 & 1 & 0 \\ 0 & 0 & 1 \end{vmatrix} \begin{vmatrix} 1 & 0 & 0 \\ 0 & 0 & 1 \\ 0 & 1 & 0 \end{vmatrix} \begin{vmatrix} 0 & 1 & 0 \\ 1 & 0 & 0 \\ 0 & 0 & 1 \end{vmatrix} \begin{vmatrix} 0 & 0 & 1 \\ 1 & 0 & 0 \\ 0 & 1 & 0 \end{vmatrix} \begin{vmatrix} 0 & 1 & 0 \\ 0 & 0 & 1 \\ 1 & 0 & 0 \end{vmatrix} \begin{vmatrix} 0 & 0 & 1 \\ 0 & 1 & 0 \\ 1 & 0 & 0 \end{vmatrix}. \tag{11-11}$$

In each case, each 1 element suppresses all of the other elements in its particular row and column via lateral inhibition. Each 0 element supports the given solution because the zero level is not capable of inhibiting the 1 elements. Starting with the matrix of Eq. (11-8), then, how can we steer the calculation so that we terminate in Eq. (11-9) rather than in one of the other stable solutions?

As Hopfield suggested, the first step is to relax the clipping characteristic so that saturation is gradual, rather than to use the sharp clipping beyond threshold that results in either 0 or 1. For this purpose, consider the sigmoid or S family of curves

$$y = \frac{\arctan Gx}{\pi}, \tag{11-12}$$

depicted in Fig. 11-17 for $G = 1, 3$, and 10. Here the threshold lies at the origin. As x becomes more positive than threshold, y *gradually* approaches the upper limit, 0.5. As x becomes more negative, y gradually approaches -0.5. The arctan function is most convenient for this purpose because it has odd symmetry and is available as an inverse key on programmable calculators. (If degrees are used rather than radians, divide by 180° instead of π.) If G approaches ∞, Fig. 11-17 yields conventional sharp saturation levels at $+0.5$ and -0.5, but in what follows, the value $G = 10$ is reasonable. The argument for gradual saturation is that it substitutes different numerical values for the 0s and 1s in the listing of stable solutions, Eq. (11-11). This allows the computer to "see" that one solution is more stable than the others.

Processing of the data follows a three-step procedure.

Step 1. Starting with a set of matrix elements, V_{new}, derive a set of inhibitory matrix elements, V_{inh}. The original V_{new} set then becomes, by definition, V_{old}.

Step 2. Apply a suitable arctan compression to a function of V_{in} and V_{inh}, where V_{in} is the input data matrix of Eq. (11-8). This yields a set of output matrix elements, V_{out}.

Step 3. From V_{out} and V_{old}, derive a new set of matrix elements, V_{new}. The process then repeats until the values converge; that is, until there is negligible change from one iteration to the next. Hopefully, this will occur when V_{out} resembles the optimum solution matrix.

Consider the calculation of inhibition, V_{inh}, for the V_{11} element of Eq. (11-7). The easiest strategy is based on the summation of horizontal elements of the same row ($V_{12} + V_{13}$) and vertical elements of the same column ($V_{21} + V_{31}$) to get

$$V_{inh\ 11} = I(V_{new\ 12} + V_{new\ 13} + V_{new\ 21} + V_{new\ 31}), \tag{11-13}$$

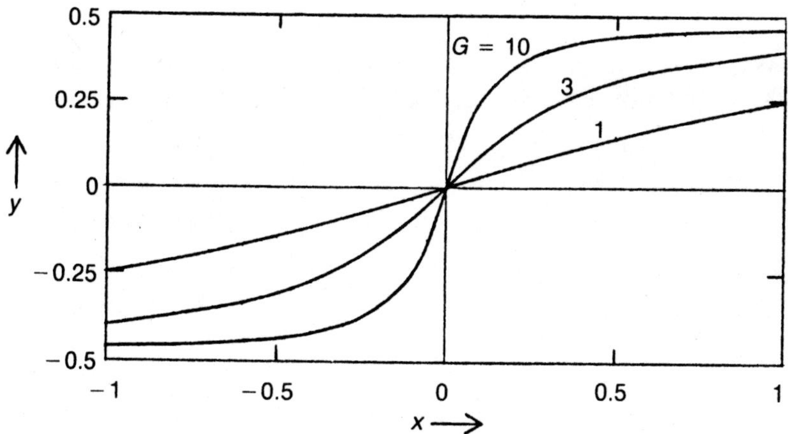

Fig. 11-17 The sigmoid or S function $y = (\arctan Gx)/\pi$ for $G = 1$, 3, and 10.

where I is an inhibitory weighting factor. Unfortunately, this simple derivation for lateral inhibition turns out to be inadequate. It allows the calculation to converge on a solution that is less than optimum. The difficulty is that Eq. (11-13) is too gentle; it does not sufficiently discriminate against less-than-optimum solutions. The remedy is to *multiply* rather than add. We cannot simply multiply all of the inhibitory contributions to get

$$V_{\text{inh } 11} = I(V_{\text{new } 12})(V_{\text{new } 13})(V_{\text{new } 21})(V_{\text{new } 31}) \tag{11-14}$$

because this leads to excessively small numerical values for V_{inh}. Instead, separately add horizontal and vertical contributions, and multiply them as follows:

$$V_{\text{inh } 11} = I(V_{\text{new } 12} + V_{\text{new } 13})(V_{\text{new } 21} + V_{\text{new } 31}). \tag{11-15}$$

To numerically compare Eqs. (11-13), (11-14), and (11-15), apply them to the V_{new} matrix when the latter is nearing the end of a computation:

$$V_{\text{new}} = \begin{vmatrix} 0.1 & 0.9 & 0.1 \\ 0.1 & 0.1 & 0.9 \\ 0.9 & 0.1 & 0.1 \end{vmatrix},$$

from which, with $I = 1$,

$$V_{\text{inh}} = \overset{\text{Eq. (11-13)}}{\begin{vmatrix} 2 & 0.4 & 2 \\ 2 & 2 & 0.4 \\ 0.4 & 2 & 2 \end{vmatrix}} \overset{\text{Eq. (11-14)}}{\begin{vmatrix} 0.0081 & 0.0001 & 0.0081 \\ 0.0081 & 0.0081 & 0.0001 \\ 0.0001 & 0.0081 & 0.0081 \end{vmatrix}} \overset{\text{Eq. (11-15)}}{\begin{vmatrix} 1 & 0.04 & 1 \\ 1 & 1 & 0.04 \\ 0.04 & 1 & 1 \end{vmatrix}}.$$

Notice that Eq. (11-13) yields a ratio of only 2/0.4 = 5 for V_{inh} values while Eq. (11-15) results in a ratio of 1/0.04 = 25.

In general, according to Eq. (11-15), for an $n \times n$ matrix,

$$V_{\text{inh } ij} = I\left[\left(\sum_{j=1}^{n} V_{\text{new } ij}\right) - V_{\text{new } ij}\right]\left[\left(\sum_{i=1}^{n} V_{\text{new } ij}\right) - V_{\text{new } ij}\right]. \tag{11-16}$$

As another example, the very first step in processing the V_{in} of Eq. (11-8) follows. Initially, $V_{\text{new}} = V_{\text{in}}$. Applying Eq. (11-15), with $I = 1$,

$$V_{\text{new}} = \begin{vmatrix} 0.4 & 0.9 & 0.7 \\ 0.8 & 0.3 & 0.6 \\ 0.6 & 0.4 & 0.1 \end{vmatrix}, \qquad V_{\text{inh}} = \begin{vmatrix} 2.24 & 0.77 & 0.91 \\ 0.90 & 1.82 & 0.88 \\ 0.60 & 0.84 & 1.30 \end{vmatrix}. \tag{11-17}$$

Notice that, although this is only the first step, it supplies a relatively large amount of inhibition against $V_{\text{new } 11}$ (0.4), $V_{\text{new } 22}$ (0.3), and $V_{\text{new } 33}$ (0.1), thus immediately eliminating

$$\begin{vmatrix} 1 & 0 & 0 \\ 0 & 1 & 0 \\ 0 & 0 & 1 \end{vmatrix}$$

as a solution. At the same time, it applies a relatively small amount of inhibition against $V_{new\,31}$ (0.6), $V_{new\,12}$ (0.9), and $V_{new\,23}$ (0.6), and this happens to be the optimum solution to the problem.

Consider next the arctan compression equation. In Fig. 11-17, y ranges between -0.5 and $+0.5$, while we want V_{out} to range between 0 and 1. This is easily accomplished by adding 0.5 to the arctan function. The equation used initially was

$$V_{out} = \frac{\arctan[G(V_{in} - V_{inh})]}{\pi} + 0.5,$$

but this failed and a trial-and-error analysis revealed that V_{inh} was too small. The correct result is obtained by doubling the V_{inh} contribution (and, as previously mentioned, we use $G = 10$):

$$V_{out} = \frac{\arctan[10(V_{in} - 2V_{inh})]}{\pi} + 0.5. \tag{11-18}$$

Applying this result to Eq. (11-17), the V_{out} matrix for the first processing step appears as

$$V_{out} = \begin{vmatrix} 0.008 & 0.049 & 0.028 \\ 0.032 & 0.010 & 0.027 \\ 0.053 & 0.025 & 0.013 \end{vmatrix} \tag{11-19}$$

Some of these values seem to be unreasonably small. Eventually, when V_{inh} approaches its final value, the corresponding V_{out} does display reasonable values.

Our calculator or computer program is now set as follows: Start with the matrix of Eq. (11-8) as V_{new}; calculate V_{inh} using Eq. (11-15), with $I = 1$; calculate V_{out} using Eq. (11-18); set V_{out} equal to V_{new} and repeat until negligible change occurs from one iteration to the next. The final result is mathematically correct, but the matrices switch violently from one stable state to another. We are trapped in a bistable pitfall (which should more properly be called a pothole):

$$\begin{matrix} V_{out} = \\ V_{new} = \end{matrix} \begin{vmatrix} 0.005 & 0.005 & 0.006 \\ 0.005 & 0.005 & 0.006 \\ 0.006 & 0.005 & 0.005 \end{vmatrix} \xrightarrow{V_{inh}} = \begin{vmatrix} 0.0001 & 0.0001 & 0.0001 \\ 0.0001 & 0.0001 & 0.0001 \\ 0.0001 & 0.0001 & 0.0001 \end{vmatrix}$$

$$\uparrow \qquad\qquad\qquad\qquad \downarrow$$

$$V_{inh} = \begin{vmatrix} 3.662 & 3.415 & 3.202 \\ 3.449 & 3.600 & 3.167 \\ 3.147 & 3.160 & 3.556 \end{vmatrix} \xleftarrow[V_{new}]{V_{out}} \begin{matrix} = \\ = \end{matrix} \begin{vmatrix} 0.922 & 0.965 & 0.955 \\ 0.960 & 0.898 & 0.947 \\ 0.947 & 0.922 & 0.750 \end{vmatrix}.$$

In the first V_{new} matrix (upper left), values are small, so the V_{inh} matrix displays practically zero inhibition. As a result, the second $V_{out} = V_{new}$ matrix (lower right) has values close to 1, which yields a V_{inh} matrix with large values of inhibition, which results in small $V_{out} = V_{new}$, and so forth.

The difficulty stems from the use of a serial computer in a problem that requires lateral interaction in a parallel processor. Fortunately, it is easy to imitate a parallel processor in this instance by reducing V_{new} changes to, say, 10% of their previous V_{old} values. In general, we demand that

$$V_{new} = V_{out}R + V_{old}(1 - R), \tag{11-20}$$

where R is a renewal factor, such as 0.1. As the calculation proceeds, and matrix entries converge toward their final values, one can increase R to 20%, 50%, and finally 100%.

With Eq. (11-20) in place, the calculations smoothly converge to

$$\begin{matrix} V_{out} = \\ V_{new} = \end{matrix} \begin{vmatrix} 0.0207 & 0.9646 & 0.0264 \\ 0.0296 & 0.0196 & 0.9471 \\ 0.9471 & 0.0212 & 0.0178 \end{vmatrix}, \qquad V_{inh} = \begin{vmatrix} 0.9680 & 0.0019 & 0.9507 \\ 0.9355 & 0.9628 & 0.0022 \\ 0.0020 & 0.9496 & 0.9426 \end{vmatrix}. \tag{11-21}$$

The three large entries in the V_{new} matrix, and their correspondingly small entries in the V_{inh} matrix, indicate that the matrix of Eq. (11-9) is the optimum solution for our task-assignment problem.

The purpose of the preceding discussion is introductory. It is hardly necessary to point out that we have not proved that the approach is optimum for task-assignment problems in general.

Back Propagation: Our third example of ANNs is a pattern-recognition circuit that uses back propagation as part of its training strategy. A small portion of one of many possible networks is depicted in Fig. 11-18. To minimize complexity, back-propagation connections are omitted in this schematic.

The circuit looks much like that of Fig. 11-5. In the latter model, however, the synaptic junction weights are built in. The circuit does not "recognize" patterns, it only dissects them into short lines and edges. In Fig. 11-18, on the other hand, the challenge is to adjust the weights, W, so that a given input pattern is recognized. (Remember that W can be negative as well as positive.)

The input signal, x, is distributed laterally by an input neuron. Only three units in each layer are shown in Fig. 11-18, but in general there are n units in an input layer and m units in an output layer. A single unit consists of several "planets" and a "sun"; these names arose from the practice of using small and large circles to represent "planet" and "sun" neurons, respectively. Notice that a "layer" actually includes a planet layer and a sun layer.

All layers between input and output are called *hidden*, which is a misleading adjective because it carries sinister implications where none are intended. Figure 11-18 only shows a single hidden layer, but the ANN may of course contain several of these layers.

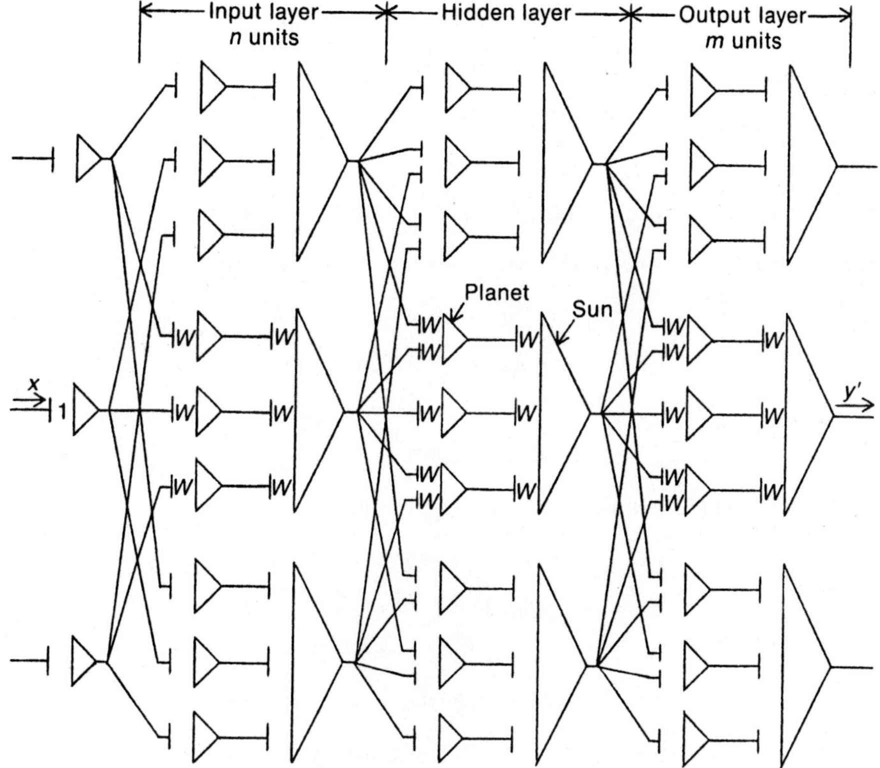

Fig. 11-18 A small portion of one of many possible ANNs that can be used for pattern recognition. To minimize complexity, back-propagation connections are omitted. The circuit shows pathways during normal operation and also during a training "forward pass."

The output is a two-dimensional signal, y'. The prime is a reminder that the output includes errors, either because W is not optimized and/or the number of m units is insufficient for fine output resolution (in which event the error can be minimized but not eliminated). The *desired* output signal is y. The error, δ, is obviously given by

$$\delta = y - y'. \tag{11-22}$$

A specific application reveals some of the problems that are encountered. Suppose that the ANN has to recognize the 10 decimal digits, $y = 0, 1, 2, \ldots, 9$, given handwritten input xs. The output layer consists of $m = 10$ units. Output unit 5 should "light up" when x is a handwritten 5; output unit 6 should turn ON when x is a handwritten 6, and so forth.

It is customary to call the signals *vectors*, so in this case y' would be a family of 10-element vectors. The unit vectors are, respectively, a, b, c,

. . . , j. During a training session, the outputs could be

$$\begin{aligned} y_0' &= 0.92a + 0.12b + \cdots - 0.08j \\ y_1' &= 0.15a + 0.90b + \cdots + 0.10j \\ &\vdots \\ y_9' &= 0.05a - 0.03b + \cdots + 0.95j. \end{aligned} \tag{11-23}$$

The desired y vector, which is without error, appears as

$$\begin{aligned} y_0 &= a + 0b + \cdots + 0j \\ y_1 &= 0a + b + \cdots + 0j \\ &\vdots \\ y_9 &= 0a + 0b + \cdots + j \end{aligned} \tag{11-24}$$

so that Eq. (11-22) indicates that the error is

$$\begin{aligned} \delta_0 &= 0.08a - 0.12b + \cdots + 0.08j \\ \delta_1 &= -0.15a + 0.10b + \cdots - 0.10j \\ &\vdots \\ \delta_9 &= -0.05a + 0.03b + \cdots + 0.05j. \end{aligned} \tag{11-25}$$

The handwritten inputs have to be read by a 10×10 grid, say, that corresponds to $n = 100$ input layer units, and x is a family of vectors having 10 rows and 100 columns whose coefficients represent the handwritten inputs.

The contraction from $n = 100$ input units to $m = 10$ output units should be gradual, and probably requires two hidden layers. The first hidden layer could have $8 \times 8 = 64$ units, and the second could have $6 \times 6 = 36$ units. An ANN such as that of Fig. 11-18 can easily contain over 1000 "synaptic" junctions.

How can we use feedback to decrease the error, δ? In a conventional system, δ is fed back to the input and, in principle, with sufficient forward gain, we can make δ as small as we please. In the preceding example, however, δ is a 10-element vector, whereas x is a 100-element vector, so it is not possible to simply feed δ back to the input. Instead, back propagation, as the name implies, is a strategy in which the direction of signal flow is reversed.

Figure 11-19 illustrates one of many possible configurations for reversing the signal flow. An ANN programmer is motivated to minimize the number of steps while maximizing accuracy, but the motivation in drawing Fig. 11-19 is to retain neuronlike elements to justify its appearance in this book. Accordingly, each neuron conducts from left to right, so the δ pathway is a "one step forward, two steps backward" locus, as a tracing of the arrows in the middle section shows.

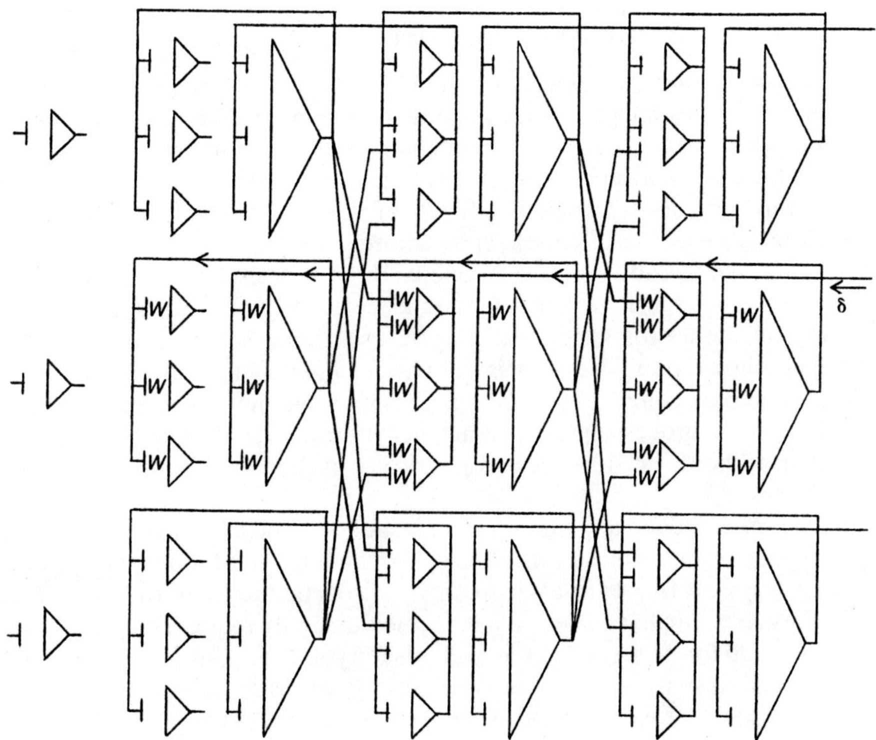

Fig. 11-19 One of many possible back-propagation configurations for the ANN of Fig. 11-18. During a training backward pass, the error signal, δ, is fed back so as to adjust the *W*s to reduce δ. Only small changes are made in *W* per training cycle.

Training the ANN, or adjusting the *W*s, follows an iterative two-step procedure. First, a forward pass is made using one of the handwritten digits as an input x vector. The x signal path contracts, say, from 100 units to 64 to 36 to 10 units. The output y' vector is subtracted from the known y vector to get δ. The latter is fed back in a backward pass. As Fig. 11-19 implies, the x input is disconnected as the δ signal path expands from 10 to 36 to 64 to 100 units.

One of the challenges for back-propagation practitioners is that the system may not converge, that it may "blow up" because the changes in *W* will increase rather than decrease δ. Since these systems are nonlinear, it is mathematically difficult to predict convergence and stability.

Although small changes in *W* per training cycle do not guarantee convergence, large changes can lead to dead ends, such as the bistable pitfall of the task-assignment problem. Therefore, training may require thousands of cycles. Although this is sometimes an unacceptable restriction, a high-speed computer can usually train the ANN in a reasonably small interval of real time.

11-6 Saccadic Motion and Compensation

We are not aware consciously of much if not most of what goes on inside the brain. In this chapter, which is devoted mainly to the visual system, it is appropriate to consider "housekeeping chores," automatic activities of which we are not aware "subjectively," in contrast to "looking at" an object, which entails a conscious effort. The remainder of this chapter is devoted to some of these housekeeping chores.

We have already considered one of these hidden activities in Chapter 10—the very important determination of color. Figure 10-20 depicts a hypothetical model for generating a two-dimensional color map. All of the neurons shown there are associated with a *single* contiguous group of red, green, and blue cones. Physically, there is no room in the retina for millions of these color processing cells. Whatever the color-determining mechanism happens to be, it must be located physically in the LGNs and/or the visual cortex.

Our next hidden activity is saccadic motion, which has been mentioned previously several times in this book. Fortunately, you the reader can now see for yourself that your eyes *do* undergo constant jerky movements. It is a remarkably simple and startling revelation that only requires that you stare at Fig. 11-20 for 30 seconds. You can blink if you must, but try to keep your eyes motionless with respect to the figure. It was discovered by F. J. Verheijen (1961). Here one can do no better than quote from Floyd Ratliff (1965):

> Fixate steadily on the black dot at the center of the grid for 30 seconds or more, and then fixate on the white dot in the center of the black square to the right and above the black dot. The negative afterimage of the entire figure can soon be seen distinctly. It is somewhat blurred, of course, because your eyes were actually moving while you attempted to fixate on the black dot. For the same reason, no matter how carefully one attempts to control fixation on the white dot, the negative afterimage of the grid will seem to move considerably with respect to the real grid.

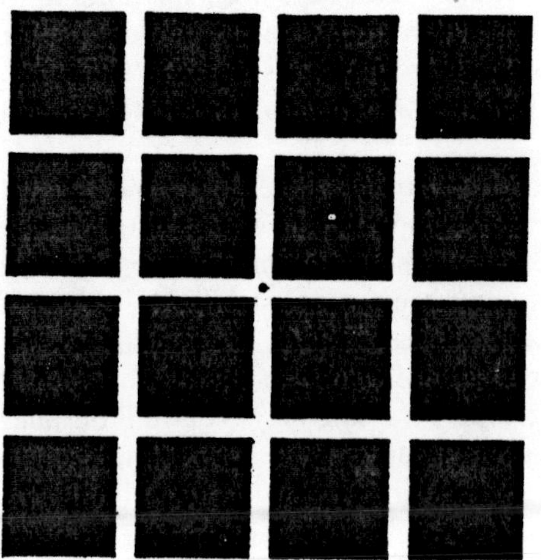

Fig. 11-20 Grid, discovered by F. J. Verheijen, that demonstrates saccadic movements of the reader's eyes by means of afterimages. Simply fixate steadily on the black dot at the center of the grid for 30 seconds, and then fixate on the white dot.

By tightly fitting a contact lens to the eye, Ratliff and Lorrin A. Riggs quantitatively measured the saccadic motion. In addition to a very small oscillation at frequencies ranging from 30 Hz to 70 Hz, we are witnessing "slow drifts of a few minutes of arc in one direction or another; and rapid jerks with an average extent of about 5 minutes of arc occurring at irregular intervals, sometimes apparently compensating for the drifts." Since the limiting resolution of the eye is 1 minute of arc, the slow drifts and rapid jerks should be noticeable, as indeed they are using Verheijen's grid.

Why does the system introduce saccadic motion? Perhaps it has no alternative. Fixation demands a steady stream of action potentials (APs) of exactly the correct frequency. Three muscle pairs are involved in each eye; the slightest increase or decrease in any of the six AP frequencies or the slightest "fatigue" in any of the muscles results in drift, and that is exactly what is observed. As the desired fixation point drifts away from the optical center, corrective feedback initiates the approximate return jerk [A. T. Bahill and L. Stark, 1979; R. Eckmiller, 1983; J. D. Enderle and J. W. Wolfe, 1987; D. H. Fender, 1964; P. Inchingolo and M. Spanio, 1985].

Since the afterimage is blurred while the real-time image is sharp, a cancellation mechanism must be operating in the visual system soon after the optic signals leave the retina. It is possible that the cancellation network can be found in the superior colliculus (SC), but a more likely location is in layers 1 and 2 of the LGNs. It is our conjecture that the six fixation muscle APs also feed into layers 1 and 2 of the LGNs, where an equal and opposite cancellation shift is introduced. It is a simple matter to devise a neural network that is capable of this lateral cancellation shift.

The detection of fixation drift has to occur before it is canceled, of course. It is also conjectured, therefore, that the difference between the desired fixation point and the actual point is also measured in LGN layers 1 and 2, and that this is the basis for a corrective burst of APs.

One can argue that the saccadic movements are not all bad. They tend to improve resolution despite the graininess associated with finite groups of red, green, and blue cones. They also tend to dilute the effects of any cones that may be damaged.

11-7 Control of Lens Opening

Control of the lens opening is also known as the pupillary reflex. In response to a sudden increase in illuminance, the lens opening of the eye, the pupil, constricts in about 30 s. The constriction serves two purposes: it guards against cone saturation in the event of excessive illuminance, and it takes advantage of a liberal amount of light to block off the outside of the lens. This improves image sharpness, and objects at various distances from the eye remain more nearly in focus. (In the case of a camera lens one would say that decreasing the aperture increases the depth of focus or the depth of field.)

After a sudden decrease in illuminance, the pupil dilates in a few minutes, as shown in Fig. 10-2. The aperture can change from a maximum diameter of 8 mm to a minimum of 1.5 mm, an area change of 30 times. The mechanism for constriction is the sphincter muscle of the iris.

We do not know all of the components of the pupillary reflex system. Somewhere, in the retina, LGNs, or cortex, a measure of illuminance is developed. In Fig. 10-20, it is hypothesized that the nervous system finds the average of the red, green, and blue cone outputs (after logarithmic compression) in each small area of the visual image. If the cone values are averaged over a large area of the visual field, we have a good measure of illuminance. Alternatively, the outputs of many horizontal cells can be averaged.

Either from the LGN or cortex, the illuminance information pathway consists of the *pretectal* region, the *Edinger–Westphal* nucleus, the ciliary ganglion, and, finally, the sphincter muscle. There is also an involvement with the sympathetic nervous system whereby stimulation of the system excites radial fibers of the iris, which in turn leads to pupillary dilation. Thus opening and closing of the lens aperture, which one would expect to be a simple system, turns out to be quite complex [A. C. Guyton, 1986; W. Krenz et al., 1985; N. Link and L. Stark, 1988].

11-8 Distance Accommodation

If a person with normal vision looks at a distant object and then switches his or her gaze to an object that is, say, 30 cm away, the ciliary muscle contracts so as to make the lens more convex in order to maintain sharp focus on the retina. This *accommodation* maneuver is discussed in Section 10-1. In a camera, the shape of the lens is constant, so accommodation is effected by changing the distance from lens to film. In a mammal, the distance from lens to retina is constant, so accommodation is accomplished by changing the shape of the lens.

We do not know how the accommodation signal originates. It probably has nothing to do with binocular vision, since closure of one eye does not cause a loss of accommodation. Two hypothetical systems can be described.

The *nonfeedback* system supposes that a "distance estimating" nucleus exists in the brain; that is, there is a group of neurons that estimates the distance to the object of interest. When the "infinite distance" neurons fire, the ciliary muscle remains relaxed; when the "distance = 8 cm" neurons fire, the ciliary muscle receives maximum stimulation, and in a child 10 years old, the focal length of the lens appropriately decreases from 15 mm for a distant object to 12.6 mm for an object 8 cm (about 3 in.) away. How does the distance-estimating nucleus arrive at a decision? If the person is using both eyes, it could be from the angle of convergence between the left and right eyes. If the person is using one eye, the estimate could be based on past experience, such as "judging from the size of the letters, this newspaper is 30 cm away." The distance-estimating nucleus could be at the end of the very sophisticated pattern-recognition apparatus of the visual system. It could be tied in with the auditory system, which also estimates distance and direction to an object of interest that is emitting sound waves.

The other hypothetical system involves feedback. We begin with a "focus detector." Since the eyes are in constant saccadic motion, the graded-potential output of many of the bipolar cells contains one or two vertical steps per second as the receptor suddenly hops from one brightness level to another. If the image is out of focus, the vertical edges become ramps. The nervous system is easily able to detect the difference between sharp and gradual time changes in the graded-potential output. Let us assume that this focus detector itself generates a graded potential, and that this voltage decreases as the focus deteriorates; it is shown as the focus-detector waveform in Fig. 11-21.

In this scheme, the ciliary muscle continually contracts and relaxes a few times per second, day in and day out, while our eyes are open. In the optimum condition, that of Fig. 11-21(*a*), the eye goes out of focus during "constriction" and "dilation"; for each cycle of constriction and dilation we get two "out of focus" cycles. In Fig. 11-21(*b*), we have insufficient contraction, so that the focus-detector output contains a component that is in phase with the ciliary muscle signal; this should feed back to the ciliary muscle and cause it to reinstate the conditions of Fig. 11-21(*a*). In Fig. 11-21(*c*), we have excessive contraction and the focus-detector signal is opposite in phase to that of the ciliary muscle; feedback should now cause the ciliary muscle to relax until Fig. 11-21(*a*) is reestablished.

It is unlikely that the preceding solution to a fairly common engineering problem is neurophysiologically correct. However, regardless of whether

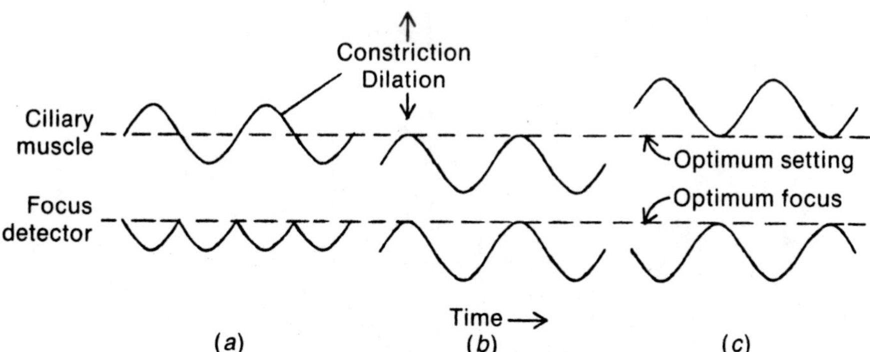

Fig. 11-21 Ciliary muscle and "focus detector" waveforms for a hypothetical distance accommodation system that involves feedback. (*a*) Ciliary muscle oscillates about its optimum setting; (*b*) with insufficient contraction; (*c*) with excessive contraction. (From Deutsch and M-Tzanakou, *Neuroelectric Systems*, New York Univ. Press, New York, 1987.)

the accommodation signal is derived with or without feedback, it seems to follow a neural pathway similar to that of the pupillary reflex of Section 11-7 [G. K. Hung, J. L. Semmlow & K. J. Ciuffreda, 1984, 1986].

11-9 Binocular Convergence

Binocular convergence requires that the eyes change their vergence angle (the angle between the optical axes) when shifting from a distant object to a nearer object or vice versa. The convergence operation is performed automatically via a built-in feedback system. Figure 11-22 illustrates a possible mechanism, similar to a model proposed by David Marr and Tomaso Poggio (1976). The vertical strips are the interleaved left- and right-eye contributions of Fig. 11-2.

In the center of Fig. 11-22 is a column of "L minus R" neurons, with dendritic branches extending to the left and right as shown. Each L and R signal corresponds to a tiny portion of the visual image, but a good strategy in understanding how the scheme operates is to use numerical values to

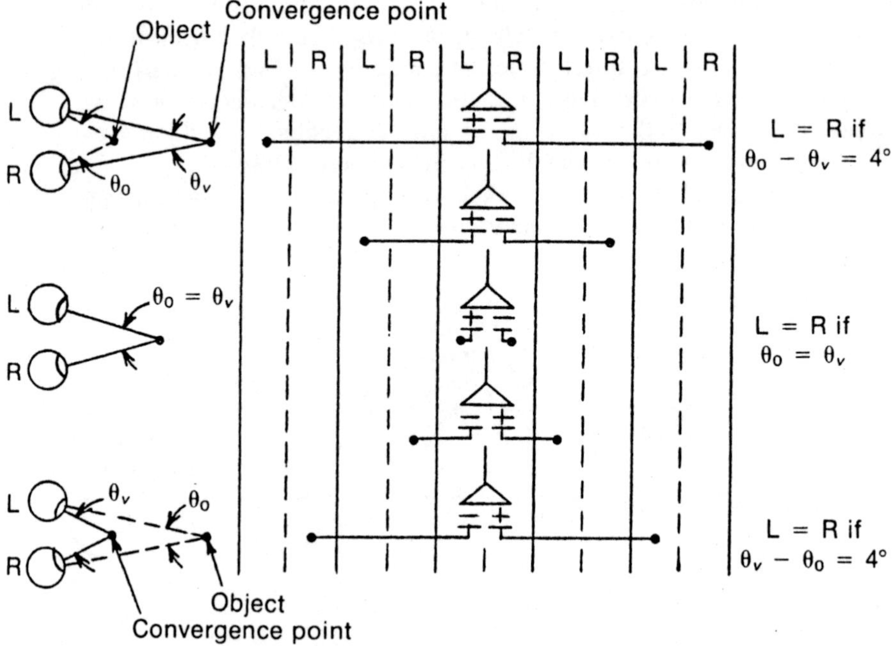

Fig. 11-22 A possible mechanism for automatic binocular convergence. The vertical strips are the interleaved left- and right-eye contributions of Fig. 11-2. The column of neurons in the center performs the operation L minus R. On the left, θ_0 is the angle formed by the eyes (dashed lines) and a visual object, while θ_v is the angle between the optical axes (solid lines). On the right, the numerical values correspond to strips that are 1° (visual angle) apart.

represent the L and R signals; that is, suppose that the visual image consists of the integers 0, 1, 2, . . . , 8. (It may be confusing to use numerical values rather than, say, the letters of the alphabet, but numerical values will lead to numerical answers, as shown below.) Also, suppose that the strips are 1° (visual angle) apart.

If vergence is correct (that is, if the eyes are looking in the correct directions), each eye sees the central group 2, 3, 4, 5, 6, so that the strip values are

```
L R   L R   L R   L R   L R
2 2   3 3   4 4   5 5   6 6
```

and the neuron at $\theta_0 = \theta_v$ generates $L - R = 0$ (no change is needed).

Now suppose that vergence is incorrect, with the L and R eyes shifted 2° to the left and right, respectively, from where they should be, so that θ_v is 4° smaller than θ_0. The optical conditions are illustrated (with angles highly exaggerated) by the sketch at the upper left of Fig. 11-22. Now the left eye sees 4, 5, 6, 7, 8 while the right eye sees 0, 1, 2, 3, 4. Interleaved, we have

```
L R   L R   L R   L R   L R
4 0   5 1   6 2   7 3   8 4 .
```

The neuron at $\theta_v - \theta_0 = 4°$ (bottom of Fig. 11-22) yields $L - R = 8$ as follows:

```
L R   L R   L R   L R   L R
4 0   5 1   6 2   7 3   8 4 .
  |<--------- −   +--------->|
```

On the other hand, the neuron at $\theta_0 - \theta_v = 4°$ (top of Fig. 11-22) has dendritic inputs that are spaced just right to yield $L - R = 0$:

```
L R   L R   L R   L R   L R
4 0   5 1   6 2   7 3   8 4 .
|<----------- +   −----------->|
```

This removes inhibition from a control neuron (not shown) that in turn stimulates the eye muscles so as to *increase* θ_v.

Next, suppose that the L and R images are each shifted 2° in the other direction from where they should be, so that θ_0 is 4° smaller than θ_v. The optical conditions are illustrated by the sketch at the lower left of Fig. 11-22. Now the left eye sees 0, 1, 2, 3, 4, while the right eye sees 4, 5, 6, 7, 8. Interleaved, we have

```
L R   L R   L R   L R   L R
0 4   1 5   2 6   3 7   4 8 .
```

Now the neuron at $\theta_0 - \theta_v = 4°$ yields $L - R = -8$:

```
L R   L R   L R   L R   L R
0 4   1 5   2 6   3 7   4 8 .
|<----------- +   −----------->|
```

But for the neuron at $\theta_v - \theta_0 = 4°$ we get L − R = 0:

$$\begin{array}{cccccccccc} L & R & L & R & L & R & L & R & L & R \\ 0 & 4 & 1 & 5 & 2 & 6 & 3 & 7 & 4 & 8 \end{array}.$$

$$|\longleftarrow\!\!-\!\!-\quad - \quad + \!\!-\!\!-\longrightarrow|$$

This removes inhibition from a control neuron, which stimulates the eye muscles so as to *decrease* θ_v.

The net stimulus to the eye muscles is a function of a summation taken from thousands of L and R strips [T. Poggio, 1984; D. Regan, K. Beverley & M. Cynader, 1979; J. Ross, 1976].

REFERENCES

D. L. Alkon, Learning in a marine snail, *Sci. Am.*, vol. 249, pp. 70–84, July 1983.

S. I. Amari, Field theory of self-organizing neural nets, *IEEE Trans. Syst., Man, Cybern.*, vol. SMC-13, pp. 741–748, Sept./Oct. 1983.

J. R. Anderson and G. H. Bower, *Human Associative Memory*. Washington, D.C.: Winston, 1973.

C. Aoki and P. Siekevitz, Plasticity in brain development, *Sci. Am.*, vol. 259, pp. 56–64, Dec. 1988.

M. A. Arbib, *The Metaphorical Brain*. New York: Wiley, 1972.

A. T. Bahill and L. Stark, The trajectories of saccadic eye movements, *Sci. Am.*, vol. 240, pp. 108–117, Jan. 1979.

G. S. Brindley and W. S. Lewin, The sensations produced by electrical stimulation of the visual cortex, *J. Physiol.*, vol. 196, pp. 479–493, 1968.

S. E. Brodie, B. W. Knight, and F. Ratliff, The responses of the Limulus retina to moving stimuli: Prediction by Fourier synthesis. The spatio-temporal transfer function of the Limulus lateral eye, *J. Gen. Physiol.*, vol. 72, pp. 129–202, 1978.

M. Constantine-Paton and M. I. Law, The development of maps and stripes in the brain, *Sci. Am.*, vol. 247, pp. 62–70, Dec. 1982.

L. N. Cooper, Distributed memory in the central nervous system: Possible test of assumptions in visual cortex, in *The Organization of the Cerebral Cortex*, eds. F. O. Schmitt et al. Cambridge, Mass.: M.I.T. Press, 1981.

J. G. Daugman, Entropy reduction and decorrelation in visual coding by oriented neural receptive fields, *IEEE Trans. Biomed. Eng.*, vol. 36, pp. 107–114, Jan. 1989.

S. Deutsch, A simplified version of Kunihiko Fukushima's neocognitron, *Biol. Cybern.*, vol. 42, pp. 17–21, 1981.

R. O. Duda and P. E. Hart, *Pattern Classification and Scene Analysis*. New York: Wiley, 1973.

R. Eckmiller, Neural control of foveal pursuit versus saccadic eye movements in primates—Single-unit data models, *IEEE Trans. Syst., Man, Cybern.*, vol. SMC-13, pp. 980–989, Sept./Oct. 1983.

G. M. Edelman, *Neural Darwinism*. New York: Basic Books, 1987.

J. D. Enderle and J. W. Wolfe, Time-optimal control of saccadic eye movements, *IEEE Trans. Biomed. Eng.*, vol. BME-34, pp. 43–55, Jan. 1987.

D. H. Fender, Control mechanisms of the eye, *Sci. Am.*, vol. 211, pp. 24–33, July 1964.

W. J. Freeman, The physiology of perception, *Sci. Am.*, vol. 264, pp. 78–85, Feb. 1991.

J. P. Frisby, *Seeing—Illusion, Brain and Mind.* Oxford: Oxford Univ. Press, 1980.

K. Fukushima, Neocognitron: A self-organizing neural network model for a mechanism of pattern recognition unaffected by shift in position, *Biol. Cybern.*, vol. 36, pp. 193–202, 1980.

K. Fukushima, Cognitron: A self-organizing multilayered neural network model, *NHK (Japan Broadcasting Corp.) Tech. Monograph No. 30*, pp. 6, 7, Jan. 1981.

S. Grossberg, Nonlinear neural networks: Principles, mechanisms, and architectures, *Neural Networks*, vol. 1, pp. 17–61, 1988.

S. Grossberg, E. Mingolla, and D. Todorovic, A neural network architecture for preattentive vision, *IEEE Trans. Biomed. Eng.*, vol. 36, pp. 65–84, Jan. 1989.

A. C. Guyton, *Textbook of Medical Physiology.* Philadelphia: Saunders, 1986.

D. O. Hebb, *The Organization of Behavior.* New York: Wiley, 1949.

R. Hecht-Nielsen, *Neurocomputing.* Reading, Mass.: Addison-Wesley, 1990.

G. E. Hinton and J. A. Anderson, eds., *Parallel Models of Associative Memory.* Hillsdale, N.J.: Erlbaum, 1981.

J. J. Hopfield and D. W. Tank, Computing with neural circuits: A model, *Science*, vol. 233, pp. 625–633, Aug. 1986.

D. H. Hubel and T. N. Wiesel, Receptive fields of single neurones in the cat's striate cortex, *J. Physiol.* (*London*), vol. 148, pp. 574–591, 1959.

D. H. Hubel and T. N. Wiesel, Brain mechanisms of vision, in *The Brain.* San Francisco: Freeman, pp. 84–96, 1979.

G. K. Hung, J. L. Semmlow, and K. J. Ciuffreda, The near response: Modeling, instrumentation, and clinical applications, *IEEE Trans. Biomed. Eng.*, vol. BME-31, pp. 910–919, Dec. 1984.

G. K. Hung, J. L. Semmlow, and K. J. Ciuffreda, A dual-mode dynamic model of the vergence eye movement system, *IEEE Trans. Biomed. Eng.*, vol. BME-33, pp. 1021–1028, Nov. 1986.

P. Inchingolo and M. Spanio, On the identification and analysis of saccadic eye movements—A quantitative study of the processing procedures, *IEEE Trans. Biomed. Eng.*, vol. BME-32, pp. 683–695, Sept. 1985.

M. Kabrinsky, *Proposed Model for Visual Information Processing in the Human Brain.* Champaign, Ill.: Univ. Illinois Press, 1966.

T. Kohonen, *Associative Memory: A System-Theoretical Approach.* New York: Springer-Verlag, 1978.

T. Kohonen, *Self-Organization and Associative Memory*, 2d ed. Heidelberg: Springer-Verlag, 1988.

W. Krenz, M. Robin, S. Barez, and L. Stark, Neurophysiological model of the normal and abnormal human pupil, *IEEE Trans. Biomed. Eng.*, vol. BME-32, pp. 817–825, Oct. 1985.

J. J. Kulikowski, S. Marcelja, and P. O. Bishop, Theory of spatial position and spatial frequency relations in the receptive fields of simple cells in the visual cortex, *Biol. Cybern.*, vol. 43, pp. 187–198, 1982.

C. G. Y. Lau and B. Widrow, eds., Special issues on neural networks, *Proc. IEEE*, vol. 78, Sept. and Oct. 1990.

B. B. Lee, K. Albus, P. Haselund, M. J. Hulme, and O. Creutzfeldt, The depth distribution of optimal stimulus orientations for neurons in cat area 17, *Exp. Brain Res.*, vol. 27, pp. 301–314, 1977.

N. Link and L. Stark, Latency of the pupillary response, *IEEE Trans. Biomed. Eng.*, vol. 35, pp. 214–218, March 1988.

D. Marr and T. Poggio, Co-operative computation of stereo disparity, *Science*, vol. 194, pp. 283–287, 1976.

C. R. Michael, Columnar organization of color cells in monkey's striate cortex, *J. Neurophysiol.*, vol. 46, pp. 587–604, 1981.

M. Minsky and S. Papert, *Perceptrons*. Cambridge, Mass.: M.I.T. Press, 1969.

V. B. Mountcastle, Modality and topographic properties of single neurons of cat's somatic sensory cortex, *J. Neurophysiol.*, vol. 20, pp. 408–434, 1957.

E. M-Tzanakou, Neural aspects of vision and related technological advances, *Proc. IEEE*, vol. 76, pp. 1130–1142, Sept. 1988.

N. J. Nilsson, *Learning Machines*. New York: McGraw-Hill, 1965.

B. R. Payne, N. Berman, and E. H. Murphy, Organization of direction preferences in cat visual cortex, *Brain Res.*, vol. 211, pp. 445–450, 1981.

T. Poggio, Vision by man and machine, *Sci. Am.*, vol. 250, pp. 106–116, Apr. 1984.

M. Porat and Y. Y. Zeevi, Localized texture processing in vision: Analysis and synthesis in the Gaborian space, *IEEE Trans. Biomed. Eng.*, vol. 36, pp. 115–129, Jan. 1989.

T. P. S. Powell and V. B. Mountcastle, Some aspects of the functional organization of the cortex of the postcentral gyrus of the monkey: A correlation of findings obtained in a single unit analysis with cytoarchitecture, *Johns Hopkins Bull.*, vol. 105, pp. 133–162, 1959.

F. Ratliff, *Mach Bands: Quantitative Studies on Neural Networks in the Retina*. San Francisco: Holden-Day, 1965.

F. Ratliff and L. A. Riggs, Involuntary motions of the eye during monocular fixation, *J. Exp. Psychol.*, vol. 40, pp. 687–701, 1950.

D. Regan, K. Beverley, and M. Cynader, The visual perception of motion in depth, *Sci. Am.*, vol. 241, pp. 122–133, July 1979.

F. Rosenblatt, *Principles of Neurodynamics*. Washington, D.C.: Spartan, 1962.

A. Rosenfeld, Computer vision: A source of models for biological visual processes? *IEEE Trans. Biomed. Eng.*, vol. 36, pp. 93–96, Jan. 1989.

J. Ross, The resources of binocular perception, *Sci. Am.*, vol. 234, pp. 80–86, March 1976.

D. E. Rumelhart, J. L. McClelland, and PDP Research Group, *Parallel Distributed Processing*. Cambridge, Mass.: M.I.T. Press, 1986.

E. Sánchez-Sinencio, and C. G. Lau, eds., *Artificial Neural Networks: Paradigms, Applications, and Hardware Implementations*. Piscataway, N.J.: IEEE Press, 1992.

F. O. Schmitt, F. G. Worden, G. Adelman, and S. G. Dennis, eds., *The Organization of the Cerebral Cortex*. Cambridge, Mass.: M.I.T. Press, 1981.

J. Szentagothai, Synaptology of the visual cortex, in *Handbook of Sensory Physiology*, ed. R. Jung. New York: Springer-Verlag, pp. 269–324, 1973.

D. W. Tank and J. J. Hopfield, Collective computation in neuronlike circuits, *Sci. Am.*, vol. 257, pp. 104–114, Dec. 1987.

I. D. Thompson and D. J. Tolhurst, Columnar organization for optimal spatial frequency in cat striate cortex, *J. Physiol. (London)*, vol. 319, p. 79, 1981.

A. Treisman, Features and objects in visual processing, *Sci. Am.*, vol. 255, pp. 114–125, Nov. 1986.

F. J. Verheijen, A simple after-image method demonstrating the involuntary multi-directional eye movements during fixation, *Opt. Acta*, vol. 8, pp. 309–311, 1961.

A. B. Watson and A. J. Ahumada, Jr., A hexagonal orthogonal-oriented pyramid as a model of image representation in visual cortex, *IEEE Trans. Biomed. Eng.*, vol. 36, pp. 97–106, Jan. 1989.

D. Wang and M. A. Arbib, Complex temporal sequence learning based on short-term memory, *Proc. IEEE*, vol. 78, pp. 1536–1543, Sept. 1990.

J. M. Wolfe, Hidden visual processes, *Sci. Am.*, vol. 248, pp. 72–85, Feb. 1983.

D. E. Wooldridge, *Sensory Processing in the Brain*. New York: Wiley, 1979.

U. Yinon and E. Auerbach, Receptive fields and firing patterns of sustained cells in the cat visual cortex, *Vision Res.*, vol. 15, pp. 1245–1250, 1975.

Problems

1. Make three photocopies of Fig. 11-4. Use these to draw the circuits and calculate values for the following inputs:
   (a) + + + + + + − − − − [the "60% white" column of Table 11-1];
   (b) − − − + + + + + + − [the fourth input column of Table 11-2(*a*)];
   (c) − − − + + + − − − − [the "30% white" column of Table 11-2(*b*)]. In each case, check your answer versus that of Tables 11-1 and 11-2.
2. Figure 11-12(*b*) depicts the topological representation of an $S_1C_1S_2C_2S_3C_3$ model in which there is 2-to-1 convergence between every S layer and its companion C layer. (a) Draw the topological layout if there is 3-to-1 convergence between every S layer and its C layer. (b) Given that the C_3 layer can recognize 64 different 8-element patterns, construct a table showing the number of patterns, terminals, and neurons versus G, S, and C layers. [Ans.: (b)

	G	S_1	C_1	S_2	C_2	S_3	C_3
Number of patterns			4		16		64
Number of terminals	40	39	13	12	4	3	1
Number of neurons	40	156	52	192	64	192	64.]

3. In Eqs. (11-2) and (11-3) it is assumed that $B = 0$. Suppose, however, that $B \neq 0$. As before, given that $I = 1.5$, $C = 2$ mV, and $E = 3$, find the values of A and B that satisfy Eq. (11-1). (These are not viable, however, because positive answers are unstable, and negative answers are not allowed.) [Ans.: (mV units): $A = 3.2$, $B = 0.1111$; $A = B = 0.5744$; $A = B = -1.741$.]
4. Given that the neural network of Fig. 11-14 is driven by the C_1 layer of Fig. 11-5: (a) Make up a work sheet using algorithms A1 to A8, similar to Table 11-3, for 14 events during the "learning" phase if the "% white" zigzags as follows: 30, 40, 50, 60, 70, 60, 50, 40, 30, 20, 10, 0, 10, and 20. (b) Using algorithms A9 to A11 and associated reasoning, find the patterns claimed by each neuron during the "organizing" phase. [Ans: (a)

% white =	30	40	50	60	70	60	50	40	30	20	10	0	10	20
Dominant neuron =	N1	N1	N2	N2	N2	N2	N3	N1	N1	N1	N4	N4	N4	N5

(b) N1 = + + − −, N2 = + + + +, N3 = + + + −, N4 = − − − −, N5 = + − − −.]

5. In the "organizing" phase of Section 11-4, various arguments are given with regard to which pattern should be claimed by each neuron. Express these arguments in the form of algorithms.
6. Given the six-neuron associative memory of Eq. (11-5) and its W table: (a) construct a table similar to Table 11-5 showing V_{in} total if the robot meets person B. (b) Repeat if the robot meets person C. [Ans.: (a) V_{in} total = 3, −3, 3, 3, −3, 3. (b) V_{in} total = 5, 5, −1, 1, −5, −5.
7. Given the six-neuron associative memory of Eq. (11-5) and its W table, the robot only gets partial data. In each case, set up a work sheet similar to Table 11-6, and derive the memory neuron signature: (a) the robot's visual cortex turns N1 and N4 ON, N2 OFF; (b) it turns N1, N2, and N4 ON; (c) it turns N1 and N2 ON; (d) it turns N1 ON and N2 OFF. [Ans.: (a) + − + + − + (B). (b) + + − + − − (C). (c) + + 00 − − (A or C). (d) + − + + − + (B).]
8. Plot the sigmoid curve $y = [(\arctan Gx)/\pi] + 0.5$ for $G = 20$.
9. In a book-shelving problem, the rates per minute are found to be:

	Physics	Chemistry	History
Jessica	1	2	3
George	8	9	4
Tim	7	6	5

After dividing the matrix by 10, apply Eq. (11-15) with $I = 1$, Eq. (11-18) with $G = 10$, and Eq. (11-20) with R small enough to avoid a bistable pitfall. Find the final $V_{out} = V_{new}$ and V_{inh} matrices to which the calculation converges, and the indicated task assignment. [Ans.:

$$V_{out} = V_{new} = \begin{vmatrix} 0.0187 & 0.0198 & 0.8964 \\ 0.0283 & 0.9646 & 0.0222 \\ 0.9545 & 0.0239 & 0.0245 \end{vmatrix},$$

$$V_{inh} = \begin{vmatrix} 0.9005 & 0.9046 & 0.0018 \\ 0.9604 & 0.0022 & 0.9144 \\ 0.0023 & 0.9637 & 0.8989 \end{vmatrix}, \begin{vmatrix} \cdot & \cdot & 3 \\ \cdot & 9 & \cdot \\ 7 & \cdot & \cdot \end{vmatrix}.$$

10. In a book-shelving problem, the rates per minute are found to be [D. W. Tank and J. J. Hopfield, 1987]:

	Geology	Physics	Chemistry	History	Poetry	Art
Sarah	10	5	4	6	5	1
Jessica	6	4	9	7	3	2
George	1	8	3	6	4	6
Karen	5	3	7	2	1	4
Sam	3	2	5	6	8	7
Tim	7	6	4	1	3	2

(a) After dividing the matrix by 10, apply Eq. (11-16), with $I = 1$, to get the initial V_{inh} matrix. (b) Assume that the smallest V_{inh} entry is a partial solution to the task-assignment problem. Delete its row and column, and repeat until the total solution is obtained. Compare with the optimum solution as given in the Tank and Hopfield article. [Ans.:

(a)

$$V_{inh} = \begin{vmatrix} 4.62 & 5.98 & 7.56 & 5.50 & 4.94 & 6.30 \\ 6.50 & 6.48 & 5.06 & 5.04 & 5.88 & 5.80 \\ 8.37 & 4.00 & 7.25 & 4.84 & 4.80 & 3.52 \\ 4.59 & 4.75 & 3.75 & 5.20 & 4.83 & 3.24 \\ 8.12 & 7.54 & 7.02 & 5.50 & 3.68 & 3.60 \\ 4.00 & 3.74 & 5.32 & 5.94 & 4.20 & 4.20 \end{vmatrix}$$

(b)

$$\begin{vmatrix} 10 & \cdot & \cdot & \cdot & \cdot & \cdot \\ \cdot & \cdot & 9 & \cdot & \cdot & \cdot \\ \cdot & \cdot & \cdot & 6 & \cdot & \cdot \\ \cdot & \cdot & \cdot & \cdot & \cdot & 4 \\ \cdot & \cdot & \cdot & \cdot & 8 & \cdot \\ \cdot & 6 & \cdot & \cdot & \cdot & \cdot \end{vmatrix} \qquad \text{Optimum solution} = \begin{vmatrix} 10 & \cdot & \cdot & \cdot & \cdot & \cdot \\ \cdot & \cdot & \cdot & 7 & \cdot & \cdot \\ \cdot & \cdot & \cdot & \cdot & \cdot & 6 \\ \cdot & \cdot & 7 & \cdot & \cdot & \cdot \\ \cdot & \cdot & \cdot & \cdot & 8 & \cdot \\ \cdot & 6 & \cdot & \cdot & \cdot & \cdot \end{vmatrix}.]$$

Total/min = 43 Total/min = 44

11. For saccadic motion: (a) draw a block diagram of the control system. Include corrective cancellation before the image reaches the visual cortex. (b) Derive a neural network that is capable of lateral shift to accomplish corrective cancellation.
12. Draw a block diagram of the control system for distance accommodation assuming that Fig. 11-21 is correct.

12

About the Brain

ABSTRACT. A typical human brain is 17 cm from front to back and contains between 50 and 100 compartments. It is topologically organized.

The cerebellum is involved with the coordination of simultaneous muscle movements. It cannot initiate any activity by itself, but acts as a helper to the motor cortex. There are five different types of neurons in the cerebellar cortex: Purkinje, granule, Golgi, basket, and stellate cells. They are arranged with stereotyped precision. Granule cell axons are called parallel fibers; their junctions with Purkinje cell dendrites may be modifiable as the cerebellum learns a particular skill.

Brain signals (EEGs) can be picked up by electrodes making contact with the scalp. Only activity in the immediate vicinity of the electrode is detected. In an awake person the EEG is due to slowly propagating graded potentials. If the person's eyes are closed, *alpha* oscillations are picked up, especially from an electrode at the back of the head. It is conjectured that the alpha rhythm is due to a visual feedback loop that goes into oscillations because gain is increased. During deep sleep the EEG displays slow waves of large amplitude, probably as neurons are rejuvenated.

Any change in the brain that represents memory is called an *engram*. The modern view is that the basis for memory is synaptic junction modification, and that engrams are stored in the same part of the brain that recognizes that particular type of pattern.

A block diagram attempt to summarize the mechanisms of memory is presented. As far as time is concerned, there is extremely-short-term, short-term, and long-term memory. There are two categories of stored information: procedural and declarative. Damage to certain central structures in the brain can result in *anterograde* or *retrograde* declarative amnesia, or various degrees of both.

The block diagram of a hypothetical simple brain is presented. A "central station" box is the "seat of consciousness." Each sensory receptor leads to a "feature extraction" box. Important patterns are stored in "pattern memory stacks," with association fibers running between the various stacks. A "thought pattern memory stack" circulates association fiber signals over and over again. Because of a random noise generator, the robot is somewhat unpredictable, and sometimes comes up with a new idea.

The determinism versus free-will controversy is reviewed. Laplacian determinism should be modified because it is impossible for us to predict the future of a single neuron, let alone 10^{12} of them. Besides, prediction requires precise measurements that violate quantum mechanics.

Consciousness is defined as an awareness of one's own existence, sensations, thoughts, and environment. It is a mysterious attribute of the brain that is undoubtedly shared by all mammals. Since honey bees are capable of a great deal of learning, perhaps they (and insects in general) also have an awareness of being.

12-1 Some Anatomy of the Brain

A discussion about the brain is a peculiar business. You, the reader, must use your brain to understand what is written about itself. The subject has been a favorite of philosophers since prehistory, undoubtedly since the human brain developed to the point where a person could ask "Why am I here, why are these animals and plants and boulders here, why does the sun move through the sky?" In this chapter, we consider some of the anatomy and electrical signals, and a crude explanation, full of conjecture, about how the brain functions [F. H. C. Crick, 1979; D. H. Hubel, 1979; W. J. H. Nauta and M. Feirtag, 1979; D. Regan, 1979].

With regard to anatomy, we will concentrate on Fig. 12-1. Since the brain is three-dimensional, some compromises have been made in this two-dimensional presentation; we have chosen to show an approximate cross section, with some regions in dashed lines for the sake of clarity. We have labeled 17 of the most important regions, plus the spinal cord leading into the base of the brain; but an anatomy book will show that names (mostly of Latin origin) have been assigned to many additional regions of the brain.

As indicated, a typical human brain is 17 cm (6.7 in.) from front to back. There are large variations in size, of course, but within the human species there is little correlation between brain size and intelligence. Species of large animals, on the other hand, tend to have relatively large brains. A chimpanzee is approximately the same size as a human, but its brain has one-fourth the volume of a human brain. (The front-to-back dimension is given by $17/(4^{1/3}) = 11$ cm.) In this case, there *is* a correlation between brain size and intelligence. In general, there is a strong correlation between the number

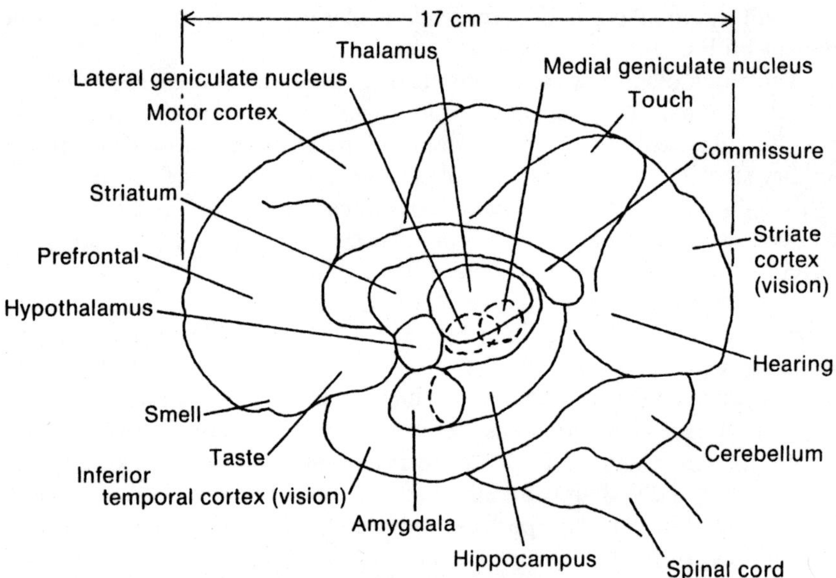

Fig. 12-1 Approximate cross section of the human brain, with some regions in dashed lines for the sake of clarity. Seventeen of the most important regions are labeled.

of neurons in a particular region of the brain and the amount of information processed by, or stored in, that tissue.

The years have seen an accumulation of case histories of patients who have suffered brain damage because of disease, accidental injury, or tumors and associated surgical intervention. These histories show that the brain is compartmentalized. It is an electrochemical processor of signals, with compartment A feeding to compartment B, or perhaps to B in parallel with C. There seem to be between 50 and 100 compartments. Unfortunately for those who dare to decipher the brain's inner workings, in most instances there is no sharp dividing line between one compartment and the next. The neurons in one compartment imperceptibly give way to similar neurons in an adjacent compartment. Furthermore, axons do not run in robust sheathed cables like telephone wires; the tissue is like gelatin, and falls apart if one tries to trace the "cables." Brain research is painstaking, frustrating, and expensive. There is no doubt, nevertheless, that some day all of the bits and pieces will fall into place, and the mysteries of the brain will be explained just as the DNA molecule has explained, to a great extent, the mysteries of individual variability.

Dysfunction associated with brain damage frequently shows what processing takes place and where. Regions tied to the sensory receptors of taste, smell, touch, hearing, and vision are labeled in Fig. 12-1. In accordance with its importance, a relatively large volume is devoted to vision. The primary visual area is the striate cortex at the rear of the brain. From here the signal processing hierarchy seems to proceed downward and forward, terminating in the inferior temporal cortex (*inferior* here means that it is situated under or beneath; *temporal* refers to the sides and base of the skull; and the *cortex* is the outer layer of an organ or part).

A relatively large area is devoted to the motor cortex. It is from here (but not exclusively) that action potentials (APs) that are directed to appropriate motoneurons originate.

The prefrontal cortex remains somewhat mysterious. It is the region where "thought is elaborated," to quote a frequent characterization by psychologists.

One of the most important clues in solving the mysteries of the brain is that it is topologically organized; that is, there is a continuous, gradual progression within a given sensory modality. In the hearing region, there is a continuous, gradual transition from low to high frequencies (tones) that is called *tonotopic* organization. In the striate cortex, there is a smooth radial transition from the foveal projections outward. The tongue is mapped in the area of taste. In the area of touch, one can smoothly and continuously map the entire outer surface of the body from head to toes. Similarly, in the motor cortex, one can smoothly and continuously map the skeletal muscles as one travels from head to toes. All of these maps are "distorted" because the more important sensory and motor regions, such as the fingers and lips, are represented by correspondingly large numbers of brain neurons, disproportionate to the surface area of the fingers and lips.

Centrally located in the brain is the thalamus. It, too, is compartmentalized because fibers from each sensory modality first travel to the thalamus before going on to their respective cortical areas. Two of the thalamic regions are shown as dotted ovals: the lateral geniculate nuclei, which receive axons from the optic nerves before sending fibers on to the striate cortex for further processing; and the medial geniculate nuclei, which indirectly receive axons from the cochlear nerves before sending fibers on to the auditory cortex for further processing.

The hypothalamus receives fibers from the thalamus and evidently compares its signals with built-in patterns that are instinctive in the sense that they are determined genetically rather than environmentally. The hypothalamus "stimulates" the glands; it controls emotional responses and the activities of the autonomic nervous system (pupils of the eyes, sweat glands, heart muscle, breathing, digestion, kidney output, sexual response, body temperature, and many others).

The cerebellum is topologically organized in accordance with the skeletal muscle system. Its neurons form part of feedback loops that stabilize muscle contraction throughout the body. The cerebellum is considered in detail in the next section.

The functions of the amygdala, hippocampus, and striatum are imprecisely known. In a subsequent model, these three are lumped, along with the thalamus, into a box called the "central station" (CS). The CS controls learning and feeds the hypothalamus. It is the "seat of consciousness."

The commissure is part of the tract of fibers passing between the two hemispheres of the brain. Sometimes, in patients desperately ill with recurrent epileptic seizures, the commissure is cut surgically in order to open the feedback path that is responsible for the oscillations. This is known as "split brain" surgery. Actually, it is not possible to completely split the brain because one would have to cut through vital neurons in the central structures. Aside from benefit to the patient, the split brain is interesting because it is possible for an experimenter to feed conflicting information into each hemisphere. Normally, each hemisphere knows what information the other is receiving because of the commissure fibers (although there is a great deal of duplication, the two hemispheres are far from identical). In the case of a split-brain patient who is receiving conflicting signals, the dominant hemisphere "decides" that its information is the true state of affairs, and it suppresses the recessive hemisphere. (If a person is right-handed, the left hemisphere is dominant in most of the dual-choice situations, and vice versa.)

12-2 The Cerebellum

A cross section of the brain shows very little. In order to highlight the neuronal circuits, the material has to be stained, somewhat reminiscent of the way a photographic negative has to be developed. With staining, one can see bits and pieces of dendrites, synaptic junctions, neuron bodies, axons, and glial (support) cells, since microscopes can only examine a thin slice taken through a three-dimensional object. The business of actually tracing the entire structure of a neuron is a long, tedious, painstaking, and expensive procedure. A dendrite, which is relatively short (Fig. 5-1 is based on a spinal motoneuron whose dendrite is 1.35 mm long), can have thousands of junctions. Usually, only a few of the axons from which the junctions arise can be traced out. Axons can be relatively long, so that in only a few cases has it been possible to trace out the complete path. Even if it is possible to draw part of a neuronal circuit, one does not know without additional work which junctions are excitatory or inhibitory, let alone any quantity that resembles the junction weighting factors. In other words, decipherment of the networks of the brain is in its infancy.

There is one relatively small volume of the brain that is an important exception—the cortex (outer layer) of the cerebellum at the end of the spinal cord in Fig. 12-1. The cerebellum is involved with the coordination of simultaneous muscle movements. Here, there are only five different types of neuron (recall that the retina, also, has five different types), and the circuits are arranged with such stereotyped precision that neuroanatomists have succeeded in uncovering the basic network. Work with experimental animals has disclosed, furthermore, which junctions are excitatory (E) and which are inhibitory (I) [A. Brodal, 1954; J. C. Eccles, M. Ito & J. Szentagothai, 1967; M. Hassul and P. D. Daniels, 1977; R. R. Llinas, 1975; F. Walberg, 1954].

A three-dimensional view—three mutually perpendicular cuts—is depicted in Fig. 12-2. At first glance this seems to be a "hopeless mess," but with a reasonable amount of patience (and a magnifying glass, perhaps) one can extract the main features of the cerebellar cortex. This extraction may be assisted by a drawing of the basic network, in Fig. 12-3, which reveals a system of medium complexity. In what follows, Figs. 12-2 and 12-3 are discussed together.

The most prominent feature of the cerebellar cortex is the Purkinje cell, named after Johannes E. Purkinje (1787–1869), who described it in 1837. Its dendrites form an extensively bifurcating fan, all of whose branches lie *in one plane* (that of the front cut in Fig. 12-2), similar to the veins of a pressed leaf. The Purkinje cell is unique in that its dendrites form around 100,000 synapses, far more than any other cell in the nervous system.

The Purkinje cell has two E and two I inputs. The *climbing fiber E* input

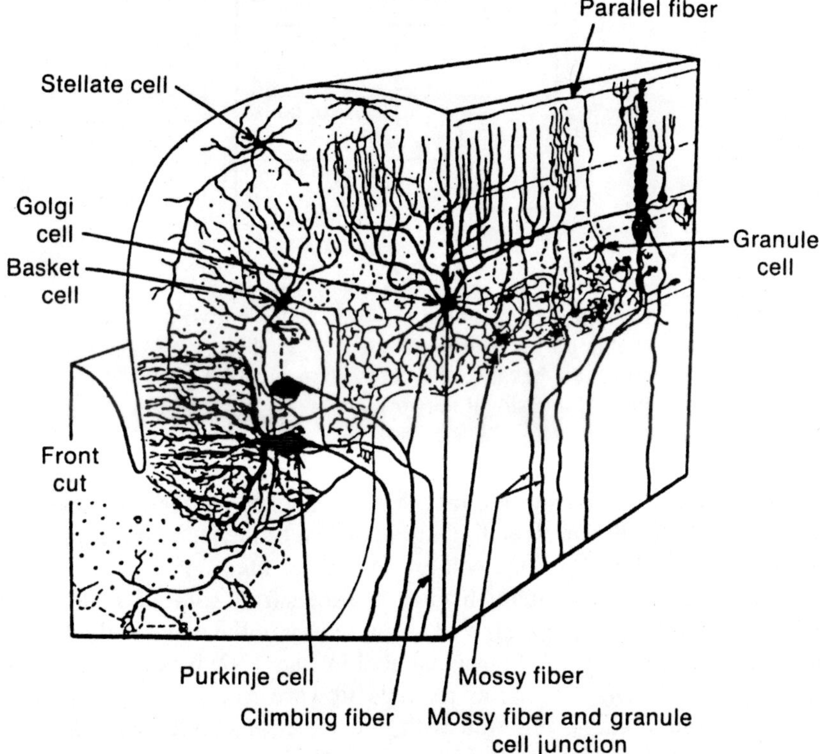

Fig. 12-2 A three-dimensional view of the cerebellar cortex as provided by three mutually perpendicular cuts. Parallel to the front cut we have dendrites of the Purkinje, basket, and stellate cells. The following are mutually perpendicular: (1) Purkinje cell dendrites; (2) parallel fibers; and (3) basket and stellate cell axons. The Golgi cell dendrites have a cylindrical architecture, like a conventional tree. (From Marr, A theory of cerebellar cortex, *J. Physiol.*, 1969.)

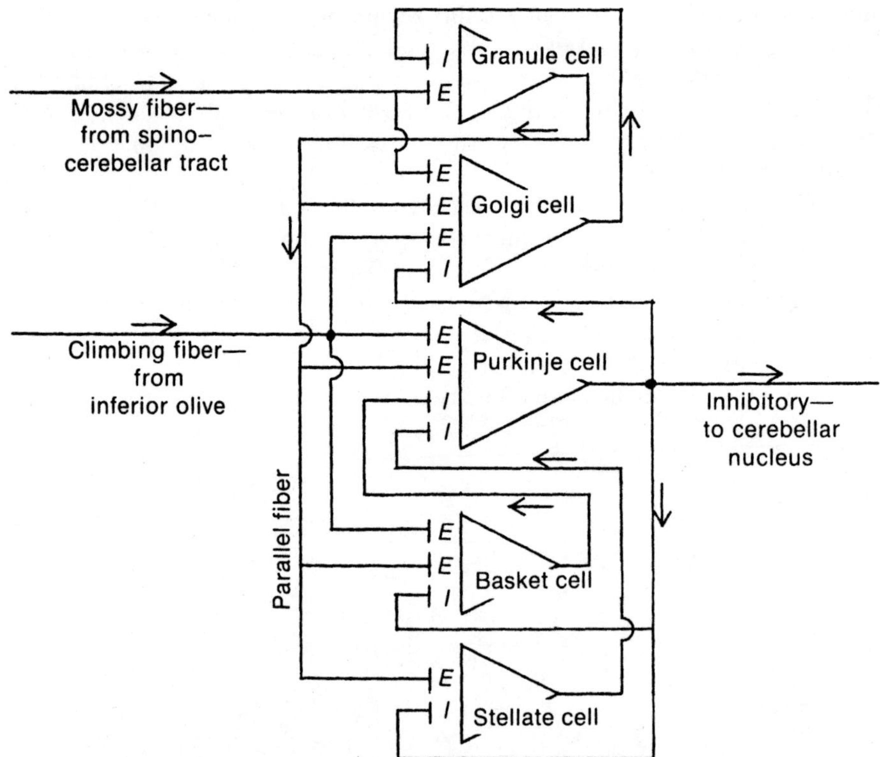

Fig. 12-3 Hypothetical model of the basic neural network of the cerebellar cortex. It is useless to analyze this circuit as shown, however, because the cerebellum must be analyzed as a three-dimensional network.

comes from the *inferior olive*, a region in the upper end of the spinal cord. The fiber gets its name from the fact that it branches and climbs around the Purkinje dendrites like a vine climbs around the branches of a tree. A single climbing fiber and its Purkinje cell are mated for life.

The other *E* input is equally bizarre. It comes from some 100,000 *parallel fibers*, which are directed *precisely at right angles* to the plane of the Purkinje dendrites. A single parallel fiber is labeled in Fig. 12-2; it is the axon of the labeled *granule cell.* The axon reaches upward and then bifurcates, as shown, forming a single junction with every Purkinje cell that it meets on each side of the bifurcation. David Marr (1969), who analyzed the system, proposed that these parallel fiber–Purkinje cell dendrite junctions are modifiable; that is, they have the weighting factors that change as the system learns.

The many black dots that appear in the front cut of Fig. 12-2 are *not* defects introduced by a photocopy machine; they are the cross sections of parallel fibers that, because of the right-angle orientation, show up as dots.

The Purkinje cell *I* inputs come from the axons of *basket* and *stellate* cells. These two neurons are similar, but are located at different depths in the cortex. They are more like the conventional neuron of Fig. 5-1, with a medium-sized dendritic tree and a single output axon that displays a relatively small number of branches. Although it may be difficult to see in Fig. 12-2, the dendrites of basket and stellate cells are flattened, and lie parallel to the plane of the Purkinje cell dendrites. In addition, the axons of basket and stellate cells are at right angles to the Purkinje cell dendrites and also to the parallel fibers. In the front cut of Fig. 12-2, Purkinje dendrites run left and right, parallel fibers run into and out of the page, while basket and stellate axons run up and down. The fact that these three axes are mutually perpendicular is vitally important in any explanation of cerebellar function.

The axons leaving the Purkinje cells constitute the *only* output of the cerebellar cortex. They go to the *cerebellar nuclei*, interneurons deep inside the cerebellum.

The source of the parallel fibers, the granule cell, has an unusual *E* dendritic input structure. As can be seen in Fig. 12-2, around four dendrite branches are connected to the cell body. A *single* synaptic junction appears at the end of each branch in the form of a clawlike appendage (which is therefore called a *claw*). Several claws, from different granule cells, grab at a roselike formation (called a *rosette*) which is one of the axon terminations of the *mossy fiber*. The latter originates in the spino–cerebellar tract (i.e., the upper end of the spinal cord). One of the mossy fiber–granule cell junctions is labeled in Fig. 12-2. It is obvious that, following mossy fiber stimulation to each claw, the postsynaptic potentials are attenuated equally in traveling to the granule cell body, in contrast with the tree structure of Fig. 5-1. In the latter case, junctions are located at varying distances from the cell body, and attenuation increases with distance.

Where the two vertical cuts in Fig. 12-2 intersect, we have a Golgi cell, named after Camillo Golgi (1844–1926), who made possible the discovery of this and other cells with his staining techniques. The Golgi cell has many bifurcating dendritic branches, but it has a cylindrical architecture, like a conventional tree, in contrast with the single-plane fan of the Purkinje cell. Its axon feeds an inhibitory input to each of the granule cells in its vicinity; since this is the *only* output of the Golgi cell, the *I* junction on each granule cell must be an important network element.

The Golgi cell receives excitatory inputs from mossy, parallel, and climbing fibers, and an inhibitory input from Purkinje cells.

Basket and stellate cell dendrites receive *E* inputs from the parallel fibers and *I* inputs from the Purkinje cell axons. Basket cells also receive *E* stimulation from the climbing fibers.

So much for anatomy and junction types. What is the purpose of the cerebellum? Much of the following is conjectural because we do not know where axons go to or from in the spino–cerebellar tract or cerebellar nucleus or inferior olive. We do know the result if an experimental animal is deprived of its cerebellum: it suffers from poor coordination of simultaneous muscle movements, and has difficulty maintaining its equilibrium.

Suppose, for example, that one wishes to walk. The motor cortex (near the top of Fig. 12-1) sends a "contract" command to the many motoneurons that are involved with the first stage of walking. The muscles contract but, alas, they tend to overshoot or undershoot. The skeletal muscle fiber is a giant, incapable of fine, accurate movements. As Fig. 8-6 shows, no less than three feedback loops are needed for fine control: one to strongly prod the motoneuron if the distance moved is grossly insufficient; one to gently nudge if the distance moved is almost correct; and one to inhibit the motoneuron if the contractile force is excessive. But walking calls upon the simultaneous contraction of many *different* muscles in the feet, legs, and back. This is where the cerebellum gets into the act. For a stereotyped activity such as walking, why should the motor cortex serve as a "common laborer" by directing every one of the many muscles as their force of contraction has to increase and decrease during each cycle of walking? The motor cortex is a member of the aristocracy, one of the higher brain centers; it has to take over when you step off a curb, or encounter a depression in the ground, or have to step over some obstacle. For any routine, repetitive activity such as walking, the cerebellum behaves as if it is reading a long list of instructions. It anticipates motion so as to prevent overshoots. However, it has to be capable of modifying each instruction. Regardless of how stereotyped a skilled activity may be, minor changes are forever occurring so that the cerebellum cannot behave as a simple idiot robot. Walking against the wind calls for somewhat different responses from walking with the wind, and so forth.

In order to do its job, the cerebellum needs two types of information: first, by way of the mossy fibers, it has to receive "walking" and other commands from the motor cortex. The cerebellum cannot initiate any activity on its own. It has learned many possible skilled movements from the previous lifetime of training to which a person has been exposed, but the higher brain centers have to call forth the particular action the person has "in mind." If the higher brain centers decide on a new activity, such as juggling balls, the person has to practice painstakingly until the relevant parallel fiber–Purkinje cell junctions are appropriately modified.

Second, the cerebellum receives proprioceptive information by way of the climbing fibers: that is, the degree of contraction and tension of every muscle group in the body. As a person walks (to use the previous example), proprioceptive feedback modifies the Purkinje cell output. The Purkinje cells must have a line, albeit an indirect line, to every motoneuron in the body. In this way, Purkinje action potential discharges, modified in accordance with mossy fiber commands and climbing fiber proprioceptive feedback, coordinate the simultaneous contraction of many different muscle groups.

The element of time is implicit in all of this. As each muscle contracts, its length and tension change with time; the updated proprioceptive information results in a corresponding update of Purkinje cell outputs. How fast can the cerebellum update these movements? In the case of walking, the distance from motoneuron to foot to cerebellum to motoneuron is 4 m. If high-speed axons are involved, the total round-trip time is some 50 ms. Add synaptic junction time and muscle contraction time to give a total of 100 ms, or 0.1 s. This is not very fast, so much of the burden of directing the muscles must rest with the motor cortex. The cerebellum may be a helper that handles most of the work, but the "boss," the motor cortex, can always override the helper when an unusual situation arises.

Why is there a precise right-angle topology in the cerebellar cortex? Because, it is conjectured, this makes it possible for every output command neuron in the motor cortex to communicate with every motoneuron, in the spinal cord or anywhere else, via the mossy fibers, granule cells, parallel fibers, and Purkinje cells. Two networks at right angles for this purpose are commonplace, in fact, in electronics communication equipment.

We could make reasonable guesses as to the values of the weighting factors, and the time taken for signals to travel from one neuron to the next, but the cerebellum must be analyzed as a three-dimensional network, with simultaneous inputs from the mossy and climbing fibers. This is far too complex a matter to allow further consideration in this book. Some day, however, the typical "switching" network will be worked out: "Parallel fiber number 11 is connected to Purkinje cells numbered 2, 4, 7, 13, . . . ; parallel fiber number 12 is connected to Purkinje cells numbered 2, 4, 6, 14, . . . ," and so forth.

12-3 Electroencephalographic (EEG) Signals

In 1924, Hans Berger connected two electrodes (small, round, metal disks) to a patient's scalp and detected a feeble current by using a very delicate galvanometer. Berger thereby opened up a window on the brain. As electronic amplifiers and recording equipment came into common use, the brain signal—called an electroencephalogram (EEG)—became a major clinical tool of the neurologist [J. D. Bronzino, 1984; R. Elul, 1972; A. S. Gevins, 1984; A. S. Gevins et al., 1975; K. A. Kooi, R. P. Tucker & R. E. Marshall, 1978; M. Matonsek et al., 1967; G. Pfurtscheller and R. Cooper, 1975; A. Remond, 1971; E. Vaadia, H. Bergman & M. Abeles, 1989].

Five EEG records taken from a normal individual are shown in Fig. 12-4 [H. H. Jasper, 1941]. The time scale is given by the 1-s line at the bottom. The voltage scales are given by the 50-μV lines at the right side. For this type of recording, the "hot" electrode is usually located on the back of the head, opposite the striate cortex, and the other on a neutral point such as an ear lobe. In order to get a good recording, one must of course make good contact with the skin. It is not necessary to shave the patient's head, but skin oil should be removed with a solvent. A conducting paste is applied to the electrode, which is held by a suitable strap or cap against the scalp. The neutral ear-lobe electrode can be held in place with a simple clip.

[S.D. writes: "When I was at the Rutgers Medical School in New Jersey, I designed equipment for telemetering 15 simultaneous EEGs to a receiver in an adjoining room, via FM, so that the animal or human patient could be

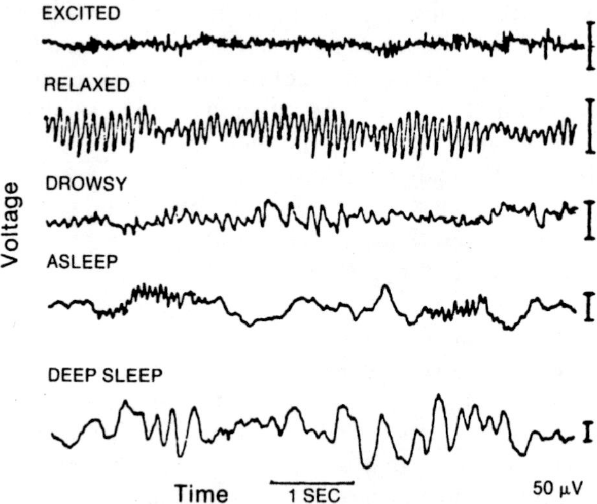

Fig. 12-4 Electroencephalographic recordings taken from a normal individual. For this type of recording, the "hot" electrode is usually located on the back of the head, the other on a neutral point such as an ear lobe. In the "relaxed" waveform the patient's eyes are closed. Notice the time and voltage scales: the 1-s line at the bottom and 50-μV lines at the right side. (From Jasper, Epilepsy and cerebral localization, Thomas, 1941.)

free to move about. In the first test of the equipment I volunteered to be the patient. My neurologist friend and colleague insisted that the normal solvent treatment was inadequate for my oily skin, and proceeded to sandpaper my ear lobe until it bled. The recording was normal, but I was the butt of jokes for the next week or two about how my ear became bitten. The moral of the story—don't let a Ph.D. connect EEG electrodes to a patient or anybody else."]

The ear-lobe electrode does not pick up appreciable signals because it is relatively far away from electrical sources such as brain cells and muscle fibers; in fact, that is why it is considered to be a neutral point. The other electrode only "sees" the electrical activity in its immediate vicinity. To begin with, because of the thickness of scalp, muscles, skull bone, and membrane surrounding the brain, the electrode is around 1 cm away from the cerebral cortex. Signals from neurons deeper in the brain are rapidly attenuated as distance increases. The traces of Fig. 12-4 therefore reveal local happenings. The voltage from an individual neuron is far too small to show up on an electrode 1 cm away. What we are seeing are the voltages from thousands of neurons whose action or graded potentials happen to be headed in the same direction at the same time. It is somewhat like being up in a satellite and trying to see individuals on earth. If everybody in Times Square suddenly headed north, say, you could possibly see the flow in one direction. Normally, however, people moving one way cancel out the people moving in the opposite direction.

In the first waveform of Fig. 12-4, the patient is "excited" (but not agitated) in the sense that he (or she) is looking at the people in the room and listening to their conversations. To avoid interference from the APs that elicit muscle contraction, the patient is not talking or moving about. We see that the EEG is a weak (10-μV), chaotic signal. From the frequencies present in the trace, we can conclude that it is not derived from high-speed (100 meters/second (m/s)) APs, which would induce relatively high frequencies as they fly by. It is not even derived from slow 1 m/s APs. The frequencies are reminiscent of the 0.1-m/s progression of a graded potential of maximum amplitude 20 mV as it is conducted along a dendrite: its neuron integrates many inputs to generate its own graded potential, and the latter is "leisurely" transmitted to another dendrite via an electrical gap junction or via a chemical agent that diffuses across a synaptic junction.

The fear voiced by some individuals that the neurologist can read their "inner thoughts" is without foundation. Certain regularities can be observed if the person is looking at, say, a pattern of vertical stripes, but the same thought repeated over and over again merely results in a random, nonrepetitive and meaningless EEG pattern [*Skeptical Inquirer*].

In the second waveform, which is "relaxed," the patient's eyes are closed. This state gives rise to a relatively strong, nonsteady oscillation at around 10 Hz, known as the *alpha* rhythm [H. T. Castello, 1983; B. H. Jansen, 1985; J. G. Okyere, P. Y. Ktonas & J. S. Meyer, 1986]. The origin of the alpha rhythm is controversial, but it is probably connected to vision because it is strongest in the vicinity of the striate cortex. Especially mysterious is the fact that it vanishes when the patient opens his or her eyes and looks at the world about. A hypothetical explanation is offered at the end of this section.

As the patient progresses (or retrogresses) from relaxed to drowsy to asleep to deep sleep, radical changes in frequency and amplitude take place. "Deep sleep" displays slow waves, only a few cycles per second, of large amplitude (100 μV). The large amplitude indicates that entire sections of tissue beneath the electrode are undergoing the same electrochemical changes in unison. The slow waves hint at the relatively slow transport of ions. Perhaps the EEG is telling us that the debris following many hours of wakefulness is being removed and "normal" material is being replenished. Alternatively, in Chapter 11, it is suggested that self-organization takes place during "quiet" periods such as sleep. These are relatively slow processes; the human cycle consists of 16 hours of alertness followed by eight hours of sleep (not all of it, of course, as "deep sleep") [A. R. Morrison, 1983].

The clinical value of the EEG comes from the fact that an abnormal waveform is correlated with abnormalcy underneath the electrode [N. I. Bachen, 1986; D. G. Childers et al., 1982; C. D. McGillem, J. I. Aunon & K.-B. Yu, 1985; N. H. Morgan and A. S. Gevins, 1986; G. Pfurtscheller,

H. Maresch & S. Schuy, 1977]. In epilepsy, many fibers discharge synchronously, yielding large (100-μV) oscillations. By employing a multielectrode recording (16, for example), the neurologist can sometimes locate the focal point of the seizure; that is, the location of scar tissue or damage that triggers the seizures. In that event, surgical excision of the tissue can cure the epilepsy. Similar procedures are used to pinpoint subsurface damage caused by disease, accidental injury, or tumors [J. Bancaud et al., 1974; A. J. Wilkins, C. E. Darby & C. D. Binnie, 1979].

We end this section by deriving highly idealized EEG waveforms. The strategy consists of examining the EEG output due to a simple geometric configuration, and trying to draw reasonable conclusions. Because of the layered organization of the cortex, a suitable configuration is a discharging strip of excitable tissue. Two orientations are of interest: first, the discharge propagates in a direction parallel to the cortical surface; second, it propagates at right angles to the surface. We first find the voltage due to an infinite slab.

Analysis of the voltage due to a discharging slab proceeds in easy stages from basic principles [J. P. Ary, S. A. Klein & D. H. Fender, 1981; S. Deutsch and E. M-Tzanakou, 1987; J. D. Kraus and K. R. Carver, 1973; P. L. Nunez, 1981; R. Plonsey, 1969]. We start, in Fig. 12-5(a), with a point current source I located at the origin of an infinite sphere. It is easy to show, from the voltage drop between a finite sphere at radius r and the infinite sphere, that the voltage at point P is given by

$$V = \frac{I}{4\pi g r}, \tag{12-1}$$

where g is the conductivity of the volume conductor. If a point current sink is combined with the source, as shown in Fig. 12-5(b), we get the voltage due to a current dipole [J. C. de Munck, B. W. van Dijk & H. Spekreijse, 1988; Y. Watanabe and Y. Sakai, 1984]:

$$V = \frac{Il \cos \theta}{4\pi g r^2}, \tag{12-2}$$

where $r \gg l$. If many of these current dipoles fill an area dA so as to constitute a current density J, as in Fig. 12-5(c), substitution into Eq. (12-2) yields

$$dV = \frac{Jl \cos \theta \, dA}{4\pi g r^2}. \tag{12-3}$$

We now take two important steps that are conceptually simplifying. First, regard dA as part of an excitable membrane that is in the process of discharging. Then membrane voltage V_m is related to J via

$$V_m = \frac{Jl}{g} \tag{12-4}$$

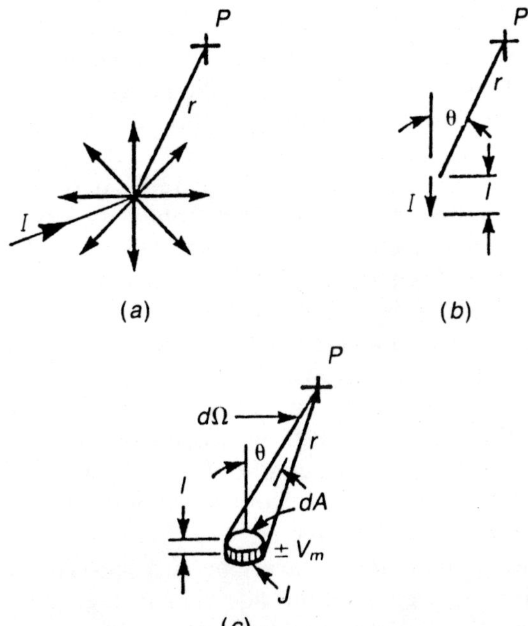

Fig. 12-5 Geometric construction used to derive the potential due to a small area of excitable membrane, dA, as a voltage source, V_m, discharges across the membrane. The membrane is embedded in an isotropic infinite volume conductor. (*a*) Potential due to a point current source. (*b*) Potential due to a current dipole. (*c*) Potential due to the discharging membrane. (From Deutsch and M-Tzanakou, *Neuroelectric Systems*, New York Univ. Press, New York, 1987.)

that Eq. (12-3) becomes

$$dV = \frac{V_m \cos\theta \, dA}{4\pi r^2} . \tag{12-5}$$

cond, solid angle $d\Omega$ in Fig. 12-5(c) is given by

$$d\Omega = \frac{\cos\theta \, dA}{r^2} , \tag{12-6}$$

that

$$dV = \frac{V_m \, d\Omega}{4\pi} . \tag{12-7}$$

other words, the voltage at point P due to a piece of excitable tissue is portional to the solid angle subtended by the tissue as seen from point For example, if point P is inside a neuron body, so that it is completely surrounded by the neuron's membrane, the solid angle is 4π, and $V = V_m$. If point P is outside the neuron, the near wall of the neuron displays $+V_m$ while the far wall displays $-V_m$, so that the net contribution is zero, $V = 0$. This is consistent with the convention that the extracellular fluid is at zero reference potential.

Next consider the slab of axons of Fig. 12-6. As shown, the slab has a height d and, in the front view, it extends from $y = -\infty$ to $+\infty$. The front view represents a cross section through cortical tissue containing many parallel, simultaneously discharging axons. As far as the contribution to any relatively distant point P is concerned, the $-+$ and $+-$ contributions of internal axons cancel each other, so that only the outside contour contributes net voltage to point P and is shown as having a membrane potential V_m.

In the side view, the axons run from $x = -\infty$ to $+\infty$. The discharge is moving from left to right. The EEG electrode is at point P. In accordance with Eq. (12-7), in the side view, contributions to point P from the top and bottom layers of the slab cancel each other except for the strip w wide along the leading edge of the bottom layer.

The voltage at point P due to the discharging slab is derived in Appendix A12-1. If the discharge region in the side view of Fig. 12-6 is a strip l units long rather than infinitely long, the derivation yields

$$V = \frac{V_m d}{2\pi a}\left\{\frac{(b + 0.5l)/a}{1 + [(b + 0.5l)/a]^2} - \frac{(b - 0.5l)/a}{1 + [(b - 0.5l)/a]^2}\right\}. \qquad (12\text{-}8)$$

We illustrate with a numerical example: $V_m = 50$ mV (this is the average

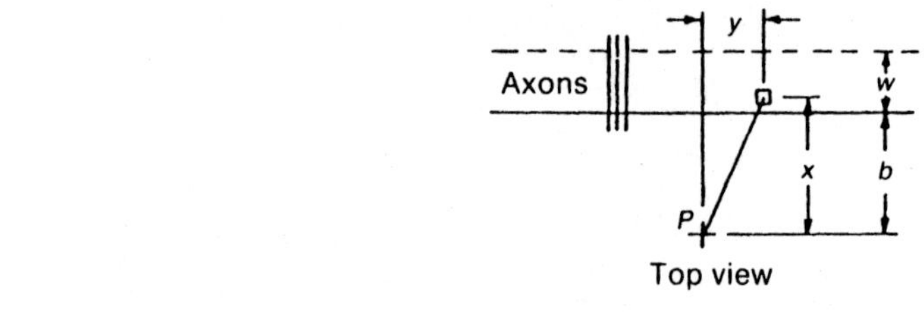

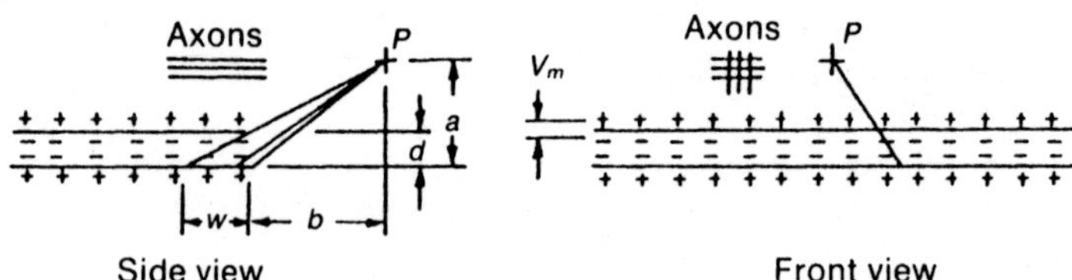

Fig. 12-6 An infinite slab of tissue, d units thick, filled with parallel, simultaneously discharging axons. In the side view, the potentials are moving from left to right, their leading edges are b units to the left of point P, and the discharge region extends an infinite distance to the left. (From Deutsch and M-Tzanakou, *Neuroelectric Systems*, New York Univ. Press, New York, 1987.)

value as the membrane capacitance discharges from 100 to 0 mV), $d = 0.1$ cm, $a = 1$ cm, and $l = 0.1$ cm (this is the discharge region calculated in connection with Fig. 4-2). The resulting EEG waveform, plotted as a function of b, is shown in Fig. 12-7. Here we have a horizontal slab of neurons oriented parallel to and 1 cm below the surface of the scalp. An EEG electrode sits at the $b = 0$ point. (The other electrode is at a neutral point, such as an ear lobe.) The slab is 1 mm thick, but is magnified by a factor of 4 to better show the movement of sodium ions. If the slab is filled with Hubel–Wiesel neurons in accordance with Fig. 11-2, which are equivalent to solidly packed cubes 27 μm on edge, in a 1-mm thickness we have 1 mm/0.027 mm = 37 neurons. A graded potential is propagating to the right along the slab, and all of the neurons are discharging synchronously. The output is 80 μV at $b = 0$, when the discharge region is directly underneath point P; the leading and trailing edges then each contribute +40 μV. Positive and negative areas are of course equal, but the negative areas are in the form of long, weak outputs as shown. The amplitude is reasonable for an epileptic seizure or for the slow "deep sleep" waves of Fig. 12-4.

At $b = 1.5$ cm in Fig. 12-7, we get $V = -9.4$ μV. Using this as a baseline, the area under the curve is equivalent to a rectangle 1.14 cm wide, which corresponds to a spatial cycle 2 × 1.14 = 2.3 cm wide. But a large-amplitude EEG spike such as 80 μV, which is caused by a broad synchronous discharge, can only represent a relatively low-frequency component, such as 5 Hz. The discharge that gives birth to the spike is therefore flying by (but 1 cm underneath the electrode) at a speed of 2.3 cm × 5 Hz = 11.5

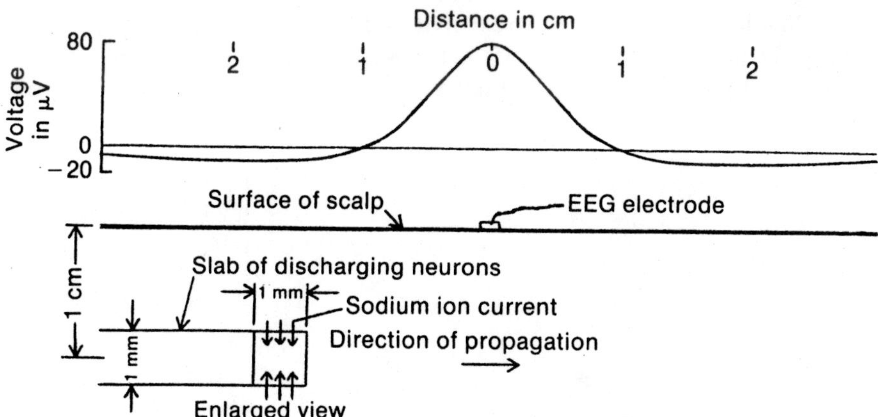

Fig. 12-7 Highly idealized EEG waveform. The surface of the scalp is a horizontal line, and the electrode sits at the distance = 0 point. A slab of neurons 1 mm thick is discharging synchronously. (An enlarged view of the discharge region, by a factor of 4, is given.) The voltage across the edges of the slab is 50 mV. The discharge propagates from left to right 1 cm below the scalp; a positive voltage bump of +80 μV is picked up.

cm/s ≅ 0.1 m/s. This would be the effective velocity when a graded potential is coupled to a dendrite, propagates without regeneration, and is coupled from the neuron axon to the next dendrite. Coupling could be by way of electrical gap junctions for which the equivalent circuit is, simply, a resistor.

The second orientation, in which the wavefront is traveling at right angles to the surface, toward (or away from) the EEG electrode, contributes a negative output. The model is depicted in Fig. 12-8. The slab of Fig. 12-6 is repeated in Fig. 12-8, but the discharge region in the side view is now on a collision course with point *P*.

If the discharge region in the side view of Fig. 12-8 is a strip l units long, the derivation in Appendix A12-1 yields for the voltage at point *P*

$$V = -\frac{V_m l d}{2\pi b^2}. \qquad (12\text{-}9)$$

We illustrate with the values used in the previous numerical example: $V_m = 50$ mV, $d = 0.1$ cm, and $l = 0.1$ cm. The resulting EEG waveform is shown in Fig. 12-9. At $b = -1$ cm away from the EEG electrode, the wavefront ends with a maximum output of -80 μV. We again conclude that this is a large-amplitude low-frequency spike caused by signal processing that proceeds at a velocity of 0.1 m/s. Although epileptic seizure recordings may be due to long-distance propagation over 1-m/s axons, the normal EEG trace is due to relatively slow graded-potential activity over local pathways.

Please note the following discrepancies between the models of Figs. 12-6 and 12-8 and a typical EEG: The models assume an infinite homogeneous volume conductor. Instead, the tissue is surrounded by a thick layer of skull bone that has a relatively high resistivity, in turn surrounded by air that has relatively infinite resistivity. Also, the analysis assumes that the EEG is picked up by a point electrode. Instead, a relatively large-diameter electrode is used to ensure a low-resistance contact. The large electrode tends to

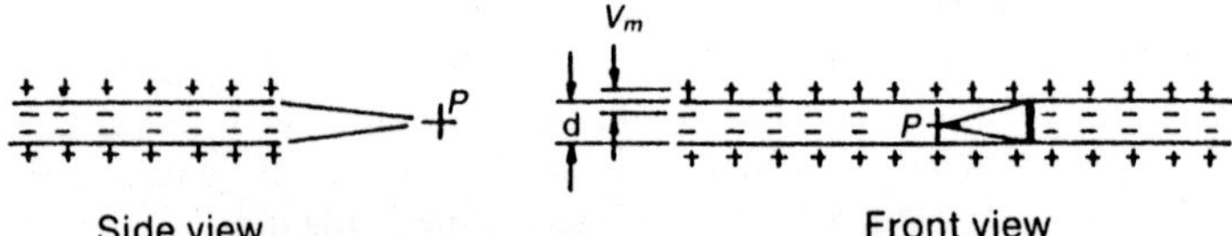

Fig. 12-8 The discharging slab of Fig. 12-6, except that the discharge region is now on a collision course with the EEG electrode at point *P*. (From Deutsch and M-Tzanakou, *Neuroelectric Systems*, New York Univ. Press, New York, 1987.)

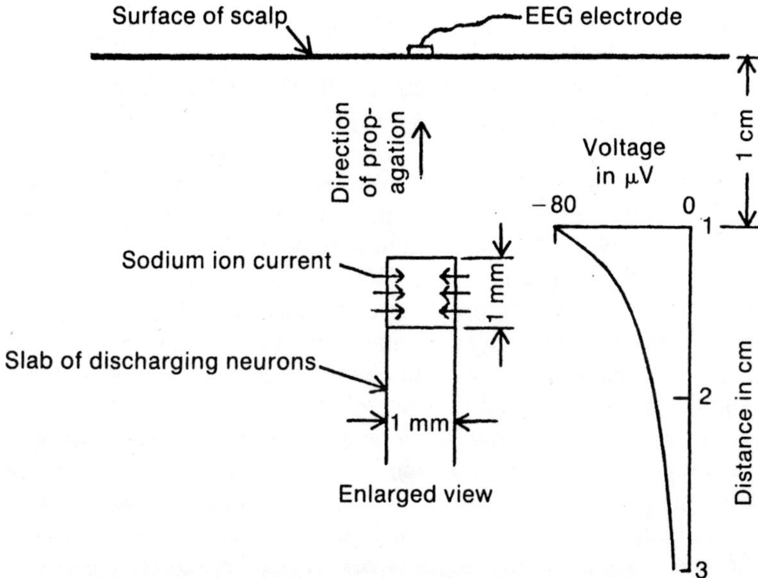

Fig. 12-9 Highly idealized EEG waveform. The surface of the scalp is a horizontal line, and the electrode sits at the distance = 0 point. A slab of neurons 1 mm thick is discharging synchronously. (An enlarged view of the discharge region, by a factor of 4, is given.) The voltage across the edges of the slab is 50 mV. The discharge propagates upward; a voltage bump of −80 μV is picked up when it is 1 cm away from the scalp.

smooth out high-frequency spatial components, which is equivalent to discrimination against high-frequency time components.

In the actual EEG outputs of Fig. 12-4, all of the discharges except the "relaxed" alpha rhythm are headed in many different directions. Positive and negative voltage bumps or waveforms follow each other randomly, with random frequency and amplitude variations, but higher frequencies correspond to more rapid propagation speeds of the underlying graded potentials.

The constant frequency of the alpha rhythm does not fit the preceding description. This type of oscillation is reminiscent of clonus, the instability discussed in Section 8-6. In the numerical example given there, a patient's wrist muscle, say, contracts; in response to the involuntary contraction, the muscle spindle generates an error signal. It requires 86 ms for the error signal to reach the motoneuron in the spinal cord. Normally, the error signal is multiplied by a synaptic junction weighting factor of 4.4 (see Fig. 8-2), but in our clonus example it is multiplied by 20 (see Fig. 8-9). The motoneuron sends a contraction signal to the wrist muscle, where it arrives 86 ms later. The contraction that it produces is exactly equal to the original contraction; in other words, feedback supplies exactly whatever is lost so as to maintain clonus indefinitely. The clonus frequency depends on the time delay, junction weighting factors, and muscle load parameters. For the numerical example just given the frequency is 0.48 Hz.

A hypothetical model for the alpha rhythm is based on the fact that there is plenty of feedback in the visual system (and almost everywhere else, for that matter), as follows: We have previously come across fibers that run from the lateral geniculate nucleus to the striate cortex. It happens that there are also fibers that extend back from the visual cortex to the thalamus. During normal viewing, the synaptic junction weights are relatively low so that the feedback system is stable. When the eyes are closed, the visual system increases sensitivity (i.e., increases the weighting factors) in order to properly process supposedly dim images. The system goes into oscillation, but it is not the wild oscillation of an epileptic seizure; the system limits the alpha amplitude to a reasonable level.

(In the early days of radio, there was a popular circuit known as a "regenerative receiver" that enabled a single vacuum tube to achieve tremendous gain via feedback that resulted in oscillations. The oscillations were not audible because they were above 20,000 Hz, but in a properly designed receiver, the weak incoming radio waves were able, nevertheless, to modulate the oscillations. These circuits were abandoned eventually because it was difficult to control the feedback and tuning.)

Why do we not "see" the alpha rhythm as a light flashing on and off 10 times a second (which, incidentally, is a good frequency for inducing epileptic seizures in a person susceptible to this type of behavior modification)? Because the alpha feedback loop is located beyond our conscious perception, wherever that happens to be. As Marvin Minsky emphasizes in his book *Society of Mind* (1986), our conscious perception encompasses very little of what is actually going on in our brain.

12-4 Memory

One of the distinguishing features of humans compared to other animals is the huge amount of information we can learn and remember. Any change that represents memory, the storage of information, is called an *engram*. A major research effort has been directed toward discovering the forms and locations of the engrams in the brain.

Because the human brain can store a tremendous amount of information, Holger Hyden proposed in 1961 that the basis is molecular; that is, a memory engram consists of molecules that are altered so as to constitute the storage element. The modern view, however, is that the basis is synaptic junction modification; that is, each part of the brain that can recognize patterns can also store the memory of new patterns via synaptic junction changes [D. L. Alkon, 1989; J. L. Gould and P. Marler, 1987; J. L. McClelland, D. E. Rumelhart & PDP Research Group, 1986; M. Mishkin and T. Appenzeller, 1987; E. T. Walters and J. H. Byrne, 1983]. In the visual cortex, for example, the area that can recognize eyes, ears, and noses has the ability to do so because these patterns are already stored therein. If a new, appreciably different nose, say, enters the visual system, this area is capable somehow of adding an engram of the new nose. Simultaneously, the auditory system stores engrams of the person's voice characteristics (frequencies, accent, and any other peculiarities), the inferior temporal cortex stores an engram of the person's facial characteristics and peculiarities, and time and place are stored in yet other locations. All of these engrams are connected via association fibers so that the "new" nose, when next encountered, calls forth all of the associated memories (or the person's voice will call forth a memory of the nose, and so forth). In this way, various aspects of memory are localized in many different areas of the brain and, in addition, each event can be widely distributed in time.

Most of what we know about memory has been gleaned from research with patients who suffer from various memory deficiencies—that is, amnesiacs. The block diagram of Fig. 12-10 is an attempt to summarize the mechanisms of memory, as follows: Sensory receptors feed neuronal circuits that seek out the salient features in the "feature extraction" block. In the preceding example, this block in the visual cortex looks for small geometrical structures such as eyes, ears, and noses. From feature extraction the signals proceed to the "pattern memory stack," which for convenience is shown as a separate block, although, as noted previously, it is embedded actually in feature extraction.

The "central station" (CS) is the "seat of consciousness," to use an expression that is popular with philosophers. It is also identified by the acronym HATS, which stands for hippocampus, amygdala, thalamus, and striatum. As Fig. 12-1 shows, these are centrally located in the brain, and their functions are poorly understood. Of vital importance for the mechanisms of memory, however, is the fact that amnesia can be traced to damage in one or more regions of the HATS ensemble.

The number of neurons in the central station is insufficient for a large amount of memory storage in addition to associated signal processing. With

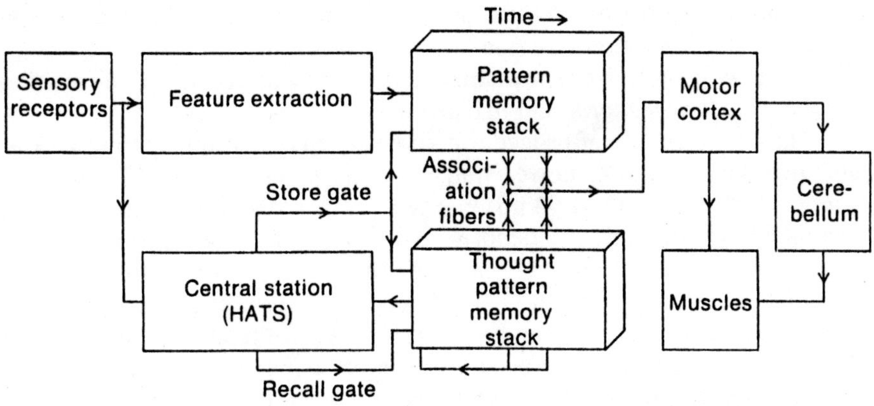

Fig. 12-10 Hypothetical block diagram that attempts to summarize the mechanisms of memory.

regard to memory, the chief function of the CS, according to Fig. 12-10, is to generate a "store gate." The CS continually examines incoming sensory information and indirectly compares it against the recognition of previously stored patterns that is taking place in the pattern memory stacks. The CS tries to answer the questions "Is it new? Is it important? Is it dangerous, unusual, or useful information?" If it *is* important, the CS generates the store gate, which leads to the memory stacks.

An analogous role, in a real brain, is played by the ascending reticular activation system (ARAS), which is located in the brain stem, in the upper end of the spinal cord in Fig. 12-1. It has been found that stimulation by the ARAS is necessary for normal wakefulness and to maintain the overall degree of activity of the brain. A brain tumor in the ARAS region can result in a coma in which the person does not respond to normal awakening stimuli. From this, it is concluded that sleep results from the absence of adequate brain excitation [V. Bloch, 1976].

In the memory stacks, with regard to time, there are several different types of memory. There is extremely short-term memory, in which several seconds of the latest sensory information are sustained, presumably via reverberation around neuronal feedback loops. If the sensory information is deemed to be important, the store gate acts as a trigger, instructing the memory stacks to permanently store the pattern. First there is some form, as yet unknown, of temporary or short-term storage. In humans, about one hour is required before this short-term memory is consolidated into permanent or long-term form. This conclusion is based on the fact that a massive trauma, such as a severe blow to the head, or electroconvulsive shock, may partially or fully erase memories that have been stored for less than an hour, but it has no effect on events experienced before that.

The form in which long-term memory is stored is not known. One theory has it that two initially excited neurons are induced to send out collateral branches that grow toward each other, establishing synaptic contact in about one hour. We know that nerve fibers that have been cut are capable of regeneration. Alternatively, the branches and junctions may be in place permanently, but it may require one hour to activate them.

Work with amnesiacs shows that there are two categories of information or knowledge: *procedural* and *declarative*.

The best example of procedural knowledge is the acquisition of a skill. This involves the motor cortex, muscles, and cerebellum at the right side of Fig. 12-10. In Section 12-2 it is pointed out that, according to one theory, "parallel fiber–Purkinje cell dendrite junctions are modifiable; that is, these are the weighting factors that change as the system learns." In learning a skill, according to this, memory is stored in the form of synaptic junction changes in the cerebellum (and also, probably, in the motor cortex). This theory would explain why procedural memory is immune to damage to the HATS ensemble; the latter can cause amnesia.

Declarative knowledge, on the other hand, is the only category lost by an amnesiac. It involves specific facts or data, such as the "new nose" seen in the previous example, along with associated information regarding time and place. This is the type of engram that is stored in pattern memory stacks. Extremely short-term memory capability remains intact; only short- and long-term memory are impaired in amnesia.

Work with amnesiacs reveals that some of them lose the ability to long-term store declarative information that is received *after* their brain injury; this is known as *anterograde* (AN′TER·O·GRADE) amnesia. Others lose the ability to recall declarative information that was stored *before* their injury; this is known as *retrograde* (RET′RO·GRADE) amnesia. There are also various degrees of both anterograde and retrograde amnesia.

A somewhat amusing interplay between procedural, declarative, anterograde, and retrograde effects sometimes occurs: A is taught to drive a car, then suffers brain damage causing anterograde amnesia, and some time later is taught to juggle balls. He learns this skill with normal aptitude, and can also remember the driving instructor, but cannot remember when, where, and from whom he took juggling lessons! B is taught to drive a car, then suffers damage causing retrograde amnesia, and some time later is taught to juggle balls. He cannot remember when, where, and from whom he took driving lessons, but perfectly well remembers the juggling instructor!

Anterograde amnesia is simply explained in the block diagram of Fig. 12-10: Because of damage somewhere in the HATS ensemble, the store gate is lost. Salient features are extracted, and the pattern memory stacks remain intact, but the signal enabling the brain to store long-term information is absent. In all other respects, the person seems to be normal. The reason for the store gate is obvious: If the gate is open all of the time, the pattern memory stacks will soon become cluttered with trivia (useless except, perhaps, for the playing of "Trivia" games). If the gate is never open, the person suffers from anterograde amnesia.

Typical are the following reports by Larry R. Squire (1986):

> Patient N.A. . . . is a pleasant man with an agreeable sense of humor, who could join in any social activity without special notice. However, he would be unable to learn the names of his colleagues, or keep up with a developing conversation, or speak accurately about public events that have occurred since his injury. He has an intelligence quotient (IQ) of 124, . . .

Also, "Our laboratory recently obtained extensive clinicopathological information from a patient who developed . . . marked anterograde amnesia, little if any retrograde amnesia, and no signs of cognitive impairment other than memory. . . . " After his death:

> Thorough histological examination revealed a circumscribed bilateral lesion of the CA1 field of the hippocampus that extended its full rostral-caudal length but not beyond. . . . Although the lesion was spatially limited, it affected an estimated 4.6 million pyramidal cells and would be expected to have a profound impact on the function of the hippocampus. A lesion in the CA1 field interrupts the essentially unidirectional flow of information that begins at the dentate gyrus

and ends in the subicular complex and entorhinal cortex. These structures are the main sources of output from the hippocampus to subcortical, limbic, and cortical structures. Thus, a CA1 lesion would significantly disrupt the interaction between the hippocampus and memory storage sites, an interaction presumed to be critical for the storage and consolidation of declarative memory.

It is more difficult to explain retrograde amnesia. In Fig. 12-10, it is conjectured that the central station generates a "recall gate" when you wish to examine the contents of a particular engram. This has to be part of a thought process since the engram has to be selected from among millions of stored memories. A "thought pattern memory stack" is accordingly depicted, including feedback from output to input. The thought pattern memory stack presumably resides in the prefrontal cortex of the brain. The input to the stack comes from the association fibers. Thoughts, in other words, circulate over and over again, modified by ever-present random noise (not shown) and by whatever is coming across from the other memory stacks via the association fibers.

The reason for the "recall gate" follows by analogy with the store gate: If the recall gate is always open, you could never concentrate on any particular event. Your thoughts would keep looping wildly from one event to another; you would be the epitome of a "scatterbrain." If afflicted with the other extreme, a closed recall gate, you could not remember any past long-term event. An entire lifetime of experiences would be stored in the pattern memory stacks, frustratingly out of reach because you have "blocked it out" via a defective recall gate.

Retrograde amnesia is common in patients with Korsakoff's syndrome. This is caused by the destruction of brain tissue due to chronic alcoholism. Of these people, Squire (1986) writes: "In amnesic patients with diencephalic lesions, the nature of anterograde and retrograde amnesia is less clear. For example, patients with Korsakoff's syndrome exhibit, instead of a temporally limited retrograde amnesia, a severe and extensive impairment of remote memory that covers most of their adult lives."

The store and recall gates are hypothetical, of course. In an analogous piece of electronic equipment, the store and recall gate paths of Fig. 12-10 would consist of single insulated wires. The nervous system would, however, employ many "wires," thus achieving reliability in exchange for redundancy, and a graded output that would allow partial storage and partial recall of information.

The memory stacks in Fig. 12-10 are portrayed as rectangular parallelepipeds in which time proceeds from left to right. Whenever a store gate is received, the salient features are entered in the stack. In a most simple and naive interpretation, zero time corresponds to birth, and the end of the stack, at the right, corresponds to (hopefully) 100 years of age with all of the person's "marbles" intact. If the brain is that of a 25-year-old, incoming patterns are stored at a location one-fourth of the distance from the left, and so forth.

From left to right, the two association fibers shown in Fig. 12-10 are located at layers that are 45 and 65 years old.

Alternatively, one can assume that new memories are stored at the left edge, and previous memories are progressively displaced to the right. This seems to be a natural arrangement whereby old memories gradually recede. A similar technique is employed by electronic calculators: When you enter a number such as 365 in a calculator, the 3 first appears at the right, then it moves to the left to make room for the 6, and finally 36 moves to the left to make room for the 5. The maneuver is controlled by a circuit known as a "shift register." In a shift register model of memory storage, the memories at birth of our 25-year-old individual will have reached a location one-fourth of the distance from the left. This must somehow occur despite the fact that neurons are stationary entities. During the embryonic stage, neurons migrate along with other cells; once in place, however, they grow dendrites and axons and never again move since, obviously, this would alter their circuits. Cells grow by dividing in two, but this would also disrupt the dendrites and axons of a neuron, so they remain set for life. (Actually, recent evidence has shown that there may be some neuron growth in the brain throughout life [F. Nottebohm, 1989].) As previously noted, neurons are probably rejuvenated during sleep when "debris" is removed and metabolic material is replenished.

The time-sequential-stack conjecture receives support when you become a tourist. After a week spent furiously dashing about from one city to the next, and trying to store so many "unforgettable" memories, it will seem as if you have been away for a month rather than a week [W. J. Friedman, 1990].

In the simple scheme previously described, memories need never fade. Would we not survive much more nobly if, after learning something once and storing it in long-term memory, it is never forgotten? Perhaps not, since traumatic experiences would forever remain fresh, and the psychologists would devote their time to sugarcoating every disturbing memory from childhood on. Nevertheless, the fading of memory with time, as if storage elements are evaporating, is probably unavoidable, and may be an important clue to solving the mysteries of memory storage. Perhaps, if appropriately selected synaptic junctions are recycled, there *are* enough of them to account for the huge memory capacity of a human without invoking Hyden's molecular hypothesis.

Personal experience shows that the acquisition of information is followed by gradual forgetting. This is illustrated by the "amount of declarative information remembered" curve of Fig. 12-11. At first, forgetting takes place rapidly; later on, more slowly. The curve is exponential, with a one-year time constant (time to drop down from 100% to 37%). The horizontal scale is proportional to the logarithm of $(1 + t)$, where t units are in years. The numerical values are very inexact because it is difficult to define and measure the amount of information stored in a pattern memory stack.

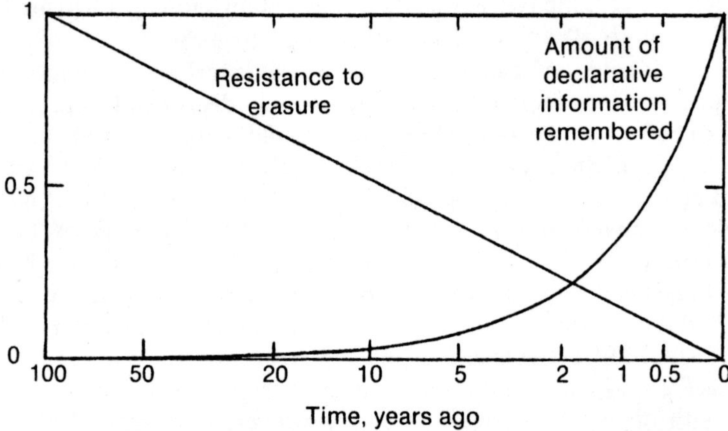

Fig. 12-11 Hypothetical "forgetting" curve (the amount of declarative information remembered), and curve showing how long-term memory becomes increasingly resistant to erasure.

[S.D. writes: "My own experience on this subject comes from some extensive traveling my wife and I undertook in 1964 and 1968. Ruth kept a diary on both occasions. When I refer back to the events that took place in a particular city, I find three categories of memory: Most of the minor details have been completely forgotten, as if they never occurred, and are read with amusement and (sometimes) disbelief. A second category is of events recalled vaguely, the kind I would testify, in a court of law, have only a 50% chance of being accurate. The third category consists of events I still remember vividly, such as the time (traveling alone) I was placed under house arrest at a Tokyo airport because I did not have a valid visa for Japan. I was confined for a day to a high-priced airport hotel (for which *I* had to pay). Resigned to my fate, I spent the day reading the *Scientific American*."]

One of the surprising results of modern research is that we gradually develop greater resistance, as time goes on, to the loss of whatever memories remain. This is shown by the "resistance to erasure" curve of Fig. 12-11. The exact shape of this curve is not known, but the intention is to show that resistance increases as the amount of remembered information decreases. The classic example is the vividness and accuracy with which a few childhood events can be recalled by older people.

12-5 A Hypothetical Brain Model

How does the brain function? The classic comment of the neurophysiologist is that we know very little about how it functions. He or she may refer you to a neurophysiology textbook that summarizes what is known about the 17 regions of Fig. 12-1, plus other regions.

In this section, a different strategy is attempted in response to the question. We will construct the block diagram of a relatively simple brain, a brain that can perform two tasks: It can read and tell us what it is reading, and it can listen to a conversation and write down what it hears. Synthesis of the block diagram is a two-stage repetitive process. First, we design a machine that can read, speak, hear, and write. Second, we modify it in accordance with what is known about an actual brain, notably from the work of Mortimer Mishkin. Then the process is repeated until the diagram stabilizes. The final result obtained here is shown in Fig. 12-12. In the upper half of the diagram, the robot reads and speaks; in the lower half, it hears and writes. You will recognize that much of the diagram is a more detailed version of the memory block diagram of Fig. 12-10; not surprisingly, much of the discussion of the previous section on "Memory" is pertinent to the hypothetical brain model [J. A. Anderson, A. Pellionisz & E. Rosenfeld, 1990; J. A. Anderson and E. Rosenfeld, 1988; M. Sharples et al., 1989].

From the retina, a layer of red, green, and blue cones feeds a "generator of color maps." This is physically located in the lateral geniculate nucleus (LGN). It generates thousands of two-dimensional color maps, one for each contiguous group of red, green, and blue cones, as modeled in the lower-right portion of Fig. 10-20. Color is therein defined in accordance with 0.9 (log red) + 0.1 (log blue) − (log green), and (log green) − (log blue). Luminance information, the log red, log green, and log blue average, is added as a third dimension. From the LGNs, a branch proceeds upward to the "visual feature extraction" box. As described in Section 11-1, "If the visual image is a face, the image is broken down into small ensembles that are first 'seen' as tiny pieces of eyes, ears, nose, and so forth; later on, neuronal ensembles see eyes, ears, and a nose; finally, the complete face of your grandmother, say, is recognized." The visual feature extraction box stands for the hierarchical progression from the striate to inferior temporal cortices.

From the cochlea, a one-dimensional array of hair-cell neurons feeds a "generator of frequency versus amplitude map." It generates a two-dimensional map as modeled in the right-hand side of Fig. 9-16. An incoming sound is therein defined in accordance with the amplitudes of its various frequency components. From the medial geniculate nucleus, a branch proceeds downward to the "auditory feature extraction" box. Here, the neuronal circuits seek out the salient feature of incoming sound. In language, for example, auditory features are the string of sounds that make a syllable, the string of syllables that make a word, and the string of words that make a short, meaningful phrase. In music, it is the string of sounds that make up a short melody, and so forth. In an actual brain, the auditory feature extraction box is in the auditory cortex.

The CS is the "seat of consciousness." Our usually unreliable agent, introspection, tells us that the CS receives visual color information, including illuminance; from the auditory system it receives frequency and amplitude data. We have suggested that CS may consist of the hippocampus, amygdala, thalamus, and striatum (HATS).

Why is there a box labeled "generator of random noise?" It is to remind

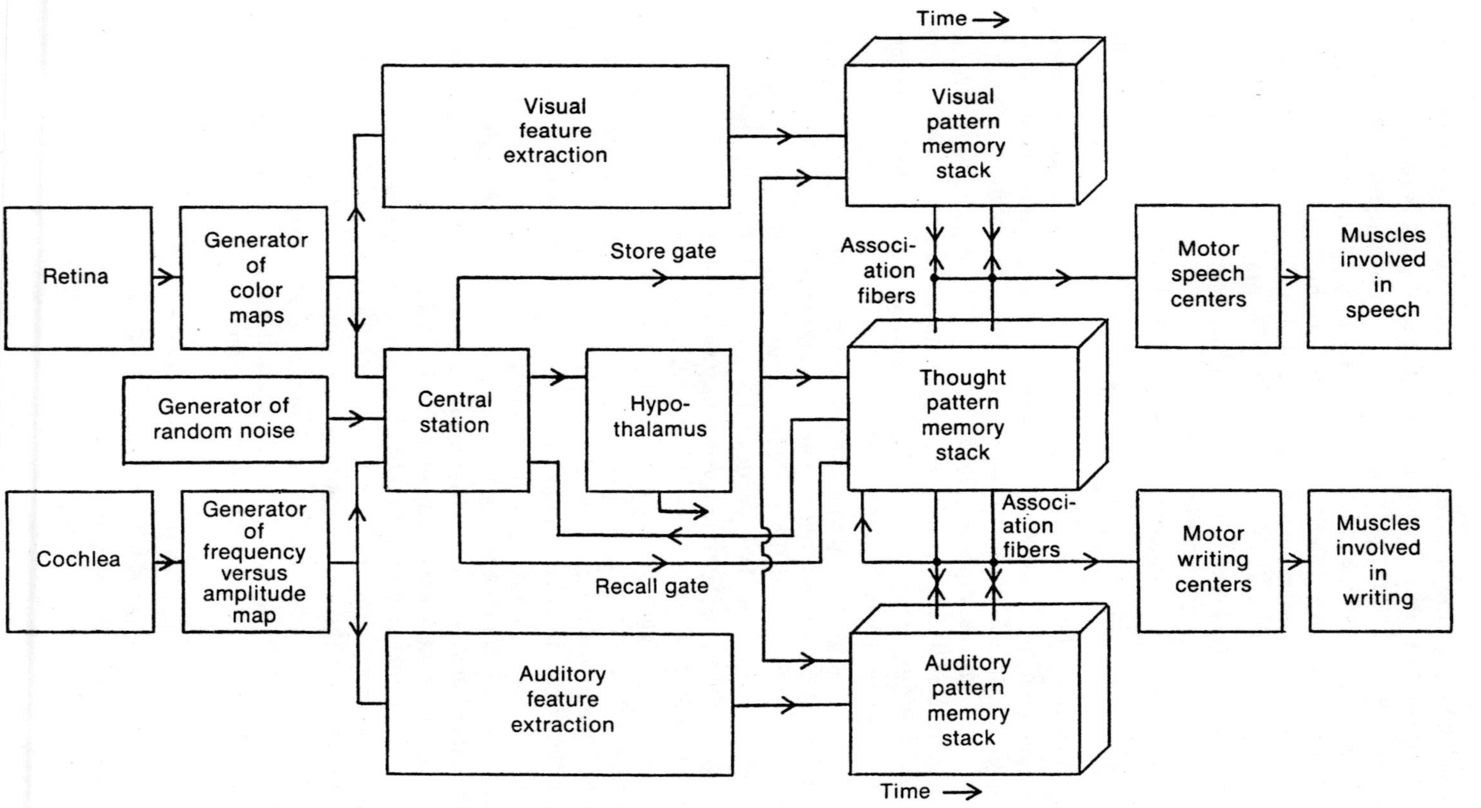

Fig. 12-12 Hypothetical block diagram of a relatively simple brain that can read and tell us what it is reading in the upper half of the diagram; listen to a conversation and write down what it hears in the lower half.

us explicitly that every train of action potentials is "corrupted by noise." Typical APs are shown in Fig. 3-3; a large gap, randomly induced, is shown in Fig. 3-4. In the discussion connected with these figures, it is pointed out that the randomness can be blamed for mistakes, but it is also responsible for much of what we call "creativity." In addition to the noise generator, the creative robot must be programmed to recognize "improvements," and to revise its memory and program in accordance with the new-found wisdom. The process is exactly analogous to the genetic mutations that can gradually give rise to a new species. This itself is a mindless process that resulted in human creativity. In other words, the robot of Fig. 12-12 is somewhat unpredictable, and sometimes comes up with a new idea (an example of robot creativity is given later on in this section).

In Chapter 11, Fig. 11-14 and Table 11-3 represent a system based on trial and error in the sense that the input comes from five binary sequences (− − − −, + + − −, + + + +, + − − −, and + + + −) taken in any order. Given certain algorithms that reward the "correct" response by increasing synaptic junction weights, and punish "incorrect" responses by decreasing junction weights, the system organizes itself. Compare this with the case of Helen A. Keller (1880–1968), who became deaf, dumb, and blind at the age of 1½ years. "At the age of six . . . she was put under the charge of Miss Annie M. Sullivan, who undertook to instruct her in the touch alphabet . . . she is now (age of 10) able to read and write with perfect facility" [*Scientific American*, 1990]. How does one communicate with a child who cannot hear, speak, or see? By a system of "rewards" and "punishments." Keller became a famous author and lecturer, communicating via braille, by touching hands, and even speaking although she could not hear herself.

Returning to Fig. 12-12, we see that three outputs leave the CS. One goes to the hypothalamus, which contains a stack of built-in, instinctive patterns; this is discussed in Section 12-1. The other two outputs are the memory store and recall gates of Section 12-4.

For convenience, the visual and auditory "pattern memory" stacks are shown as separate blocks although they are embedded actually in visual and auditory "feature extraction," respectively.

The "thought pattern memory stack" is discussed in Section 12-4. The following thought is added to this: It is sometimes claimed, as one of the peculiarities of a brain, that it can only "concentrate on one thing at a time." The explanation is that only a single pattern can scan through a memory stack at one time; in other words, we can concentrate on whatever has been extracted by visual or auditory or other sensory cortices, and whatever is being fed back to the "thought pattern memory stack" input.

At this point, it is appropriate to illustrate how the brain model functions with a simple example. Let us assume that the retina sees

Burglar alarm went off!

printed exactly as shown here. The visual image, in full color, is projected into the CS. Visual feature extraction dissects the image into short lines, alphabetical characters, words, and finally the entire phrase. The image "Burglar alarm went off!" rapidly scans through the visual pattern memory stack. The pattern is recognized at many locations in the stack. More or less simultaneously four actions take place:

(a) From the recognition centers, the pattern goes to the "thought" and "auditory" pattern memory stacks via association fibers. The "thought" stack has previously learned, many times, that this message is associated with Danger. The "thought" stack sends a Danger signal to CS, which in turn alerts the hypothalamus. The latter mobilizes the autonomic nervous system.
(b) The central station sends a store gate to all of the memory stacks. All patterns derived from sensory inputs are stored at a location corresponding to the time of occurrence (i.e., the age of the robot). Associated "thought" patterns are likewise stored.
(c) The association fibers reaching the "auditory" stack evoke two sound patterns: a rising and falling siren, and the clanging of bells. These patterns flow to CS to tell it that, associated with the visual message, one may hear a siren or clanging bells.
(d) The robot is programmed to speak what it reads. A collective output of the association fibers feeds the "Burglar alarm went off!" pattern to the "motor speech centers." Here the memory stack (not shown) has long ago learned the complicated muscle sequence needed to recite "Burglar alarm went off!" The motoneurons are properly stimulated, and the robot speaks as called for in the design.

What is the effect of the random noise generator? The robot once made a mistake—it said "Bugler alarm went off." The auditory system of the robot, monitoring itself, sent this pattern through its memory stack. The "thought pattern memory stack" received the word "bugler" and the phrase "alarm went off," but had no stored linkage between the two. It accordingly sent CS a "repeat" command, CS directed the motor centers to reread the message more slowly, and the robot correctly voiced the message the second time around.

On another occasion, the robot was creative. In the context of reading a spy thriller, it said "Bungled—the alarm went off." The mistake was subsequently discovered, but it was felt that, under the circumstances, it was an improvement over the hackneyed phrase "Burglar alarm went off."

Suppose that, instead of reading the message, the robot hears another robot (or person) say "Burglar alarm went off!" This activates the lower half of Fig. 12-12. It is not necessary for us to follow the sequence of events in detail since it is mainly a paraphrase of the earlier discussion regarding activities of the top half.

Is it possible to construct the robot of Fig. 12-12? A few years ago we would scoff at the idea. Today we are not so sure that it could not be done. Many details have been omitted, and perhaps some major reshuffling of blocks is necessary but, *in principle*, there is no reason why it cannot be built given sufficient time and money. Artificial intelligence people are forever improving their robots. Many of the blocks are already operational; some even are commonplace.

The retina could be a video camera, but the processing has to be different. In a conventional system the image is processed one element at a time, from left to right and top to bottom. Our robot should use instead parallel processing in which each element has its own simple but private computer. The "generator of color maps" is a luxury; our first model could certainly be a night watchman that relies exclusively on black-and-white information.

The world abounds in "visual feature extractors," but most of them consist of an image sampled one element at a time and processed by a fast computer that has a large memory and a program dedicated to a single simple geometric shape. Our robot requires parallel processing in which each element is compared with the surrounding elements so that the image can be dissected into short lines, as discussed in connection with Fig. 11-2. After a hierarchical buildup of more complicated architectures, we are still a long way from acceptable feature extraction. The acid test is this: Can the robot recognize a particular face? There are many features to be extracted, more or less independent of image size and partially independent of the direction in which the person is facing. We need skin texture, color (or shades of gray), shape of hair, forehead, eyes, nose, lips, chin, scars, overall contour, and so forth. All of this has to be coded (numerical values corresponding to each feature are fine for a robot) and fed to the memory stack. The coding of "Burglar alarm went off!" may have to be done in two readings ([burglar alarm] and [went off]), but is simple compared to that of a face.

The auditory section is less complicated. We already have standard equipment that supplies the frequency spectrum versus time of a microphone signal. It is a relatively simple matter to modify existing equipment to generate the frequency spectrum versus amplitude map at a rate, say, of 10 maps per second. The human auditory apparatus does it differently, however. It in effect employs parallel processing—the auditory spectrum is dissected into 3000 channels. Ideally, for uniform coverage from 20 to 15,000 Hz, the frequency ratio between one channel and the next is $(15{,}000/20)^{1/3000} = 1.0022$. The first channel is tuned to 20 Hz, the next to $20 \times 1.0022 = 20.044$ Hz, and so forth. A hypothetical tuning circuit is given in Fig. 9-13. The channel frequencies are given by 20, 20.044, 20.088, . . . , 500, 501.1, . . . , 14,967, 15,000 Hz. Each channel yields a continuous output amplitude that corresponds to the frequency components within the narrow range to which it is tuned. The amplitude of the strongest frequency components is the raw data fed to the auditory feature extraction box.

What are the salient auditory features? It is easier to answer this if we consider that the acid test for the human auditory apparatus is understanding a phrase such as "Burglar alarm went off!" regardless of who says it, child or adult, male or female, Brooklyn or Jerusalem accent, sober or drunk, soft or loud. The salient features consist of a sequence of many simultaneous frequencies, each rising and falling in amplitude. It is the *ratios* between frequencies and amplitudes that remain the same regardless of who the speaker happens to be. The color-determining scheme of the eye is faced with exactly the same problem: color is defined by the ratios between the APs derived indirectly from red, green, and blue cones. Since the calculation of ratios seems to be difficult for the nervous system, a much simpler engineering solution is suggested in Fig. 10-20: First, generate the logarithms, which is quite natural for a nonlinear device such as a neuron; second, find the difference (subtract), which is easy with excitatory and inhibitory junctions.

A simple numerical example will illustrate the technique: Suppose that, in the process of speaking, a deep male voice generates a complex sound consisting of 25-, 50-, and 250-Hz vibrations. The putative operations are summarized as follows:

Frequency components	25		50		250
Natural logarithm	3.22		3.91		5.52
Difference		0.69		1.61	

Exactly the same output is obtained if a high-pitched voice utters the *same* complex sound with components 10 times higher in frequency than that of the deep male voice:

Frequency components	250		500		2500
Natural logarithm	5.52		6.21		7.82
Difference		0.69		1.61	

The hypothesis, then, is that 3000 axons enter the auditory feature extraction box of Fig. 12-12. This is the same as the frequency component extractor box of Fig. 9-16. Major frequency components are plotted vertically, say, using a logarithmic scale (as in Fig. 9-16), while the amplitude of each component is plotted horizontally using a logarithmic scale (the decibel sound pressure level (dB SPL) scale). The maps for the deep male and high-pitched voices are identical patterns, but displaced vertically and horizontally because of frequency and amplitude differences, respectively. The temporal (in regard to time) changes that take place in this map constitute the final auditory feature extraction output. The requirements for music are exactly suited to those for voice. In remembering and recognizing a melody, it is again the *ratios* between frequencies and amplitudes that are invariant. The key in which the music is played, or how loud it happens to be, are of peripheral importance in determining the melody.

The memory stacks of Fig. 12-12 can be of the "random access" variety: Visual, thought, and auditory patterns are stored in the form of 0s and 1s. By *random access* we mean that the incoming pattern can scan through the stack until it finds a matching pattern. There are usually many approximately matching patterns in the case of our robot. Every syllable, word, and short phrase that has been learned is connected by association fibers running between the memory stacks. Its visual features appear in visual memory; its sound appears in auditory memory; and its contextual environment and usage appear in thought memory. The facial features of every person the brain is capable of recognizing are similarly connected, via association fibers, to that person's voice and, in thought memory, to events surrounding the acquaintanceship. It all adds up to a huge amount of information, again pointing, perhaps, to a molecular form of memory storage.

Incidentally, the hypothalamus can be implemented by a "read only" memory, or ROM. A familiar example is a phonograph record, which can only be played back, or "read." Except for inadvertent scratches, one cannot alter the stored information. This is in contrast with the ubiquitous reel or floppy disk of magnetic tape, which can survive many cycles of erase, write, and read. Instincts are stored in the hypothalamus, never to be modified or erased, and learning never takes place.

The "motor speech centers" and "muscles involved in speech" boxes can be implemented by electronic equipment that synthesizes speech. Similarly, the "motor writing centers" and "muscles involved in writing" can be implemented by electronic typewriters.

The "generator of random noise" in Fig. 12-12 is a standard piece of test equipment, long available, that is used to check the noise characteristics of electronic devices.

12-6 Determinism versus Free Will

It is not possible to consider *how* the nervous system functions, as we do in the preceding pages of this book, without getting caught up in the determinism versus free-will controversy.

> *Determinism:* The doctrine that every event is the inevitable consequence of antecedent physical conditions; applied to humans, it is the doctrine that every act and decision is the inevitable consequence of antecedent physical, psychological, and environmental conditions, independent of the human will.
>
> *Free Will:* The belief that a human's choices can be voluntary, and not determined by external causes.

Now that we have constructed the robot of Fig. 12-12, at least conceptually, in what sense does it differ from a living human being? The robot never gets tired, it is immune to drug addiction, it continues to operate as long as it is plugged into its wall outlet, its memory never fades, and it does not age (or, if surgery is required, one simply replaces the defective part with a new component taken off the shelf). The robot makes mistakes, some of which, upon rare occasions, turn out to be creative innovations. It has a hypothalamus, which is a read-only memory stack consisting of permanently stored instincts. Instincts guide the robot exclusively when it is first turned on (innately guided learning), but its subsequent behavior is modified gradually by what it learns, by what it has stored in its random-access memory stacks. With a bit of clever programming based on the hypothalamus, one can even endow the robot with emotional responses, a love of music and art, and so forth. Many lesser robots, nowadays, are programmed to engage in long conversations with human counterparts. Some of them even do an excellent job of playing chess!

Over 200 years ago, the great mathematician Pierre-Simon Laplace (1749–1827) launched Laplacian determinism by declaring that, given the position and velocity of every particle in the universe, one could predict the future for all time. His statement was based on the fact that every mechanical motion followed the simple equations put forth by Sir Isaac Newton. At the time Laplace lived, little was known about atoms, molecules, living systems, organic chemistry, and the nervous system. The basis of reproduction, the DNA molecule, was of course a remote dream. Nevertheless, undaunted, Laplace was certain that it would be explained some day, and that all was predetermined. In his view, the writing of this sentence, its publication, and your reading of it were all predestined, and could have been predicted by a supercomputer that was given the location and motion of every particle in the universe at the beginning of time. When tragedy strikes, there is no justification for "guilt feelings" since there was no way to avoid the tragedy and, in fact, it could have been predicted by the supercomputer. (It goes without saying that we do not have access to this supercomputer.)

Laplace was well aware that the laws of chance and theories of probability must be applied frequently because we have insufficient data:

> . . . ignorance of the different causes involved in the production of events, as well as their complexity, taken together with the imperfection of analysis, prevents our reaching . . . certainty about the vast majority of phenomena. Thus there are things that are uncertain for us, things more or less probable, and we seek to compensate for the impossibility of knowing them by determining their different degrees of likelihood. [J. P. Crutchfield et al., 1986]

The reader's attention is called again to Figs. 3-3 and 3-4, which show the randomness associated with a *single* discharging neuron. It is impossible to predict the future of this neuron for more than a few milliseconds. How, then, can one predict the future of a human, which is orchestrated by 10^{12} neurons? What Laplace meant, of course, was that it could be predicted in principle.

Laplacian determinism received a psychological boost when James Clerk Maxwell (1831–1879) showed that the laws of electromagnetism also were simple. Electrical as well as mechanical phenomena could now be described in uncomplicated terms. The laws of the universe were beautifully sparse and unadorned.

With the twentieth century, alas, determinism has received some scientific blows. So much randomness is involved in everything about us that, for all practical purposes, one cannot determine the future. Jules Henri Poincaré (1854–1912) wrote that " . . . it may happen that small differences in the initial conditions produce very great ones in the final phenomena. A small error in the former will produce an enormous error in the latter. Prediction becomes impossible, and we have the fortuitous phenomenon'' [J. P. Crutchfield et al., 1986]. (Fortuitous here refers to chance or accidental.) For the human brain, as previously pointed out, prediction is impossible.

''Besides,'' cried the physicists, ''the future cannot be predicted even in principle because it violates quantum mechanics.'' Quantum mechanics gets its name from the fact that energy is quantized; it is as if you could have a 60- or 61-W bulb, but anything in between, such as 60.3 W, is impossible. (The electric bill that you get every month is quantized to the nearest penny.) One of the central tenets of quantum mechanics is that there is a fundamental limit to the accuracy with which the position and velocity of a particle can be measured, so it would be impossible for Laplace or anybody else to accumulate the data needed to predict the future location and motion of the particle. This, like much of quantum mechanics, does not make sense, but it certainly does not imply that ''free will'' is valid simply because Laplacian determinism should be modified. The future is somewhat chaotic, but not completely so.

[S.D. writes: ''My own views on the controversy come from an occurrence in the 1950s. The Polytechnic Institute of Brooklyn received a relay computer; that is, a computer that used relays rather than vacuum tubes ('twas in the days before transistors). The relay issues an audible click each time the armature is pulled down and released. It is much slower than a vacuum tube, of course, but the relay was wonderful as a device for learning computer techniques. One evening I spent several hours solving for x in an equation of the form $ax^6 + bx^5 + \cdots g = 0$. Some of the values of x were real but some were imaginary. The next day, in order to have an independent check on my work, I asked the programmer to solve the problem on the relay computer. Here I was privileged to be subjected to a most eerie experience: As I sat in the computer room the robot was happily clicking away, with orange trouble-shooting lights blinking at me, as it went through the same calculations my own brain went through the night before! Wherein was there a difference? The thought processes were the same and the numerical answers were the same. Free will was an illusion! My brain was programmed the same as the computer, and went through the same predetermined gyrations. I became converted to Laplacian determinism. (To which A.D. interjects: 'But didn't you have more fun figuring out the answer?') Subsequently, the *Scientific American* and the *Proceedings of the Institute of Radio Engineers* published letters in which I criticized statements that violated determinism. In reaction to these letters I received a barrage of mail pointing out that determinism was invalid because of quantum mechanics. In discussions with my colleagues it turns out that they cannot accept the proposition that their clever equations, turned out in Ph.D. dissertations and subsequent research, are anything short of God-like emanations. They have very healthy egos, and the human animal in general has a need to believe in his or her omniscience. At the other extreme, those who subscribe to determinism can show step-by-step how *their* equations come from previously stored patterns scanning over and over again through memory stacks until something sensible turns up. But one should not forget that, alas, the sun is slated to become unstable 5 billion years from now, and all of our efforts will come to naught!"]

12-7 Consciousness

The dictionary defines consciousness as "The state of having an awareness of one's own existence, sensations, and thoughts, and of one's environment." It is a mysterious attribute of the brain. As before, we point out that any discussion about consciousness is a peculiar business because the brain, an electrochemical ensemble of neuron bodies, dendrites, and axons, is trying to analyze itself [D. C. Dennett, 1991; R. Jackendoff, 1987].

One constructs a robot, as in Fig. 12-12, that is crudely equivalent to a human brain or part thereof. It is assumed that the CS is the seat of consciousness. The CS receives sensory inputs in the form of coded equivalents of what the eyes see, of frequencies and amplitudes entering the ears, of pressure transducers "telling" it that it is sitting in a chair, of tastes and smells, and so forth. The central station has access to an extremely short-term memory of signals occurring during the past few seconds. The central station gets a constant barrage from the association fibers, such as "The man who just walked by is Mr. A, who speaks in a very high-pitched voice and is a political conservative," or "That voice sounds like Mrs. B, who is friendly and even older than I am." The association fibers also convey the coded equivalents of what is circulating in the "thought pattern memory stack." All of this is modified by ever-present noise, as modeled by a random noise generator, so that the robot is somewhat unpredictable and, hopefully, some new and useful patterns are created once in a while.

One day, after the robot has reached maturity and has become quite sophisticated, you ask it "Do you have a consciousness? Do you have an awareness of being?" The robot answers "Of course I do." The crux of the problem is this: How can an electromechanical gadget, constructed out of transistors and relays and similar paraphernalia, have an "awareness of being?" This robot can think—it has a "thought pattern memory stack," it can learn, it can solve reasonably difficult mathematical problems. It says that it has a consciousness, but how can you believe it? Laplace claimed that free will is an illusion, but our consciousness is real, and we fail to see how a robot can be aware of its own existence. How can a machine feel the subjective beauty of vision, the subjective impression of a red flower or green grass or blue sky; of music played on a piano; of hot and cold; of the fragrance of a perfume; of tastes good and bad? Something terribly important is missing.

The machine continues: "I can tell the difference between red or green or blue; I can tell the difference between low and high notes on a piano; I can tell temperature; I can give you the relative atomic constituents of anything I smell or taste." You have no way of knowing what consciousness means to a machine or, indeed, what it means to anybody but yourself. We await the invention of an electronic consciousness meter.

In the meantime, the greatest mystery of all time remains—why do we have an awareness of our own existence? It is appropriate to end this book with a few brief comments on the subject of consciousness:

(a) Perhaps we can receive some guidance from past mysteries that have been solved. Consider that the human race has been "cut down to size" by these developments: The universe no longer revolves around the earth; the heart is merely a pump, and can be replaced by an electromechanical surrogate if need be; and the DNA molecule of a plant or a worm uses the same four nucleotide bases as does the DNA of a human (for an *Escherichia coli* bacterium, the blueprint is 4.7 million bases long; for a human, it is 3000 million bases long). This recitation implies that consciousness is so commonplace that insects also have an "awareness of being."

(b) You have heard about emotionally disturbed people who block out reality. Here is an opportunity for you to prove to yourself that you can do the same. Stare at the two circles of Fig. 12-13, but let your gaze look beyond so that the two circles coincide. The + and × will also, of course, coincide, but please concentrate on keeping the *circles* superimposed. Almost immediately you will lose the +, or the ×, or bits and pieces of the + and ×. The reason is that there is a conflict between the left eye's + and the right eye's ×. (Some people have difficulty seeing this optical illusion.) According to Fig. 11-1, the blocking-out has to occur in the optic nerve, before the bifurcation leading to the LGN. Here, then, we have thousands of fibers blocked by some kind of inhibitory process. A similar mechanism must exist for blocking out other parts of the body in order to avoid conflict. Perhaps this has some connection with the *unconscious psyche* of psychology.

(c) Another example of the fragility of consciousness: The definition

Fig. 12-13 Visual stimulus used to demonstrate the "blocking out of reality." Stare at the two circles, but let your gaze look beyond so that the two *circles* coincide. Almost immediately you will lose the +, or the ×, or bits and pieces of the + and ×. This is an example of binocular rivalry.

of consciousness implies that it is lost during sleep. Well, after 10 minutes in Professor X's class, many students lose consciousness

despite a . . .

desperate . . .

desire to . . .

desist from . . .

dozing off.

[S.D. writes: "I went to school 20 years at night (1935–1955) and became an expert on keeping awake. I am somewhat embarrassed to admit that the best solution, for me, was cigarette smoke. In those days we were allowed to smoke in the classroom."]

(d) At what point in the development of living organisms did consciousness begin? The timetable (including the end when the sun becomes unstable) is something like this:

Big bang	15 billion years ago
Solar system starts to form	5 billion years ago
Earth forms	4.5 billion years ago
Life on earth begins	3 billion years ago
Emergence of animals	0.57 billion years ago
Emergence of mammals	0.2 billion years ago
Earth is destroyed	5 billion years in the future.

Animals move about from place to place in search of food, to escape from predators, find a mate, escape from a hostile environment, and so forth. Movement requires a nervous system, and the integration of movement requires a brain. The simplest of animals has sensory inputs and the equivalents of a hypothalamus and motor center. One can conceive of a completely instinctive creature, without sensory feature extraction and pattern memory stacks. Most environments are hostile, however, in that food is scarce and predators are plentiful. In that event, an animal that cannot learn, that cannot store memories, is at a severe disadvantage. In a relatively short time after animals emerged, 570 million years ago, the rudimentary equivalent of pattern memory stacks and consciousness must have developed.

(e) At what point in embryonic development does consciousness begin? We start with simple materials: carbon, oxygen, nitrogen, hydrogen, and a sprinkling of sulphur, iron, and a few other elements. Properly assemble around one million of these atoms and we have one of the smallest of living creatures, a virus (the complete structure of a poliovirus has been deciphered [J. M. Hogle, M. Chow & D. J. Filman, 1987]). With many more atoms and a much more complicated structure we can get a robot. With yet another arrangement of atoms we get a fertilized human egg. As directed by its DNA molecule, given the proper raw materials, amino acids and protein molecules are synthesized. The egg starts to grow by dividing repeatedly in two: 1, 2, 4, 8, 16, . . . cells. Eventually a "neural plate" forms, cells proliferate in localized regions, the immature neurons migrate to their final residences, they aggregate and differentiate to form the various parts of the brain, they mature and form connections with other neurons. Starting with almost nothing, with a few simple building blocks, a brain is thus created. Somewhere along the way, probably after birth when a certain minimum number of connections have been completed, the human embryo becomes aware that it exists. (For a cat, the circuits between the retina and visual cortex are not completed until eight weeks after birth.) [C. Aoki and P. Siekevitz, 1988; R. E. Kalil, 1989]

(f) Do other animals have an awareness of being? Because all mammalian brains are anatomically similar, the answer, unquestionably, is that all mammals have a consciousness. How about insects? Experiments with honey bees show that they are capable of a great deal of learning [R. Menzel and J. Erber, 1978]. The bees collect nectar and pollen; they land instinctively on objects that look like flowers. If the object rewards a bee with nectar or pollen, the bee will store five types of information related to that flower: (1) odor; (2) color; (3) a low-resolution picture of the flower pattern; (4) the approximate time of day, and (5) direction and distance to the food. Bees have been taught to store information for as many as nine different flowers in one day. The following day the bee visits each flower, according to its "appointment book," until a particular flower is no longer available. A human would find it very difficult to memorize all of these details!

Referring to Fig. 12-12, the bee's block diagram contains a visual feature extraction box and pattern memory stack, as well as an odor feature extraction box and pattern memory stack. The memory stacks are one day (daylight) in length. The major conceptual difference between a mammalian and bee brain is that the insect probably does not have a "thought pattern memory stack." If consciousness resides in the CS, a bee (and insects in general) should have an awareness of being, but if consciousness requires a "thought" stack, as implied by the dictionary definition, then the insect does not have an awareness of being. The brain that belongs to you, the reader, will have to decide this question for itself [D. R. Griffin, 1991; J. Horgan, 1989].

A12-1 VOLTAGE DUE TO A DISCHARGING STRIP

To calculate the voltage at point P due to the discharging slab, we start with the contribution from the square $dx\ dy$ in the top view of Fig. 12-6. Substituting into Eq. (12-5),

$$dV = \frac{V_m a\ dx\ dy}{4\pi(a^2 + x^2 + y^2)^{3/2}}. \tag{12-10}$$

A strip dx wide contributes

$$dV = \frac{V_m a\ dx}{2\pi}\int_0^\infty \frac{dy}{(a^2 + x^2 + y^2)^{3/2}} = \frac{V_m a\ dx}{2\pi(a^2 + x^2)}. \tag{12-11}$$

Integrating in the x direction, we get

$$V = \frac{V_m a}{2\pi}\int_b^{b+w} \frac{dx}{a^2 + x^2} = \frac{V_m}{2\pi}\left(\arctan\frac{b + w}{a} - \arctan\frac{b}{a}\right). \tag{12-12}$$

For the range of interest, where $d \ll a$, this simplifies to

$$V = \frac{V_m d}{2\pi a}\left(\frac{b/a}{1 + (b/a)^2}\right). \tag{12-13}$$

As the potential propagates from left to right in the side view of Fig. 12-6, the output is at first negative because b is negative; it reaches its maximum negative value at $b = a$, then rapidly swings to a positive peak when the leading edge is a units to the right of point P, and so forth.

A change is now made in Eq. (12-13) to convert it into a useful waveform equation: Let the discharge region in the side view be a strip l units long rather than infinitely long; in this event, the contribution from the trailing edge subtracts from that of the leading edge. The following equation, derived from Eq. (12-13), is symmetrical with respect to $b = 0$:

$$V = \frac{V_m d}{2\pi a}\left\{\frac{(b + 0.5l)/a}{1 + [(b + 0.5l)/a]^2} - \frac{(b - 0.5l)/a}{1 + [(b - 0.5l)/a]^2}\right\}. \tag{12-14}$$

For Fig. 12-8, substituting into Eq. (12-5), the voltage at point P due to the discharging slab appears as

$$V = \frac{V_m bd}{2\pi}\int_0^\infty \frac{dy}{(b^2 + y^2)^{3/2}} = \frac{V_m d}{2\pi b}. \tag{12-15}$$

As before, this is converted into a useful waveform equation by letting the discharge region in the side view be a strip l units long. The following equation is symmetrical with respect to $b = 0$ (that is, at $b = 0$, point P is in the center of the discharge region). For $b \gg l$,

$$V = -\frac{V_m l d}{2\pi b^2}. \tag{12-16}$$

REFERENCES

D. L. Alkon, Memory storage and neural systems, *Sci. Am.*, vol. 261, pp. 42–50, July 1989.

J. A. Anderson, A. Pellionisz, and E. Rosenfeld, *Neurocomputing 2*. Cambridge, Mass.: M.I.T. Press, 1990.

J. A. Anderson and E. Rosenfeld, *Neurocomputing*. Cambridge, Mass.: M.I.T. Press, 1988.

C. Aoki and P. Siekevitz, Plasticity in brain development, *Sci. Am.*, vol. 259, pp. 56–64, Dec. 1988.

J. P. Ary, S. A. Klein, and D. H. Fender, Location of sources of evoked scalp potentials: Corrections for skull and scalp thickness, *IEEE Trans. Biomed. Eng.*, vol. BME-28, pp. 447–452, June 1981.

N. I. Bachen, Detection of stimulus-related (evoked response) activity in the electroencephalogram (EEG), *IEEE Trans. Biomed. Eng.*, vol. BME-33, pp. 566–571, June 1986.

J. Bancaud, J. Talairach, R. Morel, M. Bresson, A. Bonis, S. Geier, E. Hermon, and P. Buser, Generalized epileptic seizures elicited by electrical stimulation of the frontal lobe in man, *Electroencephalogr. Clin. Neurophysiol.*, vol. 37, pp. 275–282, 1974.

V. Bloch, Brain activation and memory consolidation, in *Neural Mechanisms of Learning and Memory*, eds. M. R. Rosenzweig and E. L. Bennett. Cambridge, Mass.: M.I.T. Press, 1976.

A. Brodal, Afferent cerebellar connections, in *Aspects of Cerebellar Anatomy*, eds. J. Jansen and A. Brodal. Oslo, Norway: Johan Grundt Tanum, pp. 82–188, 1954.

J. D. Bronzino, Quantitative analysis of the EEG—General concepts and animal studies, *IEEE Trans. Biomed. Eng.*, vol. BME-31, pp. 850–856, Dec. 1984.

H. T. Castello, A cardiac hypothesis for the origin of EEG alpha, *IEEE Trans. Biomed. Eng.*, vol. BME-30, pp. 793–796, Dec. 1983.

D. G. Childers, P. A. Bloom, A. A. Arroyo, S. E. Roucos, I. S. Fischler, T. Achariyapaopan, and N. W. Perry, Jr., Classification of cortical responses using features from single EEG records, *IEEE Trans. Biomed. Eng.*, vol. BME-29, pp. 423–438, June, 1982.

F. H. C. Crick, Thinking about the brain, *Sci. Am.*, vol. 241, pp. 219–232, Sept. 1979.

J. P. Crutchfield, J. D. Farmer, N. H. Packard, and R. S. Shaw, Chaos, *Sci. Am.*, vol. 255, pp. 46–57, Dec. 1986.

J. C. de Munck, B. W. van Dijk, and H. Spekreijse, Mathematical dipoles are adequate to describe realistic generators of human brain activity, *IEEE Trans. Biomed. Eng.*, vol. 35, pp. 960–966, Nov. 1988.

D. C. Dennett, *Consciousness Explained*. Boston: Little, Brown, 1991.

S. Deutsch and E. M-Tzanakou, *Neuroelectric Systems*. New York: New York Univ. Press, 1987.

J. C. Eccles, M. Ito, and J. Szentagothai, *The Cerebellum as a Neuronal Machine*. Berlin: Springer-Verlag, 1967.

R. Elul, The genesis of the EEG, *Int. Rev. Neurobiol.*, vol. 15, pp. 227–272, 1972.

W. J. Friedman, *About Time*. Cambridge, Mass.: M.I.T. Press, 1990.

A. S. Gevins, Analysis of the electromagnetic signals of the human brain: Milestones, obstacles, and goals, *IEEE Trans. Biomed. Eng.*, vol. BME-31, pp. 833–850, Dec. 1984.

A. S. Gevins, C. L. Yeager, S. L. Diamond, J. P. Spire, G. M. Zeitlin, and A. H. Gevins, Automated analysis of the electrical activity of the human brain (EEG): A progress report, *Proc. IEEE*, vol. 63, pp. 1382–1399, Oct. 1975.

J. L. Gould and P. Marler, Learning by instinct, *Sci. Am.*, vol. 256, pp. 74–85, Jan. 1987.

D. R. Griffin, Animal thinking, *Sci. Am.*, vol. 265, p. 144, Nov. 1991.

M. Hassul and P. D. Daniels, Cerebellar dynamics: The mossy fiber input, *IEEE Trans. Biomed. Eng.*, vol. BME-24, pp. 449–456, Sept. 1977.

J. M. Hogle, M. Chow, and D. J. Filman, The structure of poliovirus, *Sci. Am.*, vol. 256, pp. 42–49, March 1987.

J. Horgan, Do bees think? Profile of D. R. Griffin, *Sci. Am.*, vol. 260, pp. 36–38, May 1989.

D. H. Hubel, The brain, *Sci. Am.*, vol. 241, pp. 44–53, Sept. 1979.

H. Hyden, Satellite cells in the nervous system, *Sci. Am.*, vol. 205, pp. 62–69, Dec. 1961.

R. Jackendoff, *Consciousness and the Computational Mind.* Cambridge, Mass.: M.I.T. Press, 1987.

B. H. Jansen, Comments on "A cardiac hypothesis for the origin of EEG alpha," by H. T. Castello, *IEEE Trans. Biomed. Eng.*, vol. BME-32, pp. 347–349, May 1985.

H. H. Jasper, *Epilepsy and Cerebral Localization*, Chap. 14, eds. W. Penfield and T. C. Erikson. Springfield, Ill.: Thomas, 1941.

R. E. Kalil, Synapse formation in the developing brain, *Sci. Am.*, vol. 261, pp. 76–85, Dec. 1989.

K. A. Kooi, R. P. Tucker, and R. E. Marshall, *Fundamentals of Electroencephalography*. Hagerstown, Pa.: Harper & Row, 1978.

J. D. Kraus and K. R. Carver, *Electromagnetics*, 2d ed. New York: McGraw-Hill, 1973.

R. R. Llinas, The cortex of the cerebellum, *Sci. Am.*, vol. 232, pp. 56–71, Jan. 1975.

D. Marr, A theory of cerebellar cortex, *J. Physiol.*, vol. 202, pp. 437–470, 1969.

M. Matonsek, J. Volavka, J. Roubicek, and Z. Roth, EEG frequency analysis related to age in normal adults, *Electroencephalogr. Clin. Neurophysiol.*, vol. 23, pp.162–167, 1967.

J. L. McClelland, D. E. Rumelhart, and PDP Research Group, *Parallel Distributed Processing*. Cambridge, Mass.: M.I.T. Press, 1986.

C. D. McGillem, J. I. Aunon, and K.-B. Yu, Signals and noise in evoked brain potentials, *IEEE Trans. Biomed. Eng.*, vol. BME-32, pp. 1012–1016, Dec. 1985.

R. Menzel and J. Erber, Learning and memory in bees, *Sci. Am.*, vol. 239, pp. 102–110, July 1978.

M. Minsky, *Society of Mind.* New York: Simon & Schuster, 1986.

M. Mishkin and T. Appenzeller, The anatomy of memory, *Sci. Am.*, vol. 256, pp. 80–89, June 1987.

N. H. Morgan and A. S. Gevins, Wigner distributions of human event-related brain potentials, *IEEE Trans. Biomed. Eng.*, vol. BME-33, pp. 66–70, Jan. 1986.

A. R. Morrison, A window on the sleeping brain, *Sci. Am.*, vol. 248, pp. 94–102, Apr. 1983.

W. J. H. Nauta and M. Feirtag, The organization of the brain, *Sci. Am.*, vol. 241, pp. 88–109, Sept. 1979.

F. Nottebohm, From bird song to neurogenesis, *Sci. Am.*, vol. 260, pp. 74–79, Feb. 1989.

P. L. Nunez, A study of origins of the time dependencies of scalp EEG: Theoretical basis and experimental support of theory, *IEEE Trans. Biomed. Eng.*, vol. BME-28, pp. 271–288, March 1981.

J. G. Okyere, P. Y. Ktonas, and J. S. Meyer, Quantification of the alpha EEG modulation and its relation to cerebral blood flow, *IEEE Trans. Biomed. Eng.*, vol. BME-33, pp. 690–696, July 1986.

G. Pfurtscheller and R. Cooper, Frequency dependence of the transmission of the EEG from cortex to scalp, *Electroencephalogr. Clin. Neurophysiol.*, vol. 38, pp. 93–96, 1975.

G. Pfurtscheller, H. Maresch, and S. Schuy, Inter- and intrahemispheric differences in the peak frequency of rhythmic activity within the alpha band, *Electroencephalogr. Clin. Neurophysiol.*, vol. 42, pp. 77–83, 1977.

R. Plonsey, *Bioelectric Phenomena*. New York: McGraw-Hill, 1969.

D. Regan, Electrical responses evoked from the human brain, *Sci. Am.*, vol. 241, pp. 134–146, Dec. 1979.

A. Remond, ed., *Handbook of Electroencephalography and Clinical Neurophysiology*, vols. 1–16. Amsterdam: Elsevier, 1971–1978.

Scientific American, 50 and 100 years ago, vol. 262, p. 12, Jan. 1990.

M. Sharples, D. Hogg, C. Hutchison, S. Torrance, and D. Young, *Computers and Thought*. Cambridge, Mass.: M.I.T. Press, 1989.

Skeptical Inquirer, magazine pub. by Committee for the Scientific Investigation of Claims of the Paranormal, Buffalo, New York.

L. R. Squire, Mechanisms of memory, *Science*, vol. 232, pp. 1612–1619, June 1986.

E. Vaadia, H. Bergman, and M. Abeles, Neuronal activities related to higher brain functions—Theoretical and experimental implications, *IEEE Trans. Biomed. Eng.*, vol. 36, pp. 25–35, Jan. 1989.

F. Walberg, Descending connections to the inferior olive, in *Aspects of Cerebellar Anatomy*, eds. J. Jansen and A. Brodal. Oslo, Norway: Johan Grundt Tanum, pp. 249–263, 1954.

E. T. Walters and J. H. Byrne, Associative conditioning of single sensory neurons suggests a cellular mechanism for learning, *Science*, vol. 219, pp. 405–408, Jan. 1983.

Y. Watanabe and Y. Sakai, A graphic method for estimating equivalent dipole of localized EEG discharge, *IEEE Trans. Biomed. Eng.*, vol. BME-31, pp. 435–439, May 1984.

A. J. Wilkins, C. E. Darby, and C. D. Binnie, Neurophysiological aspects of pattern sensitive epilepsy, *Brain*, vol. 102, pp. 1–25, 1979.

Problems

1. For the cerebellar network of Fig. 12-3, given the following symbol definitions:

$$V_M = \text{Mossy fiber input}$$

$$V_C = \text{Climbing fiber input}$$

V_R = Granule cell output

V_G = Golgi cell output

V_P = Purkinje cell output

V_B = Basket cell output

V_S = Stellate cell output

Assume that each of the *E*s has a junction weight of 2, and each of the *I*s has a weight of 3. Substituting into Eq. (11-1), write the equation for each output voltage. [Ans.: $V_P = \frac{1 + 2V_C + 2V_R}{1 + 3V_B + 3V_S} - 1$, V_P positive, otherwise 0; etc.]

2. Carry out the step-by-step derivation of Eq. (12-7).
3. Given Eq. (12-5), carry out the step-by-step derivation of Eq. (12-13).
4. Given Eq. (12-13), carry out the step-by-step derivation of Eq. (12-14).
5. In Fig. 12-7, given that the slab of neurons is 0.5 mm thick, the discharge region is 1 mm wide, the voltage across the edges of the slab is 25 mV, and the discharge propagates from left to right 5 mm below the scalp. Plot *V* picked up by the electrode. [Ans.: V_{peak} = 78.79 μV.]
6. In Fig. 12-7, (a) find the area of the curve above the baseline at the −1.5 and +1.5 cm points; (b) find the width of the equivalent rectangle; (c) if an action potential is flying by at a speed of 100 m/s, find the equivalent frequency of the rectangle. [Ans.: (b) 1.14 cm; (c) 4390 Hz.]
7. Given Eq. (12-5), carry out the step-by-step derivation of Eq. (12-16).
8. In Fig. 12-9, given that the slab of neurons is 0.5 mm thick, the discharge region is 1 mm wide, the voltage across the edges of the slab is 25 mV, and the discharge propagates upward to within 5 mm of the scalp, plot *V* picked up by the electrode. [Ans.: V_{peak} = 79.58 μV.]
9. Plot a curve similar to that of Fig. 12-11, y = "Amount of declarative information remembered," for a *bee* given the following assumptions: The time scale is uniform; $y = \epsilon^{-1}$ at $t = 1$ day; $y = 0$ at $t = 2$ days. [Ans.: $y = 1.5122\epsilon^{-0.5413t} - 0.5122$.]

Index